Uberveillance and the Social Implications of Microchip Implants:

Emerging Technologies

M.G. Michael
University of Wollongong, Australia

Katina Michael
University of Wollongong, Australia

A volume in the Advances in Human and
Social Aspects of Technology (AHSAT)
Book Series

An Imprint of IGI Global

Managing Director: Lindsay Johnston
Production Manager: Jennifer Yoder
Development Editor: Joel Gamon
Acquisitions Editor: Kayla Wolfe
Typesetter: Christina Henning
Cover Design: Jason Mull

Published in the United States of America by
    Information Science Reference (an imprint of IGI Global)
    701 E. Chocolate Avenue
    Hershey PA 17033
    Tel: 717-533-8845
    Fax: 717-533-8661
    E-mail: cust@igi-global.com
    Web site: http://www.igi-global.com

Library of Congress Cataloging-in-Publication Data

Uberveillance and the social implications of microchip implants: emerging technologies / M.G. Michael and Katina Michael, editors.
     pages cm
  Includes bibliographical references and index.
  Summary: "This book presents case studies, literature reviews, ethnographies, and frameworks supporting the emerging technologies of RFID implants while also highlighting the current and predicted social implications of human-centric technologies"-- Provided by publisher.
   ISBN 978-1-4666-4582-0 (hardcover) -- ISBN 978-1-4666-4583-7 (ebook) -- ISBN 978-1-4666-4584-4 (print & perpetual access) 1. Electronic surveillance--Social aspects. 2. Privacy, Right of--Social aspects. 3. Radio frequency identification systems. 4. Implants, Artificial. I. Michael, M. G. II. Michael, Katina, 1976-
   HM846.U24 2014
   323.44'82--dc23
                                   2013020727

This book is published in the IGI Global book series Advances in Human and Social Aspects of Technology (AHSAT) (ISSN: 2328-1316; eISSN: 2328-1324)

British Cataloguing in Publication Data
A Cataloguing in Publication record for this book is available from the British Library.

For electronic access to this publication, please contact: eresources@igi-global.com.

Advances in Human and Social Aspects of Technology (AHSAT) Book Series

ISSN: 2328-1316
EISSN: 2328-1324

MISSION

In recent years, the societal impact of technology has been noted as we become increasingly more connected and are presented with more digital tools and devices. With the popularity of digital devices such as cell phones and tablets, it is crucial to consider the implications of our digital dependence and the presence of technology in our everyday lives.

The **Advances in Human and Social Aspects of Technology (AHSAT) Book Series** seeks to explore the ways in which society and human beings have been affected by technology and how the technological revolution has changed the way we conduct our lives as well as our behavior. The AHSAT book series aims to publish the most cutting-edge research on human behavior and interaction with technology and the ways in which the digital age is changing society.

COVERAGE

- Activism & ICTs
- Computer-Mediated Communication
- Cultural Influence of ICTs
- Cyber Behavior
- End-User Computing
- Gender & Technology
- Human-Computer Interaction
- Information Ethics
- Public Access to ICTs
- Technoself

IGI Global is currently accepting manuscripts for publication within this series. To submit a proposal for a volume in this series, please contact our Acquisition Editors at Acquisitions@igi-global.com or visit: http://www.igi-global.com/publish/.

Titles in this Series

For a list of additional titles in this series, please visit: www.igi-global.com

Gamification for Human Factors Integration Social, Education, and Psychological Issues
Jonathan Bishop (Centre for Research into Online Communities and E-Learning Systems, Belgium)
Information Science Reference • copyright 2014 • 378pp • H/C (ISBN: 9781466650718) • US $175.00 (our price)

Emerging Research and Trends in Interactivity and the Human-Computer Interface
Katherine Blashki (Noroff University College, Norway) and Pedro Isaias (Portuguese Open University, Portugal)
Information Science Reference • copyright 2014 • 580pp • H/C (ISBN: 9781466646230) • US $175.00 (our price)

Creating Personal, Social, and Urban Awareness through Pervasive Computing
Bin Guo (Northwestern Polytechnical University, China) Daniele Riboni (University of Milano, Italy) and Peizhao
Hu (NICTA, Australia)
Information Science Reference • copyright 2014 • 440pp • H/C (ISBN: 9781466646957) • US $175.00 (our price)

Gender Divide and the Computer Game Industry
Julie Prescott (University of Bolton, UK) and Jan Bogg (The University of Liverpool, UK)
Information Science Reference • copyright 2014 • 334pp • H/C (ISBN: 9781466645349) • US $175.00 (our price)

User Behavior in Ubiquitous Online Environments
Jean-Eric Pelet (KMCMS, IDRAC International School of Management, University of Nantes, France) and Pan-
agiota Papadopoulou (University of Athens, Greece)
Information Science Reference • copyright 2014 • 325pp • H/C (ISBN: 9781466645660) • US $175.00 (our price)

Uberveillance and the Social Implications of Microchip Implants Emerging Technologies
M.G. Michael (University of Wollongong, Australia) and Katina Michael (University of Wollongong, Australia)
Information Science Reference • copyright 2014 • 368pp • H/C (ISBN: 9781466645820) • US $175.00 (our price)

Innovative Methods and Technologies for Electronic Discourse Analysis
Hwee Ling Lim (The Petroleum Institute-Abu Dhabi, UAE) and Fay Sudweeks (Murdoch University, Australia)
Information Science Reference • copyright 2014 • 546pp • H/C (ISBN: 9781466644267) • US $175.00 (our price)

Advanced Research and Trends in New Technologies, Software, Human-Computer Interaction, and Communicability
Francisco Vicente Cipolla-Ficarra (ALAIPO – AINCI, Spain and Italy)
Information Science Reference • copyright 2014 • 696pp • H/C (ISBN: 9781466644908) • US $175.00 (our price)

DISSEMINATOR OF KNOWLEDGE

www.igi-global.com

701 E. Chocolate Ave., Hershey, PA 17033
Order online at www.igi-global.com or call 717-533-8845 x100
To place a standing order for titles released in this series, contact: cust@igi-global.com
Mon-Fri 8:00 am - 5:00 pm (est) or fax 24 hours a day 717-533-8661

To the cherished memory of George Michael (1924-2013).

Editorial Advisory Board

Table of Contents

Section 1
The Veillances

Chapter 1

> M. G. Michael, University of Wollongong, Australia

Interview 1.1

> *Interview with Professor Roger Clarke*
> *Interview conducted by MG Michael on 4 May 2013, Canberra, Australia*

Chapter 2

> Steve Mann, University of Toronto, Canada

Chapter 3

> Alexander Hayes, University of Wollongong, Australia

Section 2
Applications of
Humancentric Implantables

Chapter 4

> Kevin Warwick, University of Reading, UK
> Mark N. Gasson, University of Reading, UK

Section 3
Adoption of RFID
Implants for Humans

Section 4
Tracking and Tracing Laws, Directives, Regulations, and Standards

Section 5
Health Implications of Microchipping Living Things

Section 6
Socio-Ethical Implications of RFID Tags and Transponders

Detailed Table of Contents

Section 1
The Veillances

Chapter 1

M. G. Michael, University of Wollongong, Australia

When or how *uberveillance* will be implemented in its full-blown manifestation is still a subject for some intriguing discussion and a topic of robust disagreement, but what is generally accepted by most of the interlocutors is that an "uberveillance society" will emerge sooner rather than later, and that one way or another this will mean an immense upheaval in all of our societal, business, and government relationships. What is apparent from the numerous qualitative and quantitative studies conducted is that microchipping people is a discernibly divisive issue. If we continue on the current trajectory, we will soon see further divisions – not just between those who have access to the Internet and those who do not, but between those who subjugate themselves to be physically connected to the Web of Things and People, and those who are content enough to simply have Internet connectivity through external devices like smart phones, to those who opt to live completely off the grid. Time will only tell how we as human-beings will adapt after we willingly adopt innovations with extreme and irreversible operations. This introduction serves to provide a background context for the term uberveillance, which has received significant international attention since its establishment.

Interview 1.1

Interview with Professor Roger Clarke
Interview conducted by MG Michael on 4 May 2013, Canberra, Australia

This chapter builds upon the concept of Uberveillance introduced in the seminal research of M. G.
Michael and Katina Michael in 2006. It begins with an overview of *sousveillance* (underwatching)
technologies and examines the *"We're watching you but you can't watch us"* hypocrisy associated with
the rise of *surveillance* (overwatching). Surveillance cameras are often installed in places that have "NO
CAMERAS" and "NO CELLPHONES IN STORE, PLEASE!" signage. The author considers the chilling
effect of this veillance hypocrisy on LifeGlogging, wearable computing, "Sixth Sense," AR Glass, and
the Digital Eye Glass vision aid. If surveillance gives rise to hypocrisy, then to what does its inverse,
sousveillance (wearable cameras, AR Glass, etc.), give rise? The opposite (antonym) of *hypocrisy* is
integrity. How might we resolve the conflict-of-interest that arises in situations where, for example, police
surveillance cameras capture the only record of wrongdoing by the police? Is sousveillance the answer
or will centralized dataveillance merely turn sousveillance into a corruptible *uberveillance* authority?

The intensification and diversification of surveillance in recent decades is now being considered within
a contemporary theoretical and academic framework. The ambiguity of the term 'surveillance' and the
surreptitiousness of its application must now be re-considered amidst the emergent concept of Uberveillance. This chapter presents three cases of organisations that are currently poised or already engaging in
projects using location-enabled point-of-view wearable technologies. Reference is made to additional
cases, project examples, and testimonials including the Australian Federal Police, Northern Territory
Fire Police and Emergency Services, and other projects funded in 2010 and 2011 by the former Australian Flexible Learning Framework (AFLF), now the National VET E-learning Strategy (NVELS).
This chapter also examines the use of location-enabled POV (point-of-view) or Body Wearable Video
(BWV) camera technologies in a crime, law, and national security context, referencing cross-sectoral
and inter-disciplinary opinions as to the perceived benefits and the socio-ethical implications of these
pervasive technologies.

Section 2
Applications of
Humancentric Implantables

In this chapter, the authors report on several different types of human implants with which the authors
have direct, first hand, experience. An indication is given of the experimentation actually carried out
and the subsequent immediate consequences are discussed. The authors also consider likely uses and
opportunities with the technology should it continue to develop along present lines and the likely social
pressures to adopt it. Included in the chapter is a discussion of RFID implants, tracking with implants,
deep brain stimulation, multi-electrode array neural implants, and magnetic implants. In each case,
practical results are presented along with expectations and experiences.

The ability to "write" data to the Internet via tags and barcodes offers a context in which objects will increasingly become a natural extension of the Web, and as ready as the public was to adopt cloud-based services to store address books, documents, photos, and videos, it is likely that we will begin associating data with objects. Leaving messages for loved ones on a tea cup, listening to a story left on a family heirloom, or associating a message with an object to be passed on to a stranger. Using objects as tangible links to data and content on the Internet is predicted to become a significant means of how we interact with the interface of things, places, and people. This chapter explores this potential and focuses upon three contexts in which the technology is already operating in order to reflect upon the impact that the technology process may have upon social processes. These social processes are knowledge browsing, knowledge recovery, and knowledge sharing.

Section 3
Adoption of RFID
Implants for Humans

This chapter presents the results of research designed to investigate differences between and among personality dimensions as defined by Typology Theory using the Myers-Briggs Type Indicator (MBTI). The study took into account levels of willingness toward implanting an RFID (Radio Frequency Identification) chip in the body (uberveillance) for various reasons including the following: to reduce identity theft, as a lifesaving device, for trackability in case of emergency, as a method to increase safety and security, and to speed up the process at airport checkpoints. The study was conducted with students at two colleges in the Northeast of the United States. The author presents a brief literature review, key findings from the study relative to personality dimensions (extroversion vs. introversion dimensions, and sensing vs. intuition dimensions), a discussion on possible implications of the findings when considered against the framework of Rogers' (1983; 2003) Diffusion of Innovation Theory (DoI), and recommendations for future research. A secondary, resultant finding reveals frequency changes between 2005 and 2010 relative to the willingness of college students to implant an RFID chip in the body. Professionals working in the field of emerging technologies could use these findings to better understand personality dimensions based on MBTI and the possible affect such personality dimensions might have on the process of adoption of such technologies as uberveillance.

This chapter describes the usefulness of RFID (Radio Frequency Identification Device) implant technology to monitor the elderly, who are aging in place in various retirement arrangements, and who need to maintain optimal functioning in the absence of available, and on location, service or care providers. The need to maintain functioning or sustainable aging is imperative for countries experiencing rapid growth as a demographic trend for the elderly. The chapter also raises some concerns including the social acceptance or rejection of RFID implant technology, despite the utility of the device. These concerns include a variety of political, social, and religious issues. Further, the chapter also attempts to show how RFID implant technology could be used in combination with other emerging technologies to maintain physical, emotional, and social functioning among the growing population of elderly. What follows is the introduction and a partial literature review on emergent elderly needs, and on the utilization of RFID and other technologies.

Section 4
Tracking and Tracing Laws, Directives, Regulations, and Standards

The European Court of Human Rights (ECtHR) ruling of S and Marper v United Kingdom will have major implications on the retention of Deoxyribonucleic Acid (DNA) samples, profiles, and fingerprints of innocents stored in England, Wales, and Northern Ireland. In its attempt to develop a comprehensive National DNA Database (NDNAD) for the fight against crime, the UK Government has come under fire for its blanket-style coverage of the DNA sampling of its populace. Figures indicate that the UK Government retains a highly disproportionate number of samples when compared to other nation states in the Council of Europe (CoE), and indeed anywhere else in the world. In addition, the UK Government also retains a disproportionate number of DNA profiles and samples of specific ethnic minority groups such as the Black Ethnic Minority group (BEM). Finally, the S and Marper case demonstrates that innocent children, and in general innocent citizens, are still on the national DNA database, sometimes even without their knowledge. Despite the fact that the S and Marper case concluded with the removal of the biometric data of Mr S and Mr Marper, all other innocent subjects must still apply to their local Metropolitan Police Service to have their fingerprints or DNA removed from the register. This is not only a time-consuming process, but not feasible.

Chapter 9

Darren Palmer, Deakin University, Australia

Ian Warren, Deakin University, Australia

Peter Miller, Deakin University, Australia

ID scanners are promoted as an effective solution to the problems of anti-social behavior and violence in many urban nighttime economies. However, the acceptance of this and other forms of computerized surveillance to prevent crime and anti-social behavior is based on several unproven assumptions. After outlining what ID scanners are and how they are becoming a normalized precondition of entry into one Australian nighttime economy, this chapter demonstrates how technology is commonly viewed as the key to preventing crime despite recognition of various problems associated with its adoption. The implications of technological determinism amongst policy makers, police, and crime prevention theories are then critically assessed in light of several issues that key informants talking about the value of ID scanners fail to mention when applauding their success. Notably, the broad, ill-defined, and confused notion of "privacy" is analyzed as a questionable legal remedy for the growing problems of überveillance.

Chapter 10

Jann Karp, C.C.C. Australia, Australia

Technology, trucking, and the surveillance of workers in the workplace: helpful or a hindrance? Technological advances are produced by the creative ideas individuals: these ideas then become selling items in their own right. Do tracking devices effectively regulate traffic breaches and criminality within the trucking industry? The data collection was conducted in the field while the authors rode as a passenger with truck drivers on long-haul trips. The complexities of tracking systems became more apparent as the authors listened to the men and placed their narratives in a broader context for a broader audience. The results of the work indicated that the Global Positioning System (GPS) has a role in the management of the industry as a logistics tool, but that there are limitations to the technology. The drivers use the devices and also feel the oppressive oversight when managers use the data as a disciplinary tool.

Chapter 11

Brigette Garbin, University of Queensland, Australia

Kelly Staunton, University of Queensland, Australia

Mark Burdon, University of Queensland, Australia

Online behavioural profiling has now become an industry that is worth billions of dollars throughout the globe. The actual practice of online tracking was once limited to individual Websites and individual cookies. However, the development of new technologies has enabled marketing corporations to track the Web browsing activities of individual users across the Internet. Consequently, it should be no surprise that legislative initiatives are afoot throughout the world including the United States (US), the European Union (EU), and Australia. These different jurisdictions have put forward different methods of regulating online behavioural profiling and Do Not Track initiatives. Accordingly, this chapter overviews legislative developments and puts forward a typology of different legislative initiatives regarding the regulation of online behavioral profiling and Do Not Track issues. Particular focus is given to the Australian situation and whether existing Australian privacy law is sufficient to protect the privacy interests of individuals against the widespread use of online behaviour profiling tools.

Chapter 12

Uberveillance of humans will emerge through embedding chips within nonhumans in order to monitor humans. The case explored in this chapter involves the development of nanotechnology and biosensors for the real-time tracking of the identity, location, and properties of livestock in the U.S. agrifood system. The primary method for research on this case was an expert forum. Developers of biosensors see the tracking capabilities as empowering users to control some aspects of a situation that they face. Such control promises to improve public health, animal welfare, and/or economic gains. However, the ways in which social and ethical frameworks are built into standards for the privacy/access, organization, adaptability, and transferability of data are crucial in determining whether the diverse actors in the supply chain will embrace nanobiosensors and advance the ideals of the developers. Further research should be done that explores the possibilities of tripartite standards regimes and sousveillance in relation to nanobiosensors in agrifood.

Section 5
Health Implications of Microchipping Living Things

This chapter reviews literature published in oncology and toxicology journals between 1990 and 2006 addressing the effects of implanted radio-frequency (RFID) microchips on laboratory rodents and dogs. Eleven articles were reviewed in all, with eight investigating mice and rats, and three investigating dogs. In all but three of the articles, researchers observed that malignant sarcomas and other cancers formed around or adjacent to the implanted microchips. The tumors developed in both experimental and control animals and in two household pets. In nearly all cases, researchers concluded that the microchips had induced the cancers. Possible explanations for the tumors are explored, and a set of recommendations for policy makers, human patients and their doctors, veterinarians, pet owners, and oncology researchers is presented in light of these findings.

Section 6
Socio-Ethical Implications of RFID Tags and Transponders

Chapter 14

This chapter analyzes some tools of pervasive surveillance in light of the growing philosophical literature regarding the nature and value of privacy. It clarifies the conditions under which a person can be said to have privacy, explains a number of ways in which particular facets of privacy are morally weighty, and explains how such conceptual issues may be used to analyze surveillance scenarios. It argues that in many cases, surveillance may both increase and decrease aspects of privacy, and that the relevant question is whether those privacy losses (and gains) are morally salient. The ways in which privacy diminishment may be morally problematic must be based on the value of privacy, and the chapter explains several conceptions of such values. It concludes with a description of how some surveillance technologies may conflict with the value of privacy.

Panel 14.1

Chapter 15

Transformations of humans through advances in bioelectronics, nanotechnologies, and computer science are leading to hybrids of humans and machines. Future brain-machine interfaces will enable humans not only to be constantly linked to the Internet, and to cyber think, but will also enable technology to take information directly from the brain. Brain-computer interfaces, where a chip is implanted in the brain, will facilitate a tremendous augmentation of human capacities, including the radical enhancement of the human ability to remember and to reason, and to achieve immortality through cloning and brain downloading, or existence in virtual reality. The ethical and legal issues raised by these possibilities represent global challenges. The most pressing concerns are those raised by privacy and autonomy. The potential exists for control of persons, through global tracking, by actually "seeing" and "hearing" what the individual is experiencing, and by controlling and directing an individual's thoughts, emotions, moods, and motivations. Public dialogue must be initiated. New principles, agencies, and regulations need to be formulated and scientific organizations, states, countries, and the United Nations must all be involved.

As cybersurveillance, datamining, and social networking for security, transparency, and commercial purposes become more ubiquitous, individuals who use and rely on various forms of electronic communications are being absorbed into a new type of cellular society. The eventual end of this project might be a world in which each individual, each cell in the electronic "body politic," can be monitored, managed, and, if dangerous to the social organism, eliminated. This chapter examines the objectives, desires, and designs associated with such a cellular biopolitics. Are individuals being incorporated into a Borg-like cyber-organism in which they no longer "own" their substance, preferences, desires, and thoughts and in which they are told what they should be doing next?

Uberveillance extends the responsibilities of faith-based organisations to the power imbalances now emerging. This is less a matter of governance and strategy, and more one of the core values of faith-based organisations. These might be regarded from an ethical or moral standpoint, but the approach taken is to focus on the constituencies of faith-based organisations and the imperatives that have been woven into their aims and values. The specific ways in which such disempowerments emerge and the functional importance of making organizational responses are considered. Acknowledgement is made of the Science and Society Council of the Churches of Scotland, who catalysed the expression and articulation of these issues.

Foreword

ONE GENERATION IS ALL THEY NEED

By the time my four-year-old son is swathed in the soft flesh of old age, he will likely find it unremarkable that he and almost everyone he knows will be permanently implanted with a microchip. Automatically tracking his location in real time, it will connect him with databases monitoring and recording his smallest behavioural traits.

Most people anticipate such a prospect with a sense of horrified disbelief, dismissing it as a science-fiction fantasy. The technology, however, already exists. For years, humane societies have implanted all the pets that leave their premises with a small identifying microchip. As well, millions of consumer goods are now traced with tiny radio frequency identification chips that allow companies to track their exact location.

A select group of people are already "chipped" with devices that automatically open doors, turn on lights, and perform other low-level miracles. Prominent among such individuals is researcher Kevin Warwick of Reading University in England; Warwick is a leading proponent of the almost limitless potential uses for such chips.

Other users include the patrons of the Baja Beach Club in Barcelona, many of whom paid about $150 (U.S.) for the privilege of being implanted with an identifying chip that allows them to bypass lengthy club queues and purchase drinks by being scanned. These individuals are the advance guard of an effort to expand the technology as widely as possible.

From this point forward, microchips will become progressively smaller, less invasive, and easier to deploy. Thus, any realistic barrier to the wholesale "chipping" of Western citizens is not technological but cultural. It relies upon the visceral reaction against the prospect of being personally marked as one component in a massive human inventory.

Today, we might strongly hold such beliefs, but sensibilities can, and probably will, change. How this remarkable attitudinal transformation is likely to occur is clear to anyone who has paid attention to privacy issues over the past quarter-century. There will be no 3 a.m. knock on the door by storm troopers come to force implants into our bodies. The process will be more subtle and cumulative, couched in the unassailable language of progress and social betterment, and mimicking many of the processes that have contributed to the expansion of closed-circuit television cameras and the corporate market in personal data.

A series of tried and tested strategies will be marshaled to familiarize citizens with the technology. These will be coupled with efforts to pressure tainted social groups and entice the remainder of the population into being chipped.

This, then, is how the next generation will come to be microchipped.

It starts in distant countries. Having tested the technology on guinea pigs, both human and animal, the first widespread use of human implanting will occur in nations at the periphery of the Western world. Such developments are important in their own right, but their international significance pertains to how they familiarize a global audience with the technology and habituate them to the idea that chipping represents a potential future.

An increasing array of hypothetical chipping scenarios will also be depicted in entertainment media, furthering the familiarization process.

In the West, chips will first be implanted in members of stigmatized groups. Pedophiles are the leading candidate for this distinction, although it could start with terrorists, drug dealers, or whatever happens to be that year's most vilified criminals. Short-lived promises will be made that the technology will only be used on the "worst of the worst." In fact, the wholesale chipping of incarcerated individuals will quickly ensue, encompassing people on probation and on parole.

Even accused individuals will be tagged, a measure justified on the grounds that it would stop them from fleeing justice. Many prisoners will welcome this development, since only chipped inmates will be eligible for parole, weekend release, or community sentences. From the prison system will emerge an evocative vocabulary distinguishing chippers from non-chippers.

Although the chips will be justified as a way to reduce fraud and other crimes, criminals will almost immediately develop techniques to simulate other people's chip codes and manipulate their data.

The comparatively small size of the incarcerated population, however, means that prisons would be simply a brief stopover on a longer voyage. Commercial success is contingent on making serious inroads into tagging the larger population of law-abiding citizens. Other stigmatized groups will therefore be targeted. This will undoubtedly entail monitoring welfare recipients, a move justified to reduce fraud, enhance efficiency, and ensure that the poor do not receive "undeserved" benefits.

Once e-commerce is sufficiently advanced, welfare recipients will receive their benefits as electronic vouchers only accessibly by having their microchips read, a policy that will be tinged with a sense of righteousness, as it will help ensure that clients can only purchase government-approved goods from select merchants, reducing the always disconcerting prospect that poor people might use their limited funds to purchase alcohol or tobacco.

Civil libertarians will try to foster a debate on these developments. Their attempts to prohibit chipping will be handicapped by the inherent difficulty in animating public sympathy for criminals and welfare recipients – groups that many citizens are only too happy to see subjected to tighter regulation. Indeed, the lesser public concern for such groups is an inherent part of the unarticulated rationale for why coerced chipping will be disproportionately directed at the stigmatized.

The official privacy arm of the government will now take up the issue. Mandated to determine the legality of such initiatives, privacy commissioners and Senate Committees will produce a forest of reports presented at an archipelago of international conferences. Hampered by lengthy research and publication timelines, their findings will be delivered long after the widespread adoption of chipping is effectively a fait accompli. The research conclusions on the effectiveness of such technologies will be mixed and open to interpretation.

Officials will vociferously reassure the chipping industry that they do not oppose chipping itself, which has fast become a growing commercial sector. Instead, they are simply seeking to ensure that the technology is used fairly and that data on the chips is not misused. New policies will be drafted.

Employers will start to expect implants as a condition of getting a job. The U.S. military will lead the way, requiring chips for all soldiers as a means to enhance battlefield command and control – and to identify human remains. From cooks to commandos, every one of the more than one million U.S. military personnel will see microchips replace their dog tags.

Following quickly behind will be the massive security sector. Security guards, police officers, and correctional workers will all be expected to have a chip. Individuals with sensitive jobs will find themselves in the same position.

The first signs of this stage are already apparent. In 2004, the Mexican attorney general's office started implanting employees to restrict access to secure areas. The category of "sensitive occupation" will be expansive to the point that anyone with a job that requires keys, a password, security clearance, or identification badge will have those replaced by a chip.

Judges hearing cases on the constitutionality of these measures will conclude that chipping policies are within legal limits. The thin veneer of "voluntariness" coating many of these programs will allow the judiciary to maintain that individuals are not being coerced into using the technology.

In situations where the chips are clearly forced on people, the judgments will deem them to be undeniable infringements of the right to privacy. However, they will then invoke the nebulous and historically shifting standard of "reasonableness" to pronounce coerced chipping a reasonable infringement on privacy rights in a context of demands for governmental efficiency and the pressing need to enhance security in light of the endless wars on terror, drugs, and crime.

At this juncture, a unfortunately common tragedy of modern life will occur: A small child, likely a photogenic toddler, will be murdered or horrifically abused. It will happen in one of the media capitals of the Western world, thereby ensuring non-stop breathless coverage. Chip manufactures will recognize this as the opportunity they have been anticipating for years. With their technology now largely bug-free, familiar to most citizens and comparatively inexpensive, manufacturers will partner with the police to launch a high-profile campaign encouraging parents to implant their children "to ensure your own peace of mind."

Special deals will be offered. Implants will be free, providing the family registers for monitoring services. Loving but unnerved parents will be reassured by the ability to integrate tagging with other functions on their PDA so they can see their child any time from any place.

Paralleling these developments will be initiatives that employ the logic of convenience to entice the increasingly small group of holdouts to embrace the now common practice of being tagged. At first, such convenience tagging will be reserved for the highest echelon of Western society, allowing the elite to move unencumbered through the physical and informational corridors of power. Such practices will spread more widely as the benefits of being chipped become more prosaic. Chipped individuals will, for example, move more rapidly through customs.

Indeed, it will ultimately become a condition of using mass-transit systems that officials be allowed to monitor your chip. Companies will offer discounts to individuals who pay by using funds accessible through their embedded chip, on the small-print condition that the merchant can access large swaths of their personal data. These "discounts" are effectively punitive pricing schemes, charging unchipped individuals more as a way to encourage them to submit to monitoring. Corporations will seek out the personal data in hopes of producing ever more fine-grained customer profiles for marketing purposes, and to sell to other institutions.

By this point, all major organizations will be looking for opportunities to capitalize on the possibilities inherent in an almost universally chipped population. The uses of chips proliferate, as do the types of discounts. Each new generation of household technology becomes configured to operate by interacting with a person's chip.

Finding a computer or appliance that will run though old-fashioned "hands-on'" interactions becomes progressively more difficult and costly. Patients in hospitals and community care will be routinely chipped, allowing medical staff—or, more accurately, remote computers—to monitor their biological systems in real time.

Eager to reduce the health costs associated with a largely docile citizenry, authorities will provide tax incentives to individuals who exercise regularly. Personal chips will be remotely monitored to ensure that their heart rate is consistent with an exercise regime.

By now, the actual process of "chipping" for many individuals will simply involve connecting certain functions of their existing chip. Any prospect of removing the chip will become increasingly untenable, as having a chip will be a precondition for engaging in key dynamics of modern life, such as shopping, voting, and driving.

The remaining holdouts will grow increasingly weary of Luddite jokes and subtle accusations that they have something to hide. Exasperated at repeatedly watching neighbours bypass them in "chipped" lines while they remain subject to the delays, inconveniences, and costs reserved for the unchipped, they too will choose the path of least resistance and get an implant.

In one generation, then, the cultural distaste many might see as an innate reaction to the prospect of having our bodies marked like those of an inmate in a concentration camp will likely fade.

In the coming years, some of the most powerful institutional actors in society will start to align themselves to entice, coerce, and occasionally compel the next generation to get an implant.

Now, therefore, is the time to contemplate the unprecedented dangers of this scenario. The most serious of these concern how even comparatively stable modern societies will, in times of fear, embrace treacherous promises. How would the prejudices of a Joe McCarthy, J. Edgar Hoover, or of southern Klansmen—all of whom were deeply integrated into the American political establishment—have manifest themselves in such a world? What might Hitler, Mao, or Milosevic have accomplished if their citizens were chipped, coded, and remotely monitored?

Choirs of testimonials will soon start to sing the virtues of implants. Calm reassurances will be forth-coming about democratic traditions, the rule of law, and privacy rights. History, unfortunately, shows that things can go disastrously wrong, and that this happens with disconcerting regularity. Little in the way of international agreements, legality, or democratic sensibilities has proved capable of thwarting single-minded ruthlessness.

"It can't happen here" has become the whispered swan song of the disappeared. Best to contemplate these dystopian potentials before we proffer the tender forearms of our sons and daughters. While we cannot anticipate all of the positive advantages that might be derived from this technology, the negative prospects are almost too terrifying to contemplate.*

This book represents the leading edge analysis of developments in implant technology. However, the volume is not only of interest to experts in such devices. Collectively, the different contributions amount to a case study in how society can move towards adopting technologies that only a short time ago appeared so intrusive that it was barely imaginable that they might be embraced. The processes by which this happens should be of interest to us all, and the editors M. G. and Katina Michael are to be commended for presenting a timely and interdisciplinary volume that will continue the debate about microchipping both things and people.

*An earlier version of this piece appeared in the 'Ideas' section of the Toronto Star, December 10, 2006.

Kevin Haggerty
University of Alberta, Canada

Kevin Haggerty *is a Professor of Criminology and Sociology at the University of Alberta, Canada. He received his Ph.D. in sociology from the University of British Columbia. Haggerty is the Editor of the Canadian Journal of Sociology. He is also a member of the Executive Team for the SSHRC Major Collaborative Research Initiative (MCRI) on "The New Transparency". Haggerty has recently won the Faculty of Arts Award for Associate Professors. He is also the book review editor for the international journal Surveillance and Society.*

Preface

INTRODUCTION

Uberveillance can be defined as an omnipresent electronic surveillance facilitated by technology that makes it possible to embed surveillance devices into the human body. These embedded technologies can take the form of traditional pacemakers, radio-frequency identification (RFID) tag and transponder implants, smart swallowable pills, nanotechnology patches, multi-electrode array brain implants, and even smart dust to mention but a few form factors. To an extent, head-up displays like electronic contact lenses that interface with the inner body (i.e. the eye which sits within a socket) can also be said to be embedded and contributing to the uberveillance trajectory, despite their default categorisation as body wearables.

Uberveillance has to do with the fundamental who (ID), where (location), and when (time) questions in an attempt to derive why (motivation), what (result), and even how (method/plan/thought). Uberveillance can be a predictive mechanism for a person's expected behaviour, traits, likes, or dislikes based on historical fact; or it can be about real-time measurement and observation; or it can be something in between. The inherent problem with uberveillance is that facts do not always add up to truth, and predictions or interpretations based on uberveillance are not always correct, even if there is direct visual evidence available (Shih, 2013). Uberveillance is more than closed circuit television feeds, or cross-agency databases linked to national identity cards, or biometrics and ePassports used for international travel. Uberveillance is the sum total of all these types of surveillance and the deliberate integration of an individual's personal data for the continuous tracking and monitoring of identity, location, condition, and point of view in real-time (Michael & Michael, 2010b).

In its ultimate form, uberveillance has to do with more than automatic identification and location-based technologies that we carry with us. It has to do with under-the-skin technology that is embedded in the body, such as microchip implants. Think of it as Big Brother on the inside looking out. It is like a black box embedded in the body which records and gathers evidence, and in this instance, transmitting specific measures wirelessly back to base. This implant is virtually meaningless without the hybrid network architecture that supports its functionality: making the person a walking online node. We are referring here, to the lowest common denominator, the smallest unit of tracking – presently a tiny chip inside the body of a human being. But it should be stated that electronic tattoos and nano-patches that are worn on the body can also certainly be considered mechanisms for data collection in the future. Whether wearable or bearable, it is the intent and objective which remains important, the notion of "people as sensors." The gradual emergence of the so-called human cloud, that cloud computing platform which allows for the Internetworking of human "points of view" using wearable recording technology (Nolan, 2013), will also be a major factor in the proactive profiling of individuals (Michael & Michael, 2011).

AUDIENCE

This present volume will aim to equip the general public with much needed educational information about the technological trajectory of RFID implants through exclusive primary interviews, case studies, literature reviews, ethnographies, surveys and frameworks supporting emerging technologies. It was in 1997 that bioartist Eduardo Kac (Figure 1) implanted his leg in a live performance titled *Time Capsule* (http://www.ekac.org/timec.html) in Brazil (Michael & Michael, 2009). The following year in an un-related experiment, Kevin Warwick injected an implant into his left arm (Warwick, 2002; K. Michael, 2003). By 2004, the Verichip Corporation had their VeriChip product approved by the Food and Drug Administration (FDA) (Michael, Michael & Ip 2008). And since that point, there has been a great deal of misinformation and confusion surrounding the microchip implant, but also a lot of build-up on the part of the proponents of implantables.

Radio-Frequency Identification (RFID) is not an inherently secure device, in fact it can be argued that it is just the opposite (Reynolds, 2004). So why someone would wish to implant something beneath the skin for non-medical reasons is quite surprising, despite the touted advantages. One of the biggest issues, not commonly discussed in public forums, has to be the increasing numbers of people who are suffering from paranoid or delusional thoughts with respect to enforced implantation or implantation through stealth. We have already encountered significant problems in the health domain- where for example, a clinical psychologist can no longer readily discount completely the claims of patients who identify with having been implanted or tracked and monitored using inconspicuous forms of ID. This will be especially true in the era of the almost "invisible scale to the naked eye" smart dust which has yet to fully arrive. Civil libertarians, religious advocates, and so-named conspiracy theorists will not be the only exclusive groups to discuss the real potential of microchipping people, and for this reason, the discussion will move into the public policy forum, all inclusive of stakeholders in the value chain.

Figure 1. Eduardo Kac implanting himself in his left leg with an RFID chip using an animal injector kit on 11 November 1997. Courtesy Eduardo Kac. More at http://www.ekac.org/figs.html.

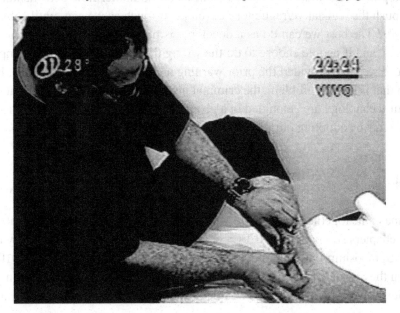

Significantly, this book will also provide researchers and professionals who are engaged in the development or implementation of emerging services with awareness of the social implications of human-centric technologies. These implications *cannot* be ignored by operational stakeholders, such as engineers and the scientific elite, if we hope to enact long-term beneficial change with new technologies that will have a positive impact on humanity. We cannot possess the attitude that says- *let us see how far we can go with technology and we will worry about the repercussions later*: to do so would be narrow-sighted and to ignore the importance of socio-technical sustainability. Ethics are apparently irrelevant to the engineer who is innovating in a market-driven and research funded environment. For sure there are some notable exceptions where a middle of the way approach is pursued, notably in the medical and educational contexts. Engineering ethics, of course exist, unfortunately often denigrated and misinterpreted as discourses on "goodness" or appeals to the categorical imperative. Nevertheless industry as a whole has a social responsibility to consumers at large, to ensure that it has considered what the misuse of its innovations might mean in varied settings and scenarios, to ensure that there are limited, if any, health effects from the adoption of particular technologies, and that adverse event reports are maintained by a centralised administrative office with recognised oversight (e.g. an independent ombudsman).

Equally, government agencies must respond with adequate legislative and regulatory controls to ensure that there are consequences for the misuse of new technologies. It is not enough for example, for a company like Google to come out and openly "bar" applications for its Glass product, such as biometric recognition and pornography, especially when they are very aware that these are two application areas for which their device will be exploited. Google is trying to maintain its brand by stating clearly that it is not affiliated with negative uses of its product, knowing too well that their proclamation is quite meaningless, and by no means legally binding. And this remains one of the great quandaries, that few would deny that Google's search rank and page algorithms have meant we have also been beneficiaries of some extraordinary inventiveness.

According to a survey by CAST, one in five persons, have reported that they want to see a Google Glass ban (Nolan, 2013). Therefore, the marketing and design approach nowadays, which is broadly evident across the universal corporate spectrum, seems to be:

We will develop products and make money from them, no matter how detrimental they may be to society. We will push the legislative/regulatory envelope as much as we can, until someone says: *Stop. You've gone too far!* The best we can do as a developer is place a warning on the packaging, just like on cigarette notices, and if people choose to do the wrong thing our liability as a company is removed completely because we have provided the prior warning and only see beneficial uses. If our product is used for bad then that is not our problem, the criminal justice system can deal with that occurrence, and if non-users of our technology are entangled in a given controversy, then our best advice to people is to realign the asymmetry by adopting our product.

INSPIRATION

This edited volume came together over a three year period. We formed our editorial board and sent out the call for book chapters soon after the IEEE conference we hosted at the University of Wollongong, the International Symposium on Technology and Society (ISTAS) on 7-10 June 2010, sponsored by IEEE's Society on the Social Implications of Technology (SSIT) (http://iibsor.uow.edu.au/conferences/ISTAS/home/index.html). The symposium was dedicated to emerging technologies and there were a great many papers presented from a wide range of views on the debate over the microchipping of people. It was such a highlight to see this sober conversation happening between experts coming at the debate from different perspectives, different cultural contexts, and different lifeworlds. A great deal of the spirit from that conversation has taken root in this book. The audio-visual proceedings aired on the Australian Broadcasting Corporation's much respected *7.30 Report* and received wide coverage in major media outlets. The significance is not in the press coverage but in the fact that the topic is now relevant to the everyday person. Citizens will need to make a personal decision- do I receive an implant or not? Do I carry an identifier on the surface of my skin or not? Do I succumb to 24x7 monitoring by being fully "connected" to the grid or not?

Individuals who were present at ISTAS10 and were also key contributors to this volume include keynote speakers Professor Rafael Capurro, Professor Roger Clarke, Professor Kevin Warwick, Dr Katherine Albrecht, Dr Mark Gasson, Mr Amal Graafstra, and attendees Professor Marcus Wigan, Associate Professor Darren Palmer, Dr Ian Warren, Dr Mark Burdon, and Mr William A. Herbert. Each of these presenters have been instrumental voices in the discussion on Embedded Surveillance Devices (ESDs) in living things (animals and humans), and tracking and monitoring technologies. They have dedicated a portion of their professional life to investigating the possibilities and the effects of a world filled with microchips, beyond those in desktop computers and high-tech gadgetry. They have also been able to connect the practice of an Internet of Things (IoT) from not only machine-to-machine but nested forms of machine-to-people-to-machine interactions and considered the implications. When one is surrounded by such passionate voices, it is difficult not to be inspired onward to such an extensive work.

A further backdrop to the book is the annual workshops we began in 2006 on the *Social Implications of National Security* which have had ongoing sponsorship by the Australian Research Council's Research Network for a Secure Australia (RNSA). Following ISTAS10, we held a workshop on the "Social Implications of Location-Based Services" at the University of Wollongong's Innovation Campus and were fortunate to have Professor Rafael Capurro, Professor Andrew Goldsmith, Professor Peter Eklund, and Associate Professor Ulrike Gretzel present their work (http://iibsor.uow.edu.au/conferences/ISTAS/workshops/index.html). Worthy of note is the workshop proceedings which are available online have been recognised as major milestones for the Research Network in official government documentation. For example, the Department of the Prime Minister and Cabinet (PM&C) among other high profile agencies in Australia and abroad have requested copies of the works for their libraries.

In 2012, the topic of our annual RNSA workshop was "Sousveillance and the Social Implications of Point of View Technologies in Law Enforcement" held at the University of Sydney (http://works.bepress.com/kmichael/249/). Professor Kevin Haggerty keynoted that event, speaking on a theme titled "Monitoring within and beyond the Police Organisation" and also later graciously contributed the foreword to this book, as well as presenting on biomimetics at the University of Wollongong. The workshop again acted to bring exceptional voices together to discuss audio-visual body-worn recording technologies including, Professor Roger Clarke, Professor David Lyon, Associate Professor Nick O'Brien, Associate Professor Darren Palmer, Dr Saskia Hufnagel, Dr Jann Karp, Mr Richard Kay, Mr Mark Lyell, and Mr Alexander Hayes.

In 2013, the theme of the National Security workshop was "Unmanned Aerial Vehicles - Pros and Cons in Policing, Security & Everyday Life" held at Ryerson University in Canada. This workshop had presentations from Professor Andrew Clement, Associate Professor Avner Levin, Mr Ian Hannah, and Mr Matthew Schroyer. It was the first time that the workshop was held outside Australian borders in eight years. While drones are not greatly discussed in this volume, they demonstrate one of the scenario views of the fulfilment of uberveillance. Case in point, the drone killing machine signifies the importance of a remote controlled macro-to-micro view. At first, something needs to be able to scan the skies to look down on the ground, and then when the target has been identified and tracked it can be extinguished with ease. One need only look at the Israel Defence Force's pinpoint strike on Ahmed Jabari, the head of the Hamas Military Wing, to note the intrinsic link between the macro and micro levels of details (K. Michael, 2012). How much "easier" could this kind of strike have been if the GPS chipset in the mobile phone carried by an individual communicated with a chip implant embedded in the body? RFID *can* be a tracking mechanism, despite the claims of some researchers that it has only a 10cm proximity. That may well be the case for your typical wall-mounted reader, but the mobile phone can act as a continuous reader if in range, as can a set of traffic lights, lampposts, or even wifi access nodes, depending on the on-board technology and the power of the reader equipment being used. A telltale example of the potential risks can be seen in the rollout of Real ID driver's licenses in the USA, since the enactment of the REAL ID Act of 2005.

In 2013, it was also special to meet some of our book contributors for the first time at ISTAS13, held at the University of Toronto on the theme of "Wearable Computers and Augmediated Reality in Everyday Life," among them Professor Steve Mann, Associate Professor Christine Perakslis, and Dr Ellen McGee. As so often happens when a thematic interest area brings people together from multiple disciplines, an organic group of interdisciplinary voices has begun to form at www.technologyandsociety.org. The holistic nature of this group is especially stimulating in sharing its diverse perspectives. Building upon these initial conversations and ensuring they continue as the social shaping of technology occurs in the real world is paramount.

As we brought together this edited volume, we struck a very fruitful collaboration with Reader, Dr Jeremy Pitt of Imperial College London, contributing a large chapter in his disturbingly wonderful edited volume entitled *This Pervasive Day: The Potential and Perils of Pervasive Computing* (2012). Jeremy's book is a considered forecast of the social impact of new technologies inspired by Ira Levin's *This Perfect Day* (1970). Worthy of particular note is our participation in the session entitled "Heaven and Hell: Visions for Pervasive Adaptation" at the *European Future Technologies Conference and Exhibition* (Paechter, 2011). What is important to draw out from this is that pervasive computing will indeed have a divisive impact on its users: for some it will offer incredible benefits, while to others it will be debilitating in its everyday effect. We hope similarly, to have been able to remain objective in this edited volume, offering viewpoints from diverse positions on the topic of humancentric RFID. This remained one of our principal aims and fundamental goals.

Questioning technology's trajectory, especially when technology no longer has a medical corrective or prosthetic application but one that is based on entertainment and convenience services is extremely important. What happens to us when we embed a device that we cannot remove on our own accord? Is this fundamentally different to wearing or lugging something around? Without a doubt, it is! And what of those technologies, that are presently being developed in laboratories across the world for microscopic forms of ID, and pinhole video capture? What will be the impact of these on our society with respect to covert surveillance? Indeed, the line between overt and covert surveillance is blurring- it becomes indistinguishable when we are surrounded by surveillance and are inside the thick fog itself. The other thing that becomes completely misconstrued is that there is actually logic in the equation that says that there is a trade-off between privacy and convenience. There is *no* trade-off. The two variables cannot be discussed on equal footing – you cannot give a little of your privacy away for convenience and hope to have it still intact thereafter. No amount of monetary or value-based recompense will correct this asymmetry. We would be hoodwinking ourselves if we were to suddenly be "bought out" by such a business model. There is no consolation for privacy loss. We cannot be made to feel better after giving away a part of ourselves. It is not like scraping one's knee against the concrete with the expectation that the scab will heal after a few days. Privacy loss is to be perpetually bleeding, perpetually exposed.

Additionally, in the writing of this book we also managed a number of special issue journals in 2010 and 2011, all of which acted to inform the direction of the edited volume as a whole. These included special issues on "RFID – A Unique Radio Innovation for the 21st Century" in the *Proceedings of the IEEE* (together with Rajit Gadh, George Roussos, George Q. Huang, Shiv Prabhu, and Peter Chu); "The Social Implications of Emerging Technologies" in *Case Studies in Information Technology* with IGI (together with Dr Roba Abbas); "The Social and Behavioral Implications of Location-Based Services" in the *Journal of Location Based Services* with Routledge; and "Surveillance and Uberveillance" in *IEEE Technology and Society Magazine*. In 2013, Katina also guest edited a volume for *IEEE Computer* on "Big Data: Discovery, Productivity and Policy" with Keith W. Miller. If there are any doubts about the holistic work supporting uberveillance, we hope that these internationally recognised journals, amongst others, that have been associated with our guest editorship indicate the thoroughness and robustness of our approach, and the recognition that others have generously provided to us for the incremental work we have completed.

It should also not go without notice that since 2006 the term uberveillance has been internationally embedded into dozens of graduate and undergraduate technical and non-technical courses across the globe. From the University of New South Wales and Deakin University to the University of Salford, from the University of Malta right through to the University of Texas at El Paso and Western Illinois University- we are extremely encouraged by correspondence from academics and researchers noting the term's insertion into outlines, chosen text book, lecture schedules, major assessable items, recommended readings, and research training. These citations have acted to inform and to interrogate the subjects that connect us. That our research conclusions resonate with you, without necessarily implying that you have always agreed with us, is indeed substantial.

OUTLINE

Uberveillance and the Social Implications of Microchip Implants: Emerging Technologies follows on from a 2009 IGI Premier Reference source book titled *Automatic Identification and Location-Based Services: from Bar Codes to Chip Implants*. This volume consists of 6 sections, and 18 chapters with 7 exclusive addendum primary interviews and panels. The strength of the volume is in its 41 author contributions. Contributors have come from diverse professional and research backgrounds in the field of emerging technologies, law and social policy, including, information and communication sciences, administrative sciences and management, criminology, sociology, law and regulation, philosophy, ethics and policy, government, political science, among others. Moreover, the book will provide insights and support to every day citizens who may be questioning the trajectory of micro and miniature technologies or the potential for humans to be embedded with electro-magnetic devices. Body wearable technologies are also directly relevant, as they will act as complementary if not supplementary innovations to various forms of implants.

Section 1 is titled "The Veillances" with a specific background context of uberveillance. This section inspects the antecedents of surveillance, Roger Clarke's *dataveillance* thirty years on, Steve Mann's *sousveillance*, and MG Michael's *uberveillance*. These three neologisms are inspected under the umbrella of the "veillances" (from the French *veiller*) which stems from the Latin *vigilare* which means to "keep watch" (Oxford Dictionary, 2012).

In 2009, Katina Michael and MG Michael presented a plenary paper titled: "Teaching Ethics in Wearable Computing: the Social Implications of the New 'Veillance'" (K. Michael & Michael, 2009d). It was the first time that surveillance, dataveillance, sousveillance, and uberveillance were considered together at a public gathering. Certainly as a specialist term, it should be noted "veillance" was first used in an important blogpost exploring equiveillance by Ian Kerr and Steve Mann (2006): "the valences of veillance" were briefly described. In contrast to Kerr and Mann (2006), Michael and Michael (2006) were pondering on the intensification of a state of uberveillance through increasingly pervasive technologies that can provide details from the big picture view right down to the miniscule personal details.

Alexander Hayes (2010), pictorialized this representation using the triquetra, also known as the trinity knot and Celtic triangle (Figure 2), and describes its application to uberveillance in the educational context in chapter 3. Hayes uses mini cases to illustrate the importance of understanding the impact of body-worn video across sectors. He concludes by warning that commercial entities should not be engaged in "techno-evangelism" when selling to the education sector but should rather maintain the purposeful intent of the use of point of view and body worn video recorders within the specific educational context. Hayes also emphasises the urgent need for serious discussion on the socio-ethical implications of wearable computers.

By 2013, K. Michael had published proceedings from *the International Symposium on Technology and Society* (ISTAS13) using the veillance concept as a theme (http://veillance.me), with numerous papers submitted to the conference exploring veillance perspectives (Ali & Mann, 2013; Hayes, et al., 2013; K. Michael, 2013; Minsky, et al., 2012; Paterson, 2013). Two other crucial references to veillance include "in press" papers by Michael and Michael (2013) and Michael, Michael, and Perakslis (2014). But what does *veillance* mean? And how is it understood in different contexts? What does it mean to be watched by a *CCTV* camera, to have one's personal details deeply scrutinized; to watch another; or to watch oneself?

Figure 2. Uberveillance triquetra (Hayes, 2010). See also Michael and Michael (2007).

Dataveillance (see Interview 1.1) conceived by Roger Clarke of the Australian National University (ANU) in 1988 "is the systematic use of personal data systems in the investigation or monitoring of the actions or communications of one or more persons" (Clarke, 1988a). According to the Oxford Dictionary, dataveillance is summarized as "the practice of monitoring the online activity of a person or group" (Oxford Dictionary, 2013). It is hard to believe that this term was introduced a quarter of a century ago, in response to government agency data matching initiatives linking taxation records and social security benefits, among other commercial data mining practices. At the time it was a powerful statement in response to the Australia Card proposal in 1987 (Clarke, 1988b) which was never implemented by the Hawke Government, despite the Howard Government attempts to introduce an Access Card almost two decades later in 2005 (Australian Privacy Foundation, 2005). The same issues ensue today, only on a more momentous magnitude with far more consequences and advanced capabilities in analytics, data storage, and converging systems.

Sousveillance (see chapter 2) conceived by Steve Mann of the University of Toronto in 2002 but practiced since at least 1995 is the "recording of an activity from the perspective of a participant in the activity" (Wordnik, 2013). However, its initial introduction into the literature came in the inaugural publication of the *Surveillance and Society* journal in 2003 with a meaning of "inverse surveillance" as a counter to organizational surveillance (Mann, Nolan, & Wellman, 2003). Mann prefers to interpret sousveillance as under-sight which maintains integrity, contra to surveillance as over-sight which equates to hypocrisy (Mann, 2004).

Whereas dataveillance is the systematic use of personal data systems in the monitoring of people, sousveillance is the inverse of monitoring people; it is the continuous capture of personal experience. For example, dataveillance might include the linking of someone's tax file number with their bank account details and communications data. Sousveillance on the other hand, is a voluntary act of logging what one might see around them as they move through the world. Surveillance is thus considered watching from above, whereas sousveillance is considered watching from below. In contrast, dataveillance is the monitoring of a person's online activities, which presents the individual with numerous social dangers (Clarke, 1988a).

Uberveillance (see chapter 1) conceived by MG Michael of the University of Wollongong (UOW) in 2006, is commonly defined as: "ubiquitous or pervasive electronic surveillance that is not only 'always on' but 'always with you,' ultimately in the form of bodily invasive surveillance" (ALD, 2010). The term entered the *Macquarie Dictionary of Australia* officially in 2008 as "an omnipresent electronic surveillance facilitated by technology that makes it possible to embed surveillance devices in the human body" (Macquarie, 2009, p. 1094). The concern over uberveillance is directly related to the *misinformation*, *misinterpretation*, and *information manipulation* of citizens' data. We can strive for omnipresence through real-time remote sharing and monitoring, but we will never achieve simple omniscience (Michael & Michael, 2009).

Uberveillance is a compound word, conjoining the German *über* meaning over or above with the French *veillance*. The concept is very much linked to Friedrich Nietzsche's vision of the *Übermensch*, who is a man with powers beyond those of an ordinary human being, like a super-man with amplified abilities (Honderich, 1995; M. G. Michael & Michael, 2010b). Uberveillance is analogous to embedded devices that quantify the self and measure indiscriminately. For example, heart, pulse, and temperature sensor readings emanating from the body in binary bits wirelessly, or even through amplified eyes such as inserted contact lens "glass" that might provide visual display and access to the Internet or social networking applications.

Uberveillance brings together all forms of watching from above and from below, from machines that move to those that stand still, from animals and from people, acquired involuntarily or voluntarily using obtrusive or unobtrusive devices (Figure 3) (K. Michael, et al., 2010). The network infrastructure underlies the ability to collect data direct from the sensor devices worn by the individual, and big data analytics ensures an interpretation of the unique behavioral traits of the individual implying more than just predicted movement, but intent and thought (K. Michael & Miller, 2013).

It has been said that uberveillance is that part of the veillance puzzle that brings together the *sur*, *data*, and *sous* to an intersecting point (Stephan, et al., 2012). In uberveillance, there is the "watching" from above component (*sur*), there is the "collecting" of personal data and public data for mining (*data*), and there is the watching from below (*sous*) which can draw together social networks and strangers, all coming together via wearable and implantable devices on/in the human body. Uberveillance can be used for good but we contend that independent of its application for non-medical purposes, it will *always* have an underlying control factor of power and authority (Masters & Michael, 2005; Gagnon, et al., 2013).

Section 2 is dedicated to applications of humancentric implantables in both the medical and non-medical space. Chapter 4 is written by professor of cybernetics, Kevin Warwick at the University of Reading and his senior research fellow, Dr Mark Gasson. In 1998, Warwick was responsible for Cyborg 1.0, and later Cyborg 2.0 in 2002. In chapter 4, Warwick and Gasson describe implants, tracking and monitoring functionality, Deep Brain Stimulation (DBS), and magnetic implants. They are pioneers in the implantables arena but after initially investigating ID and location interactivity in a closed campus environment using humancentric RFID approaches, Warwick has begun to focus his efforts on medi-

Figure 3. From surveillance to uberveillance (K. Michael, et al., 2009b)

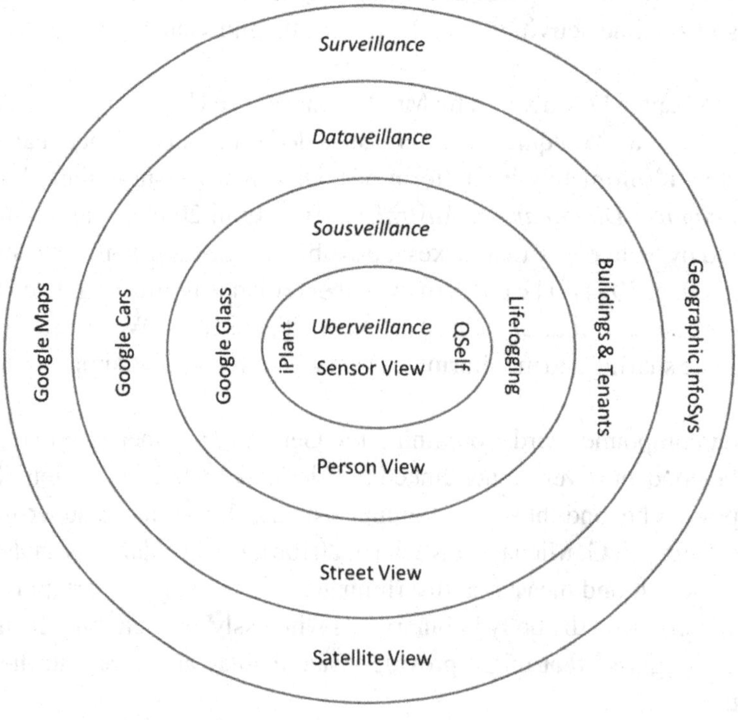

cal solutions that can aid the disabled, teaming up with Professor Tipu Aziz, a neurosurgeon from the University of Oxford. He has also explored person-to-person interfaces using the implantable devices for bi-directional functionality.

Following on from the Warwick and Gasson chapter are two interviews and a modified presentation transcript demonstrating three different kinds of RFID implant applications. Interview 4.1 is with Mr Serafin Vilaplana the former IT Manager at the Baja Beach Club who implemented the RFID implants for club patronage in Barcelona, Spain. The RFID implants were used to attract VIP patrons, perform basic access control, and be used for electronic payments. Katina Michael had the opportunity to interview Serafin after being invited to attend a Women's in Engineering (WIE) Conference in Spain in mid-2009 organised by the Georgia Institute of Technology. It was on this connected journey that Katina Michael also met with Mark Gasson during a one day conference at the London School of Economics for the very first time, and they discussed a variety of incremental innovations in RFID.

In late May 2009, Mr Gary Retherford, a Six Sigma black belt specialising in Security, contacted Katina to be formally interviewed after coming across the Michaels' work on the Internet. Retherford was responsible for instituting the Citywatcher.com employee access control program using the VeriChip implantable device in 2006. Interview 4.2 presents a candid discussion between Retherford and K. Michael on the risk versus reward debate with respect to RFID implantables. While Retherford can see the potential for ID tokens being embedded in the body, Michael raises some very important matters with respect to security questions inherent in RFID. Plainly, Michael argues that if we invite technology into the body, then we are inviting a whole host of computer "connectedness" issues (e.g. viruses, denial-of-service-attacks, server outages, susceptibility to hacking) into the human body as well. Retherford believes that these are matters that can be overcome with the right technology, and predicts a time that RFID implant maintenance may well be as straightforward as visiting a Local Service Provider (LSP).

Presentation 4.3 was delivered at IEEE ISTAS10 by Mr Amal Graafstra and can be found on the Internet here: http://www.youtube.com/watch?v=kraWt1adY3k. This chapter presents the Do-It-Yourselfer perspective, as opposed to getting an implant that someone else uses in their operations or commercial applications. Quite possibly, the DIY culture may have an even greater influence on the diffusion of RFID implantables than even the commercial arena. DIYers are usually circumspect of commercial RFID implant offerings which they cannot customise, or for which they need an implant injected into a predefined bodily space which they cannot physically control. Graafstra's published interview in 2009, as well as his full-length paper on the RFID subculture with K. Michael and M.G. Michael (2010), still stand as the most informative dialogue on the motivations of DIYers. Recently, in 2012, Graafstra began his own company DangerousThings.com touting the benefits of RFID implantables within the DIY/hacking community. Notably, a footer disclaimer statement reads: "Certain things sold at the Dangerous Things Web shop are dangerous. You are purchasing, receiving, and using the items you acquired here at your own peril. You're a big boy/girl now, you can make your own decisions about how you want to use the items you purchase. If this makes you uncomfortable, or you are unable to take personal responsibility for your actions, don't order!"

Chapter 5 closes section 2, and is written by Maria Burke and Chris Speed on applications of technology with an emphasis on memory, knowledge browsing, knowledge recovery, and knowledge sharing. This chapter reports on outcomes from research in the Tales of Things Electronic Memory (TOTeM) large grant in the United Kingdom. Burke and Speed take a fresh perspective of how technology is influencing societal and organisational change by focusing on Knowledge Management (KM). While the chapter does not explicitly address RFID, it rather explores technologies already widely diffused under the broad category of tagging systems, such as quick response codes, essentially 2D barcodes. The authors also do not fail to acknowledge that tagging systems rely on underlying infrastructure, such as wireless networks and the Internet more broadly through devices we carry such as smartphones. In the context of this book, one might also look at this chapter with a view of how memory aids might be used to support an ageing population, or those suffering with Alzheimer's disease for example.

Section 3 is about the adoption of RFID tags and transponders by various demographics. Christine Perakslis examines the willingness to adopt RFID implants in chapter 6. She looks specifically at how personality factors play a role in the acceptance of uberveillance. She reports on a preliminary study, as well as comparing outcomes from two separate studies in 2005 and 2010. In her important findings, she discusses RFID implants as lifesaving devices, their use for trackability in case of an emergency, their potential to increase safety and security, and to speed up airport checkpoints. Yet the purpose of the Perakslis study is not to identify implantable applications as such but to investigate differences between and among personality dimensions and levels of willingness toward implanting an RFID chip in the human body. Specifically, Perakslis examines the levels of willingness toward the uberveillance trajectory using the Myers Briggs Type Indicator (MBTI).

Interview 6.1 Katina Michael converses with a 16-year-old male from Campbelltown, NSW, about tattoos, implants, and amplification. The interview is telling with respect to the prevalence of the "coolness" factor and group dynamics in youth. Though tattoos have traditionally been used to identify with an affinity group, we learn that implants would only resonate with youth if they were functional in an advanced manner, beyond just for identification purposes. This interview demonstrates the intrinsic connection between technology and the youth sub-culture which will more than likely be among the early adopters of implantable devices, yet at the same time remain highly susceptible to peer group pressure and brand driven advertising.

In chapter 7, Randy Basham considers the potential for RFID chip technology use in the elderly for surveillance purposes. The chapter not only focuses on adoption of technology but emphasises the value conflicts that RFID poses to the elderly demographic. Among these conflicts are resistance to change, technophobia, matters of informed consent, the risk of physical harm, Western religious opposition, concerns over privacy and GPS tracking, and transhumanism. Basham who sits on the Human Services Information Technology Applications (HUSITA) board of directors provides major insights to resistance to change with respect to humancentric RFID. It is valuable to read Basham's article alongside the earlier interview transcript of Gary Retherford, to consider how new technologies like RFID implantables may be diffused widely into society. Minors and the elderly are particularly dependent demographics in this space and require special attention. It is pertinent to note, that the protests by CASPIAN led by Katherine Albrecht in 2007 blocked the chipping of elderly patients who were suffering with Alzheimer's Disease (Lewan, 2007; ABC, 2007). If one contemplates on the trajectory for technology crossover in the surveillance atmosphere, one might think on an implantable solution with a Unique Lifetime Identifier (ULI) which follows people from cradle-to-grave and becomes the fundamental componentry that powers human interactions.

Section 4 draws on laws, directives, regulations and standards with respect to challenges arising from the practice of uberveillance. Chapter 8 investigates how the collection of DNA profiles and samples in the United Kingdom is fast becoming uncontrolled. The National DNA Database (NDNAD) of the UK has more than 8% of the population registered with much higher proportions for minority groups, such as the Black Ethnic Minority (BEM). Author Katina Michael argues that such practices drive further adoption of what one could term, national security technologies. However, developments and innovations in this space are fraught with ethical challenges. The risks associated with familial searching as overlaid with medical research, further compounds the possibility that people may carry a microchip implant with some form of DNA identifier as linked to a Personal Health Record (PHR). This is particularly pertinent when considering the European Union (EU) decision to step up cross-border police and judicial cooperation in EU countries in criminal matters, allowing for the exchange of DNA profiles between the authorities responsible for the prevention and investigation of criminal offences (see Prüm Treaty).

Chapter 9 presents outcomes from a large Australian Research Council-funded project on the night time economy in Australia. In this chapter, ID scanners and uberveillance are considered in light of trade-offs between privacy and crime prevention. Does instituting ID scanners prevent or minimise crime in particular hot spots or do they simply cause a chilling effect and trigger the redistribution of crime to new areas. Darren Palmer and his co-authors demonstrate how ID scanners are becoming a normalized precondition of entry into one Australian nighttime economy. They demonstrate that the implications of technological determinism amongst policy makers, police and crime prevention theories need to be critically assessed and that the value of ID scanners needs to be reconsidered in context. In chapter 10, Jann Karp writes on global tracking systems in Australian interstate trucking. She investigates driver perspectives and attitudes on the modern practice of fleet management, and on the practice of tracking vehicles and what that means to truck drivers. Whereas chapter 9 investigates the impact of emerging technology on consumers, chapter 10 gives an employee perspective. While Palmer et al. question the effectiveness of ID scanners in pubs and clubs, Karp poses the challenging question- is locational surveillance of drivers in the trucking industry helpful or is it a hindrance?

Chapter 11 provides legislative developments in tracking, in relation to the "Do Not Track" initiatives written by Mark Burdon et al. The chapter focuses on online behavioral profiling, in contrast to chapter 8 that focuses on DNA profiling and sampling. US legislative developments are compared with those in the European Union, New Zealand, Canada and Australia. Burdon et al. provide an excellent analysis of the problems. Recommendations for ways forward are presented in a bid for members of our communities to be able to provide meaningful and educated consent, but also for the appropriate regulation of transborder information flows. This is a substantial piece of work, and one of the most informative chapters on Do Not Track initiatives available in the literature.

Chapter 12 by Kyle Powys Whyte and his nine co-authors from Michigan State University completes section 4 with a paper on the emerging standards in livestock industry. The chapter looks at the benefits of nanobiosensors in livestock traceability systems but does not neglect to raise the social and ethical dimensions related to standardising this industry. Whyte et al. argue that future development of nano-biosensors should include processes that engage diverse actors in ways that elicit productive dialogue on the social and ethical contexts. A number of practical recommendations are presented at the conclusion of the chapter, such as the role of "anticipatory governance" as linked to Science and Technology Studies (STS). One need only consider the findings of this priming chapter, and how these results may be applied in light of the relationship between non-humancentric RFID and humancentric RFID chipping. Indeed, the opening sentence of the chapter points to the potential: "uberveillance of humans will emerge through embedding chips within nonhumans in order to monitor humans."

Section 5 contains the critical chapter dedicated to the health implications of microchipping living things. In chapter 13, Katherine Albrecht uncovers significant problems related to microchip-induced cancer in mice or rats (2010). A meta-data analysis of eleven clinical studies published in oncology and toxicology journals between 1996 and 2006 are examined in detail in this chapter. Albrecht goes beyond the prospective social implications of microchipping humans when she presents the physical adverse reactions to implants in animals. Albrecht concludes her chapter with solid recommendations for policy-makers, veterinarians, pet owners, and oncology researchers, among others. When the original report was first launched (http://www.antichips.com/cancer/), Todd Lewan (2007) of the Associated Press had an article published in the *Washington Post* titled, "Chip Implants Linked to Animal Tumors." Albrecht is to be commended for this pioneering study, choosing to focus on health related matters which will increasingly become relevant in the adoption of invasive and pervasive technologies.

The sixth and final section addresses the emerging socio-ethical implications of RFID tags and transponders in humans. Chapter 14 addresses some of the underlying philosophical aspects of privacy within pervasive surveillance. Alan Rubel chooses to investigate the commercial arena, penal supervision, and child surveillance in this book chapter. He asks: what is the potential for privacy loss? The intriguing and difficult question that Rubel attempts to answer is whether privacy losses (and gains) are morally salient. Rubel posits that determining whether privacy loss is morally weighty, or of sufficient moral weight to give rise to a right to privacy, requires an examination of reasons why privacy might be valuable. He describes both instrumental value and intrinsic value and presents a brief discussion on surveillance and privacy value.

Panel 14.1 is a slightly modified transcription of the debate over microchipping people recorded at IEEE ISTAS10 (https://www.youtube.com/watch?v=dI3Rps-VFdo). This distinguished panel is chaired by lawyer William Herbert. Panel members included, Rafael Capurro, who was a member of the European Group on Ethics in Science and New Technologies (EGE), and who co-authored the landmark Opinion piece published in 2005 "On the ethical aspects of ICT implants in the human body." Capurro, who is the director for the International Center for Information Ethics, was able to provide a highly specialist ethical contribution to the panel. Mark Gasson and Amal Graafstra, both of whom are RFID implantees, introduced their respective expert testimonies. Chair of the Australian Privacy Foundation Roger Clarke and CASPIAN director Katherine Albrecht represented the privacy and civil liberties positions in the debate. The transcript demonstrates the complexity and multi-layered dimensions surrounding humancentric RFID, and the divisive nature of the issues at hand: on whether to microchip people, or not.

In chapter 15 we are introduced to the development of brain computer interfaces, brain machine interfaces and neuromotor prostheses. Here Ellen McGee examines sophisticated technologies that are used for more than just identification purposes. She writes of brain implants that are surgically implanted and affixed, as opposed to simple implantable devices that are injected in the arm with a small injector kit. These advanced technologies will allow for radical enhancement and augmentation. It is clear from McGee's fascinating work that these kinds of leaps in human function and capability will cause major ethical, safety, and justice dilemmas. McGee clearly articulates the need for discourse and regulation in the broad field of neuroprosthetics. She especially emphasises the importance of privacy and autonomy. McGee concludes that there is an urgent need for debate on these issues, and questions whether or not it is wise to pursue such irreversible developments.

Ronnie Lipschutz and Rebecca Hester complement the work of McGee, going beyond the possibilities to making the actual assumption that the human will assimilate into the cellular society. They proclaim "We are the Borg!" And in doing so point to a future scenario where not only bodies are read, but minds as well. They describe "re(b)organization" as that new phenomenon that is occurring in our society today. Chapter 16 is strikingly challenging for this reason, and makes one speculate *what* or *who* are the driving forces behind this cyborgization process. This chapter will also prove of special interest for those who are conversant with Cartesian theory. Lipschutz and Hester conclude by outlining the very real need for a legal framework to deal with hackers who penetrate biodata systems and alter individual's minds and bodies, or who may even kill a person by tampering with or reprogramming their medical device remotely.

Interview 16.1 (see chapter 16 Appendix) directly alludes to this cellular society. Videographer Jordan Brown interviews Katina Michael on the notion of the "screen bubble." What is the screen culture doing to us? Rather than looking up as we walk around, we divert our attention to the screen in the form of a smart phone, iPad, or even a digital wearable glass device. We look down increasingly, and not at each other. We peer into lifeless windows of data, rather than peer into one another's eyes. What could this mean and what are some of the social implications of this altering of our natural gaze? The discussion between Brown and K. Michael is applicable to not just the implantables space, but to the wearables phenomenon as well.

The question of faith in a data driven and information-saturated society is adeptly addressed by Marcus Wigan in the Epilogue. Wigan calls for a new moral imperative. He asks the very important question in the context of "who are the vulnerable now?" What is the role of information ethics, and where should targeted efforts be made to address these overarching issues which affect all members of society- from children to the elderly, from the employed to the unemployed, from those in positions of power to the powerless. It is the emblematic conclusion to a book on uberveillance.

Katina Michael
University of Wollongong, Australia

M.G. Michael
University of Wollongong, Australia

REFERENCES

ABC. (2007). Alzheimer's patients lining up for microchip. *ABCNews*. Retrieved from http://abcnews.go.com/GMA/OnCall/story?id=3536539

Albrecht, K. (2010). Microchip-induced tumors in laboratory rodents and dogs: A review of the literature 1990–2006. In *Proceedings of IEEE International Symposium on Technology and Society (ISTAS10)*. Wollongong, Australia: IEEE.

Ali, A., & Mann, S. (2013). The inevitability of the transition from a surveillance-society to a veillance-society: Moral and economic grounding for sousveillance. In *Proceedings of IEEE International Symposium on Technology and Society (ISTAS13)*. Toronto, Canada: IEEE.

Australian Privacy Foundation. (2005). Human services card. *Australian Privacy Foundation*. Retrieved 6 June 2013, from http://www.privacy.org.au/Campaigns/ID_cards/HSCard.html

Clarke, R. (1988a). Information technology and dataveillance. *Communications of the ACM, 31*(5), 498–512. doi:10.1145/42411.42413

Clarke, R. (1988b). Just another piece of plastic in your wallet: The 'Australian card' scheme. *ACM SIGCAS Computers and Society, 18*(1), 7–21. doi:10.1145/47649.47650

Gagnon, M., Jacob, J. D., & Guta, A. (2013). Treatment adherence redefined: A critical analysis of technotherapeutics. *Nursing Inquiry, 20*(1), 60–70. doi:10.1111/j.1440-1800.2012.00595.x PMID:22381079

Graafstra, A. (2009). Interview 14.2: The RFID do-it-yourselfer. In K. Michael, & M. G. Michael (Eds.), *Innovative automatic identification and location based services: from bar codes to chip implants* (pp. 427–449). Hershey, PA: IGI Global.

Graafstra, A., Michael, K., & Michael, M. G. (2010). Social-technical issues facing the humancentric RFID implantee sub-culture through the eyes of Amal Graafstra. In *Proceedings of IEEE International Symposium on Technology and Society (ISTAS10)*. Wollongong, Australia: IEEE.

Hayes, A. (2010). *Uberveillance (triquetra)*. Retrieved 6 May 2013, from http://archive.org/details/Uberveillancetriquetra

Hayes, A., Mann, S., Aryani, A., Sabbine, S., Blackall, L., Waugh, P., & Ridgway, S. (2013). Identity awareness of research data in veillance and social computing. In *Proceedings of IEEE International Symposium on Technology and Society (ISTAS13)*. Toronto, Canada: IEEE.

Kerr, I., & Mann, S. (n.d.). Exploring equiveillance. *ID TRAIL MIX*. Retrieved 26 September 2013 from http://wearcam.org/anonequiveillance.htm

Levin, I. (1970). *This perfect day: A novel*. New York: Pegasus.

Lewan, T. (2007, September 8). Chip implants linked to animal tumors. *Washington Post*. Retrieved from http://www.washingtonpost.com/wp-dyn/content/article/2007/09/08/AR2007090800997_pf.html

Macquarie. (2009). Uberveillance. In S. Butler (Ed.), *Macquarie dictionary* (5th ed.). Sydney, Australia: Sydney University.

Mann, S. (2004). Sousveillance: Inverse surveillance in multimedia imaging. In *Proceedings of the 12th Annual ACM International Conference on Multimedia*. New York, NY: ACM.

Mann, S., Nolan, J., & Wellman, B. (2003). Sousveillance: Inventing and using wearable computing devices for data collection in surveillance environments. *Surveillance & Society, 1*(3), 331–355.

Masters, A., & Michael, K. (2005). Humancentric applications of RFID implants: The usability contexts of control, convenience and care. In *Proceedings of the Second IEEE International Workshop on Mobile Commerce and Services*. Munich, Germany: IEEE Computer Society.

Michael, K. (2003). The automatic identification trajectory. In E. Lawrence, J. Lawrence, S. Newton, S. Dann, B. Corbitt, & T. Thanasankit (Eds.), *Internet commerce: Digital models for business*. Sydney, Australia: John Wiley & Sons.

Michael, K. (2012). Israel, Palestine and the benefits of waging war through Twitter. *The Conversation*. Retrieved 22 November 2012, from http://theconversation.com/israel-palestine-and-the-benefits-of-waging-war-through-twitter-10932

Michael, K. (2013a). High-tech lust. *IEEE Technology and Society Magazine, 32*(2), 4–5. doi:10.1109/MTS.2013.2259652

Michael, K. (Ed.). (2013b). Social implications of wearable computing and augmediated reality in every day life. In *Proceedings of IEEE Symposium on Technology and Society*. Toronto, Canada: IEEE.

Michael, K., McNamee, A., & Michael, M. G. (2006). The emerging ethics of humancentric GPS tracking and monitoring. In *Proceedings of International Conference on Mobile Business*. Copenhagen, Denmark: IEEE Computer Society.

Michael, K., & Michael, M. G. (Eds.). (2007). *From dataveillance to überveillance and the realpolitik of the transparent society*. Wollongong, Australia: Academic Press.

Michael, K., & Michael, M. G. (2009a). *Innovative automatic identification and location-based services: From bar codes to chip implants*. Hershey, PA: IGI Global. doi:10.4018/978-1-59904-795-9

Michael, K., & Michael, M. G. (2009c). Predicting the socioethical implications of implanting people with microchips. *PerAda Magazine*. Retrieved from http://www.perada-magazine.eu/view.php?article=1598-2009-04-02&category=Citizenship

Michael, K., & Michael, M. G. (2009d). Teaching ethics in wearable computing: The social implications of the new 'veillance'. *EduPOV*. Retrieved June 18, from http://www.slideshare.net/alexanderhayes/2009-aupov-main-presentation?from_search=3

Michael, K., & Michael, M. G. (2010). Implementing namebers using implantable technologies: The future prospects of person ID. In J. Pitt (Ed.), *This pervasive day: The potential and perils of pervasive computing* (pp. 163–206). London: Imperial College London.

Michael, K., & Michael, M. G. (2011). The social and behavioral implications of location-based services. *Journal of Location-Based Services*, *5*(3-4), 121–137. doi:10.1080/17489725.2011.642820

Michael, K., & Michael, M. (2013). No limits to watching? Communications of the ACM, 56(11), 26-28. Association for Computing Machinery. doi:10.1145/2527187

Michael, K., Michael, M. G., & Abbas, R. (2009b). *From surveillance to uberveillance (Australian Research Council Discovery Grant Application)*. Wollongong, Australia: University of Wollongong.

Michael, K., Michael, M. G., & Ip, R. (2008). Microchip implants for humans as unique identifiers: A case study on VeriChip. In *Proceedings of Conference on Ethics, Technology, and Identity*. Delft, The Netherlands: Delft University of Technology.

Michael, K., Michael, M. G., & Perakslis, C. (2014). Be vigilant: There are limits to veillance. In J. Pitt (Ed.), *The computer after me*. London: Imperial College Press.

Michael, K., & Miller, K. W. (2013). Big data: New opportunities and new challenges. *IEEE Computer*, *46*(6), 22–24. doi:10.1109/MC.2013.196

Michael, K., Roussos, G., Huang, G. Q., Gadh, R., Chattopadhyay, A., & Prabhu, S. (2010). Planetary-scale RFID Services in an age of uberveillance. *Proceedings of the IEEE*, *98*(9), 1663–1671. doi:10.1109/JPROC.2010.2050850

Michael, M. G. (2000). For it is the number of a man. *Bulletin of Biblical Studies*, *19*, 79–89.

Michael, M. G., & Michael, K. (2009). Uberveillance: Microchipping people and the assault on privacy. *Quadrant, 53*(3), 85–89.

Michael, M. G., & Michael, K. (2010). Towards a state of uberveillance. *IEEE Technology and Society Magazine, 29*(2), 9–16. doi:10.1109/MTS.2010.937024

Minsky, M. (2013). The society of intelligent veillance. In *Proceedings of IEEE International Symposium on Technology and Society* (ISTAS13). Toronto, Canada: IEEE.

Nolan, D. (2013, June 7). The human cloud. *Monolith*. Retrieved from http://www.monolithmagazine. co.uk/the-human-cloud/

Oxford Dictionary. (2012). *Dataveillance*. Retrieved 6 May 2013, from http://oxforddictionaries.com/ definition/english/surveillance

Paechter, B., Pitt, J., Serbedzijac, N., Michael, K., Willies, J., & Helgason, I. (2011). Heaven and hell: Visions for pervasive adaptation. In *Fet11 essence*. Budapest, Hungary: Elsevier. doi:10.1016/j. procs.2011.12.025

Paterson, N. (2013). Veillances: Protocols & network surveillance. In *Proceedings of IEEE International Symposium on Technology and Society* (ISTAS13). Toronto, Canada: IEEE.

Pitt, J. (Ed.). (2012). *This pervasive day: The potential and perils of pervasive computing*. London: Imperial College London.

Pitt, J. (2014). *The computer after me*. London: Imperial College Press.

Reynolds, M. (2004). Despite the hype, microchip implants won't deliver security. *Gartner*. Retrieved 6 May 2013, from http://www.gartner.com/DisplayDocument?doc_cd=121944

Rodotà, S., & Capurro, R. (2005). Ethical aspects of ICT implants in the human body. *Opinion of the European Group on Ethics in Science and New Technologies to the European Commission, 20*.

Shih, T. K. (2013). Video forgery and motion editing. In *Proceedings of International Conference on Advances in ICT for Emerging Regions*. ICT.

Stephan, K. D., Michael, K., Michael, M. G., Jacob, L., & Anesta, E. (2012). Social implications of technology: Past, present, and future. *Proceedings of the IEEE, 100*(13), 1752–1781. doi:10.1109/ JPROC.2012.2189919

(1995). Superman. InHonderich, T. (Ed.), *Oxford companion to philosophy*. Oxford, UK: Oxford University Press.

(2010). Uberveillance. InALD (Ed.), *Australian law dictionary*. Oxford, UK: Oxford University Press.

Warwick, K. (2002). *I, cyborg*. London: Century.

Wordnik. (2013). *Sousveillance*. Retrieved 6 June 2013, from http://www.wordnik.com/words/sousveillance

Acknowledgment

Piglet noticed that even though he had a Very Small Heart, it could hold a rather large amount of Gratitude. -A.A. Milne, Winnie-the-Pooh

Thank you to *Diane Walton Editorial Consulting* for her copyediting. Diane you were so enthusiastic throughout the project, and without you, this volume may never have reached completion. It was on more than one occasion that authors mentioned your conscientiousness.

Thank you to Jordan Brown who transcribed Gary Retherford's interview and who also transcribed his own video documentary interview with Katina Michael. Jordan, for one so young, you have an exceptional sense of direction when it comes to the social implications of technology. To Sarah Kepa, formerly of Indiana University, thank you for the transcription and translation of the interview with Serafin Vilaplana. Thank you to *The Transcription People* for the remaining transcriptions, and to Jessica Brown for addressing some last minute requirements.

To the awesome foursome who graduated during the completion of this book—Sarah Jean Fusco, Dr. Anas Aloudat, Dr. Roba Abbas, and Dr. Jennifer Heath—we love you dearly. You have been on this journey with us and you have each inspired.

A very special thank you to our PhD student, Alexander Hayes, who has supported us at every event we have hosted in the last 3 years – and whose own PhD on wearable technologies will indeed be groundbreaking.

To the University of Wollongong's Faculty of Informatics *Soaring Eagle* program administered by Professor Philip Ogunbona and Lyn Bosanquet in 2012, thank you! To our Dean of Engineering and Information Sciences, Professor Chris Cook, we so appreciate that you've never shied away from social policy discussions on technical matters.

To our communications team at UOW led by Bernie Goldie for all the exposure our work has received. We extend acknowledgment to the Australian Science Media Centre (AusSMC), to the science and technology editors at *The Conversation*, *ABC Science Online*, and *The Drum*, and investigative reporters at the Australian Broadcasting Corporation. To Michael Organ, UOW's archivist, for always keeping up-to-date with our research output. Thanks also to Vicky Wallace for her diligence in reporting on UOW research and for facilitating Katina's socio-ethics presentation at TEDxUWollongong's 2012 on the theme of "Medical Bionics."

To our amazingly gifted colleagues at IEEE SSIT amongst an extended list Professors Keith Miller, Luis Kun, Joe Herkert, Ken Foster, Michael Loui, Clinton Andrews, Karl Stephan, Jeff Robbins, and Ms Janet Rochester. We are especially indebted to Terri Bookman, Managing Editor of *T&S,* who has been a rich reservoir of knowledge and a true support throughout Katina's editorship. Our IEEE SSIT extended family continues to be a unique communal voice with a socially responsible backbone untethered by the demands to innovate aimlessly.

To the team at the *Macquarie Dictionary* who entered *uberveillance* into the official Australian dictionary in 2009, after the term was nominated for "word of the year" and went on to win the technology category in 2008. To the folks at the *Australian Law Dictionary* for their inclusion of the term uberveillance in the Oxford University Press volume, and to Australia's Les Murray, arguably one of the world's greatest living poets, for his essay titled: "Infinite Anthology: Adventures in Lexiconia" in *The Monthly,* citing uberveillance in 2010. Also to novelist Sam Yarney for his noteworthy use of the word, along with its broader implications in his international suspense thriller *The Banjo Player* (2010). All these are landmark inclusions in the growing body of supplementary literature surrounding the study of embedded surveillance devices.

A big thank you to Professor Kevin Haggerty for agreeing to contribute the foreword to the book long before it was completed – this was a special show of trust and it meant more to us than you might imagine.

To our long list of international expert Editorial Advisory Board members, you never tired of our questions and correspondence. You have not only been there for us through the development of this edited book, but right through to the international conferences and workshops and special issues we have hosted and edited in the past. The quality of this book is also a reflection of your diligence and your own expert research into emerging technologies. Thank you for your tireless reviews and for always being willing to give of your time in a voluntary capacity.

To our individual author contributors, some of whom we never knew before this project began. A massive thank you for your immense patience and for dedicating time from your busy lives to go through two sets of revisions – one to address reviewer comments, and then to address currency in terms of changes to your manuscripts reflecting real-world incremental innovations that had taken place since 2010. So much has happened over the last three years, thank you for sticking by us throughout! You have discerned what *uberveillance* is all about, tested its implications further in your own research, and have challenged us to additional explorations. And many thanks, of course, to each one of our interviewees. Your time, effort, and trust in our motivation was greatly appreciated and, you enriched this book.

To the publishers at IGI and our editorial directors, Joel Gamon and Christine Smith. Thanks for hanging in there when it seemed we just kept pushing all the deadlines!

To Professors Roger Clarke and Marcus Wigan, a special thank you for being there from the beginning. You have been great teachers and inspirational colleagues.

To Matthew Fedele-Sirotich, who asked me *that* question during the lecture that got me to searching for the "right" word: "So, then, where *is* all of this [surveillance] heading?" You were a fantastic student and we are incredibly proud of your progress in industry.

To our three little children who have been ever so patient with our busy writing schedules, we say thank you forever and ever. You are a big part of the reason we believe our research into the social implications of microchipping humans is so enormously important.

To our parents, we thank you greatly for the blessings and prayers. Without your support and succor, little of this research would have been possible.

This book is dedicated to you, dad. *Thank you so much for everything*. May you rest in peace (George Michael, 1924-2013) http://veillance.me/blog/2013/3/10/memories.

M.G. Michael
University of Wollongong, Australia

Katina Michael
University of Wollongong, Australia

Section 1
The Veillances

Chapter 1
Introduction:
On the "Birth" of Uberveillance

M. G. Michael
University of Wollongong, Australia

ABSTRACT

When or how uberveillance will be implemented in its full-blown manifestation is still a subject for some intriguing discussion and a topic of robust disagreement, but what is generally accepted by most of the interlocutors is that an "uberveillance society" will emerge sooner rather than later, and that one way or another this will mean an immense upheaval in all of our societal, business, and government relationships. What is apparent from the numerous qualitative and quantitative studies conducted is that microchipping people is a discernibly divisive issue. If we continue on the current trajectory, we will soon see further divisions – not just between those who have access to the Internet and those who do not, but between those who subjugate themselves to be physically connected to the Web of Things and People, and those who are content enough to simply have Internet connectivity through external devices like smart phones, to those who opt to live completely off the grid. Time will only tell how we as human-beings will adapt after we willingly adopt innovations with extreme and irreversible operations. This introduction serves to provide a background context for the term uberveillance, which has received significant international attention since its establishment.

Ultimately, the big choices must be made by citizens, who will either defend their freedom or surrender it, as others did in the past. -David Brin (1998), The Transparent Society.

INTRODUCTION

The conception of the word *uberveillance* came about during question time at the conclusion of a class I was guest lecturing on the "Consequences of Innovation" in May of 2006 at the University

DOI: 10.4018/978-1-4666-4582-0.ch001

of Wollongong. In that enthusiastic group of young men and women were a number of my former students. One of these, who was at the time completing his honors research project with Katina Michael, asked the key question: "So then, where is all this [surveillance] heading?" I pondered for a moment searching for a word or term which would summarize what I was 'seeing' in my mind's eye and what I had been reflecting upon for a long time: a coming together of Big Brother, dataveillance, microchip implants, RFID, GPS, A-LBS, Apocalypse (Rev 13), and Übermensch. There was nothing I could think of that would capture all of these indispensable components and hybrid architectures of the trajectory of electronic surveillance and information gathering, including the wider implications of the "technological society" as I had understood it from my study of Jacques Ellul and his analysis of "technique" (Ellul 1967). If technique is that component of technology which has *maximum efficiency*, that is, "the totality of methods" as its primary goal, then uberveillance can be understood in similar terms insofar as surveillance is concerned.

BACKGROUND

So here is something of the background that led to the birth of *Uberveillance* and a summary of the fundamental components of the term.

During my preparation for the class which would also include readings from Martin Heidegger (1982) *The Question Concerning Technology*, Paul Feyerabend (1978) *Against Method*, Everett M. Rogers (1995) *Diffusion of Innovations*, and Richard S. Rosenberg (2004) *The Social Impact of Computers*, I came across one of Franz Kafka's fascinating letters to Milena Jesenská on "intercourse with ghosts". A powerful albeit little critique on the underlying structures of industrial technology and the resulting consequences on communication, "[t]he ghosts won't starve, but we will perish" (Kafka 1999). Fyodor Dostoevsky

another of the great students on the conditions of bureaucracy and isolation, held similar reservations and concerns with a designed utopia, represented at the time by his experience of Saint Petersburg. *Notes from Underground*, still has much to say to 'Technological Man' in pursuit of the "golden dream" (Dostoevsky, 1992). Writers with these sorts of sensibilities and philosophical intelligence, such as Kafka and Dostoevsky, have fascinated me since my undergraduate days when I first stumbled upon them after reading Nietzsche with Paul Crittenden (2008) at Sydney University, more than thirty years ago. This genre of writing, roughly categorized "existentialist", awakened in me deep-seated sensitivities to do with abuses in bureaucracy and in the practices of the ruling elite.

I had also spent time thinking on Ray Kurzweil's (2005) "singularity" and the connection of exponential growth on the future prospects of surveillance. Later, having arrived in this 'place' after the convergence of a number of interrelated subjects, I would continue to discover many more intuitive and forward thinking authors in the emerging fields of privacy advocacy and surveillance studies. Authors, who would both inform and challenge me with their cutting-edge work. This is a long and imposing list from which I have had the privilege in a number of instances to have together presented at conferences, to have co-authored with, or to have published papers as an editor. One of the key texts that I would discover from that time was the seminal publication of David Lyon's *The Electronic Eye: The Rise of Surveillance Society* (1994).

Two other important works which have not dated on account of their continuing significance and which I turned to in the earlier years before the great and imposing flood of later surveillance and privacy literature are Simon Davies' (1992) *Big Brother: Australia's Growing Web of Surveillance*, and Anne Wells Branscomb's (1994) *Who Owns Information?* I would go back to these books, better still *testimonies*, when I feared that I might be reading too much into what I was finding or had been overly reliant on my intuition. The ultimate questions which Davies and Branscomb pose to do with the *gathering, storing,* and *distribution* of information, not only remain paramount but are even more momentous today.

The week before I delivered that guest lecture, and for reasons not necessarily connected to my presentation, I also revisited Anton Chekhov's discerning short story, *The Bet* (1999). Once more, I was drawn to the concluding paragraphs of this story. And in one place which will be patently obvious, I allowed for my imagination to go on a flight of fancy. I considered that Chekhov had also outside his penetrating critique of unfettered materialism and greed, by "accident" looked ahead to genetic engineering gone awry. And not unimportantly, we should remember, Anton Chekhov was also a physician:

And I despise your books. I despise all the blessings of this world, all its wisdom. Everything is worthless, transient, illusory and as deceptive as a mirage. You may be proud, wise and handsome, but death will wipe you from the face of the earth, together with the mice under the floorboards. And your posterity, your history, your immortal geniuses will freeze or be reduced to ashes, along with the terrestrial globe. You've lost all reason and are on the wrong path. You mistake lies for the truth and ugliness for beauty. You'd be surprised if apple and orange trees suddenly started producing frogs and lizards instead of fruit, or if roses smelt of sweaty horses. I'm amazed at people who have exchanged heaven for earth. I just don't want to understand you.

Like most things passed off as "new", surveillance itself is not new. Its antecedents are as old as our earliest creation accounts when the gods gazed out onto the cosmos they had created and "saw everything" (Gen 1:31). It is ironic that surveillance via low earth orbiting satellites is back up in the "heavens" scanning the entirety of

the earth in an astonishing 90 minutes. And so in response to our student's above-noted question, I was about to give up and settle for the stock response: 'Big Brother on a microchip' when at the last moment I came out with "uberveillance". This coining of the word at that particular moment is the least complicated part of the larger story of how the term itself emerged. The trajectory of surveillance technologies together with the underlying technological, social, and religious implications and impacts has been an ongoing interest of mine for the better part of my adult life. At least since the early 1980s when I first started to look for and collect materials of a diverse provenance connected to the idea of a *centralized* or *distributed* surveillance and of the technologies propagating these innovations. Therefore, it is enormously important to stress that the term itself was certainly not without an informed context or without a rich bibliography of previous research spanning from my initial tertiary studies in 1981 to the present times.

I have written two dissertations on the Book of Revelation- one of these examining the infamous "666" conundrum- and that cannot but be an influence when I look into some of the more unwelcome trajectories of mass surveillance and political terrorization. I am glad, however, to have been in a position to introduce the apocalyptic genre (with its mixed bag of eschatological anxieties and apprehensions) as a credible "sociological" resource in the ICT bibliography outside its more commonly held fundamentalist ascriptions (Michael, 1999; 2000). Using the historical-critical method for my studies in ancient history, I brought with me significant lessons that were to prove very useful in computing studies (Michael, 1998; 2002; 2010). I should add that a perceptive interviewer once quipped that "uberveillance" not only suggests sinister overtones, but that the word itself gives resonance to them. He was close to the mark. It is enough to read Primo Levi (1919-1987) and Alexander Solzhenitsyn (1918-2008), for instance, to get a realistic idea of what murderous regimes can do with surveillance and identity management systems (Levi, 1989; Solzhenitsyn, 2003).

As all good students do, they invite questions toward clarification, and so our very keen undergraduate went on to subsequently ask: "What *is* this uberveillance?" I remember replying something along the lines of it being a "super surveillance" technology which comprised of all that we had come to expect of George Orwell's big brother in *1984* (1949); Roger Clarke's dataveillance (1988); computer technology in terms of both unprecedented processing power (e.g. predictive analytics) and rapid miniaturization; Kevin Warwick's (2002) microchip implant experiments in 1998 (Cyborg 1.0) and 2002 (Cyborg 2.0); and the fact that apocalyptic scenarios were now no longer the exclusive domain of "fundamentalist" interpretations.

Later on when I looked further into the inventive research and writings of Steve Mann (2001), particularly to do with "wearables", sousveillance came into the equation. In fact, one of our doctoral candidates at UOW, Alexander Hayes (2012; 2013), has been studying the union of surveillance, dataveillance and sousveillance, and has used the triquetra to depict the underlying intersection into uberveillance. The three-looped triquetra with its ü intersection is a good summary of the interaction between the fundamental components when considered in the context of embedded surveillance devices (ESDs).

DEFINING UBERVEILLANCE

Uberveillance was, at least as I had originally understood it and presented it on that day, the disturbing technological scenario of putting all of this hybrid architecture on a microchip, and beyond just wearing it, having it *implanted* beneath the skin. In layman terms, uberveillance is a kind of CCTV on the "inside" looking out rather than

on the "outside" looking down: an "above and beyond", an exaggerated 24/7 surveillance embedded inside the human body. Locating, tracking, and real-time monitoring without any cease, that is, constant and unending. This condition not only begged the question of "the death of privacy" and of autonomous action but that of data integrity as well (*misinformation, misinterpretation,* and *information manipulation*). In addition, my contention that 24/7 surveillance would give rise to a new class of functional mental disorders and exacerbate already existing ones, is now commonly considered a legitimate concern and a subject of funded research (Michael & Michael, 2011).

It was only after I discussed this interesting exchange during the guest lecture with my fellow collaborator, Katina Michael, that I became convinced after listening to her technical references, that the word might indeed have legs and that we needed to explore the implications of this neologism further. By this time, Katina and I had been collaborating in one way or another for the greater part of 10 years, the culmination of some heavy-duty cross-disciplinary activities formalized in 2006 with a special issue which we were invited to guest edit on the social implications of national security in *Prometheus* (Michael & Michael, 2006). And also the launch of a long term workshop series on the *Social Implications of National Security* now in its eighth year, which was funded by the Australian Research Council's Research Network for a Secure Australia, (*RNSA*), between 2006-2013. I co-edited the proceedings of the first three volumes, the second of which in 2007 was entitled: "From Dataveillance to Uberveillance and the Realpolitik of the Transparent Society" (Michael & Michael, 2007). It was during the course of this workshop that Roger Clarke delivered the keynote address: "What 'uberveillance' is and what to do about it" (Clarke, 2007).

In 2009, around three years from its conception and after a great deal of hard work, many presentations, and open scrutiny, "uberveillance" was entered into the *Macquarie Dictionary* after having been earlier nominated for "Word of the Year" in 2008. The *Macquarie Dictionary* entry reads as follows: "an omnipresent electronic surveillance facilitated by technology that makes it possible to embed surveillance devices in the human body" (Butler, 2009). Another significant milestone was the inclusion of the term in the *Australian Law Dictionary* published by Oxford University Press. Mann and Blunden (2012) define uberveillance as: "ubiquitous or pervasive electronic surveillance that is not only 'always on' but 'always with you', ultimately in the form of bodily invasive surveillance".

Überveillance is a compound word, conjoining the German *über* meaning over or above with the French *veillance*. On the question of the umlaut, it was decided early on to omit the *u-mutation* principally for the purposes of search engine retrievals.

The concept is very much linked to Friedrich Nietzsche's vision of the Übermensch, who is a man with powers beyond those of an ordinary human being, like a super-man with amplified abilities. For example, heart, pulse, and temperature sensor readings emanating from the body in binary bits wirelessly, or even through amplified eyes such as contact "glass" that might provide visual display and access to the Internet or social networking applications. It has been said that uberveillance centralizes all the forms of watching (from above, from below, by collectives, by individuals) because the sensor devices carried or embedded in the body are the lowest common denominator in tracking elements- the individual. The network infrastructure underlies the ability to collect data direct from the sensor devices worn by the individual and big data analytics ensures an interpretation of the unique behavioral traits of the individual implying more than just predicted movement, but intent and thought. Uberveillance as having "to do with the fundamental who (ID), where (location), and when (time) questions in an attempt to derive why (motivation), what (result), and even how (method/plan/thought)" (Michael & Michael, 2010).

ON THE FLUIDITY OF UBERVEILLANCE IN COMPETING NARRATIVES

After the word had entered academic discourse, a number of researchers suggested the term "uber surveillance" as a more precise alternative to "uberveillance". Uber surveillance, I believe, would have been redundant in a "dictionary" which already included the "synonyms" of mass-surveillance, wholesale surveillance, or total surveillance. At the same time Bentham's famous "all-seeing" panopticon certainly captures the chief elements of our collective efforts to describe newer forms of surveillance. Uber surveillance, then, if I can put it this way, is one of the results of uberveillance. They are *not* the same thing. A good illustration of this is the translation of uberveillance into supervigilancia by some South American writers. This does not carry over the deeper content and underlying narrative of the word, for instance the Nietzschean, Orwellian and Apocalyptic components of uberveillance, but rather it endeavors to translate "uber surveillance" instead. By the way, I do not think that "supervigilancia" entirely expresses the idea of "uber surveillance" either, but it does tease out other significant implications such as supervision and control.

One of the things we now look forward to doing is to continue to demonstrate and to dig deeper into how the term has been used, and continues to be used, by other researchers worldwide towards scientific work across various disciplines including- health, business, ethics, management and military. And to also document its increasing reference in blogs, the media and popular culture. We remain passionate on questions dealing with the applied ethics in relation to surveillance, especially in places where the application of uberveillance would violate the body, that is to say, the "sacred space" or where surveillance in general would impact upon our abilities to act as free agents outside any "visible" or "invisible" coercion. My own personal interest has also been wound up by multi-national and corporate involvement (chiefly in the area of consumption analysis) in the creation of demand, supply, and merchandizing of "spy wear" and the inter-play of these global entities with governments.

Ultimately, it is not the word as such which matters, that is neither here nor there, but I consider its value and its usefulness by the broad impact it has had and the global discussion that it has ignited. For some readers tutored in social semiotics, "uberveillance" might largely function as a 'sign', something in the vein of Sartre's "nausea". I never set out to 'pre-emptively' create a word and though I could argue for its etymological credentials, it would under the circumstances be a redundant exercise. Given my original training in the humanities, which also included the study of linguistics under the legendary Michael (Mak) Halliday (2013), I have a great love and respect for words and a good understanding of their power, especially as to their heuristic and representational function in language. At the same time we know that new realities – "new" but still very much informed and illuminated by history- borne from fresh contexts and innovative applications, need to be defined and described in original vocabulary. And often enough this will mean 'breaking' the rules. The fundamental thing is that words are *not* meaningless. This is ultimately determined, not even by linguists, but by common usage and popular consent.

One of the vital positions of our research into uberveillance has been the conscious and ongoing effort to reach out to prominent researchers who have invested decades of their lives whether in the critique of new technologies or to their develop-ment and advancement. The contributors to this volume are wholly demonstrable of this ethos. Perhaps the more visible case in point here is the contribution of Katherine Albrecht (2006) and Kevin Warwick (2002). Both Albrecht and Warwick, who are not rarely set up as polar opposites on their respective position on RFID implantables and often enough 'demonized' by their criticizers, are long-time friends of the present editors and significant collaborators on other projects. Albrecht's documented concerns are considerable and unquestionably real and Warwick's pioneering research into implantables has brought (and will bring) healing to sufferers of severe and debilitating disease. Yet, he too understands the multitude of dangers to the abuse of his work and has not kept his apprehensions secret, even hiring an ethics expert to challenge his trajectories.

Certainly, the subject of *technotherapeutics* raises a new chapter of weighty questions and serious apprehensions to do with power and authority. It would be a great mistake in the current debate to marginalize such knowledgeable voices which shift our comfort zones and challenge our perspectives. One of the open-ended problems that we have faced and will continue to face is what Haggerty and Samatas (2010) point to in a marvelous essay, that generalizations particularly when it comes to surveillance are fraught with problems "given the dynamics and normative implications of different surveillance practices." The irony would be that in fighting 'big brother' we inadvertently give rise to privacy groups and organizations which become a 'law and state' unto themselves. And this remains one of my great anxieties and fears; only beneath my trepidation for *fully blown* uberveillance. We need to be highly discerning to what lies behind the electrified glass tubes which flash WYSIATI "what you see is all there is" (Kahneman, 2011). It is just not true above all when it comes to embedded surveillance devices (ESDs) and we will suffer the awful consequences if we do not ask for "more information".

CONCLUSION

Competing narratives and criticisms of the research underpinning Uberveillance are more than welcome. This is a complex and controversial field and we need to update and inform each other whether this has to do with new technologies, amendments to legislation, or simple and plain correction. Those of us who are genuinely concerned with the quickening erosion of our right to privacy as a fundamental component of our natural rights on which John Locke (2003) and the social contract thinkers had some momentous things to say, will achieve little if anything- at least in the long term- if we go about it alone or side-line and downgrade colleagues who might approach the debate a little differently or inform it from another perspective.

The question remains, why do researchers who believe that trajectories mapped out by engineers given the principle of exponential growth will invariably be realized in ubiquitous surveillance, continue to spend time and resources on the subject? The answer need not be intricate. It is because a large group of these researchers believe that ultimately *whatever* the cost individuals will still possess the freedom to decide to what extent they integrate themselves into the electronic grid. Additionally, philosophers who have contemplated on the question of technology and its impact on society such as Martin Heidegger, Ivan Illich, Jacques Ellul, and those from the Frankfurt School, have argued that technology must be vigorously critiqued for the worst of all possible outcomes would be the de-humanization of the individual and the loss of dignity resulting in a "standardized subject of brute self-preservation." One of the fundamental elements of such literature is the profound comprehension that technology has not only to do with *building* but that it is also a *social process*. Charlie Chaplin's "culturally significant" *Modern Times* (1936) is an unmatched visual accompaniment, the classic scene of the iconic Little Tramp caught up into the cogs of the giant machine, of the unintended consequences of the efficiencies of modern industrialization. A decade earlier Fritz Lang's futuristic *Metropolis* (1926) the story of a mechanized underground city set in a dystopian society, would likewise leave its indelible mark. It was a prescient summary of what was to follow, the troubling link between teleology and technology.

It is fitting to conclude with a recent citation from authors Lisa Shay et al. (2012) of the Cyber Research Center of the United States Military Academy in West Point, New York, which points to the maturation of the term uberveillance and to the realization of its potential consequences:

Roger Clarke's concept of dataveillance and M.G. Michael and Katina Michael's more recent uberveillance serve as important milestones in awareness of the growing threat of our instrumented world.

REFERENCES

Albrecht, K. (2006). *Spychips: How major corporations and government plan to track your every purchase and watch your every move*. New York: Plume.

Branscomb, A. W. (1994). *Who owns information?* New York: Basic Books.

Brin, D. (1998). *The transparent society*. Boston: Perseus Books.

Chekhov, A. (1999). *Later short stories, 1888-1903*. New York: Modern Library.

Clarke, R. (1988). Information technology and dataveillance. *Communications of the ACM, 31*(5), 498–512. doi:10.1145/42411.42413

Clarke, R. (2007). What 'überveillance' is, and what to do about it. In *Proceedings of 2nd RNSA Workshop on the Social Implications of National Security - From Dataveillance to Überveillance*. Retrieved from http://www.rogerclarke.com/DV/RNSA07.html

Clarke, R. (2012). The regulation of point-of-view surveillance in point of view technologies in law enforcement. In *Proceedings of the Sixth Workshop on the Social Implications of National Security*. Sydney, Australia: Wollongong University. Retrieved from http://www.rogerclarke.com/DV/PoVSR.html

Crittenden, P. (2008). *Changing orders: Scenes of clerical and academic life*. Sydney, Australia: Brandl & Schlesinger.

Davies, S. (1992). *Big brother: Australia's growing web of surveillance*. Sydney: Simon & Schuster.

Dostoevsky, F. (2000). *Notes from underground*. New York: W.W. Norton & Company.

Ellul, J. (1967). *The technological society* (J. Wilkinson, Trans.). New York: Vintage.

Feyerabend, P., & Hacking, I. (2010). *Against method*. London: Verso Books.

Haggerty, K., & Samatas, M. (2010). *Surveillance and democracy*. New York: Routledge-Cavandish.

Halliday, M. (2013). *Interviews with M.A.K. Halliday: Language turned back on himself*. London: J.R.Martin, Bloomsbury Academic.

Hayes, A. (2012). *Uberveillance triquetra*. Retrieved from http://archive.org/details/Uberveillancetriquetra

Hayes, A. (2013). Cyborg cops, googlers, and connectivism. *IEEE Technology and Society Magazine, 32*(1), 23–24. doi:10.1109/MTS.2013.2247731

Heidegger, M. (1977). *The question concerning technology, and other essays*. New York: Harper & Row.

Kafka, F. (1999). *Letters to milena*. New York: Vintage Classics.

Kahneman, D. (2011). *Thinking, fast and slow*. London: Penguin.

Kurzweil, R. (2005). *The singularity is near: When humans transcend biology*. New York: Penguin.

Levi, P. (1989). *The drowned and the saved*. New York: Vintage.

Locke, J. (2003). *Two treatises of government and a letter concerning toleration* (I. Shapiro, Ed.). New Haven, CT: Yale University Press.

Lyon, D. (1994). *The electronic eye: The rise of surveillance society*. Minneapolis, MN: University of Minnesota Press.

Mann, S., & Niedzviecki, H. (2001). *Cyborg: Digital destiny and human possibility in the age of the wearable computer*. Toronto, Canada: Doubleday.

Mann, T., & Blunden, A. (2012). Uberveillance. In *Australian law dictionary*. Oxford, UK: Oxford University Press.

Michael, K., & Michael, M. G. (2007). *From dataveillance to überveillance and the realpolitik of the transparent society*. Wollongong, Australia: University of Wollongong.

Michael, K., Roussos, G., Huang, G. Q., Gadh, R., Chattopadhyay, A., Prabhu, S., & Chu, P. (2010). Planetary-scale RFID services in an age of uberveillance. *Proceedings of the IEEE, 98*(9), 1663–1671. doi:10.1109/JPROC.2010.2050850

Michael, M. G. (1998). The number of the beast, 666 (revelation 13: 16-18), background, sources, and interpretation. New South Wales, Australia: Macquarie University.

Michael, M. G. (1999). The genre of the apocalypse: What are they saying now? *Bulletin of Biblical Studies*, *18*, 115–126.

Michael, M. G. (2000). For it is the number of a man. *Bulletin of Biblical Studies*, *19*, 79–89.

Michael, M. G. (2002). *The canonical adventure of the apocalypse of John: An Eastern Orthodox perspective*. New South Wales, Australia: Australian Catholic University.

Michael, M. G. (2010). Demystifying the number of the beast in the book of revelation: Examples of ancient cryptology and the interpretation of the 666 conundrum. In *Proceedings of IEEE International Symposium on Technology and Society*. IEEE.

Michael, M. G., Fusco, S. J., & Michael, K. (2008). A research note on ethics in the emerging age of überveillance. *Computer Communications*, *31*(6), 1192–1199. doi:10.1016/j.comcom.2008.01.023

Michael, M. G., & Michael, K. (2006). National security: The social implications of the politics of transparency. *Prometheus*, *24*(4), 359–364. doi:10.1080/08109020601029912

Michael, M. G., & Michael, K. (2007). *A note on überveillance: From dataveillance to überveillance and the realpolitik of the transparent society*. Wollongong, Australia: University of Wollongong.

Michael, M. G., & Michael, K. (2009). Uberveillance: Microchipping people and the assault on privacy. *Quadrant*, *53*(3), 85–89.

Michael, M. G., & Michael, K. (2010a). Surveillance and uberveillance. *IEEE Technology and Society Magazine*, *29*(2).

Michael, M. G., & Michael, K. (2010b). Towards a state of uberveillance. *IEEE Technology and Society Magazine*, *29*(2), 9–16. doi:10.1109/MTS.2010.937024

Michael, M. G., & Michael, K. (2011). The fallout from emerging technologies: On matters of surveillance, social networks and suicide. *IEEE Technology and Society Magazine*, *30*(3), 15–18. doi:10.1109/MTS.2011.942312

Nietzsche, F. (1990). *The twilight of the idols and the anti-Christ* (R. J. Hollingdale, Trans.). London: Penguin Classics.

Orwell, G. (1983). 1984. New York: Plume.

Rogers, E. M. (1995). *Diffusion of innovations: Modifications of a model for telecommunications*. New York: The Free Press.

Rosenberg, R. S. (2004). *The social impact of computers*. Los Angeles, CA: Elsevier.

Semple, J. (1993). *Bentham's prison: A study of the panopticon penitentiary*. Oxford, UK: Clarendon Press. doi:10.1093/acprof:oso/9780198273875.001.0001

Shay, L. A., Conti, G., Larkin, D., & Nelson, J. (2012). A framework for analysis of quotidian exposure in an instrumented world. In *Proceedings of IEEE Carnahan Conference on Security Technology*. IEEE.

Solzhenitsyn, A. (2003). *The Gulag archipelago*. London: The Harvill Press.

Stephan, K. D., Michael, K., Michael, M. G., Jacob, L., & Anesta, E. (2012). Social implications of technology: Past, present, and future. *Proceedings of the IEEE*, *100*(13), 1752–1781. doi:10.1109/JPROC.2012.2189919

(2009). Uberveillance. In Butler, S. (Ed.), *Macquarie Dictionary* (5th ed.). Sydney, Australia: Sydney University.

Warwick, K. (2002). *I, cyborg*. London: Century.

Weiss, J. H. (1982). *The making of technological man: The social origins of French engineering education*. Cambridge, MA: The MIT Press.

ADDITIONAL REFERENCES

Abbas, R., Michael, K., Michael, M. G., & Aloudat, A. (2011). Emerging Forms of Covert Surveillance Using GPS-Enabled Devices. *Journal of Cases on Information Technology, 13*(2), 19–33. Available at http://works.bepress.com/kmichael/224 doi:10.4018/JCIT.2011040102

ABC. (2007). You're being watched right now: in an era of 'Internet everywhere' everyone is being tracked all the time, *ABCAmericaOnline News*. http://abcnews.go.com/Business/FunMoney/Story?id=3937203&page=2

Allenby, B. R., & Sarewitz, D. R. (2011). *The Techno-Human Condition*. USA: The MIT Press.

Asher, J. (2009). Humans will be implanted with microchips, *ninemsn*, 30 Jan. Available from http://news.ninemsn.com.au/technology/735519/humans-will-be-implanted-with-microchips

Bendavid, Y., & Cassivi, L. (2010). Bridging the gap between RFID/EPC concepts, technological requirements and supply chain e-business processes. *Journal of Theoretical and Applied Electronic Commerce Research, 5*(3). http://www.jtaer.com/dec2010/bendavid_cassivi_p1.pdf doi:10.4067/S0718-18762010000300002

Capurro, R., Eldred, M., & Nagel, D. (2012). Digital Whoness: Identity, Privacy and Freedom in the Cyberworld, Berlin: ontos verlag. 310pp.

Cincotta, K., & Ashford, K. (2011). The New Privacy Predators. *Women's Health*, November.

Clarke, R. (2012). A Framework for Surveillance Analysis, February, http://www.rogerclarke.com/DV/FSA.html.

Dann, S. (2008). A Leximancer analysis of social marketing definitions versus social marketing literature. *Australia and New Zealand Marketing Academy Conference*, ed. D. Spanjaard, S. Denize, N. Sharma, Australian and New Zealand Marketing Academy (ANZMAC), Sydney.

de Almeida Amazonas, J. R. Opportunities, Challenges for Internet of Things Technologies. In (Eds.) Robert, J. Vermesan, O., & Friess, P. Internet of Things - Global Technological and Societal Trends From Smart, Aalborg, Denmark: River Publishers, 195-240.

Ekholm, K., & Karhula, P. (2012). Sleepwalking toward a control society? Ten Must-Know Trends, *IFLA*, http://www.ifla.org/publications/sleepwalking-toward-a-control-society-ten-must-know-trends

Ema, A., & Fujigaki, Y. (2011). How far can child surveillance go? Assessing the parental perceptions of an RFID child monitoring system in Japan. *Surveillance & Society, 9*(1/2), 132–148.

Ferenczi, P.M. (2009). You are Here. *Laptop*, February, 98-102.

Friggieri, A., & Michael, K. (2009). The legal ramifications of microchipping people in the United States of America- A state legislative comparison. ISTAS '09. *IEEE International Symposium on Technology and Society*, 18-20 May, Tempe: Arizona.

Gagnon, M., Jacob, J. D., & Guta, A. (2013). Treatment adherence redefined: a critical analysis of technotherapeutics. *Nursing Inquiry, 20*(1), 60–70. doi:10.1111/j.1440-1800.2012.00595.x PMID:22381079

Goldie, B. Charting unknown cyber-seas and the dangers for privacy, UOW Latest News, Sep. 2007. http://media.uow.edu.au/news/UOW037156.html

Goldie, B. (2009). Uberveillance Exhibition and Book Launch, University of Wollongong Latest News. Sep. 2009. http://media.uow.edu.au/news/UOW066120.html

Goldsmith, A., & Michael, K. (2009). Police Accountability in an Age of Uberveillance, *Australia and New Zealand Society of Criminology Conference: Crime and Justice Challenges in the 21st Century: Victims, Offenders and Communities*. Perth, Western Australia, Australia. Nov.

Gretzel, U. (2011). Intelligent systems in tourism: A Social Science Perspective. *Annals of Tourism Research*, *38*(3), 757–779. doi:10.1016/j.annals.2011.04.014

Gürses, S., Troncoso, C., & Diaz, C. (2011). *Engineering privacy by design*. Computers, Privacy & Data Protection.

Guta, A., Gagnon, M., & Jacob, J. D. (2012). Using Foucault to Recast the Telecare Debate. *The American Journal of Bioethics*, *12*(9), 57–59. doi:10.1080/15265161.2012.699140 PMID:22881861

Harfield, C. (2010). Transnational Criminal Investigation and Modes of Governance. *Policing. Journal of Policy Practice*, *5*(1), 3–14. doi:doi:10.1093/police/paq039

Harfield, C. (2012). E-Policing and the Social Contract. *IEEE Technology and Society Magazine*, *31*(1), 33–41. doi:10.1109/MTS.2012.2185272

Harrison, K., & Rainey, B. (2013). *The Wiley-Blackwell Handbook of Legal and Ethical Aspects of Sex Offender*. West Sussex: Wiley-Blackwell. doi:10.1002/9781118314876

Hayes, H., Mann, S., Aryani, A., Sabine, S., Blackall, L., Waugh, P., & Ridgway, S. (2013). Identity Awareness of Research Data in Veillance & Social Computing. *IEEE Symposium on Technology and Society (ISTAS13)*, http://veillance.me, Ed. Katina Michael, University of Toronto, Canada, 27-29 June.

Heath, J. (2011). Consumers, ALRC Privacy Principles and the 2010 Healthcare Identifiers Act, *Telecommunications Journal of Australia*, *61*(3), 46.1-46.8.

Herold, R. (2007). 6 Scary Stuff Privacy Terms IT, Info Sec and Privacy Folks Should Know, *Realtime IT Compliance*. Available from: http://www.realtime-itcompliance.com/privacy_and_compliance/2007/11/6_scary_stuff_privacy_terms_it.htm

Hjorth, L. (2013). Relocating the mobile: A case study of locative media in Seoul, South Korea. *Convergence*, *19*(2), 237–249. doi:10.1177/1354856512462360

Hufnagel, S., Harfield, C., & Bronitt, S. (2011). *Cross-Border Law Enforcement: Regional Law Enforcement Cooperation*. Sydney: Routledge.

Iglesias, R., Nicholls, R., & Travis, A. (2012). Private Clouds with No Silver Lining: Legal Risk in Private Cloud Services. *Communications & Strategies*, *85*, 125–140.

Iqbal, M. U. (2009). *Location privacy in automotive telematics*. PhD Thesis. School of Surveying & Spatial Information Systems, UNSW, Australia.

Iqbal, M. U., & Lim, S. (2007). Privacy Implications of Automated GPS Tracking and Profiling, *The Second Workshop on the Social Implications of National Security*, Research Network for a Secure Australia, Wollongong: University of Wollongong, 225-240. http://works.bepress.com/kmichael/51/

Iqbal, M. U., & Lim, S. (2007). Designing privacy-aware mobility pricing systems based on user perspective. *Journal of Location Based Services*, *1*(4), 274–299. doi:10.1080/17489720802183415

Ivanov, S., Webster, C., & Mladenovic, A. (2012). *The Microchipped Tourist: Implications for European Tourism*. Rochester: SSRN Working Papers Series.

Janssen, C. (2010). Uberveillance. *Techopedia*. http://www.techopedia.com/definition/4150/uberveillance

Kargl, F., Lawrence, E., Fischer, M., & Lim, Y. Y. (2008). Security, Privacy and Legal Issues in Pervasive eHealth Monitoring Systems, *7th International Conference on Mobile Business*, Barcelona, Spain, 296-304. http://medien.informatik.uni-ulm.de/~frank/research/icmb2008.pdf

Kerr, J., Marshall, S., Raab, F., Godbole, S., Chen, J., Schipperijn, J., & Doherty, A. (2012). Using technology to better assess active commuting and sedentary behavior, Active Living Research, *SenseCam2012*, http://activelivingresearch.org/files/2012_UsingTechnology_Marshall-Schipperijn-Foster-Kerr.pdf

Klugman, K. (2013). All spies and no responsibility. Civil Liberties Australia. http://www.cla.asn.au/Article/2013/Rpt%20NGOs%20130624.pdf

Krishnan, A. (2009). *Killer Robots: Legality and Ethicality of Autonomous Weapons*. Great Britain: Ashgate.

Kulesza, J. (2013). International law challenges to location privacy protection. *International Data Privacy Law*, *3*(3), 158–169. doi:10.1093/idpl/ipt015

Kurkovsky, S., Syta, E., & Casano, B. (2010). Continuous RFID-enabled authentication and its privacy implications. *IEEE International Symposium on Technology and Society (ISTAS10)*, 7-9 June, Wollongong: University of Wollongong, 103-110.

Kurosawa, S. (2009). Setjetters glamping it up. *The Australian*. 31 January 2009.

Labrador, M., Michael, K., & Kupper, A. (2008). Advanced Location-Based Services. *Computer Communications*, *31*(6), 1053–1054. doi:10.1016/j.comcom.2008.01.033

Leskinen, S. (2012). Veterinarian work, enhanced by mobile technology – an empirical study, 45 Hawaii International Conference on Systems Science (HICSS), 1403-1412.

Lien, T. H. (2011). Automatic identification technology: Tracking weapons and ammunition for the Norwegian Armed Forces. Naval Postgraduate School, California. http://www.dtic.mil/dtic/tr/fulltext/u2/a547879.pdf

Mann, S. (2013). Veillance and Reciprocal Transparency: Surveillance versus Sousveillance, AR Glass, Lifeglogging, and Wearable Computing. *IEEE Symposium on Technology and Society (ISTAS13)*, http://veillance.me, Ed. Katina Michael, University of Toronto, Canada, 27-29 June.

Mann, S., & Hrelja, M. (2013). Perakslis, C. (2013 Praxistemology: Early childhood education, engineering education in a university, and universal concepts for people of all ages and abilities, *IEEE Symposium on Technology and Society (ISTAS13)*, http://veillance.me, Ed. Katina Michael, University of Toronto, Canada, 27-29 June.

McPhee, L. (2009). Twitterverse, zombie debt crack Macquarie. *The West Australian*, 28 October.

Michael, K. (2011). Event Report: The IEEE Symposium on Technology and Society 2010 (7-10 June), The Social Implications of Emerging Technologies. *Journal of Cases on Information Technology*, 2(13), 80–87. http://www.igi-global.com/Files/Ancillary/10.4018_jcit.2011040106.pdf

Michael, K. (2012). Self-guided bullets won't stuff up, but what about the grunts and drones firing them? *The Conversation*, 22 May, http://theconversation.com/self-guided-bullets-wont-stuff-up-but-what-about-the-grunts-and-drones-firing-them-7104

Michael, K., & Clarke, R. (2013). Location and Tracking of Mobile Devices: Überveillance Stalks the Streets, *Computer Law and Security Review*, 29(2), 216-228. Available at: http://works.bepress.com/kmichael/305

Michael, K., & Hayes, A. (2013). Are We Ready to Live In An Uberveillance Society? *UOW Latest News* Apr. 2013 http://media.uow.edu.au/news/UOW147704.html

Michael, K., McNamee, A., & Michael, M. G. (2006). The Emerging Ethics of Humancentric GPS Tracking and Monitoring, *International Conference on Mobile Business*, Copenhagen, Denmark: IEEE Computer Society. Available at: http://works.bepress.com/kmichael/7

Michael, K., McNamee, A., Michael, M. G., & Tootell, H. (2006). Location-Based Intelligence – Modeling Behavior in Humans using GPS, *IEEE International Symposium on Technology and Society*, New York, United States. Available at: http://works.bepress.com/kmichael/6

Michael, K., & Michael, M. G. (2007). The Rise of Homo Electricus, University in the Brewery, An Australian Innovation Festival Event Lecture: 16th May, http://www.uow.edu.au/research/uni-brewery/UOW009417.html

Michael, K., & Michael, M.G. (2009). Controlling Technology, *Illawarra Mercury*, September, 23.

Michael, K., & Michael, M. G. (2009). Predicting the socioethical implications of implanting people with microchips. *PerAda Magazine*. http://www.perada-magazine.eu/view.php?article=1598-2009-04-02&category=Citizenship

Michael, K., & Michael, M.G. (2009). Teaching ICT Ethics Using Wearable Computing: the Social Implications of the New 'Veillance'. *AUPOV09-Australian Point of View Technologies Conference*. Wollongong, Australia. Jun.

Michael, K., & Michael, M. G. (2009). *Innovative Automatic Identification and Location-based Services: From Bar Codes to Chip Implants*. New York: Information Science Reference. doi:10.4018/978-1-59904-795-9

Michael, K., & Michael, M.G. (2010). Implementing Namebers Using Implantable Technologies: The Future Prospects of Person ID. In J. Pitt (Ed.), *This Pervasive Day: The Potential and Perils of Pervasive Computing*. London: Imperial College London.

Michael, K., & Michael, M.G. (2012). Converging and coexisting systems towards smart surveillance. *Awareness Magazine: Self-Awareness in Autonomic Systems*. http://www.awareness-mag.eu/view.php?article=003989-2012-06-19&category=Networks+%26+Infrastructure

Michael, K., & Michael, M. G. (2012). Commentary on: Mann, Steve (2012), Wearable Computing, *Encyclopedia of Human-Computer Interaction*. Eds. Soegaard, Mads and Dam, Rikke Friis. Aarhus, Denmark: The Interaction-Design.org Foundation. Available at: http://works.bepress.com/kmichael/272

Michael, K., & Michael, M. G. (2013). The future prospects of embedded microchips in humans as unique identifiers: the risks versus the rewards. *Media Culture & Society*, *35*(1), 78–86. doi:10.1177/0163443712464561

Michael, K., & Miller, K. W. (2013). Big Data: New Opportunities and New Challenges. *IEEE Computer*, *46*(6), 22–24. doi:10.1109/MC.2013.196

Michael, K., Stroh, B., Berry, O., Muhlbauer, A., & Nicholls, T. (2006). The AVIAN Flu Tracker - a Location Service Proof of Concept. *Recent Advances in Security Technology* (1 ed). Ed. P. Mendis, J. Lai, E. Dawson. Canberra, Australia: Australian Homeland Security Research Centre, 244-258. Available at: http://works.bepress.com/kmichael/3

Michael, M. G. (2007). Uberveillance in *29th International Conference of Data Protection and Privacy Commissioners*. Privacy Horizons: Terra Incognita, Location Based Tracking Workshop. Montreal, Quebec, Canada. Sep.

Mickhail, G. (2007). The Agora-Pnyx Paradox, *The Second Workshop on the Social Implications of National Security*, Research Network for a Secure Australia, Wollongong: University of Wollongong, 169-179. http://works.bepress.com/kmichael/51/

Minsky, M., Kurzweil, R., & Mann, S. (2013) The Society of Intelligent Veillance, *IEEE Symposium on Technology and Society (ISTAS13)*, http://veillance.me, Ed. Katina Michael, University of Toronto, Canada, 27-29 June. http://wearcam.org/sensularity.pdf

Moore, C. (2011). The magic circle and the mobility of play. *Convergence*, *17*(4), 373–387. doi:10.1177/1354856511414350

Moore, C. Identities and Überveillance, Civil Society Fourth Estate Digital Journalism Citizen Journalism Public Sphere, *ALC215 Globalisation and the Media 2013 W.5*. http://prezi.com/wrzooe_n_z7x/alc2013-w5-identities-and-uberveillance/

Moran, S. (2011). User Perceptions of System Attributes in Ubiquitous Monitoring: Toward a Model of Behavioural Intention. PhD Thesis, School of Construction Management and Engineering and Informatics Research Centre, University of Reading, Reading, England.

Mulligan, E. (2007). Spying at Home, *Illawarra Mercury: Higher Education IQ*, 27 November.

Murray, L. (2010). Infinite Anthology: Adventures in Lexiconia. *The Monthly*. 59, http://www.themonthly.com.au/issue/2010/august/1280988123/les-murray/infinite-anthology

Nayak, D., Venkata Swamy, M., & Ramaswamy, S. (2013). Supporting Location Information Privacy in Mobile Devices. *Distributed Computing and Internet Technology. Lecture Notes in Computer Science*, *7753*, 361–372. doi:10.1007/978-3-642-36071-8_28

Nellis, M. (2013). Implant technology and the electronic monitoring of offenders in eds A. Crawford & A. Hucklesby. *Legitimacy and Compliance in Criminal Justice*, New York: Routledge, 159-179.

Nicholls, R. (2012). Right to Privacy: Telephone Interception and Access in Australia. *IEEE Technology and Society Magazine*, *31*(1), 42–49. doi:10.1109/MTS.2012.2185274

Nicholls, S., & McKenny, L. (200). With Words That Last. *Sydney Morning Herald*, Dec., 27.

NYTimes. (2009). Schotts Vocab: Uberveillance. *The New York Times*, Feb. http://schott.blogs. nytimes.com/2009/02/04/uberveillance/?_r=0

Offman, C. (2007). You are tagged, *The National Post*. Available from: http://www.nationalpost. com/scripts/story.html?id=139966

Pauli, D. (2010). Grandkids to welcome 'uber-veillance': security and freedom a fallacy for microchip implants. *Computerworld*, http://www. computerworld.com.au/article/349457/grand-kids_welcome_uberveillance_/

Pauli, D. (2010). Govt launches security research network. *ZDNet*. http://www.zdnet.com/govt-launches-security-research-network-1339306227/

Perakslis, C. (2013). Millennials' Increasing Openness to Microchip Implants in Humans: A Confluence of Factors, *IEEE Symposium on Technology and Society (ISTAS13)*, http://veillance. me, Ed. Katina Michael, University of Toronto, Canada, 27-29 June.

Perakslis, C., & Michael, K. (2012). Indian Millennials: Are microchip implants a more secure technology for identification and access control? *IEEE International Symposium on Technology and Society (ISTAS12)*. Ed. Michael Arnold. Singapore: IEEE. Available at: http://works.bepress. com/kmichael/288

Resnyansky, L. (2008). *Technology in foreign policy and national security: a factor, a tool, and a mediator, Australia and the New Technologies: Evidence Based Policy in Public Administration, Research Network for a Secure Australia* (pp. 123–130). Wollongong: University of Wollongong.

Roberts, J. M. (2009). *The competent public sphere: global political economy, dialogue and the contemporary workplace*. Great Britain: Palgrave Macmillan. doi:10.1057/9780230244535

Russo, P. (2012). *The antecedents, objects, and consequents of user trust in location-based social networks*. Polytechnic Institute of New York University, Doctor of Philosophy.

Schmidt, L. (2009). Profile Susan Butler. *The Age*, http://www.theage.com.au/news/ business/money/planning/profile-susan-but-ler/2009/01/26/1232818337607.html

Schultz, T. (2012). Is it Cyberstalking? *Sunshine Coast Daily*, 28 August, 22.

Sedlmayr, M., & Münch, U. (2012). Smart Objects in Healthcare: Impact on Clinical Logistics, *Critical Issues for the Development of Sustainable E-health Solutions* Healthcare Delivery in the Information Age Series. *Springer Science, LLC*, 293–312.

Shay, L., Conti, G., & Hartzog, W. (2013). Beyond Sunglasses and Spray Paint: A Taxonomy of Surveillance Countermeasures. *IEEE Symposium on Technology and Society (ISTAS13)*, http:// veillance.me, Ed. Katina Michael, University of Toronto, Canada, 27-29 June.

Slane, C. (2013). Google Glass Privacy Issues Could Be a Buzzkill. *BestTechie*, 15 Mar. http:// www.besttechie.com/2013/03/15/google-glass-privacy-issues-could-be-a-buzzkill/

Smith, R. E. (2007). New ways your privacy is being invaded, *Yahoo!Canada Finance*. http:// ca.pfinance.yahoo.com/ca_finance_general/435/ new-ways-your-privacy-is-being-invaded

Smith, R. E. (2007). Scary stuff, *Forbes.com*. Available from: http://www.forbes.com/opin-ions/2007/11/21/privacy-surveillance-technolo-gy-oped-cx_res_1126privacy.html

Smith, R. E. (2008). Removing Unwanted Images from the 'Net. *Privacy Journal*, *34*(3), 6.

Tootell, H. (2007). Auto-ID and Location-Based Services in National Security: Social Implications, *The Second Workshop on the Social Implications of National Security*, Research Network for a Secure Australia, Wollongong: University of Wollongong, 201-224. http://works.bepress.com/kmichael/51/

Wamba, S. F., & Ngai, E. W. T. (2011). Unveiling the Potential of RFID-Enabled Intelligent Patient Management: Results of a Delphi Study, *44th Hawaii International Conference on System Sciences (HICSS)*, 4-7 January, 1-10.

Wang, J. L., & Loui, M. (2009). Privacy and ethical issues in location-based tracking systems. *IEEE International Symposium on Technology and Society*, 18-20 May, 1-4.

Whelan, A. (2011). Digital Media: The cultural politics of information. In B. M. Z. Cohen (Ed.), *Being Cultural*. Auckland: Pearson.

Whitehouse, D., & Duquenoy, P. (2009). Applied Ethics and eHealth: Principles, Identity, and RFID. In (Eds.) Matyáš, V., Fischer-Hübner, Cvrcek, D., Švenda, P. The Future of Identity in the Information Society: 4th IFIP WG 9.2, 9.6, 11, 298, 43-55.

Wicktionary (2013). Uberveillance. http://en.wiktionary.org/wiki/uberveillance

Wigan, M. (2007). Owning Identity- one or many- do we have a choice? *The Second Workshop on the Social Implications of National Security*, Research Network for a Secure Australia, Wollongong: University of Wollongong, 61-70. http://works.bepress.com/kmichael/51/

Wigan, M. (2012). Contestability, Democracy, and Trust in the Anti-Terror Age. *IEEE Technology and Society Magazine*, *31*(1), 26–32. doi:10.1109/MTS.2012.2185730

Yarali, A., & Hung, C. (2011). Wireless Services and Intelligent Vehicle Transportation Systems. Wireless Communications, 1-3 June, Vancouver, BC, Canada. DOI: 10.2316/P.2011.730-011

Yarney, S. (2010). *The Banjo Player*. United Kingdom: Story Bay Press.

KEY TERMS AND DEFINITIONS

Apocalypse: Is abbreviated Apoc and in the Greek means to uncover, translated literally to mean 'a disclosure of knowledge'. *The Apocalypse of John* is the last book of the New Testament, commonly known as *The Book of Revelation.*

Choice: The power, right, or liberty to choose. One of several options.

Consequences: Something that logically or naturally follows from an action. It is the effect of something.

De-Humanisation: To deprive of human qualities such as individuality, compassion, or civility, reducing an individual to an ID number or an ID card, or even a mechanical part (e.g. mobile phone).

Embedded: An embedded system is a special-purpose system in which the computer is completely encapsulated by the device it controls. Embedded systems can also be firmly affixed in a surrounding mass, like implants that are injected beneath the skin in the human body.

Human: A human is a member of the genus Homo and especially of the species H. Humans are to be distinguished separately from animals and mechanical apparatus.

Innovation: The act of introducing something new. Typically the process from invention to diffusion. It is the application of new solutions that meet new requirements or existing market needs. Innovations can take the form of product or process solutions.

RFID: Radio-frequency identification is the use of radio-frequency electromagnetic fields to transfer an identification number wirelessly, for the purposes of automatically tracking objects.

GPS: The Global Positioning System is a satellite navigation system that provides location and time information, anywhere on the Earth where there is an unobstructed line of sight to four or more GPS satellites. The system provides critical capabilities to military, civil and commercial users around the world. It is maintained by the United States government and is freely accessible to anyone with a GPS receiver.

Privacy: Right to be let alone. The ability of an individual or group to seclude themselves or information about themselves and thereby reveal themselves selectively. The boundaries and content of what is considered private differ among cultures and individuals, but share basic common themes.

Sensors: Also known as a detector because it converts a physical quantity into a signal that can be read by an electronic instrument. Sensors are embedded in various systems, including mobile phones. Example sensors include: accelerometer, gyroscope, altimeter, magnetometer, and thermometer.

Singularity: The quality or condition of being singular. A variety of meaning in mathematics and the natural sciences recently popularised by Ray Kurzweil.

Surveillance: A term stemming from the French meaning to *watch over* ("sur" means "from above" and "veiller" means "to watch"). It is the monitoring of a person's or group's activities, with the purpose of influencing, managing, directing, protecting or oppressing them.

Technology: The application of science to commercial objectives. Commonly used to describe high-tech electronic or digital products, systems and services.

Technotherapeutics: Technologically enhanced pharmaceuticals like humancentric embedded microchips.

Ubermensch: A person with great powers and abilities.

Uberveillance: An omnipresent electronic surveillance facilitated by technology that makes it possible to embed surveillance devices in the human body.

Ubiquitous: Being or seeming to be everywhere at the same time; omnipresent. Being omnipresent is not the same as having omniscience.

Interview 1.1
Dataveillance – Thirty Years On
Professor Roger Clarke, Canberra, Australia
Interview conducted by MG Michael on 4 May 2013.

MG MICHAEL: When and for what reasons did your interest in privacy begin?

ROGER CLARKE: In 1971, I was working in the (then) computer industry, and undertaking a 'social issues' unit towards my degree. A couple of chemical engineering students made wild claims about the harm that computers would do to society. After spending time debunking most of what they said, I was left with a couple of points that they'd made about the impact of computers on privacy that were both realistic and serious. I've been involved throughout the four decades since then, as consultant, as researcher and as advocate.

There are various levels at which the privacy need evidences itself (Clarke, 1997; 2006b). Many people tend to focus on the psychological aspects. People need private space, not only behind closed doors and drawn curtains, but also in public. At a social level, people need to be free to behave, and to associate with others, subject to broad social mores, but without the continual threat of being observed. Otherwise we reduce ourselves to the inhumanly constraining impositions suffered by people in countries behind the Iron Curtain and the Bamboo Curtain. There is also an economic dimension, because people need to be free to invent and innovate. International competition is fierce, so countries with high labour-costs need to be clever if they want to sustain their standard-of-living; and cleverness has to be continually reinvented.

My strongest personal motivations, however, have been at the political level. People need to be free to think, and argue, and act. Surveillance chills behaviour, association and speech, and hence threatens democracy. Our political well-being depends not only on the rule of law and freedom of speech, but also on privacy. The term 'dissidentity' draws together seemingly disparate ideas about the sanctity of the ballot box, whistleblower protections, and the political deviants who intellectually challenge established doctrines, policies, institutions or governments (Clarke, 2008).

MG MICHAEL: How do you define or understand privacy?

ROGER CLARKE: Privacy is a human right, and the decades of attempts by US business interests to reduce it to a mere economic right have failed, and must fail. On the other hand, definitions that are built around 'rights' are very difficult to apply. And of course the definitions in data protection statutes around the world are so narrow and legalistic that they provide no useful basis for meaningful discussions.

Privacy is most usefully defined as an 'interest' – the interest that individuals have in sustaining a 'personal space', free from interference by other people and organisations (Morison, 1973; Clarke, 1997; Clarke, 2006b).

Privacy has multiple dimensions (Clarke, 1997). Data privacy is only one of them, but it attracts almost all of the attention. Particularly when discussing surveillance, it's essential that all of the dimensions be considered, not just the protection of personal data.

Personal communications are under observation as never before. And longstanding protections have been obliterated by a combination of technological accidents and the fervour of national security extremists. The freedom of personal behaviour is greatly undermined by many different forms of monitoring.

The intrusiveness of some techniques is now reaching the point that we need to start talking about the privacy of personal experiences (e.g. what you read), and even of attitudes and thoughts. Finally, privacy of the physical person is not only negatively affected by the inroads made in the other dimensions, but is also under direct attack.

MG MICHAEL: How do you define dataveillance? How did its definition come about?

ROGER CLARKE: By the middle of the 1980s, I was frustrated with the ongoing superficiality of the conversation. The discussion was still about 'computers and privacy', even though it was clear that both technologies and their applications were developing rapidly, and that deeper analysis was essential.

I coined 'dataveillance' with several purposes in mind:

- To switch the primary emphasis towards the threat rather than the asset;
- To draw surveillance notions into the conversation;
- To expand the scope beyond personal data to human communications and behaviour; and
- To broaden the scope of technologies under discussion.

Hence dataveillance was, and is, "the systematic monitoring of people's actions or communications through the application of information technology" (Clarke, 1986; 1988; 2003).

It was clear in the 1980s that the monitoring of people through the data trails that they generate is far more economically efficient than previous forms of surveillance, and that it would be adopted rapidly and widely. As we know all too well, it has been.

The basic theory of dataveillance needed to be articulated in a number of areas, in particular identity (Clarke, 1994b; 1999b), the digital persona (Clarke, 1994a; 2013c) and national identification schemes (Clarke, 1987; 2006a). Because industry initiatives in such areas as digital signatures, identity authentication and identity management areas have embodied serious misunderstandings about key concepts, it has also proven necessary to express the consolidated model of entity, identity and (id)entity authentication and authorisation (Clarke, 2009a).

MG MICHAEL: Do you understand 'dataveillance' a little differently today than when you first introduced the idea to the world? What are the principal links between dataveillance and uberveillance?

ROGER CLARKE: The definition of 'dataveillance' appears to me to be as appropriate now as it was 25 years ago. On the other hand, the technologies have changed, and the organisations with a vested interest in promoting and conducting dataveillance have grown in number, in size, and in institutional and market power. So the importance of the discussion is vastly greater even than it was at that time.

In relation to surveillance more generally, I've had to adapt my thinking over time. Back in the mid-1980s, I distinguished:

- Physical surveillance (visual and aural observation by a person);
- Augmented physical surveillance (assisted by telescopes, directional microphones, etc.);
- Communications surveillance (mail covers, wire-taps, etc.); and
- Dataveillance.

There have been a number of technological developments that weren't anticipated in my original dataveillance paper, and I've consequently had to broaden my scope (Clarke, 2009b; 2010a). Here are some of the important aspects that have emerged during the last 25 years:

- Visual surveillance technologies have made advances in both technical and economic terms;
- Visual surveillance technologies have become a source of data, and have thereby further expanded the sources to support dataveillance;
- Experience surveillance, of what one reads, hears and views has become feasible (Clarke, 2010a). Particularly if this is combined with communications surveillance, an observer can infer a great deal about an individual's attitudes and motivations;
- Technologies for the surveillance of human bodies have emerged (Clarke, 2010a), variously:
- By associating tracking tools with the body (e.g. RFID-enabled anklets, mobile phones);
- By recognising individuals through their natural physical features (biometrics); and
- By embedding tracking-tools in the body, which is one of the two senses of the term 'ueberveillance' – although it might also be called endo-surveillance (internal) or auto-surveillance (by the self, of the self);
- The collection of location data, and hence person-tracking has proven to be feasible, with the result that it may be best to highlight it as a special case of dataveillance (Clarke, 1994b; 1999c; Clarke & Wigan, 2011; Michael & Clarke, 2013);
- Sousveillance, a notion introduced by Steve Mann in 1995, reflects the adoption of surveillance technologies not only by those powerful in society but also by those on whom surveillance is imposed (Mann et al., 2003);
- Equiveillance and inequiveillance reflect the degree of balance between the impacts of sur- and sous-veillance (Mann, 2005).

As surveillance forms proliferated and were integrated, it became important to develop an overarching concept. This emerged as 'ueberveillance' (Michael, 2006; Michael & Michael, 2006). There are several elements within the notion (Clarke, 2010a). The apocalyptic – but regrettably, not unrealistic – interpretation is of surveillance that applies across all space and all time (omni-present), and supports some organisation that is all-seeing (omnivident) and even all-knowing (omniscient), at least relative to some person or object. Another sense of 'ueber' is 'exaggerated' or 'disproportionate', perhaps because the justification for the intrusion is inadequate, or the scope is excessive. The third interpretation is as 'meta', 'supra' or 'master', implying that information from multiple surveillance streams is consolidated, coordinated or centralised.

MG MICHAEL: Where do you stand on the "dystopian" genre? i.e. *Brave New World* or *1984* etc. Are these works a genuine and helpful critique of our political system and technological fixation?

ROGER CLARKE: I've drawn heavily on many aspects of dystopian and sci-fi literatures, not only in the surveillance arena, but in other areas as well (Clarke, 1993a; 1993b; 2005; 2009c). For example, I'm currently re-reading 'The Diamond Age' (Stephenson, 1995), because it's easily the most effective experiment yet performed on the impact of miniaturised 'aero-stats', and is consequently valuable pre-reading when preparing contributions to the current debates on drones.

The single most important work is 'We' (Zamyatin, 1922). Orwell's '1984' was heavily derivative from Zamyatin, but it built on experience of both Stalin's regime and the UK's war-time Ministry of Information. In recent decades, the cyberpunk genre (of which Stephenson is a mature exponent) has been a rich source of ideas.

When reading such literatures, it's very important to keep the art and the hard-headed policy analysis distinct from one another. Art isn't constrained by the laws of physics, nor by the 'laws' of economics. It provides fore-warning, and pre-thinking, and in many cases extrapolation to the consequences of extremist forms of a technology or an application, or exploitation of them by extremist corporations, governments and political movements.

The extrapolation aspect once seemed to me to be of limited value, because I thought I lived in a world in which extremism would be largely kept in check. The last few decades have shown that sentiment to be naive. Parliaments have retreated from their responsibility to regulate both the business sector and the Executive. And a handful of successful terrorist strikes has caused once relatively free nations to eat themselves, by implementing measures that were previously associated only with repressive regimes. So, unfortunately, even some of the more extreme ideas in the dystopian and sci-fi literatures have relevance.

MG MICHAEL: I know you are not a fan of casting the net too far into the future, but do you see a time when the microchipping of humans, whether it be for medical or commercial or national security purposes, will become routine? And if yes, how would such a state of affairs impact on privacy?

ROGER CLARKE: I describe my philosophical stance as 'positively agnostic', by which I mean that there are questions that humans simply cannot even begin to answer. The source of matter, and of the various forms it takes, and the existence of G/god(s), are beyond me, and, I contend, are beyond other mere humans as well. Everyone is welcome to their spiritual framework and beliefs; but policy matters are much better dealt with in closer contact with the world we're living in. Along with metaphysical questions, I see all of us as being incapable of making judgements about distant human futures.

On the other hand, some technologies, and some applications, are readily extrapolated from the present, and some of the simpler disjunctions and shifts can be at least mulled over. In 1992-93, I had to enter into a lengthy discussion with the Editor of *Information Technology & People*, in order to be permitted to retain in a paper a passage that she regarded as alarmist and technically unrealistic:

"It has been technically feasible for some time, and is increasingly economically practicable, to implant micro-chips into animals for the purpose of identification and data-gathering. Examples of applications are to pets which may be stolen or lost, breeding stock, valuable stock like cattle, and endangered species like the wandering albatross. As the technology develops, there will doubtless be calls for it to be applied to humans as well. In order to discourage uncooperative subjects from removing or disabling them, it may be necessary for them to be installed in some delicate location, such as beside the heart or inside the gums" (Clarke, 1994b).

My understanding is that the first chip-implantation in animals was in 1991, and the first in humans (at that stage, and seemingly as late as 2013, only voluntarily) was in 1998. As short a lag as 6 years doesn't justify even the term 'prescient', let alone 'visionary'; yet I had to argue strongly to get the sentence published.

Chips with moderate computational power, storage and communications capabilities are already embedded in a great many devices. Already some of those devices are embedded in humans. Pacemakers are being joined by replacement limbs and joints, which contain chips that, at the very least, perform functions relating to balance. Many products will carry chips in order to support maintenance (e.g. aircraft components) and product recall (e.g. potentially dangerous consumer goods), and it's highly likely that materials used in operating theatres (e.g. swabs) and endo-prostheses and endo-orthoses (e.g. titanium hips, stents) will carry them as well.

In addition to medical applications, two other contexts stand out. One is fashion, not only for technophiles but also for, say, night-club patrons. The other is control over the institutionalised. Anklets with embedded chips are applied to felons in gaols, to parolees, and even to people on bail and to people who have served their time but are deemed dangerous or undesirable. In a few short years, the practice leapt from an unjustifiably demeaning imposition to an economically motivated form of retribution and stigma. Chipping of the mentally handicapped, the aged and the young can be justified with no more effort than was needed to apply it to people on bail. The migration from anklet to body will be straightforward, based no doubt on the convenience of the subject, dressed up as though it were a choice they have made.

There are virtually no intrinsic controls over such developments, and virtually no legal controls either. For example, the US Food and Drug Administration's decision that there was no health reason to preclude the insertion of chips in humans was rapidly parlayed by the industry into 'approval' for chip implantation. There are many serious concerns about imposition by corporations and the State on individuals' data, communications, behaviour and bodies; but few are as directly intrusive as the prospect of the insertion of computing and communication devices in humans becoming normalised.

My work on cyborgisation was originally predicated on the assumption that, for the reasonably foreseeable future, our successors would be technology-enhanced, but very much humans (Clarke, 2005). Regrettably, it's become necessary to recognise a strong tendency for technology to be applied so as to demean, to deny self-determination, and to impose organisational dictates on individuals (Clarke, 2011b).

MG MICHAEL: Do you believe those engaged in auto-ID technology distinguish between the locating and tracking of objects and bodies, and the monitoring the mind?

ROGER CLARKE: I've previously used the term 'sleep-walking' for the manner in which people have overlooked the march of surveillance technologies: "the last 50 years of technological development has delivered far superior surveillance tools than Orwell imagined. And we didn't even notice" (Clarke, 2001). Others prefer the 'warm frog' or zombies metaphors.

During the Holocaust, 1940-45, each of the successive impositions was grossly inhuman, but cumulative, and culminated in vast numbers of people trudging into their place of execution. In this case, on the other hand, the features are being initially pitched as being exciting, fashionable and convenient. The innately human qualities of respecting apparent authority, and of blindly trusting, results in people becoming numbed, inured, and accepting.

The sceptic, and the analyst, have no trouble recognising that location data becomes tracking data, and enables inferences to be drawn about each person's trajectory through physical space, including their likely destinations (Clarke, 1999c). Similarly, some categories of people understand that rich data-sets enable inferencing about a person's interests, social networks, and even attitudes, intentions and beliefs. Some have even noticed the morphing of digital books and newspapers, aggressive exploitation of intellectual property laws, and migration of electronic content from personal possessions to cloud-storage, into 'experience surveillance'.

For most people, however, such things are the stuff of novels, not part of their world, and an unwelcome suggestion because it intrudes into their enjoyment of a world that they assume is here to stay and theirs to exploit. Most social issues are of interest only to an intelligentsia, and abstract 'surveillance and privacy' matters are more difficult to convey in a graphic manner than, say, refugee families behind razorwire, street-people, and indigenous young people living in squalor.

Because of my background in technology, I'm uncomfortable using terms such as 'mind-monitoring' to refer to this rapidly developing aspect of surveillance. There is no doubt, however, that we must find ways to convey to a much broader public how much insight organisations and their data-gorging inference engines are gaining into our individual psyches.

MG MICHAEL: Are you an optimist insofar as privacy surviving? There are those who are already speaking of privacy in terms of it being dead. What is your response to this?

ROGER CLARKE: There are a number of variants of the 'privacy is dead' meme (Clarke, 2003; 2009c):

- 'It's great that privacy is dead' – associated with the 'original sin' mentality of moral minorities.
- 'You have zero privacy anyway. Get over it' – associated with Scott McNealy; and the corollaries expressed by Eric Schmidt "If you have something that you don't want [Google and its customers] to know, maybe you shouldn't be doing it in the first place", and by Mark Zuckerbeg that "if [I] were to create Facebook again today, user information would by default be public".
- 'Privacy through secrecy is dead' – associated with Brin (1998).
- 'The new generation has a completely different approach to privacy' – associated with investors in social media services, whose business model is predicated on voyeurism, self-exhibitionism and the exposure of others, and 'the default is social'.

There are straightforward responses to all of these schools of thought. Underlying them are the facts that privacy is a human need, and that human needs don't 'survive'; they just are. For any of these arguments to be accepted, a fundamental change in the human psyche would have had to occurred just because CCTV, search engines and/or social networking services had been invented.

People who believe that 'we're all sinners' would welcome a post-privacy era on the basis that 'you only need privacy if you have something to hide'. But there's a problem with that proposition. Quite simply, everyone has things to hide (APF, 2006). Not least, when anti-privacy advocates put their heads over the parapet, they tend to need to hide such things as their bank accounts, their blogs and Twitter accounts, and even their whereabouts.

Similarly, McNealy (previously of Sun), Schmidt (Google) and Zuckerberg (Facebook), like celebrities everywhere, continue to keep their passwords and PINs to themselves, and for the most part obscure their whereabouts and their travel plans. The fact that they do so successfully suggests that 'you have zero privacy' is something of an over-statement.

David Brin called for ubiquitous transparency, on the grounds that he sees it as a better form of protection of freedoms than secrecy (Brin, 1998). His is essentially an argument that sousveillance will solve all ills. It's an attractive thesis, but it's based on some key assumptions. It would require the enforcement of open practices on law enforcement agencies and operators of surveillance apparatus such as CCTV control rooms. In effect, we need the powerful to be convinced, or forced, to do what they have seldom done: not exercise their power. That idea is, quite simply, fanciful. It's one thing to switch the focus from hiding data to providing people with control over their own data; but it's quite another to suggest that the privacy protections that have been achieved should be abandoned in favour of a naive notion that 'the weak shall become powerful'.

As to the fourth form of the 'privacy is dead' meme, the myth that privacy attitudes of the new generation are different has arisen from a very basic misunderstanding. Middle-aged people look at young people, perceive them to be different from middle-aged people, and conclude that therefore the new generation is different. What's needed is a comparison of young people with what middle-aged people were like when they were young, and preferably (although with greater difficulty) with what middle-aged people would have been like when they were young if they had experienced the same conditions as young people do now.

Every new cohort of the young takes risks. Every cohort becomes progressively more risk-averse as it gets older and gains more things to hide. These include assets, such as possessions worth stealing and informational assets from which they extract advantages. Things worth hiding also include liabilities, such as financial and psychic debts, and informational liabilities such as a stigmatised medical condition, a criminal record, a failed study-course, a failed relationship.

Because of the naive use of social media since about 2004, many people are being bitten by the exposure of embarrassing information and images, or are gaining vicarious experience of other people's misfortunes. So the reasonable expectation is that the iGeneration, i.e. those born since the early 1990s, will be more privacy-sensitive than their predecessors, not less (Clarke, 2009d).

Privacy isn't 'surviving'. It just 'is', and will continue to 'be'.

MG MICHAEL: Turning now to the issue of how we save at least some impression of privacy. How would most privacy experts understand "ethics" in their work? How important is the question of ethics in this debate, or is this a question that will become increasingly redundant?

ROGER CLARKE: From the very outset, privacy protection has been conceived so as to primarily serve corporate and government interests, rather than human values (Clarke, 2000). The agenda was set by Columbia University economist, Alan Westin (Westin, 1967; 1971; Westin & Baker, 1974). He developed the notion of 'fair information practices' (often referred to as FIPS). FIP-based privacy regimes have been described as an 'official response' which was motivated by 'administrative convenience', and which legitimated dataveillance measures in return for some limited procedural protections (Rule, 1974; Rule et al., 1980).

The institution that published the first international set of Data Protection Principles was formed to address economic not social issues. Its highly influential document (OECD, 1980) consequently codified FIPS, and embedded the dominance of institutional over individual interests. The EU strengthened the protections; whereas the US cut the OECD Guidelines down even further, in the form of a 'safe harbor' for American corporations (USDOC, 2000), and has been seeking to weaken them further by using Australia and other members of the Asia-Pacific Economic Cooperation to create an APEC Privacy Framework (APEC, 2005) that is even weaker than the seriously inadequate 'safe harbor' formulation.

It would therefore be a very welcome new development if privacy protections were to be conceived on the basis of an ethical analysis that put people's interests before those of governments and corporations. Unfortunately, such a change appears highly unlikely.

In any case, ethics is seen by most people as being primarily confined to abstract judgements about good and evil. Ethical analyses are valuable as a component of ex post facto evaluations of actions that have been already taken, and reviews of institutional structures, processes and decision criteria. But there are doubts as to whether ethics ever have volitional or motivational power, and hence influence the behaviour of organisations (Clarke, 1999d).

MG MICHAEL: Is there a place for religious sensitivities in the dataveillance and uberveillance debate? The taboo of not making mention, for instance, of the anxiety over the 'branding' or the microchipping of humans has to a large extent been lifted and writers are engaging with this question from not only a civil libertarian point of view but also from a religious point of view. Is this contribution to the debate to be welcomed?

ROGER CLARKE: The *Book of Revelation* is expressed in mystical style, and the notion of 'the mark of the beast' can be interpreted in a great many ways. Viewed from a secular perspective, the intensity of the expression, and of its interpretation by many Christians, reflects the revulsion felt by many people about physical intrusions into their selves, and the exercise of power over their behaviour by a malevolent force.

Personally, I feel discomfort when people use '666' symbolism. I prefer to focus on evidence from our experience of the physical world, rather than ascientific assertions. I also doubt whether many of the uncommitted are won over by such arguments. But I can't and don't deny the legitimacy of approaches other than my own. The horror of impositions on our physical selves will be evidenced in many ways, and communicated in many ways.

MG MICHAEL: If we are to save at least some impression of privacy, are we largely dependent on legislation, self-regulation, or some sort of "default" in the technology?

ROGER CLARKE: There's no doubt that we need a network of interacting protections – natural, organisational, technical and regulatory – designed so as to be mutually reinforcing.

Many of the natural protections have been undermined by such changes as the digitisation of content, increases in transmission bandwidths, and greatly reduced costs. Nonetheless, some remain, such as the self-interest of organisations, and competition among them. A variety of organisational protections suit the needs of companies and government agencies as well as individuals, such as data integrity and data security safeguards.

A great many technical protections exist, and more are being developed all the time. The problem is that most of the developers are employed by organisations that seek to invade privacy and exploit personal data. So for every consumer-protective safeguard that's produced, there are scores of privacy-invasive features and products, and many countermeasures against the safeguards. W3C designed far more serious privacy-invasive features into HTML5 than it did privacy-protective features. And Mozilla and Firefox are similarly marketer-friendly and consumer-hostile.

As an antidote to this malaise, I've argued that privacy advocacy organisations need to publish their own Standards, to compete with the Standards written by industry and government to satisfy their own needs (Clarke, 2010b). A document such as the 'Policy Statement re Visual Surveillance' of the Australian Privacy Foundation (APF, 2009) could provide a basis for a Civil Society Standard. A generic set of Meta-Principles for Privacy Protection is enunciated in APF (2013).

With market failure so evident, it would be expected that regulation would be imposed. In many countries, however, legislatures have been derogating their duty to protect the public against excesses by corporations and government agencies. The chimera of 'self-regulation' has been invoked, as though it could be effective as well as efficient. It has uniformly failed to satisfy the interests of the public. Wolves self-regulate for the benefit of wolves, not sheep. And in any case there are many loner wolves out of the reach of the associations whose 'industry codes' create the pretence of a privacy-protective regime.

Faced with increasing scepticism about self-regulation, partnerships between government and business have invented the term 'co-regulation'. The concept has merit, but only if it is implemented in a meaningful manner, including a legislative framework that stipulates the privacy principles, delegated codes that bind all parties, and a watchdog agency with enforcement powers and resources (Clarke, 1999a).

Some countries, primarily in Europe, have Data Protection Commissioners that have some enforcement powers at least, and that can therefore be regarded as regulators – although they have little or no coverage of privacy of personal communications or behaviour, nor of privacy of the physical person. Some countries, such as Australia, have a 'watch dog' oversight agency rather than a 'guard dog' regulator, because the organisation lacks power, and in many cases resources as well. Such agencies typically fail to even fully exercise the influence that they could have, e.g. by failing to operationally define public expectations in even the most straightforward areas such as data security safeguards (APF, 2012; Clarke, 2013b). Some countries, such as the USA, lack even oversight.

Legislatures will only impose requirements on organisations, and will only empower regulators (and force them to do their jobs) to the extent that the public makes clear that that's what they expect. Consumer and privacy advocacy organisations need to mobilise much more activity, coordinate it, and project it through the media, in order to achieve visibility.

MG MICHAEL: Ultimately, 'who will guard the guards themselves'?

ROGER CLARKE: 'Quis custodiet ipsos custodes? Ecce, ipsi quos custodiunt custodes' 'Who will guard the guardians themselves? Lo, the very ones whom the guardians guard.'

On the one hand, this can be interpreted as an argument for the merits of sousveillance. On the other, it underlines the networked nature of effective democratic systems, whereby all powers are subject to controls, but those controls are a web rather than a simple hierarchy, and 'the public' are part of that web, through such means as periodic elections, petitions, citizen initiatives, recalls, referenda, civil disobedience, demonstrations, and revolutions.

I've long used an aphorism that is a distant relation of the Latin dictum above:

Privacy doesn't matter until it does.

By this, I mean that most people, most of the time, accept what happens. If what happens is unfriendly, their acceptance is a bit sullen. While their disenchantment remains below some critical threshold, they simply bear a bit of a grudge. When that threshold is exceeded, however, action happens. And when the feeling is held by many people, the action is like a dam-wall breaking – swift, vicious and frequently decisive (Clarke 1996, 2006c). Large numbers of corporate privacy disasters attest to that (Clarke 2013a), as do multiple failed national identification schemes, and hundreds of knee-jerk privacy laws throughout the USA and its constituent States.

MG MICHAEL: With respect to dataveillance, what is the role of privacy advocacy groups such as Privacy International and the Australian Privacy Foundation in the debate with political and corporate entities?

ROGER CLARKE: A substantial set of privacy advocacy organisations exists around the world (Bennett 2008, PAR 2013), together with powerful voices such as the Washington-based Electronic Privacy Information Center (EPIC) and London-based Privacy International.

Advocacy organisations aim for a future in which all organisations collect, use, retain and disclose personal data only in ways that are transparent, justified and proportionate, and subject to mitigation measures, controls and accountability (APF, 2013). In order to achieve that condition, a number of enabling measures are necessary (Clarke, 2012). Evaluation of proposals is essential, in accordance with the accumulated knowledge about Privacy Impact Assessments (de Hert & Wright, 2012; Clarke 2011a).

To contribute to evaluations and achieve privacy-positive outcomes, advocacy organisations need to conduct research, establish policy statements, write submissions, give evidence, and advocate the public interest verbally in meetings and in the media. That requires all of the accoutrements of a civil society organisation, including an appropriate constitution, governance, business processes, and resources. Where necessary, an advocacy organisation must conduct campaigns for enhanced laws, or (far too often) against projects such as national identification schemes, unjustified cyber-surveillance and visual surveillance, and excessive media uses of surveillance. It's seriously challenging to attract enough people with sufficient expertise and energy.

Experience has shown, however, not only in the USA, the UK, Germany, Canada and Australia, but also in many other counties, that advocacy organisations that are run with professionalism and vigour can have very substantial impacts on policy debates, enthuse the media, mobilise the public, and cause politicians to ask hard questions, and to act.

REFERENCES

APEC. (2005). *APEC privacy framework*. Retrieved from http://www.apec.org/Groups/Committee-on-Trade-and-Investment/~/media/Files/Groups/ECSG/05_ecsg_privacyframewk.ashx

APF. (2006). *If I've got nothing to hide, why should I be afraid?*. Australian Privacy Foundation. Retrieved from http://www.privacy.org.au/Resources/PAS-STH.html

APF. (2009). *Policy statement re visual surveillance*. Retrieved from http://www.privacy.org.au/Papers/CCTV-1001.html

APF. (2012). *Policy statement on information security*. Retrieved from http://www.privacy.org.au/Papers/PS-Secy.html

APF. (2013). *Meta-principles for privacy protection*. Retrieved from http://www.privacy.org.au/Papers/PS-MetaP.html

Bennett, C. (2008). *The privacy advocates: Resisting the spread of surveillance*. Cambridge, MA: MIT Press.

Brin, D. (1998). *The transparent society*. Reading, MA: Addison-Wesley.

Clarke, R. (1986). Information technology and 'dataveillance'. In *Proceedings of the Symp. on Comp., & Social Responsibility*. Academic Press.

Clarke, R. (1987). Just another piece of plastic for your wallet: The Australia card. *Prometheus, 5*.

Clarke, R. (1988). Information technology and dataveillance comm. *ACM, 31*(5).

Clarke, R. (1993a). *A future trace on dataveillance: Trends in the anti-utopia / science fiction genre*. Xamax Consultancy Pty Ltd. Retrieved from http://www.rogerclarke.com/DV/NotesAntiUtopia.html

Clarke, R. (1993b). Asimov's laws of robotics implications for information technolog. *IEEE Computer, 26*(12), 53-61.

Clarke, R. (1994a). The digital persona and its application to data surveillance. *The Information Society, 10*(2).

Clarke, R. (1994b). Human identification in information systems: Management challenges and public policy issues. *Info. Technology & People, 7*(4).

Clarke, R. (1996). *Privacy, dataveillance, organisational strategy*. Retrieved from http://www.rogerclarke.com/DV/PStrat.html

Clarke, R. (1997). *Introduction to dataveillance and information privacy, and definitions of terms*. Xamax Consultancy Pty Ltd. Retrieved from http://www.rogerclarke.com/DV/Intro.html

Clarke, R. (1999a). Internet privacy concerns confirm the case for intervention. *Communications of the ACM, 42*(2), 60-67.

Clarke, R. (1999b). Anonymous, pseudonymous and identified transactions: The spectrum of choice. In *Proceedings of IFIP User Identification & Privacy Protection Conference*. Retrieved from http://www.rogerclarke.com/DV/UIPP99.html

Clarke, R. (1999c). Person-location and person-tracking: Technologies, risks and policy implications. In *Proceedings of 21st International Conf. Privacy and Personal Data Protection*. Retrieved from http://www.rogerclarke.com/DV/PLT.html

Clarke, R. (1999d). Ethics and the internet: The cyberspace behaviour of people, communities and organizations. In *Proceedings of the 6th Annual Conf. Aust. Association for Professional and Applied Ethics*. Retrieved from http://www.rogerclarke.com/II/IEthics99.html

Clarke, R. (2000). *Beyond the OECD guidelines: Privacy protection for the 21st century*. Xamax Consultancy Pty Ltd. Retrieved from http://www.rogerclarke.com/DV/PP21C.html

Clarke, R. (2001). While you were sleeping.. Surveillance technologies arrived. *Australian Quarterly, 73*(1), 10-14.

Clarke, R. (2003). *Dataveillance - 15 years on*. Retrieved from http://www.rogerclarke.com/DV/DVNZ03. html, together with a slide-set)

Clarke, R. (2005). *Human-artefact hybridisation: Forms and consequences*. Retrieved from http://www. rogerclarke.com/SOS/HAH0505.html

Clarke, R. (2006a). *National identity schemes - The elements*. Xamax Consultancy Pty Ltd. Retrieved from http://www.rogerclarke.com/DV/NatIDSchemeElms.html

Clarke, R. (2006b). *What's privacy?*. Retrieved from http://www.rogerclarke.com/DV/Privacy.html

Clarke, R. (2006c). Make privacy a strategic factor - The why and the how. *Cutter IT Journal, 19*(11).

Clarke, R. (2008). Dissidentity: The political dimension of identity and privacy. *Identity in the Information Society, 1*(1), 221-228.

Clarke, R. (2009a). A sufficiently rich model of (id)entity, authentication and authorization. In *Proceedings of IDIS 2009 - The 2nd Multidisciplinary Workshop on Identity in the Information Society*. LSE. Retrieved from http://www.rogerclarke.com/ID/IdModel-090605.html

Clarke, R. (2009b). *A framework for surveillance analysis*. Xamax Consultancy Pty Ltd. Retrieved from http://www.rogerclarke.com/DV/FSA.html

Clarke, R. (2009c). *Surveillance in speculative fiction: Have our artists been sufficiently imaginative?*. Xamax Consultancy Pty Ltd. Retrieved from http://www.rogerclarke.com/DV/SSF-0910.html

Clarke, R. (2009d). *The privacy attitudes of the igeneration*. Xamax Consultancy Pty Ltd. Retrieved from http://www.rogerclarke.com/DV/MillGen.html

Clarke, R. (2010a). What is überveillance? (And what should be done about it?). *IEEE Technology and Society, 29*(2), 17-25.

Clarke, R. (2010b). Civil society must publish standards documents. In *Proceedings of Human Choice & Computers* (HCC'10). IFIP.

Clarke, R. (2011a). An evaluation of privacy impact assessment guidance documents. *International Data Privacy Law, 1*(2).

Clarke, R. (2011b). Cyborg rights. *IEEE Technology and Society, 30*(3), 49-57.

Clarke, R. (2012). The challenging world of privacy advocacy. *IEEE Technology & Society*. Retrieved from http://www.rogerclarke.com/DV/PAO-12.html

Clarke, R. (2013a). *Vignettes of corporate privacy disasters*. Xamax Consultancy Pty Ltd. Retrieved from http://www.rogerclarke.com/DV/PrivCorp.html

Clarke, R. (2013b). *Information security for small and medium-sized organizations*. Xamax Consultancy Pty Ltd. Retrieved from http://www.xamax.com.au/EC/ISInfo.pdf

Clarke, R. (2013c). Persona missing, feared drowned: the digital persona concept, two decades later. *Information Technology & People, 26*(3).

Clarke, R., & Wigan, M. (2011). You are where you've been the privacy implications of location and tracking technologies. *Journal of Location Based Services, 5*(3-4), 138-155.

De Hert, P., & Wright, D. (Eds.). (2012). *Privacy impact assessments: Engaging stakeholders in protecting privacy*. Berlin: Springer.

Mann, S. (2005). Equiveillance: The equilibrium between surveillance and sous-veillance. In *Proceedings of Computers, Freedom & Privacy*. Retrieved from http://idtrail.org/files/Mann,%2520Equiveillance.pdf

Mann, S., Nolan, J., & Wellman, B. (2003). Sousveillance: Inventing and using wearable computing devices for data collection in surveillance environments. *Surveillance & Society, 1*(3), 331-355.

Michael, M.G. (2006). *Consequences of innovation*. Unpublished Lecture Notes No. 13 for IACT405/905 - Information Technology and Innovation, School of Information Technology and Computer Science, University of Wollongong, Australia.

Michael, K., & Michael, M.G. (2006). Towards chipification: The multifunctional body art of the net generation. In *Proceedings of Conf. Cultural Attitudes Towards Technology and Communication*, (pp. 622-641). Tartu, Estonia: Academic Press.

Michael, K., & Clarke, R. (2011). Location and tracking of mobile devices: Überveillance stalks the streets. *Computer Law & Security Review, 29*(3).

Morison, W.L. (1973). *Report on the law of privacy*. Sydney, Australia: Govt. Printer.

OECD. (1980). *Guidelines on the protection of privacy and transborder flows of personal data*. Retrieved from http://www.oecd.org/internet/ieconomy/oecdguidelinesontheprotectionofprivacyandtransborderflowsofpersonaldata.htm

PAR. (2013). *Privacy advocates register*. Retrieved from http://privacyadvocates.ca/

Rule, J.B. (1974). *Private lives and public surveillance: Social control in the computer age*. Schocken Books.

Rule, J.B., McAdam, D., Stearns, L., & Uglow, D. (1980). *The politics of privacy*. New York: New American Library.

Stephenson, N. (1995). *The diamond age*. New York: Bantam Books.

USDOC. (2000). *Safe harbor*. U.S. Department of Commerce. Retrieved from http://export.gov/safeharbor/

Westin, A.F. (1967). *Privacy and freedom*. New York: Atheneum.

Westin, A.F. (Ed.). (1971). *Information technology in a democracy*. Cambridge, MA: Harvard University Press.

Westin, A.F., & Baker, M.A. (1974). *Databanks in a free society: Computers, record-keeping and privacy*. New York: Quadrangle.

Wright, D. (Ed.). (2011). *Privacy impact assessments: Engaging stakeholders in protecting privacy*. Academic Press.

Zamyatin, E. (1922). *We*. New York: Penguin.

KEY TERMS AND DEFINITIONS

Cyborgisation: A human who has certain physiological processes aided or controlled by mechanical or electronic devices.

Data Protection: Safeguards for individuals relating to personal data stored on a computer.

Dataveillance: The systematic monitoring of people's actions or communications through the application of information technology.

Ethics: Involves systematizing, defending and recommending concepts of right and wrong conduct.

Human Rights: Are "commonly understood as inalienable fundamental rights to which a person is inherently entitled simply because she or he is a human being.

Microchipping: An identifying integrated circuit device or RFID transponder encased in silicate glass and implanted in the body of a human being or animal or non-living thing.

Privacy: Is the interest that individuals have in sustaining a 'personal space', free from interference by other people and organisations.

Surveillance: Is the systematic investigation or monitoring of the actions or communications of one or more persons.

Transparency: Claims to offers everyone access to the vast majority of information that is collected by government or business.

Chapter 2
Veillance:
Beyond Surveillance, Dataveillance, Uberveillance, and the Hypocrisy of One–Sided Watching

Steve Mann
University of Toronto, Canada

ABSTRACT

This chapter builds upon the concept of Uberveillance introduced in the seminal research of M. G. Michael and Katina Michael in 2006. It begins with an overview of sousveillance (underwatching) technologies and examines the "We're watching you but you can't watch us" hypocrisy associated with the rise of surveillance (overwatching). Surveillance cameras are often installed in places that have "NO CAMERAS" and "NO CELLPHONES IN STORE, PLEASE!" signage. The author considers the chilling effect of this veillance hypocrisy on LifeGlogging, wearable computing, "Sixth Sense," AR Glass, and the Digital Eye Glass vision aid. If surveillance gives rise to hypocrisy, then to what does its inverse, sousveillance (wearable cameras, AR Glass, etc.), give rise? The opposite (antonym) of hypocrisy is integrity. How might we resolve the conflict-of-interest that arises in situations where, for example, police surveillance cameras capture the only record of wrongdoing by the police? Is sousveillance the answer or will centralized dataveillance merely turn sousveillance into a corruptible uberveillance authority?

INTRODUCTION: SURVEILLANCE

It is often said that we live in a "surveillance society" (Lyon, 2001), but what does "surveillance" mean? The primary definition of the word "surveillance" is:

sur-veil-lance [ser-vey-luh ns] noun

1. *A watch kept over a person, group, etc., especially over a suspect, prisoner, or the like ...*

[examples] The suspects were under police surveillance. (Random House Dictionary, © Random House, Inc. 2013, accessed through dictionary. com)

DOI: 10.4018/978-1-4666-4582-0.ch002

The word "surveillance" is from the French word "surveiller" which means "to watch over". Specifically, the word "surveillance" is formed from two parts:

1. The French prefix "sur" which means "over" or "from above"; and
2. The French word "veillance" which means "watching" or "monitoring"

The word "Veillance" comes from the French word "veiller" which means "to watch". It derives from the word "vigil".

The HarperCollins Complete and Unabridged English Dictionary defines "vigil" as:

1. A purposeful watch maintained, esp at night, to guard, observe, pray, etc.;
2. The period of such a watch

[from Old French vigile, from Medieval Latin vigilia watch preceding a religious festival, from Latin: vigilance, from vigil alert, from vigēre to be lively].

Thus "surveillance" is "watchful vigilance from above". So a literal translation of "surveillance" into English gives "overwatching" – not a real English word. The closest existing English word is the word "oversight" (dictionary.com, Random House, 2013), which emerged around the year 1300. In fact Google Translate returns the French word "surveillance" when presented with the English word "oversight". But in current English usage, the word "oversight" has a somewhat different and in fact, double, meaning, compared with "surveillance". Specifically, "oversight" can mean:

1. An omission or error due to carelessness. My bank statement is full of oversights. or;
2. Supervision; watchful care: a person responsible for the oversight of the organization. (Random House, 2013).

The fact that the English word "oversight" has two meanings perhaps explains why we use the French word "surveillance", i.e. why we use the term *"surveillance cameras"* rather than *"oversight cameras"*.

A surveillance camera is a camera that typically watches from above, e.g.:

*   The "eye-in-the-sky" afforded by an aerial surveillance "drone";
*   Cameras on property (real-estate), i.e. land or buildings.
    *   Cameras are affixed to land by way of watchtowers or poles or masts.
    *   Cameras are affixed to buildings in weatherproof enclosures, or affixed inside building interiors by way of "ceiling domes of wine-dark opacity" (Patton, 1995).

SOUSVEILLANCE

There has been a recent explosion of interest in body-borne camera systems to help people see better, as well as for body-centered sensing of the body itself and the environment around it. Such systems include the self-gesturing neckworn sensor-camera of Figure 1, when fitted also with a 3D data projector to project onto the real world (See Figure 2).

Another related invention is the MannGlassTM HDR (High Dynamic Range) welding glass that evolved out of experiments in photographically mediated visual reality in the 1970s (pre-dating 3M's SpeedGlassTM welding glass of 1981, which only provided global auto-darkening across the whole field of view), leading eventually to the general-purpose "Digital Eye Glass" (EyeTap) (See Figure 3).

It is clear from the foregoing that a portable, wearable, or implantable camera borne by an individual person is not a surveillance camera. It is still a veillance camera, i.e. it is still "watching"

Figure 1. Neckworn lifeglogging sensor-cameras. A lifeglog is a lifelog (or lifelong weblog) that does not require conscious thought or effort to generate (Mann, 2001).

Figure 2. Neckworn Augmented Reality pendant comprised of a wearable camera and projector together on the pendant (Mann, 2001). The wearer can control the apparatus by self-gesturing. Mann referred to the wearable computer in this sense as a "Sixth Sense", and used the term "Synthetic Synesthesia of the Sixth Sense" (Mann, 2001; Geary, 2003) (Source: Wikimedia Commons).

(veillance), but not necessarily "sur" (from above, physically or hierarchically).

What kind of veillance is it, though, if it is not *sur*veillance? The French prefix "sur" means "on" or "over" or "above". Its opposite is "sous" which means "under" or "below" as in "sous-chef" or "sous la table" (under the table). Thus body-borne sensing is widely referred to as "sousveillance", or just "veillance" (Mann, 2002; Kerr and Mann, 2006; Dennis, 2008; U.C. Berkeley, 2009; Bakir, 2010; Fletcher et al, 2011).

Sousveillance is a fundamental and universal concept that is widely understood across all ages and cultures (See Figure 4).

Figure 3. Digital Eye Glass (EyeTap) and the evolution of wearable computing in everyday life (top row). Mann's Digital Eye Glass (1999) and Google Glass (2012) (Source: Wikimedia Commons).

Figure 4. Six-year-old's drawing depicts Surveillance (overwatching) as compared with Sousveillance (underwatching). As a student in a French Immersion school, she thinks of Surveillance as the central-ized veillance of large entities (governments and corporations), vs. Sousveillance as the distributed "crowdsourced" veillance of the masses == Eyes AND ears down at street level, rather than high on a watchtower or streetlight pole.

VEILLANCE INTEGRITY AND VEILLANCE HYPOCRISY

To a six-year-old who's told she's not allowed to take pictures in a place where she's surrounded by surveillance cameras, it would seem to her that property (land and buildings) has more rights than people – that property almost always has the right to bear cameras, but people often don't. She does not fully comprehend the more complex sociopolitical landscape in which it is really the unseen behind-the-scenes governments, police, and property owners who have those rights, not the land or buildings themselves. Moreover, in this day and age of AR (Augmediated Reality) (Mann, 2001), the hypocrisy of one-sided veillance is accentuated by the ubiquitous appearance of 2-dimensional camera-ready QR (Quick Read) barcodes through business establishments, including those where cameras are prohibited (See Figure 5).

To the six-year old who's being taught that it is impolite to stare, how do we explain the impoliteness of the surveillance cameras all around us? The "unblinking eye" of surveillance is said to "never avert its gaze". And if she's told to stop taking pictures, should she hide her camera and keep taking pictures? Surveillance cameras are often hidden, and if you ask a security guard or staff person what's inside those mysterious ceiling domes of wine-dark opacity, they'll usually avoid giving an honest answer. In fact the security industry often advises that cameras should be hidden, or concealed, to avoid, for example, vandalism (see for example what is happening with the Camover social movement http://www.youtube.com/watch?v=9GCsd2TJKjQ). If we ask whether or not we're under surveillance, a typical answer is that we're forbidden from knowing, "for security purposes".

So should not sousveillance enjoy the same rights to secrecy? For the safety of the individual wearing a camera, should the camera not be hid-

Figure 5. Hypocrisy of surveillance: Many establishments use surveillance cameras but prohibit sousveillance. Signs in business establishments read "NO CAMERAS/VIDEO", "NO CELL PHONES", "NO CELL PHONE IN STORE PLEASE" – yet, ironically, a cell phone is needed to read the in-store pre-product Quick Read (QR) codes on boxes of watermelons and other merchandise displays.

den? Social movements like Camover merely amount to destruction of property, but social movements like "Stop the Cyborgs" could lead to violence against persons, not just property; violence against sousveillance can result in injury or death of a person. Is human life not worth more than physical property? And if so, should not sousveillance be granted the same, or more, "rights" to secrecy than surveillance?

Authorities who have the only recordings of incidents, including those depicting wrongdoing by the authorities, are being trusted to be keepers and curators of the only copy of evidence that might show their wrongdoing. Consider, for example, the Brazilian electrician who was shot and killed by police in a London subway, when they mistook him for someone else. A simple case of mistaken identity turned into a case of lost surveillance recordings. Police seized four separate surveillance system recordings and reported that all four systems were blank and failed to record the incident:

And now police say there are no CCTV pictures to reveal the truth. So why did plainclothes officers shoot young Jean Charles de Menezes seven times in the head, thinking he posed a terror threat?

Evidence of this hold-up should have been provided by CCTV footage from dozens of cameras covering the Stockwell ticket hall, escalators, platforms and train carriages.

However, police now say most of the cameras were not working. (Guardian 2005)

A similar situation occurred more recently, but this time with sousveillance:

On Memorial Day 2011, Narces Benoit witnessed and filmed a group of Miami police officers shooting and killing a suspect ... He was then confronted by officers who handcuffed him and smashed his cell phone, but Benoit was able to sneakily preserve

the video ... he discreetly removed the [memory] card and placed it in his mouth. (NBC News and the Miami Herald)

In this case Benoit was not sousveilling for personal gain, or even for personal safety or personal protection. He was sousveilling for the public good, and in some sense his effort was a noble one. Thus a person wearing a camera perhaps has a moral and ethical imperative to (1) record wrongdoing by authorities; and (2) to conceal the recording apparatus, and its associated functioning.

THE VEILLANCE CONTRACT ANALOGY: RECIPROCAL RECORDING RIGHTS

Whereas there may exist certain places like change rooms where recording is not appropriate, it has been suggested that in any place where surveillance is used, sousveillance must also be permitted.

The justification for such a reciprocal recording right can be understood by way of the "contract analogy" ("veillance contract analogy").

Imagine A and B enter into a written contract but that only A has a copy of the contract. If B were to carelessly lose its copy of the contract, the contract is still valid. But if the reason B does not have a copy of the contract is that A prohibited B from having a copy, then the contract is not valid. The reason for this rule is to prevent falsification.

Let's suppose we have a 50 page contract to which A and B both agreed, with their signatures on page 50. Later, A could go back and change page 49 (one of the non-signature pages). But if A and B both had copies, the copies would differ, and the courts would place higher scrutiny on the remaining parts, possibly examining the papers by microscope or other forensics to determine which copy was falsified.

By prohibiting these checks and balances (i.e. by prohibiting B from having a copy of the contract), A is creating a potential conflict-of-interest, and a possibility (maybe even an incentive) for falsification of the contract.

In today's world we live a social contract of the oral and action-based variety. Much of what we do is spoken or acted out, and not written. An oral contract is still legally binding. So if one entity insists on having the only copy of what was said or agreed upon, A is creating the possibility to falsify (whether by editing or simply by omission, i.e. by deleting some pictures and keeping others) the recorded evidence.

Such a monopoly on sight can create "surveillance curation", i.e. the person doing the surveillance "curates reality" by selecting certain "exhibits" to keep, and others to delete.

In response to such a proposal, Paul Banwatt, a lawyer at Gilbert's LLP, has suggested that:

1. Surveillance cannot be secret, or else individuals will be unable to tell when their right exists, or if one assumes the right is assumed to exist then; and
2. Those who sousveil must be informed that they are NOT being recorded in order to form the necessary basis for a demand to stop sousveillance. (Personal communication by way of author S. Mann's Veillance Group on LinkedIN.com).

A practical solution to this problem is to at least agree that when a person is prohibited from recording their own side of an interaction (i.e. prohibited from recording their own senses), the person who prohibited should have their side also removed from admissibility in any court of law. Such a "veillance contract" does not require either party to know whether or not their actions are being recorded!

Under the proposed rule, an organization installing a "no photography" sign, or otherwise discouraging people from keeping their own copy of the "veillance contract", would make their own surveillance recordings inadmissible in a court of law.

DEFINING SUR/SOUSVEILLANCE

The higher you look up the ladder, the more difficult it becomes to tell the good guys from the bad guys. — Reese, Person of Interest

Surveillance is the veillance of large entities in a position of authority, whereas sousveillance is the veillance of smaller entities not in a position of authority (Mann, 2002; Kerr and Mann, 2006). And to the extent that authorities want to watch, but not be watched, the veillance of authority – surveillance – has, associated with it, an inherent conflict-of-interest.

As mentioned previously, a simple and often-used definition of surveillance is cameras borne by property (land or buildings), and of sousveillance is cameras borne by people. But a hierarchical definition is also possible: surveillance is the veillance of large powerful entities on the higher rungs of the social "ladder", and sousveillance is the veillance of the masses on the lower rungs of the social "ladder". This leads to obvious problems with conflict-of-interest and possible corruption (See Figure 6). It is often said that putting surveillance cameras in one part of a city will merely "move" or "push" crime away to other parts of the city. But putting cameras everywhere down at street-level gives crime nowhere to "move" but up the social ladder toward corruption.

Moreover, not only can surveillance cameras "push" crime "up", but they can also be used to perpetrate crimes. Authorities can use surveillance cameras to determine when someone is away (notifying accomplice robbers), or use surveillance cameras to stalk a victim:

A SECURITY guard at one of Edinburgh's best-known visitor attractions used CCTV cameras to stalk a young female worker and spy on the public. James Tuff used the camera system at Our Dynamic Earth, Edinburgh, to track his victim and then radio her with lewd comments.

Figure 6. The "Ladder Theory" of Veillance. A society with surveillance-only (i.e. where sousveillance is prohibited) is a society in which there is an inherent conflict-of-interest that tends to favour higher-level corruption over low-level street crime (i.e. tends to "push" or "move" crime from being under veillance to rising above the veillance).

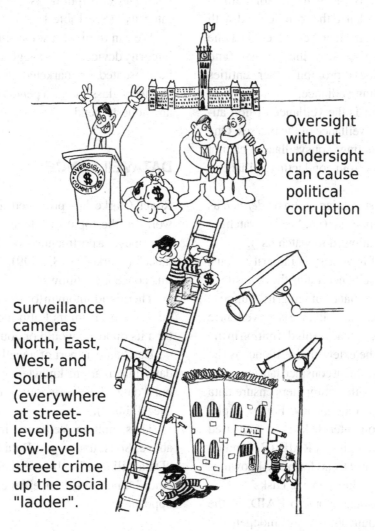

Oversight without undersight can cause political corruption

Surveillance cameras North, East, West, and South (everywhere at street-level) push low-level street crime up the social "ladder".

He even trained the cameras on members of the public milling about outside, in one case saving footage of two girls kissing to show to colleagues.

Tuff eventually sexually assaulted Dora Alves ... He was fined and placed on the sex offenders register for three years. ... She said: 'At first it was just the odd comment about my body; he would say things about me having a real woman's body ... But soon after he would appear out of nowhere when I was cleaning in the toilets. ...' as she walked to the canteen on her break and stopped to collect something from her locker. 'Mr Tuff came out of his office and grabbed me from behind. ...' She said CCTV footage which could have proved the incident took place had gone missing. (News Scotsman.com. 'Cleaner says she was stalked on CCTV by security guard', March 2011)

Thus there is obviously a conflict-of-interest among higher authorities that can give rise to an incentive to promote surveillance while prohibiting sousveillance.

This gives us a possible definition for sur/sousveillance:

- **Surveillance:** Is the veillance of authority, i.e. the veillance that can be used with a prohibition on other veillances, i.e. one entity conducting surveillance can (and sometimes does) prohibit other entities from conducting veillance;
- **Sousveillance:** Is the veillance of non-authority, i.e. the veillance of the masses. One person conducting sousveillance cannot prohibit others from conducting veillance.

In other words, surveillance can be *defined* as the veillance of hypocrisy (i.e. "we're watching you but you're not allowed to watch us").

The antonym of hypocrisy is integrity. Thus, by mere logic, sousveillance, being the opposite of surveillance, is the veillance of integrity. Suppose we accept simply that "surveillance is the veillance of integrity"? Let us "unpack" this definition in the following way: to the extent that veillance is the collection of data, we might consider data integrity. One of the best ways that computers ensure data integrity is through redundancy. For example, data integrity on computer disks is best ensured through such technologies as RAID (Redundant Array of Independent Disks, formerly known as "Redundant Array of Inexpensive Disks").

Sousveillance is analogous to RAID, in the sense that it is a Redundant Array of Independent Veillances, i.e. we could abbreviate it as "RAIV". A RAIV is therefore a possible way of ensuring veillance integrity through a decentralized veillance.

Consider, for example, a situation in which any member of a society can commit an act of wrongdoing. Centralized veillance would allow the possibility of wrongdoing by an authority to falsify self-incriminating veillance data, whereas distributed veillance is less likely to fall totally under the control of a single entity that might falsify the data.

The opposite of integrity, when referring to human-nature, is hypocrisy. The opposite of integrity when referring to data is corruption, i.e. "data corruption" is the opposite of "data integrity" (See Table 1).

We can think of a sousveillance device as an integrity device, e.g. as eyeglass, the device might be presented or marketed or understood as an "honesty glass", or as a pendant, it might become an "honesty pendant".

DATAVEILLANCE

Roger Clarke has produced a seminal body of work on the topic of "dataveillance" which he describes as a portmanteau of "data" and "surveillance" (Clarke, 1998, 1999). We can generalize this concept as follows:

There is no antonym to "Data" in the dictionary, but if we go from its Latin meaning of "given", then its opposite would or could be "taken".

But "data" is part of a well-known hierarchy: data, information, knowledge, wisdom, and all but "data" have previously defined opposites (See Table 2).

Thus, putting Veillance in the middle, Sur above, Sous, under, and Data beside it, we have "Dataveillance", "Infoveillance", "Knowveillance", and "Wisdomveillance" offset by their opposites.

Table 1. *Hypocrisy of surveillance, when it becomes (or is defined as) the veillance of a single entity (i.e. the "authority") that prohibits other veillances, as compared to the integrity of distributed veillance*

	Surveillance (Centralized)	Sousveillance (Distributed)
Human-nature	Hypocrisy	Integrity (RAIV)
Data	Corruption	Integrity (RAID)

Table 2. "DI KNOWise" veillances

Understanding	"Overstanding"	Veillance
Data	Taken	Dataveillance
Information	Misinformation	Infoveillance
Knowledge	Ignorance	"Knowveillance", "Knoveillance", "Knowleveillance"
Wisdom	Stupidity	"Wisdomveillance" ("Wisdoveillance", "Wisveillance")

UBERVEILLANCE

The seminal work of M.G. Michael and K. Michael define the concept of Uberveillance as embedded surveillance (Masters & Michael, 2005; Masters & Michael, 2007; Michael et al., 2008; Michael & Michael, 2007; Michael & Michael, 2009; Michael & Michael, 2010; Perusco & Michael 2007). The definition appears in some dictionaries, e.g. "An omnipresent form of 24/7 surveillance of humans based on widespread electronic devices, and especially computer chips embedded into bodies." – Wiktionary.

Table 3 outlines the etymology of Uberveillance in the context of its language mixtures.

As veillance encroaches closer on the body, this research on Uberveillance highlights the dangers faced by a corporeal technology that itself becomes more like surveillance than sousveillance.

Table 3. Veillances in 3 languages, French, German, English

French	Sur	Sous	Veillance
German	Über	Unter	Wachung
English	Over	Under	Watching (monitoring, sight)

FROM CROWD VEILLANCE TO CLOUD VEILLANCE: GOOVERNANCE OF THE GOOLAG, OR GAOLBROKEN GOOGLASS?

The recent introduction of Google Glass has suggested a possible reversal of sousveillance, i.e. what happens when sousveillance is commercialized by a large corporation that re-centralizes what was formerly a distributed form of computation, sensing, and "crowd veillance".

Google has decided that it will pre-approve all apps offered to glasses users, unlike its more wide open market for Android phones and tablets. 'It's so new, we decided to be more cautious,' Schmidt said. 'It's always easier to open it up more in the future.' (Reuters, 'Google's Schmidt says talking to glasses can be weird, inappropriate' Reporting by Aaron Pressman; Editing by Lisa Shumaker, Reuters, Thu Apr 25, 2013 6:21pm EDT, http://www.reuters.com/article/2013/04/25/us-google-harvard-idUSBRE93O1FF20130425

Google has also threatened to remotely sabotage anyone's eyeglass if they sell it or even lend it to others. This remote self-destruct and sabotage capability is built into the product.

This "deconomics" ("sabotage economics"), at the hardware level, suggests a form of centralized control rather than the freedom envisioned for sousveillance. Thus it has been questioned whether Google's vision is one of sousveillance or surveillance (Corey Manders, IEEE ISTAS13, Toronto, Ontario, Canada, 2013 June 28).

Table 4. Corporeality vs. existentiality

		Corporeality	
Existentiality	Low	Surveillance	Uberveillance
	High	??? ("Unterveillance")	Sousveillance

We might therefore consider the "Wearability" (Corporeality) and Existentiality axes as suggested in "Can Humans Being Clerks make Clerks be Human? - Exploring the Fundamental Difference between UbiComp and WearComp", by S. Mann, in Oldenbourg Electronic Journals, ISSN 0944-2774, Informationstechnik und Technische Informatik, Volume 43, Issue 02, pages 97-106 http://wearcam.org/itti/

If we put Uberveillance on these axes, we see a missing quadrant (See Table 4) which we might call "unterveillance".

UNTERVEILLANCE

Just as sousveillance can be co-opted by a large corporation, we might ask the question, "can surveillance also be reversed"? For example, what if the video captured by cameras in the environment (e.g. cameras on land and buildings) could become "existential", i.e. distributed, rather than monopolized.

This suggests that it is possible to have a reversal of surveillance, e.g. the new camera-based streetlights could catch a corrupt police chief or mayor, as easily as they might capture a robber or lone thief.

CONCLUSIONS

In this chapter we have considered Veillance (Sur and Sous), and Uberveillance (the corporatization or over-governance of sousveillance). Recent developments suggest a possible reversal of sousveillance into uberveillance, and surveillance into unterveillance.

Much needs to be explored, but one thing is certain: we have moved from living in a "surveillance society" to living in a "veillance society" where there is a plurality of veillances, not just one!

REFERENCES

Bakir, V. (2010). *Sousveillance, media and strategic political communication*. London, UK: Continuum.

Burgess, J. G., Warwick, K., Ruiz, V., Gasson, M. N., Aziz, T. Z., Brittain, J., & Stein, J. (2010). Identifying tremor-related characteristics of basal ganglia nuclei during movement in the Parkinsonian patient. *Parkinsonism & Related Disorders*, *16*(10), 671–675. doi:10.1016/j.parkreldis.2010.08.025 PMID:20884273

Choney, S. (2011). *NBC news*. Retrieved from http://www.nbcnews.com/technology/technolog/memory-card-mouth-saves-police-shooting-video-122903

Clarke, R. (1988). Information technology and dataveillance. *Communications of the ACM*, *31*(5), 498–512. doi:10.1145/42411.42413

Clarke, R. (1999). *Introduction to dataveillance and information privacy, and definitions of terms*. Roger Clarke's Dataveillance and Information Privacy Pages.

Dennis, K. (2008). Viewpoint: Keeping a close watch–the rise of self-surveillance and the threat of digital exposure. *The Sociological Review*, *56*(3), 347–357. doi:10.1111/j.1467-954X.2008.00793.x

Fletcher, G., Griffiths, M., & Kutar, M. (2011). *A day in the digital life: A preliminary sousveillance study*. Retrieved from ttp://ssrn.com/abstract=1923629

Foster, K. R., & Jaeger, J. (2007). RFID inside. *IEEE Spectrum*, *44*, 24–29. doi:10.1109/MSPEC.2007.323430

Gasson, M. N., Hutt, B. D., Goodhew, I., Kyberd, P., & Warwick, K. (2005b). Invasive neural prosthesis for neural signal detection and nerve stimulation. *International Journal of Adaptive Control and Signal Processing*, *19*(5), 365–375.

Gasson, M. N., Wang, S. Y., Aziz, T. Z., Stein, J. F., & Warwick, K. (2005a). Towards a demand driven deep-brain stimulator for the treatment of movement disorders. In *Proceedings of 3rd IEE International Seminar on Medical Applications of Signal Processing*, London, UK: IEE.

Geary, J. (2003). *The body electric: An anatomy of the new bionic senses*. New Brunswick, NJ: Rutgers University Press.

Hameed, J., Harrison, I., Gasson, M. N., & Warwick, K. (2010). A novel human-machine interface using subdermal magnetic implants. In *Proceedings of the IEEE International Conference on Cybernetic Intelligent Systems*. IEEE.

Hayes, A. (2013). Cyborg cops, googlers, and connectivism. *IEEE Technology and Society*, *32*(1), 23–24. doi:10.1109/MTS.2013.2247731

Kerr, I., & Mann, S. (2006). *Exploring equiveillance*. Retrieved from http://www.anonequity.org/weblog/archives/2006/01/exploring_equiv_1.php

Lyon, D. (2001). *Surveillance society*. Buckingham, UK: Open University Press.

Mann, S. (2001). *Intelligent image processing*. New York: John Wiley-IEEE Press. doi:10.1002/0471221635

Mann, S. (2002). Sousveillance, not just surveillance. *Metal and Flesh, 6*(1).

Masters, A., & Michael, K. (2005). Humancentric applications of RFID implants: The usability contexts of control, convenience and care. In *Proceedings of Mobile Commerce and Services*. IEEE. doi:10.1109/WMCS.2005.11

Masters, A., & Michael, K. (2007). Lend me your arms: The use and implications of humancentric RFID. *Electronic Commerce Research and Applications*, *6*(1), 29–39. doi:10.1016/j.elerap.2006.04.008

Michael, K., & Clarke, R. (2013). Location and tracking of mobile devices: Überveillance stalks the streets. *Computer Law and Security Review, 29*(2).

Michael, K., & Michael, M. G. (2008). *Innovative automatic identification and location-based services: From bar codes to chip implants*. Hershey, PA: IGI Global.

Michael, K., & Michael, M. G. (2009). *Innovative automatic identification and location-based services: From bar codes to chip implants*. Hershey, PA: IGI Global. doi:10.4018/978-1-59904-795-9

Michael, M. G., Fusco, S. J., & Michael, K. (2008). A research note on ethics in the emerging age of überveillance. *Computer Communications*, *31*(6), 1192–1199. doi:10.1016/j.comcom.2008.01.023

Michael, M. G., & Michael, K. (2007). A note on überveillance. In *From dataveillance to uberveillance and the realpolitik of the transparent society*. Wollongong, Australia: Academic Press.

Michael, M. G., & Michael, K. (2010). Toward a state of überveillance. *IEEE Technology and Society Magazine*, *29*(2), 9–16. doi:10.1109/MTS.2010.937024

Patton, P. (1995). Caught-You used to watch television: Now it watches you. *Wired, 3*(1), 124–127.

Perusco, L., & Michael, K. (2007). Control, trust, privacy, and security: Evaluating location-based services. *IEEE Technology and Society Magazine, 26*(1), 4–16. doi:10.1109/MTAS.2007.335564

Thompson, T., & Phillips, T. (2005). *Death in Stockwell: The unanswered questions*. Retrieved from http://www.guardian.co.uk/uk/2005/aug/14/july7.terrorism

UC Berkeley School of Information. (2009). *Info 290: Surveillance, sousveillance, coveillance, and dataveillance*. Retrieved from http://www.ischool.berkeley.edu/courses/i290-sscd

Warwick, K., Gasson, M. N., Hutt, B., Goodhew, I., Kyberd, P., & Andrews, B. et al. (2003). The application of implant technology for cybernetic systems. *Archives of Neurology*, *60*(10), 1369–1373. doi:10.1001/archneur.60.10.1369 PMID:14568806

KEY TERMS AND DEFINITIONS

Access Control: In information security, access control is the selective restriction of access to a place or other resource.

Advocacy: Is considered a political process usually by a group of people with the aim to influence public-policy. An advocate is usually affiliated with a non-government organisation, undertaking media campaigns, public speaking, commissioning and publishing research.

Behaviour: Is the range of actions undertaken by a human in their physical or non-physical environment. It is the response of the system or organism to various stimuli or inputs, whether internal or external, conscious or subconscious, overt or covert, and voluntary or involuntary.

Biometrics: The measurement of physical characteristics or traits, such as fingerprints, handprints, DNA, iris or retinal patterns, gait or voice recognition for use in verifying the identity of individuals.

Chilling Effect: A discouraging or deterring effect, usually resulting from a restrictive law or regulation. For example, introducing ID scanners at licensed venues to minimize patron violence, or even introducing a curfew on a particular zone.

Ethics: A set of principles of right conduct. The rules or standards governing the conduct of a person or the members of a profession.

Geographic Information Systems: A computer application used to store, view, and analyze geographical information, especially maps. A GIS can contain both vector and raster images.

Google Glass: A digital glass head up display (HUD) device that carries many smartphone functions. Its capabilities include identifying current locations (augmented reality), voice and video calls, GPS navigation, sending/receiving messages, taking notes, as well as snapping and sharing photos and video.

Harm: To cause physical or psychological injury or damage.

Human Cloud: The emerging cloud revolutionised by wearable technologies, allegedly making users feel more intelligent and in control of their lives.

Identity Cards: An identity document (also called a piece of identification, ID, or colloquially as one's 'papers') is any document which may be used to verify aspects of a person's personal identity. If issued in the form of a small, mostly standard-sized card, it is usually called an identity card (IC). Countries which do not have formal identity documents may require informal documents.

Chip Implants: A medical device manufactured to replace a missing biological function or to enhance an existing one. Medical implants are man-made devices made of a biomedical material such as titanium, silicone or apatite. In some cases implants contain electronics e.g. artificial pacemaker and cochlear implants. Some implants are bioactive, such as subcutaneous drug delivery devices in the form of implantable pills or drug-eluting stents.

iPlant: Is the potential for an invention of an Internet-enabled chip implant required to login to online applications. The word "iPlant" conjoins the words "internet" and "implant" and is a play on the internet-based applications developed by Apple Inc. and NTT Docomo (e.g. iTunes, iPod, iLife, iSight, iPhone, iPad, imode, i-shot, and iD).

Interdisciplinary: Of, relating to, or involving two or more academic disciplines that are usually considered distinct.

Legislation: A proposed or enacted law or group of laws.

Lifeloggers: Wear body-worn video recorders in order to capture their entire lives digitally, or large portions of their lives.

National Security: A collective term encompassing both national defense and foreign relations. Internationally has been used to denote an all-hazards approach to security, ensuring the safety of the citizen.

Omnipresent: Present everywhere simultaneously.

Omniscience: Having total knowledge; knowing everything.

Pacemakers: Any of several usually miniaturized and surgically implanted electronic devices used to stimulate or regulate contractions of the heart muscle.

Profiling: To draw or shape a profile of an individual. Increasingly proactive profiling in real-time is occurring, especially in shopping malls, using consumer-owned mobile phones.

Nanotechnology: The science and technology of building devices, such as electronic circuits, from single atoms and molecules.

Regulation: The act of regulating or the state of being regulated. A principle, rule, or law designed to control or govern conduct.

Smart Dust: Is a system of many tiny micro electromechanical systems (MEMS) such as sensors, robots, or other devices, that can detect, for example, light, temperature, vibration, magnetism, or chemicals.

Social Policy: A policy for dealing with social issues.

StreetView: Google Street View is a technology featured in Google Maps and Google Earth that provides panoramic views from positions along many streets in the world. It was launched on May 25, 2007, in several cities in the United States, and has since expanded to include cities and rural areas worldwide.

Transhumanism: (Abbreviated as H+ or h+) Is an international cultural and intellectual movement with an eventual goal of fundamentally transforming the human condition by developing and making widely available technologies to greatly enhance human intellectual, physical, and psychological capacities.

Value Chain: A chain of activities that a firm operating in a specific industry performs in order to deliver a valuable product or service for the market. Can also be considered a chain of firms required to offer a complex service.

Wearable Computers: Also known as body-borne computers are miniature electronic devices that are worn by the bearer under, with or on top of clothing. This class of wearable technology has been developed for general or special purpose information technologies and media development. Wearable computers are especially useful for applications that require more complex computational support than just hardware coded logics.

Chapter 3
Uberveillance:
Where Wear and Educative Arrangement

Alexander Hayes
University of Wollongong, Australia

ABSTRACT

The intensification and diversification of surveillance in recent decades is now being considered within a contemporary theoretical and academic framework. The ambiguity of the term 'surveillance' and the surreptitiousness of its application must now be re-considered amidst the emergent concept of Uberveillance. This chapter presents three cases of organisations that are currently poised or already engaging in projects using location-enabled point-of-view wearable technologies. Reference is made to additional cases, project examples, and testimonials including the Australian Federal Police, Northern Territory Fire Police and Emergency Services, and other projects funded in 2010 and 2011 by the former Australian Flexible Learning Framework (AFLF), now the National VET E-learning Strategy (NVELS). This chapter also examines the use of location-enabled POV (point-of-view) or Body Wearable Video (BWV) camera technologies in a crime, law, and national security context, referencing cross-sectoral and inter-disciplinary opinions as to the perceived benefits and the socio-technical implications of these pervasive technologies.

INTRODUCTION

[The Emperor] said: "It is all useless, if the last landing place can only be the infernal city, and it is there that, in ever-narrowing circles, the current is drawing us."

And Polo said: "The inferno of the living is not something that will be; if there is one, it is what is already here, the inferno where we live every day, that we form by being together. There are ways to escape suffering it. The first is easy for many: accept the inferno and become such a part of it

DOI: 10.4018/978-1-4666-4582-0.ch003

that you can no longer see it. The second is risky and demands constant vigilance and apprehension; seek and learn to recognize who and what, in the midst of the inferno, are not inferno, then make them endure, give them space." (Calvino, 1972, *Invisible Cities*)

BACKGROUND

Memory

Picture a domestic setting in the 1970s.

Sydney, Australia throbbed under the self-determination of the tune in, dropout culture and the soapbox debates strayed left and right as far as public tolerance would allow. Telephones were wired to the wall, spin dialled and publicly coin dependent. Dogs roamed free, unidentifiable until someone visited the local pound. The faux-wood panelled television set peddled sitcom have-it-now culture. Police officers wore two-way radios and carried Smith & Wessons.

Fast forward to 2012.

Parents know more of their family's lives through an online website designed originally to unite college sweethearts. Everyone owns a mobile phone or two and sometimes even three. Dogs are chipped, de-sexed, voice-boxed and confined to yards as are children confined to their living rooms. Long division is a practice lost to the electronic calculator.

DIY drone hobbyists gather on local town ovals.

Police officers and security agency personnel wear high definition location enabled video recorders for evidence gathering, as do teachers in educational organisations.

Have we progressed as a society over the last 30 years or have we lost the ability to think outside of the networked grid? Amidst our hyper-connectivity does anyone give himself or herself long enough to review what has been, what is happening and where we want to be?

SURVEILLANCE

We could, upon reflection, conclude that we now live in a society besieged by a technological omnipresence born of dystopia and intense paranoia.

We might also draw conclusions that communities in all parts of the world are constantly teetering between peaceful citizenship and utter chaotic anarchy, as if in a state of schizophrenia so acute that the very architectures they inhabit have become cells of their own Orwellian incarceration.

In many countries camera surveillance has become commonplace and ordinary citizens and consumers are increasingly aware that they are under surveillance in everyday life. Camera surveillance is typically perceived as the archetype of contemporary surveillance technologies and processes. While there is sometimes fierce debate about their introduction, many others take the cameras for granted or even applaud their deployment. Yet what the presence of surveillance cameras actually achieves is still very much in question. International evidence shows that they have very little effect in deterring crime and 'in making people feel safer' but they do serve to place certain groups under greater official scrutiny and to extend the reach of today's 'surveillance society'. (Doyle et al., 2011)

We could also, as optimists, consider that we have as a society developed a better sense of who we are as humans as a result of surveillance technologies, by being able to observe others at work, play and in public places volunteered to the interweb.

Some would say this developed "awareness" of self, our moral conscience before evil and malcontent, as impossible to attain given the depravities of inconsistency that humans exhibit when subject to a constancy of "unseen" supervision.

One of the goals of moral education is to cultivate a conscience – the little voice inside telling us that we should do what is right because it is right. As surveillance becomes increasingly ubiquitous, however, the chances are reduced that conscience will ever be anything more than the little voice inside telling us that someone, somewhere, may be watching. (Westacott, 2011)

By feeding the interweb's insatiable appetite for knowledge, this technology is now "us" - a manifestation of our biological milieu. This matrix of connection is what we have come to depend upon for everything we live by as humans and likewise systems and processes that we need to survive depend on the presence and connectivity of this global entity.

As ordinary citizens we frequent numerous private and public spaces that all legitimize electronic surveillance for a myriad of reasons. We acknowledge and ignore its gaze when it suits us, an omnipresence that evokes all types of emotions, substantiated or otherwise.

Technology is a wonderful thing ... With advances in technology increasing at a staggering rate there are ever more options available for public safety agencies to increase operational efficiency. (Kay, 2007, pp. 49-50)

This paper presents recent examples of organisations employing the use of these technologies in the pursuit of educational excellence, posits considerations that need to be made by anyone seeking operational efficiency using surveillance in an educational or other context, and examines the greater social impacts these technologies may be having upon learners, educators and workforce trainers.

SOUSVEILLANCE

Undoubtedly, this ever present state of electronic monitoring, static and now mobile, has altered the manner of how we can engage with each other in a social setting, how we interact with each other in the broader community, in vehicles as we travel, in our homes and now in places of educative arrangement.

Mobile phones equipped with high definition video recording have enabled citizen reporters to spread stories and realise an audience quicker than ever before. The panopticon of views support, debunk, twist and re-shape the way we understand things happening in our community, across the country and around the world.

Sousveillance activities benefit (as do surveillant activities) when networked connectivity permits synchronous participation, where the "smart-mob" of "know-where" massage an otherwise passive public with ambiguity and contradiction.

As we enter an age of swift and monumental access to information at speeds and in volumes unthinkable a few years ago, we are now also subject to a shift in the very core of humankind as we embrace technologies as inseparable necessities beyond convenience.

One of the characteristics of change is that often we do not realise it has happened until we have had time to look back on it. Change can be gradual. But a series of small, incremental events can over time amount to a fundamental shift. Sometimes change can be swift and monumental, with an impact so profound that we quickly forget what life was like beforehand. (Howarth & Ledwidge, 2011, p.1)

Until recently, in human terms, we had been subject only to the "eye of providence", governed by spiritual beliefs and religious inculcation. As the unification of interaction in apparent real time is realised by the Internet, so too has the matrix of gaze, the digitisation and repeatability of what is seen, heard and traded.

Steve Mann, Professor at the Department of Electrical and Computer Engineering, University of Toronto, is attributed with coining the term "sousveillance", which typically involves community-based recording from first person perspectives. These formative activities do not necessarily have a political purpose whereas inverse-surveillance is known as a form of sous-veillance that is typically directed at, or used to collect data to analyse or study, surveillance or its proponents.

As a 'cyborg' in the sense of long-term adaptation, body-borne technologies, etc., one encounters a new kind of existential self-determination and mastery over one's own destiny, that can be learned, in the postmodern (posthumanism) context one might think of as the 'cyborg age' in which many of us now live. (Mann, 2011)

We have, by choice and by design, broken the homo-social frameworks of ancient society and in the discordance of postmodernity emptied our fears into server farms bearing the insignia of clouds, mirrors to ourselves, heaven and hell now wrapped in plastic.

In essence, the creation of a cyberspace has made it possible for humans to connect, create and repeat patterns for mutual benefit as much as for malevolent purpose.

Users interactions, locational whereabouts and other transactional data is now of great interest to service providers interested in selling advertising space, governments protecting national interests and corporations seeking to influence the behaviours of consumers.

We have tapped into the resonance of electronic transmission and in doing so have evolved as humans now immersed, connected and embedded. As we look at what is presented from others in online social media spaces, as they "shoot back" recording civil disobedience, appreciative inquiry and the seemingly banal, we see our own future and the rapid difference of society in the past.

Sousveillance activities broaden the process of digitisation, as mobile activities are captured and then transmitted live to the internet. In replication, we better understand our own contributions as nodes in architectures of networked participation. Examining these behaviours allows us to better understand our own motivations for sharing and may shape the understanding of others as to who we are as human beings.

Welcome to the Social Media Classroom and Collaboratory. It's all free, as in both 'freedom of speech' and 'almost totally free beer.' We invite you to build on what we've started to create more free value. The Social Media Classroom (we'll call it SMC) includes a free and open-source (Drupal-based) web service that provides teachers and learners with an integrated set of social media that each course can use for its own purposes—integrated forum, blog, comment, wiki, chat, social bookmarking, RSS, microblogging, widgets, and video commenting are the first set of tools. (Rheingold, 2011)

As the Rheingold example elucidates, engendering reflective and proactive digital literacy into existing curriculum activities to bring about awareness for those who may seek recourse in uncertain futures is now a core consideration of educational organisations worldwide. Given the myriad of laws and by-laws that defend the right for businesses and organisations to retain such data flows, it is the responsibility of educators to at least attempt to inform their learners of the dangers of their disclosures to the web.

Therein lays, at the core, a privacy concern that transcends generations who currently post drunken soirees to the web with little or no recourse to delete such compromising data in the not so distant future.

To what extent must we re-consider our educative practices and policies given this interconnected and hybrid state of singularity? Have we, as Mann suggests, entered a dawn of posthumanism? And is it hive-mind formed as a "cyborg" if we begin to consider where technology exists in our everyday human ecology? To what extent can we choose to engage in activities of the everyday, of the private sanctuary, the cultural spaces and places we declare taboo without the ever-presence of the "other"?

DATAVEILLANCE

"The Digital Persona and its Application to Data Surveillance" (1994) by Roger Clarke states:

The 'digital persona', as a tool in the analysis of behaviour on the 'net'. It applies the tool, together with data surveillance theory, to predict the monitoring of the 'real-life' behaviour of individuals and groups through their net behaviour. (Clarke, 1994)

Our personal data, our identity, and our navigation may become "their" data and our choices rapidly influence those of others outside our preferred filter bubble, none more evident than location-enabled push-services we subscribe to through our mobile cell phones.

We have entered an age where technologies are as important to an individual's identity as culture is to a nation fighting to be recognised amidst the carnage of others' attempts at co-sovereignty. We are now, as consumers, nodes in a web of algorithms, as citizens in a state of constant transmission and as people of many nations unified in our geographic impermanence.

Our relationship with the Internet is, as it is with the retailer - a journey, connection dependent and increasingly intertwined in a marriage of networks. Checkout operators have given way to surveillance assistants; shopping transactions now have become a simple robotic process of swipe, pay, pack and go.

Attitudes to interruption and preference permissions to digital communication have shifted, in the behaviours of humans of all ages and in places where the tolerance for such perfusion previously did not exist. The open circuit of the mobile device positions telecommunication providers as the new lawmakers, their customers wallowing in the quagmire of their own acquiescence.

No longer is an idle conversation with the taxi driver an exposition of friendly satire unheard. Our banter with unknown baristas, airport terminal staff and with service providers in call centres in far flung countries all become part of the larger cacophony of networked, recorded and very often data-mined conversation.

UBERVEILLANCE

Uberveillance is, in essence, an embodiment of all "veillances", in totality.

At its core, is an apex of composites - triquetra - that of surveillance and all its nuances, that of dataveillance and its multitude of feeds and that of sousveillance with its manifestations of recalcitrance.

The emergent term coined in May 2006 by UOW Honorary Associate Professor Dr M.G. Michael is described as:

... an above and beyond, an exaggerated, an omnipresent 24/7 electronic surveillance. It is a surveillance that is not only 'always on' but 'always with you' (it is ubiquitous) because the technology that facilitates it, in its ultimate implementation, is embedded within the human body. The problem with this kind of bodily invasive surveillance is that omnipresence in the 'material' world will not always equate with omniscience, hence the real concern for misinformation, misinterpretation, and information manipulation. (Michael, 2007)

Michael & Michael (2009) in their seminal publication titled *Innovative Automatic Identification and Location Based Services: From Bar Code To Chip Implants* cite Steve Mann's (2001) Axis of Existentiality as a fundamental diagrammatic depiction of the trajectory of human interaction with technologies and the subsequent consequentiality of collision between the wearability of technology with that of the control one has over that technology.

This emergent concept gained entry into the *Macquarie Dictionary* in 2008 and has since challenged all who consider its emergent themes, as we all, irrespective of our role in the community, consider the implications of any action that encourages, supports or indeed advocates sub-dermal infusion of technologies.

In the context of educative arrangement, where students/learners/people from all walks of life frequent architectures of knowledge accreditation, there are increasing examples in Australia of the use of wearable technologies that capture from the first person perspective, are hands-free, continuous and in some cases automated and remotely modifiable.

This gradual but incremental shift in tolerance to the constancy of the worn technology as part of education activities, workforce practices and social interactions needs now to be at the centre of consideration by organisations as they articulate and defend the privacy, security and ethical dimensions of the identity of employees and learners.

We now must contemplate a near future that positions the carriage of our identities and our privacy as more heavily mediated by consortiums and in doing so, we need to determine what "part" we are playing in that future by our present advocacy of hurried and non-reflective advocacy of these technologies.

Uberveillance is what we all "know" as the inevitable, as metaphorically present already in technologies such as heart pacemakers and infused prosthetics that permit mobility simply by thinking or as real and as present as humans chipped by choice claiming DIY autonomy.

It is apparent that our right to remain "unmarked" or "unfound" amidst a sea of veillances fades into a distant utopia or dystopia depending on what we see. Uberveillance has already "become" us when we deliberately stop for a moment or two and think deeply about our inability to inhabit this earth without some form of electronic mediation.

Location Enabled Wearable Technologies

As consumers, we are fed a soup of service, technology access and interconnectivity for those who can afford it, amidst marketing organisations refining their ability to know where an individual frequents in order to present products and services as seamlessly as possible.

Michael & Michael (2009) interrogate the social implications of the Auto-ID trajectory, the role location based services are having as part of that development and the myriad of technologies that are encompassed within the scope of this techno-social paradigm.

Irrespective of the developed or developing status of any nation, mobile telecommunication or wireless services have exploded around land bound Internet access and mobile phones have become Internet enabled mini-computers. These pervasive and networked technologies endear themselves to humans as they navigate, communicate and contemplate, and it is in this attachment that marketers consider the location-enabled suite of services that can be ubiquitously sewn into these wearable devices.

Dependency on access, connection status, range and re-powering of personal mobile telephones has permeated all practical daily activities; so too has the plethora of applications that permit Global Information Systems (GIS) services to inform and enhance the capabilities of the device.

The consideration of how to embrace the shift to a "hand-held curriculum" gave rise in the late 1990s and early 2000s to the short-lived mobile learning, or m-learning, fraternity world wide as educators across all sectors grappled with the disruptive effects of an always-on generation of learners.

The rapid adoption of cell phone-enabled social media platforms in the last decade as a means to communicate en masse has robbed educational technologists of a substantiation through academic discourse of separation in learner style seen only in younger people enamoured of such technologies. Sold short on intergenerational discontent, pro-active educational organisations now embrace these devices in educational settings, signalling recognition that the cell phone is a socially acceptable wearable prosthetic, as much as it is a vital source for many curriculum activities.

With this shift, mobile learning or m-learning, as a moniker of differentiation or "new" methodological intervention, has now lost its catch-cry and also its potency amongst aspiring educational technologists.

Point-of-View (POV) Technology

The use of point-of-view (POV) technologies across the entire primary, secondary, vocational and tertiary Australian education sector over the last decade has developed from DIY prototypes to a recognisable and integrated workforce practice. This device type is also known as body worn video (BWV) in the policing context.

The term "POV" originated from an expression by cinematographers to denote the capture of perspective from the "third-eye" or "first-person" of the wearer.

Reference to the use of POV in cinematography is well documented, and evidenced in works by Alfred Hitchcock.

One of the prime examples of Hitchcock's use of optical point-of-view shots is his 1954 film, Rear Window. There are two main purposes for his use of optical point-of view shots in Rear Window. One has to do with the story itself. The point-of-view shots help to pull the audience into the film and to identify more with the characters, most notably the main character, L. B. Jeffries (Jimmy Stewart). The second reason is much more universal, having to do with the nature of film itself, and the essence of cinema. (Charnick, 2012)

Its most controversial use in contemporary history is the production of pornography, closely followed by its place in armed services and community policing history.

The new generation of police recruits are highly adept at using new technologies. Indeed, there is evidence that some police carry their own personal audio and video recorders and use them to provide independent evidence of 'difficult' interactions with citizens. Indeed, some jurisdictions are now trialling the use of miniaturised wearable point-of-view (POV) cameras attached to police officers' uniforms. (Bronit et al., 2010, pp. 3)

Application of these technologies in a variety of contexts and many differing sectors suggest that the concept of first person perspective digital capture and location enabled data tagging are becoming more of an accepted and integrated activity for knowledge or skill acquisition. The "forensic" nature of evidence gathering takes place within the context of accreditation for competencies or outcomes that meet a learning objective.

The very same nomenclature also exists in parallel across policing, the biological sciences and agricultural sectors, among many.

Geoff Lubich, Automotive Lecturer at TAFE WA Pilbara College, is acknowledged as a lead innovator in custom created video camera and digital video recorder devices to capture first-person processes for educational purpose. His innovation in this area commenced as early as 2004. Flexible learning workshops and conferences funded by the Department of Education and Training Western Australia quickly ignited an interest in other educators, including Sue Waters, Challenger TAFE Western Australia, to adapt the concepts for other training areas including aquaculture. The Australian Flexible Learning Framework (AFLF), an Australian federally funded FLAG initiative (Anon, 2012a), now the National VET

E-learning Strategy (NVELS) also provided funding to support projects including the "TxtMe" (Bateman, 2004) project lead by Swan TAFE Western Australia, "Digital Outback" by Pilbara TAFE, Western Australia and the "Engageme" project at TAFE NSW, Sydney Australia. These projects explored the use of mobile messaging systems amidst other infrastructural capability development for vocational learner engagement.

The 2009 AUPOV conference in Wollongong, NSW Australia provided a valuable insight into the challenges faced by organisations seeking secure and sustainable data management particularly as new and emergent video technologies increased in use. This conference provided a timely reflection for cross-sectoral educators, trainers, workplace assessors and representatives from a broad field of sectors, including the Australian armed forces.

The adoption and applied use of this technology, as a result of these two main support initiatives, spread quickly across vocational settings in the Australian trade areas of refrigeration mechanics, bricklaying, roof carpentry and hairdressing, and has continued to grow at an exponential rate.

The rapid uptake of body worn, location enabled mobile network accessible solutions for rich media creation and connection in extreme sports, military and medical sectors is now also challenging the mobile learning / distance education stereotype. The re-purposed application of these technologies in the education and training sector is now opening up new domains for connecting learners with educators, which in turn poses substantial challenges for organisations as they grapple with the implications that this technology undeniably imbues. (Hayes, 2010, p.7)

A limited "snapshot" review of the use of these technologies by Hayes (2010) has further informed project outcomes from all Australian states supported again by funding from the Australian Flexible Learning Framework.

In the two years since that review, in many workplaces across Australia, people are now employed by educational organisations to engage with learners and to collect evidence of recognised prior learning or current competencies using more informal, repeatable and accessible forms of interaction; in many cases without having to attend the workplace at all.

Educative Arrangements

The Australian education sector, in all its forms and sectoral permutations, has over the last decade transformed itself through osmosis of these information communication (ICT) technologies.

The Internet is becoming an open source library, especially for the young. All that is left to be opened is the classroom. For some they can pursue knowledge on their own, others need structure to attain it. As more of the self-taught and self-motivated involve themselves with networking, traditional schools and their certifications should become less important. What will be important is a reputation for integrity, action, creativity and applied knowledge. (Blackall, 2007, p.153, cited, Peter Allen)

Our educational institutions are no stranger to the power of this connection and harness the Internet as one of its core tools in engaging with a cohort sometimes spread far and wide across continents and remote communities.

The traditional classroom setting has remained until recently a sacrosanct zone, a last moral bastion where the teacher didactically engaged with learners and prepared young people for the servitude of timetables or the freedoms of knowledge unfettered. This engagamement has largely been unrecorded yet this priviliege is also changing.

Open the door in a contemporary classroom setting now and note the lack of apparent teacher's desk, the multiplicity of digital screens, absent learners and distant crescendos of mobile ringtones. As the architectures for educative arrangement have crumbled, so too has the manner in which knowledge engagement joined the zillion-headed electrophorus, communicable, networked as a core learning dependency (Blackall, 2011).

The shift in these previously physical architectures of participation has had a profound impact on the manner of educator and learner engagement, moving from an unmediated role differential to that of "connectivism", a state in which connections form networks and forge new curricula activity unlike groups seeking conformity and new dogma.

Behaviorism, cognitivism, and constructivism are the three broad learning theories most often utilized in the creation of instructional environments. These theories, however, were developed in a time when learning was not impacted through technology. Over the last twenty years, technology has reorganized how we live, how we communicate, and how we learn. Learning needs and theories that describe learning principles and processes, should be reflective of underlying social environments. (Siemens, 2005)

Despite the apparent groundswell of acceptance amongst educationalists worldwide of Siemens' position on contemporary learning theory, Kop & Hill (2008) question whether the proposition of Connectivism as a 21st century theory of learning in a post-constructivist paradigm is valid and noteworthy.

The role of the educator is mooted as that of facilitator, where the networked connection of the individual, tethered by peers to a multitude of feeds and unseen audiences, influences the shape and chronology of events that unfold in a learning setting. As the manner in which learners engage with organisations changes, so to do the visions of the academic fraternity, more anxious than ever to shipshape connection as the king of accreditation over what once was revered content.

Contemporary educational philosophers Mann (2011) Blackall (2011) Siemens (2005) posit the "learn-by-being" Kierkegaardian state where the individual is at the center of consideration for pedagogical development, where the organisation must make way for the dis-organisation and open state of knowledge aggregation.

Highly charged discussions take place across educational organisations in Australia, questioning whether connected classrooms are nothing more than a hybridised and consortium travesty intermediated by a nationalist curriculum. Without a doubt, in Australia, the nomenclature of educators, teachers, and trainers is now as much at threat of extinction as are past teaching practices that demanded uninterrupted didactic attention.

To what extent have we developed a curriculum that uses connectivity as a means to broaden the learning horizon, build life skills and honour the unique abilities of an individual over that of grade-driven productivity? What state of mind will exist in people post our current landscape of technological prosthetic?

The Digital Education Revolution

Educational institutions find themselves placing ICTs at the very core of curriculum, diversifying the learning experience beyond the confines of the organization's traditional catchment profile, as this omnipresence affects the community at large.

New tactics beget new technologies and vice versa, and this has informed the digital education revolution in Australia.

The Digital Education Revolution (DER) aims to contribute sustainable and meaningful change to teaching and learning in Australian schools that will prepare students for further education, training and to live and work in a digital world ... In this context, the Australian Government is investing over $2.4 billion to support the effective integration of information and communication technology (ICT) in Australian schools in line with the Government's broader education initiatives, including the Australian Curriculum. (Anon, 2012b)

As the cases in this paper evidence, workplace settings in the vocational education and training sector now stretch between cooking classes in metropolitan secondary schools through to remote mining camps in the arid deserts of Western Australia.

To meet such widely dispersed cohorts, educational organisations unify learner digital identity with flexible modes of delivery and, in doing so provide the opportunity to gain accreditation where geographical or circumstantial challenges otherwise prevented access to accreditation.

Policy makers in educational organisations now pay close attention to technology market forces on an international stage, expounding the rhetoric of catering for individualisation, equity of access, privilege and knowledge nation economics.

Meanwhile local communities struggle with the shift from facility oriented learning settings to workplace and home-based virtual attendance. As the place for learning diversifies so too has the manner in which organisations now clamour to monetise interactions as content. Connected, conferenced, multiplicity of "place" is now a prerequisite skill for the educator to demonstrate to maintain employment in the contemporary training landscape.

The premise for an educative experience has undoubtedly shifted, and the boundaries vaporized, as exclusivity has shifted away from traditional centres of excellence.

A recent discussion with an Australian independent e-learning consultant provides evidence of this shift across the Australian Vocational Education & Training (VET) sector:

In the last 2 years in my role as eMentor for practitioners in VET and ACE [Adult Community Education] there has been an increase in the use of mobile devices for connectivity with web-based courseware. More organisations are exploring the uptake of e-learning using such devices as tablets and cell phones. This has meant a deeper consideration of the type of learning management

system (LMS), social media and communication tools to be included in the blend for learners using iPads, as an example. A change is happening in the instructional design of learning experiences for the learners on the move and another change is being embraced in the accessibility of learning through massive online open courses. A new breed of self-directed learners, clamouring for free, open and networked learning experiences, are emerging who prefer the benefits of cloud computing for most of their professional development. In the field of e-portfolios this has become an important issue for the ownership and portability of their evidence of learning and employability. (McCulloch, 2012)

The "call-to-arms" discussion by McCulloch also included the future "shape" of learning and teaching:

There is a need to expand the user's skills in contexts that we have yet to experience - a swing away from teacher led instruction to self-managed networked and collaborative learning. The all-pervasive 'hive mind' approach can now tap into this new terrain of mobile devices, cloud computing and learner curated content in ways we never previously imagined. (McCulloch, 2012)

Is there a danger of losing learners to the web or in fact do educators now need to consider their role more closely because of it? What effect is an always-on expectation having on the quality of an educator's output in a blended delivery curriculum?

Case 1

Title

Point of View (POV) Cameras: Assessing their Validity and Reliability as an Adjunct for Formative Assessment of Remote Medicine Vocational Trainee Doctors.

Researchers

- **Principal Investigator:** Amber Thornburrow
- **Supervisor:** Professor Stephen Margolis
- **Additional Investigators:** Professor Sabina Knight: Director - Mt Isa Centre for Rural and Remote Health James Cook University; Dr. Stephanie DeLaRue: Deputy Director - Mt Isa Centre for Rural and Remote Health James Cook University; Dr. Pat Giddings - CEO - Remote Vocational Training Scheme

Collaboration

Mount Isa Centre for Rural and Remote Health (MICRRH), James Cook University; Remote Vocational Training Scheme (RVTS)

Funding

Primary Health Care Research, Evaluation and Development (PHCRED)

Description

Our research project will utilise POV cameras to capture a range of clinical skills by medical vocational trainee doctors. The doctors will be performing a series of clinical skills in a controlled environment, and will be assessed using standard summative assessment rating forms on-site (face to face) by an accredited medical educator. The video footage from the POV cameras will then be sent off-site to a remote accredited medical educator, who will assess the clinical skills using the same standard summative assessment rating forms. The resulting data will be analysed using correlation coefficient and Cronbach's alpha.

We hypothesise that high definition POV cameras are a reliable adjunct assessment tool and will be valid for formative assessment of remote vocational trainee doctors. Currently, remote vo-

cational doctors are disadvantaged when it comes to assessing their clinical skills.

Doctors who are geographically isolated have limited access to accredited medical educators, usually seeing a medical educator once every 3-6 months. The clinical skills are then usually assessed in a simulated environment, and discussed. This causes a clear gap in the ability for trainee doctors to be signed off on their clinical skills in a timely manner. In comparison, doctors from tertiary centres can be assessed in "real time" and their supervisor is able to give feedback immediately.

The advantage of high definition POV cameras is that the technology is lightweight and sturdy enough to be used in outback conditions. The cameras that we are using are dust resistant, shock resistant and weather resistant. These cameras are also able to capture images in low-level light, which means that the trainee doctors can take them to a large number of emergencies and use them in adverse conditions. If our hypothesis proves to be correct, a field trial will be exercised across select RVTS sites, before eventual rollout across Australia.

Notes

Amber Thornburrow, Principal Investigator stated in conversation with the Author on January 13, 2012 that the project focus was upon "… proving or not proving whether remote assessment using these technologies is as effective as face-to-face assessment, cognisant that most assessment activities occur within a blended delivery framework of learner engagement."

Case 2

Title

Angurugu School

Context

Angurugu School is located on remote Groote Eylandt, Northern Territory, Australia. This school is 1 of 4 schools that sit under the Ngakwurra Langwa College model (the other 3 being Alyangula in the special purpose mining town; Umbakumba (a 1 hour drive away) and Milyakburra on Bickerton Island (a 10 minute flight from Angurugu). The College is committed to the creation of professional learning communities and to share expertise on a regular basis.

Setting

Angurugu School has an enrolment of 326 students from pre-school through to senior secondary. Students are Warnandilakwayn people who speak Andindilyakwa as a first language. English is at best a second language.

The school has a strong partnership with Groote Eylandt Bickerton Island Enterprises (GEBIE), Groote Civil and Construction (GCC) and the social and economic development arm of the peak Indigenous body (the Anindilyakwan Land Council) who are dedicating a teacher and offering contextualized literacy and numeracy programs to students and GEBIE and GCC employees.

Funding

Funding for this project came from the Smart School Awards 2011 won by Angurugu for Excellence in Partnering. More information about this funding body is available at http://www.det.nt.gov.au/events/schoolsawards

Technology Use

The planned uses of Point of View (POV) technologies in the school setting include:

- Senior students many of whom are completing vocational education and training components, including Conservation and Land Management.
- School based traineeships and GEBIE and GCC employees.
- Filming work processes in the (training) workshop and out in the community.
- Creating training materials and resources in the Anindilyakwa language.
- Creating a portfolio of evidence for students if assessor is unable to visit (isolation factor).

Notes

Pamela McGowan, Senior Teacher of Information Communication Technologies stated in a conversation conducted by web conference on the January 12, 2012 with the Author that the Angurugu School was:

… well positioned to take advantage of these technologies, as the access to technologies that have industry equivalence is very important in the development of skills that the school students can then apply in their employment.

During the conversation Pamela spoke of the prospects of this technology being used in a community based setting and the appreciation that the Indigenous community had for:

- **Ownership of the Creation Process:** Access to the technology first-hand.
- **Learning Resources that are Technology Based:** Accessible, practical technologies.
- **Learning in their First Language:** Relevancy of learning content and experience.
- **Privacy, Security and Cognisance of Cultural Context:** Data created and retained by the Community.

Case 3

Title

Body Worn Video

Context

The following case draws upon an article written by an officer serving in the Australian Police Force at the time of publication. The article also refers to technologies employed in policing in the United Kingdom and the United States where officers wear a video recorder as part of their operational duties.

This case, based upon this article, brings to light operational practices in the Australian Police Services which squarely position technology at the core of corroboration, at the centre of debate and perhaps at the periphery a much bigger socio-technological discussion yet to unfold.

Lyell (2010) coins the term 'body worn video of BWV in preference to that of POV or point-of-view video. The difference between PoV and BWV and the manner in which it is used, is perhaps a sector driven distinction, to create a distinction or in some cases a direct comparison with hand-held to pocket-worn technologies.

Setting

What is clear is the message Sgt. Mark Lyell makes regarding the use of these technologies in core operational policing in the state of Queensland, Australia.

Body Worn Video (BWV) describes a device or system that captures images and audio recordings and is worn by the officer. BWV is a technology that offers important benefits to the Queensland Police Service (QPS) and individual police officers and significantly contributes to the achievement of the core function of the QPS. After briefly reviewing the use of technology by the QPS this article will *advance five reasons why the QPS should issue BWV to all operational officers. (Lyell, 2010, p. 29)*

Lyle paints a picture of a police force using all manner of evidence gathering technologies, with access only on an ad hoc basis to current or emergent audio and video recording technologies.

Individual officers have over the last ten years at their own expense purchased tape recorders, and more recently digital audio recorders to assist in the collection and gathering of evidence and to protect themselves from false complaints. (Lyell, 2010, p. 30)

Lyle also notes that the QPS have been instrumental in supporting the installation of CCTV in community settings, vehicular recording devices in taxis and GPS technologies in their own vehicles to enhance presence, thwart criminal behavior and improve operational efficiencies.

At the same time it can be acknowledged that the QPS has a good record of embracing and implementing technologies in other areas such as OC spray, Taser, Comfit, Livescan and DNA collection and analysis. (Lyell, 2010, p. 30)

This article poses five main reasons as substantive claims for supporting the roll-out of BWV across the entire Queensland Police Service, including evidence quality, protection against false complaints, modifying behavior (officer safety), professionalism and accountability as well as its place as an effective training tool.

Issues noted as needing to be examined and addressed include limitations of the video and audio range, cost of the device, storage of data collected, privacy as well as perspective and perception. "The experience of police using BWV has been overwhelmingly positive. Officers report saving a significant amount of time preparing their own evidence for Court." (Lyell, 2010, p. 31)

Lyle provides comprehensive examples of pilot trials in the UK and the US, cites statistics from studies into reduced aggression as a result of in-car and body worn video by officers and provides numerous points in support of police officers being equipped as part of their operational duties with BWV.

Lyle concludes with claims that:

BWV provides a significant tool that can assist police in performing their core function of law enforcement through preventing offences and detecting and prosecuting offenders. Additionally it protects officers from vexacious complaints, deter offenders from abusing and assaulting police and increase public confidence in the integrity and professionalism of police officers. (Lyell, 2010, p. 35)

Interestingly, within the conclusion Lyle reinforces in smaller italic font the following:

In the meantime it remains a decision for individual officers whether they purchase their own BWV digital recorders to assist them in discharging their duty and to protect themselves from false complaints. (Lyell, 2010, p. 35)

CONCLUSION

This state of awareness of the omnipresence of technology and its plethora of permutations in all parts of our lives presents society at large with some very real challenges. As this paper reveals, technology provides avenues to protect its citizens on the one hand and on the other provides information about those citizens to networked corporations and consortiums unseen.

The lack of totality to the increasing array of surveillant, sousveillant and dataveillant technolo-gies that make up our community fills some people with an Orwellian dread and for those who have the foresight to investigate Uberveillance, an even more urgent course of investigation.

In some contexts, surveillance helps keep us on track and thereby reinforces good habits that become second nature. In other contexts, it can hinder moral development by steering us away from or obscuring the saintly ideal of genuinely disinterested action. And that ideal is worth keeping alive. (Westacott, 2011)

Educators now find themselves in a position of making or even endorsing meaning amongst the accounts of others, paradoxically navigating around in the same maddening array of digital spaces and places that learners inhabit, perhaps with fewer skills than those they seek to educate.

Academically we interrogate the effects of technology as it widens the scope for possibility in an ever-changing world and acknowledge the anxiety that the gaze of the network causes for its inhabitants as they grow up inside this human made machine.

Like an autistic child, we create a pattern of movements to control or hold closer that of which we have little understanding and we occasionally arrive upon a state of peace. That peace is an understanding that in living, we are part of a greater form that we cannot control. This understanding occasionally presents discords to the continuity of an otherwise regular existence as we go about our daily routines.

By avoiding techno-evangelist complacency, as POV or BWV inhabits educational settings educators must now question their own purposeful intent of the use of these technologies in an educational context and acknowledge the broader social implications that other sectors more thoroughly interrogate.

REFERENCES

Anon. (2012a). Australian flexible learning framework. *Australian Flexible Learning Framework (NVELS)*. Retrieved from http://www.flexible-learning.net.au/

Anon. (2012b). Digital education revolution. *Australian Government Digital Education Revolution*. Retrieved from http://www.deewr.gov.au/Schooling/DigitalEducationRevolution/Pages/default.aspx

Bateman, C. (2004). Txt me: Supporting disengaged youth using mobile technologies, Australia. *Australian Flexible Learning Framework*. Retrieved from http://www.google.com/url?q=http://www.flexiblelearning.net.au/projects/media/txt_me_evaluation_report.pdf&sa=U&ei=3egTT_3OJezxmAX2m72BCg&ved=0CAYQFjAB&client=internal-uds-cse&usg=AFQjCNFcOS-RnC92uRBwbLBPqaub0A5ffYQ

Blackall, L. (2007). *The future of learning in a networked world.* Retrieved from http://learning-networkedworld.blogspot.com 1

Blackall, L. (2011). Leigh Blackall: What is a definition of networked learning? *Leigh Blackall*. Retrieved from http://leighblackall.blogspot.com/2011/06/what-is-definition-of-networked.html

Bronit, S., Harfield, C., & Michael, K. (2010). The social implications of covert policing. Wollongong, Australia: University of Wollongong Press - The Centre for Transnational Crime Prevention (CTCP) - Faculty of Law.

Calvino, I. (1972). *Invisible cities*. Italy: Giulio Einaudi.

Charnick, J. (2012). *Upstart film collective.* Retrieved from http://www.upstartfilmcollective.com/portfolios/jcharnick/essays/rear-window.html

Clarke, R. (1994). The digital persona and its application to data surveillance. *The Information Society, 10*(2), 77–92. doi:10.1080/01972243.1994.9960160

Doyle, A., Lippert, R., & Lyon, D. (2011). *Eyes everywhere: The global growth of camera surveillance*. Retrieved from http://www.routledge.com/books/details/9780415668644/

Hayes, A. (2010). National snapshot: Current use of POV technologies in an Australian educational context. In *MobilizeThis Symposium*. Retrieved from http://www.alexanderhayes.com/publications

Howarth, B., & Ledwidge, J. (2011). *A faster future - The future of broadband: What it means for business, society and you*. Sydney, Australia: Five Senses Education.

Kay, R. (2007). Head mounted cameras for security operations: What the officer sees the jury sees. *Security Solutions, 7*, 49–50.

Kop, R., & Hill, A. (2008). Connectivism: Learning theory of the future or vestige of the past? *International Review of Research in Open and Distance Learning, 9*(3).

Leinonen, T. (2010). *Digital learning tools: Methodological insights*. Aalto, Finland: Aalto University, School of Arts Design and Architecture.

Luckin, R., Bligh, B., Manches, A., Ainsworth, S., Crook, C., & Noss, R. (2012). *Decoding learning: The proof, promise and potential of digital education*. London: NESTA.

Lyell, M. (2010). *To fight crime and win* (pp. 29–37). Police Association News.

Mann, S. (2001). Can humans being clerks make clerks be human? Exploring the fundamental difference between UbiComp and WearComp. *Oldenbourg Electronic Journals, 43*(2), 97–106.

Mann, S. (2011). Learning by being: Thirty years of cyborg existemology. *Wearcam*. Retrieved from http://www.wearcam.org/existed/existed.ps

McCulloch, C. (2012). *About me - Baranduda blog*. Retrieved from http://coachcarole.wordpress.com/about/

Michael, K., & Michael, M. G. (2009). *Innovative automatic identification and location based services: From bar code to chip implants*. Hershey, PA: IGI Global. doi:10.4018/978-1-59904-795-9

Michael, K., & Michael, M.G. (2011). The social and behavioral implications of location-based services. *Journal of Location-Based Services, 5*.

Michael, M. (2007). A note on uberveillance. In From Dataveillance to Uberveillance and the Realpolitik of the Transparent Society, pp. 9-26.

Rheingold, H. (2011). Social media classroom. *Invitation to the Social Media Classroom and Collabatory*. Retrieved from http://socialmediaclassroom.com/

Scholz, T. S. (2011). *Learning through digital media: Experiments on technology and pedagogy*. Institute for Distributed Creativity (IDC). Retrieved from http://www.learningthroughdigitalmedia.net

Siemens, G. (2005). Connectivism: A learning theory for the digital age. *ITDL*. Retrieved from http://www.itdl.org/Journal/Jan_05/article01.htm

Wagner, D., & Berger, J. (1985). Do sociological theories grow? *American Journal of Sociology*, *90*(4), 697–728. doi:10.1086/228142

Westacott, E. (2011). Does surveillance make us morally better? *Philosophy Now*. Retrieved from http://www.philosophynow.org/issues/79/Does_Surveillance_Make_Us_Morally_Better

KEY TERMS AND DEFINITIONS

Australian Flexible Learning Framework: A former Australian national e-learning strategy (2008–2011) now known as the National Vocational E-learning Strategy (NVELS) managed by the Flexible Learning Advisory Group (FLAG), a key policy advisory group on national directions and priorities for information and communication technologies in the VET sector.

Body-Worn-Video (BWV): A specialised form of technology used in a range of occupations where the user wears a video (and audio) capture device that is clipped to or attached in some way to clothing to record from first-person perspective.

Cases: Sector specific examples.

Community: A term used to define a segment of greater society; a mixed cultural assembly or a group defined by certain geographical parameters.

Education: Includes the primary, secondary, vocational and higher education sectors as well as non-accredited training, non-formal and peer learning experiences.

Ethical Restraint: A demonstration of consideration with accompanying shift in behaviours or actions.

Location-Enabled: A technology assisted enhancement that provides global positioning system capabilities. E.g. a GPS sensor embedded in digital eyewear.

Point-of-View: A unique perspective; applied in the context of point-of-view camera technologies the term refers to photo or video capture using technologies that contain a camera function embedded in the head-worn device that capturing up to a 160 degrees from the first-person perspective of the wearer.

Police: Law enforcement in the geographical region of Australasia who provide service to protect citizens and uphold the law of that land.

Social Media: Interactions between humans that are internet enabled via online communities and platforms which allow the sharing and exchange of data kin many different forms.

Socio-Ethical Implications: The consideration for and evaluation of activities that may have positive or negative impact on others and the longer term implications that these activities or technologies may have on the user/subject.

Socio-Technical: A systems orientation or theoretical evaluation of interconnected relationship between humans and technology; social and technical relatedness.

Training: Educational knowledge acquisition, activity undertaken to gain accreditation or the act of teaching in differing vocational settings.

Veillance: The domain within which subsets of monitoring activity are defined e.g. surveillance or sousveillance.

Vocational Education and Training: Known as VET, is a discrete sector most closely aligned with trade practices and related educational activities.

Wearables: Technology that is neither handheld nor embedded in the human body; technology that can be worn. E.g. a pedometer.

Section 2
Applications of Humancentric Implantables

Chapter 4
Practical Experimentation with Human Implants

Kevin Warwick
University of Reading, UK

Mark N. Gasson
University of Reading, UK

ABSTRACT

In this chapter, the authors report on several different types of human implants with which the authors have direct, first hand, experience. An indication is given of the experimentation actually carried out and the subsequent immediate consequences are discussed. The authors also consider likely uses and opportunities with the technology should it continue to develop along present lines and the likely social pressures to adopt it. Included in the chapter is a discussion of RFID implants, tracking with implants, deep brain stimulation, multi-electrode array neural implants, and magnetic implants. In each case, practical results are presented along with expectations and experiences.

INTRODUCTION

For many years, science fiction has looked to a future in which robots are intelligent and cyborgs – a human/machine merger – are commonplace. Movies such as *The Terminator*, *The Matrix*, *Blade Runner* and *I, Robot* are all good examples of this. But until recently, any serious consideration of what this future might actually mean was not necessary because it was largely considered science fiction and not scientific reality. Now, however, science has not only caught up but, in bringing about some of the ideas initially thrown up by science fiction, has introduced wild card practicalities that which extend further than the original story lines and even beyond current fiction.

It should be clear from the start that the authors of this paper are scientific experimenters who like to look outside the box. From a background of artificial intelligence, robotics and biomedicine, the authors have been an integral part of each of these experiments and the need for discussion and

DOI: 10.4018/978-1-4666-4582-0.ch004

debate on the issues raised is recognised. The material here is presented with a view to contributing to the area in order to provide a concrete basis for what has actually been achieved and, hence, what might be possible in the future.

In each case an outline and explanation of the experimentation is given. Related academic papers are referenced, where appropriate, in order to provide more in-depth details.

Each experiment is described in its own self-contained section. Although there is clear technical overlap between the sections, they throw up individual considerations which the authors have not wished to blur. Following a description of each investigation, the authors have attempted to raise some pertinent issues on that topic. As can be seen, points have been raised with a view to near term technical advances and what these might mean in a practical scenario. This is not intended as an attempt to present a fully packaged, conclusive document. Rather, the aim is to open up the research carried out and its implications to ethical scrutiny and assessment.

RFID IMPLANTS

The first experiment to be considered is the use of implant technology, for example, the implantation of a Radio Frequency Identification Device (RFID) as a token of identity. In its simplest form, such a device transmits by radio a sequence of pulses which represent a unique number. The number can be pre-programmed to act rather like a PIN number on a credit card. So, with an implant of this type in place, when activated, the code can be checked by computer and the identity of the carrier specified.

Such implants have been used as a sort of fashion item, to gain access to night clubs in Barcelona and Rotterdam (The Baja Beach Club), as a high security device by the Mexican Government or as a medical information source (having been approved in 2004 by the U.S. Food and Drug Administration which regulates medical devices in the USA, see Graafstra, 2007; Foster & Jaeger, 2007). In the latter case, information on an individual's medication, for conditions such as diabetes, can be stored in the implant. Because it is implanted, the details cannot be forgotten, the record cannot be lost, and it will not be easily stolen.

An RFID implant does not have its own battery. It has a tiny antenna and microchip enclosed in a silicon or glass capsule. The antenna picks up power remotely when passed near to a larger coil of wire which carries an electric current. The power picked up by the antenna in the implant is employed to transmit by radio the particular signal encoded in the microchip. Because there is no battery, or any moving parts, the implant requires no maintenance whatsoever – once it has been implanted it can be left there.

The first such RFID implant to be put in place in a human occurred on 24 August 1998 in Reading, England. It measured 22 mm by 4 mm diameter. The body selected was the first author of this Chapter. The doctor involved burrowed a hole in the upper left arm, pushed the implant into the hole and closed the incision with a couple of stitches.

The main reason for selecting the upper left arm for the implant was that we were not sure how well it would work. We reasoned that, if the implant was not working, it could be waved around until a stronger signal was transmitted. It is interesting that most present day RFID implants in humans are located in a roughly similar place (the left arm or hand), even though they do not have to be. Even in the James Bond film, *Casino Royale* (the new version), Bond himself has an implant in his left arm.

The RFID implant allowed the author to control lights, open doors and be welcomed "Hello" when he entered the front door at Reading University (Warwick, 2000; Warwick & Gasson, 2006). Such an implant could be used in humans for a variety of identity purposes – e.g. as a credit card, as a car key or (as is already the case with some other animals) a passport, or at least a passport supplement.

The use of such implant technology for monitoring people opens up a considerable range of issues. Tracking individual people in this way, possibly by means of an RFID or, alternatively, for more widespread application and coverage via a Global Positioning System by means of a Wide Area Network or even via the cell phone network, is now a realistic concept. Ethically though, it raises considerable questions when it is children, the aged (e.g. those with dementia) or prisoners who are subjected to tracking, even though for some people this might be deemed to be beneficial.

In the case of a missing child, who might have been abducted, a tracking implant could be activated in order to enable them to be immediately located, possibly thereby saving their life – certainly saving a lot of police time and parental anguish. But is it appropriate for children to be given implants in this way? Shouldn't they be given the choice? In many countries children are, at a very early age, injected with vaccines we still do not fully understand, and that potentially have several side effects. This is deemed ethically acceptable, so why not a tracking implant to keep them safe? The author is regularly asked by parents (globally) to provide such a technology specifically for their children – because of the worries of the parent (and in some cases the children as well). It is not intended here to delve into a detailed ethical and social comparative analysis, but merely to pose the question, such that an analysis, if desired, can be carried out. The main point is to introduce the technical possibilities.

The use of implant technology as an extra identity device has been with us now for some time. As yet, however, no credit card company has offered a major incentive in terms of extra security or lower costs. It is suspected that if a company did so, the take up might well be considerable.

However, the broad discussion on security and privacy issues regarding mass RFID deployment has gathered momentum, and security experts are now specifically warning of the inherent risks associated with using RFID for the authentication of people, see Michael (2008) for an overview.

Implants for tracking people are still at the research stage. While the idea that RFIDs can be used to covertly track an individual 24-7 betrays a fundamental misunderstanding of the current limitations of the technology, there are genuine concerns to address. As they do become available, there are numerous (special) cases where there are distinct drivers. For example, there is potential gain for a person to be tracked and their position monitored in this way, especially where it could be deemed to either save or considerably enhance their life – as could be the case for an individual with dementia.

RFID IMPLANTS REVISITED

The core technology of RFID devices has continued to develop, and although non-implantable RFID devices in general remain more advanced than implantable, glass capsule types, these too continue to evolve and this opens up new possibilities, and new issues. To explore this further, the earliest experiments with an implanted RFID device conducted in 1998 were revisited using the latest in implantable RFID technology. On 16 March 2009, this Chapter's co-author had a glass capsule HITAG S 2048 RFID device implanted into his left hand (See Figure 1) (Gasson, 2010).

While containing a unique identifier number, similarly to older devices, the device also has memory to store data and the option of 48-bit secret key based encryption for secure data transfer. These are clear advances over the older implantable technology, which could only broadcast a fixed identifier, thus enabling new applications to be realized. However, in a series of experiments it was shown that a compromised computer system

Figure 1. An RFID tag is injected into the left hand of the author by a surgeon (left), shown in close up (top right). Two x-ray images taken post-procedure (bottom right) show the position of the tag in the hand near the thumb (Gasson, 2010)

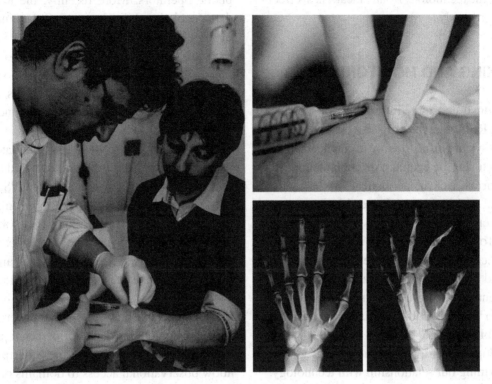

could infect the implanted device with a computer virus, and how this could be done in such a way that simply reading the infected device could further transmit the virus to other systems.

While a clear demonstration of how implantable devices are becoming more complex, capable and potentially vulnerable, being susceptible to a computer virus also raises interesting questions linked to the concept of the body. As functions of the body are restored or further enhanced by implanted devices, the boundaries of the body become increasingly unclear.

Previous recipients of RFID implants echo the sentiments of many cochlear implant and heart pacemaker users - the implant quickly becomes perceived as being part of the body and what the user understands to be their body includes the technological enhancement. In essence, the boundaries between man and machine simply become theoretical.

This development in our traditional notion of what constitutes our body and its boundaries leads to two notable repercussions. Firstly, it becomes possible to talk in terms of a human (albeit a technologically enhanced human) becoming infected by a computer virus. Thus, in that light, the simple experiments conducted gave rise to the world's first human to be infected by a computer virus. Secondly, this development of our concept of the body impacts on certain human rights, in particular the right to bodily integrity. Bodily integrity constitutes a right to do with one's body whatever one wants (a right to self-determination) and it implies the right to prevent one's body from being harmed by others.

In this context, a computer virus infecting an implanted device constitutes an infringement on the right to bodily integrity. Should the protection afforded to implantable devices from a legal perspective be different and distinct to that of any given piece of computing equipment? Surely if I consider my implant to be part of my body then attacking the technology by any means is in effect to attack my body.

TRACKING AND MONITORING

We previously considered an application scenario in which RFID may be utilised as part of an established system infrastructure to enable tracking and monitoring of people. An implanted RFID device forms a clear, permanent link with the individual and if this can be read in some way to persistently identify the person, such as via a handheld or wearable device which has additional communication capabilities, then this type of system could be realised. Here we briefly describe experimentation using mobile phone devices, with a view to the applicability to systems utilising implantable devices.

Developments in mobile phone devices are rapidly reforming our relationship with technology. The changes are not just technological - they are driving changes in cultural and social paradigms, and further empowering the consumer to seek new experiences and services (García-Montes, 2006). The drive from industry to stay at the cutting edge has seen mobile phones turned into feature packed computing devices within a decade. With Internet capabilities, high resolution cameras, GPS and growing selections of third party software applications, these devices are no longer simply mobile phones. Indeed they are now more like mobile computers on which we can make phone calls.

It appears that mobile handsets are the first wave of successful "wearable computers", at least in the sense that they comprise a relatively powerful computing device which people habitually carry. As it stands, new generations of handset are heralding a new era of information access and disclosure.

Advances in mobile technologies have meant that being able to track or locate people has been

possible for some time. However, such information is usually only readily available to mobile phone operators. More recently, the advent of data enabled mobile phones, and the emergence of popular social networking internet sites has realized a dramatic increase in the volume of information people willingly disclose about themselves. In many cases the disclosure is to large numbers of complete strangers (Parameswaran & Whinston, 2007).

In February 2009, Nokia forecast that 50% of its handsets sold in 2009 would include a GPS unit. The 3G iPhone, with integrated GPS, reportedly held a 4-6% share of the handset market in the UK toward the end of 2008 (Gasson et. al., 2011). New services are likely to appear which encourage people to reveal where they are at any time in the name of safety, convenience or for social use. The danger here is that this begins to remove the divide between safety, convenience or entertainment and invasion of personal privacy.

Modern mobile handsets, or "Smartphones", allow observational access to domains of behavioural data not previously available even with constant observation and self "diary" reports. Data collection which requires little or no user input, coupled with the strong, unique relationship between the handset and its user makes for behavioural data access on an unprecedented scale. GPS enabled phones have helped generate a new era of information disclosure, and new services are likely to appear which encourage people to reveal their whereabouts for safety, convenience or social use.

While these services will undoubtedly be useful, the seemingly harmless data may not just reveal where you are or have been – it could expose aspects of your private life that at first glance may not be apparent. It is possible to aggregate pieces of information over time and use data mining techniques to extract a "behavioral profile" from the data. Problems could arise if, for example, this information is used by third parties to vary their services or prices specific to the individual, in some cases to their detriment. This is especially so given that the end "user" may have no idea that this is happening, or indeed how it is happening.

To further explore this emerging technology, and to assess its potential impact on the privacy of the user, four people from three different European Member States took part in an experiment during April/May 2009 (Gasson, 2011). Over this period they were persistently tracked via GPS enabled mobile phones and their location data was stored in a central database for automated and manual processing. The aim of this processing was a first attempt to mine new information from the data relevant to forming behavioural profiles of the individuals.

The information was based on places they visited, people they associated with or other aspects of their complex routines determined through persistent tracking, and was used to show that simple profiles of individual or group behaviour can be drawn over short periods and that the location data can be highly privacy invasive. Pertinent to this Chapter, one of the participants was authenticated by the phone using an implanted RFID tag, and an RFID tag reader attached to the phone. In this way it was possible to have stronger confidence that the GPS information was related to that user since the phone also held a record of when it was in the locale of that particular user.

The study showed that an enormous amount of information about the individual is buried in the data available from persistently tracking people. From this short study, results of profiling ranged from intuitive through special cases to insightful, in terms of the regular times that specific places are visited. While some information such as residential address, place of work, social and business relationships between individuals could easily be uncovered, it is clear that personal and sensitive information such as family life, health, religion, personal habits and preferences could be inferred from data collection over months, and years.

It is possible that services offering customized information based on the results of such behav-ioural profiling could become commonplace. However, it may not be immediately apparent to the user that a wealth of their information potentially unrelated to the service, can be revealed. Further issues occur if the user agreed, while subscribing to the service, for data to be passed to third parties where it may be used to their detriment.

We will no doubt see a steady use of such technology within society over the years ahead. But is this simply something that society should accept? If questions should be raised, who is responsible for raising them and how will this be achieved? Clearly there are enormous commercial possibilities, should those companies be, some-how, made to pay an individual for any data they collect about that individual or is information on an individual's habits and whereabouts considered to be "common knowledge"?

But the use of such implanted technology for tracking and monitoring raises many issues with regard not only to technical aspects but also in terms of freedom of the individual and ethical concerns (Michael el. al., 2006). Indeed the same can be said for implants that dramatically change the landscape offering completely new services or enhancements. What is perhaps much less of concern ethically is the use of implants for thera-peutic purposes. Here a person has a problem of some kind and the implant is used more with the aim of restoring that individual towards the norm. Such is the case with implants used to overcome the maleffects of some neurological disorders as we will consider in the next section.

DEEP BRAIN STIMULATION

There now exist several different types of brain computer interfaces, employed either for research purposes or for standard medical procedures. The number actually in position and operating at any one time is steadily growing, a trend that is likely to increase.

As a case example, the number of Parkinson's Disease (PD) patients is estimated to be 120-180 out of every 100,000 people, although the percentage (and hence the number of people affected) is increasing rapidly as life expectancies increase. For decades, researchers have exerted considerable effort to understand more about the disease and to find methods to limit its symptoms successfully (Pinter et al., 1999), which are most commonly periodic (and frequently acute) muscle tremor and/or rigidity. Many other symptoms such as hunched stooping may however occur in later stages of PD.

Several approaches exist to treat this disease. In its early stages, the drug levodopa (L-dopa) has been the most common drug used to treat PD since 1970. However, the effectiveness of L-dopa decreases as the disease worsens and severity of the side effects increases, especially seen when PD affects a younger person.

Surgical treatment, such as lesioning, is an alternative when drug treatments have become ineffective. Lesioning can alleviate symptoms thus reducing the need for drug therapy altogether. An alternative treatment of PD, by means of Deep Brain Stimulation (DBS), only became possible when the relevant electrode technology became available from the late 1980s. Many neurosurgeons have moved to implanting neurostimulators connected to deep brain electrodes positioned in the thalamus, sub-thalamus or globus pallidus for the treatment of tremor, dystonia and pain.

A typical DBS device contains an electrode lead with four or six cylindrical electrodes at equally spaced depths, attached to an implanted pulse generator (IPG) which is surgically positioned below the collar bone. DBS has many advantages, including reversibility. It is also potentially much less dangerous than lesioning and is, in many cases, highly effective. However, it presently utilizes a continuous current simulation at high frequency resulting in the need for regular battery replacement (every 24 months or so). The cost of battery replacement, the time-consuming surgery involved and the trauma of repetitive surgery for

battery replacement severely limits patients who can benefit, ruling out those who are frail, have problems with their immune system or are not particularly wealthy.

The obvious solution, namely remote inductive battery recharging, is fraught with problems such as the size of passive coil that needs to be implanted and detrimental chemical discharges that occur within the body – even then the mean time between replacements is only marginally improved. Another solution to prolong battery life is simply to improve battery technology. However, the link between battery price and battery life is clear. If we are considering here a battery that could potentially supply the stimulation currents required over a ten or twenty year period then the technology to achieve this in a low cost, implantable, durable form, is not on the horizon.

Ongoing research, involving the author, is aimed at developing an "intelligent" stimulator (Gasson et al., 2005; Burgess, et al., 2010; Wu, et al., 2010). The idea of the stimulator is to produce warning signals before the tremor starts so that the stimulator only needs to generate stimulation signals occasionally instead of continuously – in this sense, operating in a similar fashion to a heart pacemaker.

Both Multilayer Perceptron (MLP) and Radial Basis Function (RBF) neural networks have been applied as the Artificial Intelligence tool, and are shown to successfully provide tremor onset prediction. In either case, data input to the neural network is provided by the measured electrical Local Field Potentials, obtained by means of the deep brain electrodes. Using these the network is trained to recognise the nature of electrical activity deep in the human brain and to predict (several seconds ahead) the subsequent tremor onset outcome. In this way, the DBS device is "intelligent" when the stimulation is only triggered by the neural network output. However, the required MLP network training process is very slow, and many questions remain with regard to the adaptability of the network, not only between patients but also with regard to the neural activity of one patient at different times. Meanwhile, the RBF network not only has universal approximation properties, but this is achieved with a relatively simple mathematical description. In both cases, much pre-processing of data is necessary along with frequency filtering to minimize the difficulty of prediction. Comparative studies are ongoing to ascertain which method (or hybrid of methods) appears to be the most reliable and accurate in a practical situation.

It is worth pointing out that false positive predictions (that is the network indicating that a tremor is going to occur when in fact this is not the case) are not a critical problem. The end result with such a false positive is that the stimulating current may occur when it is not strictly necessary. In any event no actual tremor would occur, which is a good outcome for the patient, however unnecessary energy would have been used. In fact, if numerous false predictions occurred, the intelligent stimulator would tend toward the present "blind" stimulator. Effectively the occasional false positive prediction is not a problem, unless it became a regular occurrence. Fortunately, results show that the network can be readily tuned to avoid the occurrence of many false positives.

Missing the prediction of a tremor onset is extremely critical, however, and is simply not acceptable. Such an event would mean that the stimulating current would not come into effect and the patient would actually suffer from tremors.

While deep brain implants are, as described, aimed primarily at stimulating the brain for therapeutic purposes, they can also have a broader portfolio in terms of their effects within the human brain. It is worth stressing that, in all cases, further implantations are forging ahead with little or no consideration given to the pervasive general technical, biological and ethical issues. It is time that such issues were given an airing.

The physical stimulator used for the treatment of Parkinson's Disease is also used, albeit in fewer

instances, for cases of Tourette's Syndrome, Epilepsy and even Clinical Depression. It is commonly considered that the use of deep brain stimulators for the treatment of Parkinson's Disease, Epilepsy or Tourette's Syndrome is perfectly acceptable because of the standard of living it can afford patients. However, long term modifications of brain organisation can occur in each case, causing the brain to operate in a completely different fashion. For example, there can be considerable long term mental side effects in the use of such technology. A stimulator, when positioned in central areas of the brain, can also cause other direct results, including distinct emotional changes. The picture is therefore far more complex than merely overcoming a medical problem.

Even the mere use of such a stimulator raises interesting questions. For example, if an individual with such a stimulator implanted in their brain was to murder another human and then deny responsibility, claiming that it was in fact a fault of the stimulator, who would be to blame for the murder? Would it be the individual, despite their protests that the implant was overriding their normal brain function? If a stray radio signal had caused the problem, could it be the person broadcasting at the time who is to blame? Perhaps it is the surgeon who put the implant in place, or a researcher who worked on the device 10 years ago? Clearly we have a potential problem.

As described here, "intelligent" deep brain stimulators are being designed. In such a case, a computer (artificial brain) is used to understand the workings of specific aspects of the human brain. The job of the artificial brain, as can be seen from the description of my own experimentation, is to monitor the normal functioning of the human brain such that it can accurately predict a spurious event, such as a Parkinson's tremor, several seconds before it actually occurs. In other words, the artificial brain's job is to out-think the human brain and stop it doing what it normally would do. Clearly the potential for this system to be applied for a broad spectrum of different uses is enormous – perhaps to assist slimming or (in some places) to control a spouse.

MULTI ELECTRODE ARRAY

With other brain-computer interfaces, the therapy versus enhancement question becomes more complex. In some cases it is possible for those who have suffered an amputation or spinal injury due to an accident, to gain control of artificial devices via their (still functioning) neural signals (Hochberg, et al., 2004). Meanwhile, stroke patients can be given limited control of their surroundings, as indeed can those who have motor neurone disease.

Even in these cases, the situation is not simple as each individual is given abilities that no normal human has – for example, the ability to move a cursor around on a computer screen from neural signals alone (Kennedy et al., 2004). The same quandary exists for blind individuals who are allowed extra sensory input, such as sonar (a bat-like sense), which doesn't repair their blindness but allows them to make use of an alternative sense.

Some of the most impressive human research to date has been carried out using the microelectrode array, shown in Figure 2. The individual electrodes are only 1.5mm long and taper to a tip diameter of less than 90 microns. A number of trials, not using humans as test subjects, have occurred but human tests are, at present, limited to two recent studies. In one of these, the array has been employed in a recording-only role, most notably as part of (what was called) the "Braingate" system. Essentially, activity from a few neurons monitored by the array electrodes was decoded into a signal to direct cursor movement. This enabled an individual to position a cursor on a computer screen, using neural signals for control combined with visual feedback. But the first use of the microelectrode array has considerably broader implications for advancing human capabilities.

Actually deriving a reliable command signal from a collection of captured neural signals is not a simple task, partly due to the complexity of signals recorded, and partly due to time constraints in dealing with the data. In some cases however, it can be relatively easy to look for and obtain a system response to certain anticipated neural

Figure 2. A 100 electrode, 4x4mm microelectrode array, shown on a UK 1 pence piece for scale

signals – especially when an individual has trained extensively with the system. In fact, neural signal shape, magnitude and waveform with respect to time, are considerably different to the other signals, so it is measurable in this situation.

The interface through which a user interacts with technology provides a distinct layer of separation between what the user wants the machine to do, and what it actually does. This separation imposes a considerable cognitive load, directly proportional to the level of difficulty experienced. The main issue is interfacing the human motor and sensory channels with the technology in a reliable, durable, effective, bi-directional way. One solution is to avoid this sensorimotor bottleneck altogether by interfacing directly with the human nervous system. It is therefore worthwhile considering what may potentially be gained from such an invasive undertaking.

An individual human so connected can potentially benefit from some of the advantages of machine/artificial intelligence. Advantages of machine intelligence over human intelligence are, for example, rapid and highly accurate mathematical abilities in terms of "number crunching", a high speed, almost infinite Internet knowledge base, and accurate long term memory. Additionally, it is widely acknowledged that humans have only five senses that we know of, whereas machines offer a view of the world which includes infra-red, ultraviolet and ultrasonic signals, to name but a few.

Humans are also limited in that they can only visualise and understand the world around them in terms of a limited 3 dimensional perception, whereas computers are quite capable of dealing with hundreds of dimensions. Perhaps most importantly, the human means of communication, essentially transferring a complex electrochemical signal from one brain to another via an intermediate, often mechanical slow and error prone medium (e.g. speech), is so poor as to be embarrassing, particularly in terms of speed, power and precision. It is clear that connecting a human brain by means of an implant with a computer network could, in the long term, open up the distinct advantages of machine intelligence, communication and sensing abilities.

As a step towards this broader concept of human-machine symbiosis, in the first study of its kind, the microelectrode array (Figure 2) was implanted into the median nerve fibres of a healthy human individual (this Chapter's first author) during two hours of neurosurgery in order to test bidirectional functionality in a series of experiments. A stimulation current directly into the nervous system allowed information to be sent to the user, while control signals were decoded from neural activity in the region of the electrodes (Warwick et al., 2003; Gasson et al., 2005b). In this way, a number of experimental trials were successfully concluded, in particular:

1. Extra sensory (ultrasonic) input was successfully implemented and used.
2. Extended control of a robotic hand across the internet was achieved, with feedback from the robotic fingertips being sent back as neural stimulation to give a sense of force being applied to an object (this was achieved between Columbia University, New York (USA) and Reading University, England).
3. A primitive form of telegraphic communication directly between the nervous systems of two humans (the author's wife assisted) was performed (Warwick, et al., 2004).
4. A wheelchair was successfully driven around by means of neural signals.
5. The colour of jewellery was changed as a result of neural signals – as indeed was the behaviour of a collection of small robots.

In most, if not all, of the above cases, the trial could be regarded as useful for purely therapeutic reasons, e.g. the ultrasonic sense could be useful for an individual who is blind or the telegraphic communication could be very useful for those with certain forms of Motor Neurone Disease. However, each trial can also be seen as a potential form of augmentation or enhancement beyond the human norm. Indeed, the recipient did not need to have the implant for medical purposes to overcome a problem. It was carried out for scientific and medical exploration. The question then arises as to the outer limts. Clearly, enhancement by means of Brain Computer Interfaces opens up all sorts of new technological and intellectual opportunities. It also throws up a raft of different ethical considerations that need to be addressed.

Such ongoing experiments involve healthy individuals where there is no reparative element in the use of a brain computer interface. The author, in carrying out such experimentation, specifically wished to investigate actual, practical enhancement possibilities (Warwick, et al., 2003; Warwick, et al., 2004).

From the trials, it is clear that extra sensory input is one practical possibility that has been successfully trialled, however, improving memory, thinking in many dimensions and communication by thought alone, are other distinct potential, yet realistic, benefits. The latter has been investigated to an extent. Clearly, all these things appear to be possible (from a technical viewpoint at least) for humans in general.

Approval for an implantation in each case (in England anyway), currently requires ethical approval from the local hospital authority in which the procedure is carried out, and, if it is appropriate for a research procedure, approval from the research and ethics committee of the establishment involved. This is quite apart from Devices Agency approval if a piece of equipment, such as an implant, is to be used on a number of individuals. Interestingly no general ethical clearance is needed from any societal body – yet the issues are complex.

Should it be possible for surgeons to place implants with which they can make the individual happy, sad or sexually excited? If it is acceptable for a person who is blind to receive an implant which allows them extra sensory input, then why cannot everyone have such an implant if they want

one? There already exists considerable literature on the treatment/therapy versus enhancement/upgrade issue. Issues raised include the dangers involved and how the vulnerable are affected. Here, however, we step into the unknown when it comes to upgrading humans. In particular, should we further develop implants that allow for brain enhancements when it may lead to non-implanted humans becoming subservient to their intellectual (implanted) superiors? On the other hand, if some individuals wish to have technical abilities way beyond the human norm then why should a "stick in the mud" person be able to stop them?

MAGNETIC IMPLANTS

The application of human implants is largely medical, and the vast majority of these devices are simple structural or mechanical devices. However, such passive devices have been utilised for many years for enhancement of healthy people through varying degrees of body modification from simple cosmetic improvements, body decoration to radical reshaping of the physical form.

In 2006 self-experimenters began to utilise small Neodymium magnet implants, typically under the skin of the fingertips, for sensory experimentation whereby the movement of the implant in the presence of magnetic fields can be felt by the individual. More recently this type of implant has been used to convert non-human sensory information, such as sonar or distance, into touch information by manipulating the implant via external electromagnets to control stimulation of sensory receptors (Hameed et. al., 2010).

Several factors need to be considered when selecting an implant magnet. Implantation is an invasive procedure and hence implant durability is a desired feature. Only permanent magnets retain their magnetic strength over a long period and are robust to various conditions. This restricts the type of magnet that can be considered for implantation. For permanent magnets materials such as Hard ferrite, Neodymium and Alnico are all readily available and provide a low cost solution.

In one application scenario, the magnetic implant was interfaced to an ultrasonic ranger for navigation assistance, possibly of potential use for the visually impaired. The distance information from the ranger was encoded as variations in frequency and output to the interface. It was found that this mechanism allowed a practical means of providing information about the surroundings for navigational assistance. The distances were intuitively understood within a few minutes of use and were enhanced by distance "calibration" through touch and sight.

In another application, the magnetic interface was utilised for communicating text messages to a human using an encoding mechanism suitable for the interface. Morse code was chosen for encoding due to its relative simplicity and ease of implementation. The application made use of a computer with software to take text input, encoded it as Morse code and sent the dots and dashes to the interface. The dots and dashes could be represented as either frequency or magnetic field strength variations. It was found that "reading" text messages using this approach was practical and intuitive. However, the achieved words-per-minute was limited due to unfamiliarity with Morse code and the frequent need to employ look up reference tables.

The use of subdermal magnets opens up exciting possibilities for implementing other application scenarios. For example "virtual surfaces" would involve the finger being drawn across a flat featureless surface and textures and/or shapes artificially generated by stimulation of the implant related to its position. Touch screen type technology could enable this, while research on capturing frequency components of surfaces has been investigated for use on haptic end effector devices in which the user 'feels' forces that can be either computer generated or computer enhanced. This is now the subject of further investigation. So too is the use of such devices for other sensory input,

such as infra-red signals, providing the opportunity to sense the heat of objects remotely.

Apart from their potential in the arts world, magnetic implants are a research topic currently receiving a lot of attention and it is still early in its research life. Clearly there are ethical implications in terms of individual choice as to whether they wish to participate. However the wider ethical and societal issues are not so profound. Using such a device to assist the disabled is clearly not an issue, as long as it is their choice, but the sensory enhancement possibilities could provide exciting commercial opportunities either purely from a fun/hobby aspect or even for serious sensing of signals presently not immediately available to humans.

CONCLUSION

In this chapter we have looked at different experimental cases in which humans merge with technology – thereby opening up a number of social and ethical considerations as well as inherent technical issues. In each case we have reported on actual practical experimentation results, rather than merely a theoretical concept.

Technological advancement is a part of our evolution, and the significant next step of forming direct bi-directional links with the human brain is moving inexorably closer. It is understandable to think that since the technology has not yet been, and may not be, perfected there is no need to address the incipient legal, ethical and social issues that the development of these devices may bring. However, the basic foundations of such advanced implant devices are being developed for clear medical purposes, and it is reasonable to assume that few would argue against this progress for such "noble", therapeutic causes.

Equally, as has been demonstrated by cosmetic surgery, we cannot assume that because a procedure is highly invasive people will refuse the procedure. So, while these procedures are not yet reality, there is clear evidence that devices capable of significant enhancement will become reality, and most probably applied in applications beyond their original purpose. Thus, consideration needs to be given now to the fundamental moral, ethical, social, psychological and legal ramifications of such enhancement technologies. It is not too soon to start a real debate.

The aim in this Chapter has been to engender reflection on the experimental material described. Some technological issues have also been considered in order to demonstrate the direction in which technological developments are heading. In each case, however, we have remained firmly in the realm of actual practical technology rather than on speculative fictional ideas. Our purpose is to hand over the baton for reflection so that our experimentation (and that of others) can be guided by the informed conclusions that result.

REFERENCES

Burgess, J. G., Warwick, K., Ruiz, V., Gasson, M. N., Aziz, T. Z., & Brittain, J. et al. (2010). Identifying tremor-related characteristics of basal ganglia nuclei during movement in the Parkinsonian patient. *Parkinsonism & Related Disorders, 16*(10), 671–675. doi:10.1016/j.parkreldis.2010.08.025 PMID:20884273

Foster, K. R., & Jaeger, J. (2007). RFID inside. *IEEE Spectrum, 44*, 24–29. doi:10.1109/MSPEC.2007.323430

García-Montes, J., Caballero-Muñoz, D., & Perez-Alvarez, M. (2006). Changes in the self resulting from the use of mobile phones. *Media Culture & Society, 28*(1), 67–82. doi:10.1177/0163443706059287

Gasson, M. N. (2010). Human enhancement: Could you become infected with a computer virus? In *Proceedings of IEEE International Symposium on Technology and Society* (pp. 61-68). Wollongong, Australia: IEEE.

Gasson, M. N., Hutt, B. D., Goodhew, I., Kyberd, P., & Warwick, K. (2005b). Invasive neural prosthesis for neural signal detection and nerve stimulation. *International Journal of Adaptive Control and Signal Processing, 19*(5), 365–375.

Gasson, M. N., Kosta, E., Royer, D., Meints, M., & Warwick, K. (2011). Normality mining: Privacy implications of behavioral profiles drawn from GPS enabled mobile phones. *IEEE Transactions on System, Man, Cybernetics. Part C, 41*(2), 251–261.

Gasson, M. N., Wang, S. Y., Aziz, T. Z., Stein, J. F., & Warwick, K. (2005). Towards a demand driven deep-brain stimulator for the treatment of movement disorders. In *Proceedings of 3rd IEE International Seminar on Medical Applications of Signal Processing* (pp.83-86). London, UK: IEE.

Graafstra, A. (2007). Hands on. *IEEE Spectrum, 44*(3), 18–23. doi:10.1109/MSPEC.2007.323420

Hameed, J., Harrison, I., Gasson, M. N., & Warwick, K. (2010). A novel human-machine interface using subdermal magnetic implants. In *Proceedings of IEEE International Conference on Cybernetic Intelligent Systems* (pp. 106-110). Reading, UK: IEEE.

Hochberg, L. R., Serruya, M. D., Friehs, G., Mukand, J. A., Saleh, M., & Caplan, A. et al. (2006). Neuronal ensemble control of prosthetic devices by a human with tetraplegia. *Nature, 442*, 164–171. doi:10.1038/nature04970 PMID:16838014

Kennedy, P., Andreasen, D., Ehirim, P., King, B., Kirby, T., & Mao, H. et al. (2004). Using human extra-cortical local field potentials to control a switch. *Journal of Neural Engineering, 1*(2), 72–77. doi:10.1088/1741-2560/1/2/002 PMID:15876625

Michael, K., McNamee, A., & Michael, M. G. (2006). The emerging ethics of humancentric GPS tracking and monitoring. In *Proceedings of IEEE International Conference on Mobile Business*. IEEE.

Michael, K., & Michael, M. G. (2008). *Innovative automatic identification and location-based services: From bar codes to chip implants*. Hershey, PA: IGI Global.

Parameswaran, M., & Whinston, A. B. (2007). Research issues in social computing. *Journal of the Association for Information Systems, 8*, 336–350.

Pinter, M. M., Alesch, F., Murg, M., Seiwald, M., Helscher, R. J., & Binder, H. (1999). Deep brain stimulation of the subthalamic nucleus for control of extrapyramidal features in advanced idiopathic Parkinson's disease. *Journal of Neural Transmission, 106*, 693–709. doi:10.1007/s007020050190 PMID:10907728

Warwick, K. (2000). The chip and I. In *The political subject: Essays on the self from art, politics and science*. London, UK: Lawrence and Wishart.

Warwick, K., & Gasson, M. N. (2006). A question of identity – Wiring in the human. In *Proceedings of IET Wireless Sensor Networks Conference* (pp. 4/1-4/6). London: IET.

Warwick, K., Gasson, M. N., Hutt, B., Goodhew, I., Kyberd, P., & Andrews, B. et al. (2003). The application of implant technology for cybernetic systems. *Archives of Neurology, 60*(10), 1369–1373. doi:10.1001/archneur.60.10.1369 PMID:14568806

Warwick, K., Gasson, M. N., Hutt, B., Goodhew, I., Kyberd, P., & Schulzrinne, H. et al. (2004). Thought communication and control: A first step using radiotelegraphy. *IEE Proceedings. Communications, 151*(3), 185–189. doi:10.1049/ip-com:20040409

Wu, D., Warwick, K., Ma, Z., Gasson, M. N., Burgess, J. G., & Pan, S. et al. (2010). Prediction of Parkinson's disease tremor onset using radial basis function neural network based on particle swarm optimization. *International Journal of Neural Systems, 20*(2), 109–118. doi:10.1142/S0129065710002292 PMID:20411594

KEY TERMS AND DEFINITIONS

Deep Brain Stimulation (DBS): Is a surgical treatment which involves the implantation of electrodes into the brain to send electrical impulses to those specific parts of the brain. DBS has proved to be very successful at treating some movement and affective disorders.

Implant: Is a human-made device to be inserted into the body for purposes of therapy or enhancement.

Monitoring: Involves closely watching the state or value of a system.

Multielectrode Arrays (MEAs): Or microelectrode arrays contain multiple shanks with electrode tips through which neural signals are monitored and/or stimulating pulses are applied.

Nanotechnology: Generally involves materials and devices with dimensions of the order from 1 to 100 nanometers.

Radio-Frequency Identification: (RFID): Involves a wireless system employing radio-frequency electromagnetic fields to transfer data to and/or from the device for the purpose of identification.

Tracking: In the sense used here implies the close monitoring of the position of an object or person.

Interview 4.1
The Baja Beach Club Implant Program
Mr. Seraphin Vilaplana
Interview conducted by Katina Michael on 4 June 2009, Barcelona, Spain

Katina Michael: I'd like to begin by asking you where the idea to implant patrons came from.

Serafin Vilaplana: Well, the owner-manager of the Baja Beach Club visited the United States, and he got the idea while traveling over there and hearing about the trial of the chip implant that was linked to electronic health records. These implants were first being used for the elderly and the sick.

Katina Michael: How did Conrad consider implementing the radio frequency identification (RFID) technology with respect to the club?

Serafin Vilaplana: Well, Conrad thought, if you can use the technology for e-health solutions, you could certainly transfer it to work for access control applications.

Katina Michael: Yes, but what was actually the main driver … why implant humans, and more to the point why implant individuals coming to visit a bar by day and a club by night?

Serafin Vilaplana: Well, you see it was all about us gaining the maximum publicity for the club that we could. Conrad believed that if we ran this implant program that clubbers in Barcelona, and even in others places in Spain, and tourists would hear about it, and be drawn to the club just to see it for themselves. It was all about the novelty of the application. In short, it was a marketing ploy, and it actually it worked well. We strongly believed that despite the high-tech application, it was good old traditional word of mouth that would drive people to come and participate in the clubbing experience.

Katina Michael: Did you charge people a fee to participate in this program?

Serafin Vilaplana: No, we did not. The VIP members traditionally spent a great deal of money at the club, so we were not about to charge them any more money. The point was to make it easier for them to spend money through the use of the chip. So convenience for our patrons was also something we thought was important.

Katina Michael: How long did the Baja Beach chip implant trial last?

Serafin Vilaplana: The VIP chip implant program began in 2004 and ended in 2009, it lasted seven years in duration.

Katina Michael: How many people were chipped in the program?

Serafin Vilaplana: Not more than one hundred people were implanted.

Katina Michael: Did you design the access control system?

Serafin Vilaplana: Yes I did.

Katina Michael: How old were you when you built the system?

Serafin Vilaplana: Twenty-six years old.

Katina Michael: How long did the system take to go from requirements gathering to operation?

Serafin Vilaplana: Roughly between two and three months.

Katina Michael: That is a pretty short roll-out time for something that had never quite been done before in terms of a commercial service. And that you worked on this program all by yourself is quite amazing. Where did you go to university?

Serafin Vilaplana: Thanks. But I must clarify something, I have never actually gone to university. Most of my information technology capability was gained on my own. I am pretty good at learning new concepts. So I just did my research, and already had the ability to program from a young teenager, so the rest was quite easy for me.

Katina Michael: You make it sound so so simple?

Serafin Vilaplana: Well, that is because it is very simple. Let me explain further, I find programming very easy, I always have. It is a hobby of mine. And now I am even a web developer with many projects on the side. I like to tinker and have a go using new technologies, all the time.

Katina Michael: Have you heard of Mr Amal Graafstra of the United States who wrote RFID Toys?

Serafin Vilaplana: No.

Katina Michael: Well, Serafin, he pretty much is like you … and just like yourself, he too is an IT Manager for an American company. It is fascinating to me that both of you are self-taught, and both of you have become pretty successful despite the fact that you do not hold university degrees. You know I teach IT students at a tertiary institution in Australia, and sometimes I wish they had the same drive as people like yourself and Amal … I don't want them to go out and put implants in their body, but sometimes I feel like students who are handed an education don't make as much of it as they could … Could I ask you Serafin, whether any of the patrons, even complained about their implant or whether they had anything negative to say about the process or the implant itself.

Serafin Vilaplana: No, never ever. In the whole history of the program we had going at Baja Beach Club, not one implanted patron ever complained about the program. Absolutely not. On the contrary, everyone who opted to take the implant was really positive about the application.

Katina Michael: Serafin, what was the main drawing point of the chip implant application?

Serafin Vilaplana: It was quite simple really … our implanted patrons would have access to special VIP areas, and when they walked into the club, their name would automatically flash up on a big screen and a loud beep could be heard echoing throughout the club … the reaction by the other clubbers was always "Oh, look, so and so is here and he is bearing an implant in his body", or "Ah, here comes that person again, he is special". It was quite interesting to sit back and watch people's reactions … it was at times quite amusing really … but the implantee patrons, really loved the extra attention, and people would instantly go up to them and talk to them and strike a conversation about anything… they were no longer anonymous, but were very approachable because people knew their name.

Katina Michael: Did people think it was cool?

Serafin Vilaplana: Oh yes, certainly. Young people love new technology. And not only that it was cool, but it brought some of our VIP patrons instant prestige. They were different and there was an attraction in that. Also, I should emphasize that our VIP patrons were pretty big spenders, they had a sizeable budget … so if your name flashed up on the screen, pretty much, people knew you were quite wealthy.

Katina Michael: Did the implant serve the patron any other functionality?

Serafin Vilaplana: Yes of course. It was not just an ID device. What I did was to create a whole purchasing program that had a back-end database with patron details. On the chip itself all that was stored was an ID number. I built a system that would read the ID number at the point of sale, and when the implantee purchased a drink or any other services, the total amount would be deducted from their stored value. They could top up their balance at any time they wished. It was really easy. We never had anything go wrong with that, ever. I had readers positioned all over the place, in strategic entry/exit points and also on all the doors, and areas leading into and out of VIP areas and of course at point of sale. I made sure to build the database on sound principles.

Katina Michael: Was it a big integration effort on your part?

Serafin Vilaplana: Not at all … it was a small customized system that I built. I did not rely on anything more than the RFID device, the readers, the antennas, a database to store the information collected, and the program I built to register patrons on the system and to enable transactions to be processed in real-time.

Katina Michael: Serafin, was it a transponder or a tag that you used for the RFID device?

Serafin Vilaplana: It was a tiny transponder.

Katina Michael: How does the technology work?

Serafin Vilaplana: Well, see this bar code [points to a bar code label on a water bottle], RFID works like this, except it is more powerful because it does not require line of sight.

Katina Michael: At what distance would a read take place? I assume the transponder was passive and not active?

Serafin Vilaplana: The read would be at very close range like within 10 centimetres and the transponder was passive.

Katina Michael: Who supplied the transponder?

Serafin Vilaplana: Well, look, there were many organizations who could have supplied the transponder. I did research on lots and lots of companies and products and found the one I liked best to meet our needs. It didn't take me that long to find a supplier. I just wanted to make sure I had the best technology for what we wanted to do.

Katina Michael: What do you consider to be the best feature of the implant system you created?

Serafin Vilaplana: Definitely it was the fact that it would keep out unwanted persons from restricted areas in the club that were meant only for the select VIPs. Of course, others wanted to be a part of those zones but had not paid or subscribed to enjoy those services. Access control was really the best outcome of the program. For instance, now we are talking here in the office deep underground, and even these doors were secured so no one could snoop around where we keep all our files and computer equipment. No one could get into this nerve centre to steal important documentation. You can use access control to exclude and include dependent on who the person is.

Katina Michael: How did you actually design this?

Serafin Vilaplana: Well, I just had to define special roles. For example, VIP patron, employee, manager, administrator. Dependent on who you were in the registered database, you enjoyed certain privileges. So it was actually an access control matrix that I defined and then the rest was easy. But you have to pay special attention to when designing this matrix, because you really do not want to make errors. For instance, you would not want a VIP patron to have access to the management area for staff only.

Katina Michael: Where did you get some of your ideas to create different zoning areas?

Serafin Vilaplana: This was naturally instinctive to me … but also I had read a lot about the military applications they are proposing, and that are currently in use in some parts of the world. Have you heard about the way that some countries are thinking to integrate RFID implants for gun control? That is, the gun can only be fired when the right person, bearing the right ID tries to fire the gun. If anyone other than the soldier picks it up to fire, they would be unable to.

Katina Michael: Serafin, when you built your implant program, had you heard of the Cincinnati trial in the state of Ohio in the US?

Serafin Vilaplana: No.

Katina Michael: Okay, basically, a number of employees in this small organization were implanted.

Serafin Vilaplana: I must stress that people should be free to join programs and not forced into anything. We were not about this at all.

Katina Michael: So you would not condone the use of blanket coverage microchip implants for government programs which enforced the whole population to get an ID number?

Serafin Vilaplana: No. There must be freedom.

Katina Michael: Do you know of any other applications on the market, where microchip implants are currently being used?

Serafin Vilaplana: Look, there are so many now. Maybe they are not large scale but there are a lot of people looking at the potential. I always like to see where things are at in this space, and often look for information related to RFID to note the developments. While the vast majority of trials have only had very few people in them, nevertheless they have tested the concept works.

Katina Michael: Were there any limitations or problems with the VIP implant program at Baja Beach Club?

Serafin Vilaplana: Not really. About the only thing that I could say was a limitation, was that the enrolment process was very messy and required all these people to come together so that just one patron could be registered. It was very time consuming in my opinion, and very disruptive to add a new patron to the program. You needed a doctor, you needed a Club employee who had been trained in the process of registration (and most of the time it was me needing to be available to do the data entry and make sure it was all working as it should be), you needed the patron to make sure they wanted to participate in the trial, and that some kind of light anesthetic would be applied so the person would not feel any pain. So for me the main problem was that you just needed too many people to be involved in the whole registration process. It is just not viable as a business process, as there are too many costs involved.

Katina Michael: Where was the actual implant zone?

Serafin Vilaplana: It was in the arm at first but to be honest we realized this was not the most convenient location and so some people had it implanted on the inside of their wrist. This is the site that gives the individual maximum mobility and is user friendly. We injected so far as the implant could lodge itself in the muscle of the person and would not go moving around inside the body to render the program useless.

Katina Michael: Serafin, you mentioned that Baja Beach Club now does not exist in any form. What happens to all those 100 people that were implanted? Do they still have the implant?

Serafin Vilaplana: Yes. We are now called the Opium and under completely new management. I have just stayed on as their IT manager.

Katina Michael: So the implants in the VIP patrons are just dormant?

Serafin Vilaplana: Yes. But they are quite useless now. Even if someone was to chop off someone's arm to use the ID device, they could not get far with it … I have changed all the ID entry access codes now- it was very simple.

Katina Michael: And nobody has ever asked you to remove the implant?

Serafin Vilaplana: No, they have not … it is just sitting in their body, it cannot do them any harm.

Katina Michael: Are there any plans to introduce a similar program under the new management in The Opium Club?

Serafin Vilaplana: No. That is now finished. Nobody will be implanted again here.

Katina Michael: Do you know of any other clubs that are now using your concept in other clubs anywhere else in the world?

Serafin Vilaplana: No, I do not think there is anyone else.

Katina Michael: Where do you think this technology is headed?

Serafin Vilaplana: Well, my hope is that one day, I do not have to carry any cards at all, that everything will be stored on centralized databases, and that my ID will be an implant so I do not have to carry around cumbersome wallets. I would feel very relieved as a consumer if all this would happen. I just want to be free of extras that are a nuisance. I just want people to be able to check my records, so I can go about my daily business without any hassles. It would save so much time … more to the point today we carry so many many cards, what is the point?! I cannot wait for this revolution to take place one day.

Katina Michael: Serafin, I am a little concerned at this statement … RFID is such an insecure device … people in the know already put sentinel jackets around their ePassport so it cannot be illegally read by another device. What do you say to this?

Serafin Vilaplana: Yes, I agree that RFID is an insecure technology, in fact it is very insecure. This is the whole point of using databases … the chip should just have an ID number, and that is it … everything else should be securely stored on a database. You would require a great deal of validation to be going on in the system, but with today's processing speeds that is achievable. So you need to check the databases, validate, and synchronize with each transaction.

Katina Michael: Okay, let us go a little further with this possibility. Say for instance, a commercial organization, like a mobile phone company, was to [tell] all their subscribers in the near future that they could take advantage of new wireless services if they got an implant.

Serafin Vilaplana: Katina, I must underscore here - consent is of utmost importance. If people want it they can adopt it, if they don't want it, then that's that. The decision must be with the consumer and not with the system creator. We never told anyone at the club - look, now you are under our patronage you must take this technology. No, never. And for the patrons, I must stress again, it was always up to them.

Katina Michael: Can you see any risks in the deployment of this type of technology for large-scale applications?

Serafin Vilaplana: No. Frankly I can only see positive benefits. There are no negatives.

Katina Michael: Is there anything else you would like to add?

Serafin Vilaplana: No. Just to say, if you need to contact me here is my business card.

Katina Michael: Thank you very much.

KEY TERMS AND DEFINITIONS

Baja Beach Club: Was an exclusive nightclub in Barcelona, Spain. In 2004 the Club began offering implantable microchips to its VIP customers for identification purposes but closed its operations in 2009. A sister club in Rotterdam, in the Netherlands also offered implantable microchips to its patrons but closed in 2011.

Chipping: The application of RFID tags/transponders, injected into the subdermal layer of the human body. To date chipping humans has happened on a voluntary basis through individual consent.

Implantee: Someone who bears a radio-frequency identification device. Can also be used to denote an implantable prosthetic device in a human. Some implants are passive (e.g. the VeriChip), others are active.

RFID: Stands for radio-frequency identification. Components of an RFID system include a reader (also known as an interrogator) which communicates with a transponder that holds digital information in a microchip, such as a unique identifier.

Risk: The potential of loss resulting from a given action, activity and/or inaction.

Sentinel jacket: Acts like a Faraday cage and shields implants from any electromagnetic waves, thus preventing reader queries from reaching them.

Transponder: Is a device that emits an identifying signal in response to an interrogating received signal.

VIP: Is a very important person.

Interview 4.2
Citywatcher.com

Mr. Gary Retherford, Six Sigma Security, US
Interview conducted by Katina Michael on 26 May 2009 via teleconference

PART ONE

Gary Retherford: A few years ago I reached out to guys who were doing Six Sigma at the University of [NAME WITHHELD] on some concepts and ideas about security and innovation in terms of methodology.

Katina Michael: Yes.

Gary Retherford: I was really fascinated that while these guys had books all over their shelves and they were certainly very intelligent guys, I was shocked at their inability to look beyond what was right in front of their face.

Katina Michael: Okay.

Gary Retherford: And it almost appeared that there was just absolutely no visionary in the room who could think beyond what they were – it was almost like they wanted to stay safe and secure within what they knew, and they were unable to really think outside the box.

Katina Michael: And that was just in the Six Sigma domain? Or implants?

Gary Retherford: No, what happened was, and this, we'll try and get into this on the question list, but my story is a little unique in the sense that my background was in working in the security industry where I was involved in video surveillance and access control systems and this was post 9/11 so you can understand it was pretty heightened.

Katina Michael: Yes.

Gary Retherford: And I was working at some chemical plants where I had met somebody who introduced me to Six Sigma. This was in 2003. And looking at Six Sigma, I began to realise that Six Sigma could have an application in security. In the area of security, you have a lot of opinions around what is considered the security that should be applied or how we are spending our money, but there's very little use of data. Being in your field, data is important and so the challenge was to take something like Six Sigma and make it applicable into a security world where you were forcing people to use the data to make decisions and start finding out, for example, what is security? So in that context, that's what happened and I started to mesh the two together, me and this other guy and we had an article published in a journal in 2004 by a company out of England on the theme of business continuity... Are you familiar with LinkedIn?

Katina Michael: Of course.

Gary Retherford: Okay, well I'm on LinkedIn and I have a LinkedIn group called Six Sigma.

Katina Michael: Okay.

Gary Retherford: And I have now people coming in there from the Middle East and India and everywhere where it's starting to catch on. And I do Six Sigma consulting now and do some of the training in Six Sigma and I have a focus that I'm working on in law enforcement and in the security realm again, but that's my involvement in Six Sigma.

Katina Michael: That's fascinating. So you're spinning security-related business processes?

Gary Retherford: Right, exactly. Because, again, there were some surveys that were done back in 2002 after 9/11 where it was towards the, I guess it might have been the Fortune 500 or the Fortune 1000 asking about how they spent their money and … on security and where they saw improvements in security and I just remember one thing out of that survey which was that 80% said they were going to … that they needed to make improvements in security and only 20% or 30% felt they would spend the money wisely. And that started the whole process of looking at: Okay, well what's the reason behind that? There were some other surveys that were done by I believe it was Bolbridge and it might have been done by the American Society of Quality, I don't remember exactly, but these surveys were basically pointing out that Six Sigma was a tool, or that these … forgive me for fixing it as a tool… but these organisations/ companies unable to really have a definitive plan and were not using any type of data to support their security decisions could apply Six Sigma principles. It was based upon basically an opinion. You know the history in security … corporate security have a history where a lot of people are in that decision making role used to have backgrounds in the area of law enforcement.

Katina Michael: Yes. This is common in Australia as well.

Gary Retherford: And law enforcement is not a good place for actually people to go into corporate security, it's a different world.

Katina Michael: Yes, I've seen that first hand. High ranking retired policemen are hired to uncover fraud in corporations.

Gary Retherford: Yes. And so that was part of it in 2000 and then in the fall of 2007 the Institute of Justice here in the U.S. gave a grant for four police departments to use Six Sigma in law enforcement to see if it would be a way for improving law enforcement and the results of that came back very positive. And so … if you look at the United Kingdom, there are initiatives there and have been for a while where they use Six Sigma in law enforcement. So it's a matter of applying data to our world to improve security. There was actually an article in *USA Today* in 2002 that talked about how the CIA was getting into using Six Sigma and I believe it's in that article where the comment was made that if the CIA had used Six Sigma, it was felt that they could have prevented 9/11. So, that's just some of the things that I was doing and then how I ended up getting into Six Sigma.

Katina Michael: So basically, that's your tool – that's your methodological approach, using Six Sigma.

Gary Retherford: Yeah, that's what I did.

Katina Michael: And if I was to ask you, taking that security context now, for you going through this process with clients, you're looking for the most foolproof technology within a particular context and that's how you came to look at implants?

Gary Retherford: Well, it's moving in that direction. What is the challenge right now has just been to get some of the basic concepts using the Six Sigma for decision makers. I'll give you an example. There's a large company, a mining company, with plants all over the world. The security department there wanted to bring Six Sigma in to align itself more with corporate goals because Six Sigma was used throughout the organisation – whether it be strategic planning or the operations department or things of that nature – and they felt that they wanted to use Six Sigma, primarily in their case just to be a part of integrating into the rest of the organisation. Where it leads to then from there is you use basically the tools of things like the project charter sypoc – which is your processes and your customer. Before you even get to the data, you're looking at the methodology and thinking of concepts- or how you're thinking about it to be able to drive a more disciplined way of thinking.

Katina Michael: Yes. Rule based.

Gary Retherford: And cutting out things that right now that you don't want.

Katina Michael: So there's a lot of reporting, a lot of seeking cost efficiencies and better effectiveness...

Gary Retherford: Yes.

Katina Michael: Basically, greater profits and reduction in risks and costs I guess and more to a rules based approach wherever you can have rules for decision-making process.

Gary Retherford: Exactly. Where it will lead is that it's going to get into better technologies.

Katina Michael: Oh, of course. It's a bit like – have you heard of on the RFID side the notion of the "Internet of Things"?

Gary Retherford: No. What is it, so I know?

Katina Michael: The "Internet of Things" is a term that was coined back in 1999 by Kevin Ashton by the Auto ID Centre at MIT and they were building a standard, basically to allow, in this new IPv6 environment, the tracking of every single object that's manufactured. And so, the notion of the Internet of Things is this notion that objects that are created, that we buy as consumers or that we sell as businesses – are tagged with passive tags or active tags depending on the context and they then become somewhat remotely 'trigger-able' if that makes sense. You know, if I have a washing machine I don't have to send a loyalty card in saying "here's my warranty status" or "here's my receipt/proof" ... they know where it's sitting. Or it can communicate if it has IP access, back to base, saying "it's time for my maintenance" or "this has gone wrong" or "here's a fault, I'm reporting it" – a bit like our desktop computers, but imagine objects like flip-flops or sandals or glasses or watches, each embedded with a passive tag that could be triggered ... say I have a reader either at a kiosk or either on my phone or somewhere in my home. And so what we're looking at now is an environment that is very much driven by hybrid wireless architectures. So you've got your Local Area Network, you might have your home network ... you may have Wi-Fi networks run by service providers or free to the public depending on the council you're part of and then you've got these mobile 2G/3G types of environments, you've also got GPS ... so it's using the different types of wireless infrastructures together, and once you can do that then things like your watch or your glasses or anything that passes in field view of a reader can be triggered and elicit a response. This response will then report back to base if that's how you have customised your services, but the Internet of Things is allowing all objects manufactured to store or have onboard an RFID device – a tag or transponder – and that can communicate back. So it's not ... I find it interesting that people continuously talk about humans being implanted – many people have argued: Why implant when you can just wear the darn things? But the thing is then, external, you know, it can be removed and so forth – we understand those issues, but that interaction with objects and subjects is what's going to become very lucrative either for Six Sigma, whether it's you're employees or whether it's the consumers you've sold things to or citizens in the context of government. But that's where I perceive the great innovations to come next – it's actually implementing that EPC (Electronic Product Code) model that MIT had put forward back in 1999 and that's now being rolled out in many places. There are big companies... Wal-Mart for example and DOD (the Department of Defence) have heavily invested in EPC oriented tagging.

Gary Retherford: See, I'm like you. The thing I see about this and when it comes to the implants is, because of everything else that's going on and I wasn't familiar with precisely with what you describe, but I have basically the concept in my head – I knew this was going to be out there. I'm like you. I really sometimes am baffled by some of the reaction to the implants when between your cell phone and everything else that you just talked about and Google ... and Google Earth.

Katina Michael: Yes.

Gary Retherford: Why are people still worried about implants? And quite frankly, I'm ... we'll get into this I've thought about the questions that you sent me and I'm a little ... There's a piece of me that doesn't really believe that we have ... there's as much of a pushback as a perceived push back because a lot of the push back, I feel, is from more of a vocal minority than a majority and because we did the thing here in Cincinnati with CityWatcher – we had hundreds, probably thousands of e-mails and I would say if you take out the ones that we believed were repeatable e-mails coming from what it sounded like either the same group or the same people, across the board it looked like the majority were in favour of what we did. Now if you look at it within inside of the pocket of the middle west of the United States where you still have an evangelical slant, you know, we have some fairly negative responses there but, by and large, it was vocal from a few but even ... but for the most part I can't say that what I realise is – and I think this is not academic type research, but it's just my research being in the role of sales, business development and you know, what I do ... What I got as feedback is that while there are people out there that might be against it, they still believe that we live in a world that it's your choice.

Katina Michael: Yes.

Gary Retherford: And as long as it's your choice, I choose not to but if you want to, I'm not going to stop you from doing what you want to do. And so in that regard, I personally don't see – I see a lot of the hurdles to doing implantable microchips to somewhat being more perceived by the industry as a hurdle than what they really would have if they would just go out and just start doing it … That there may be some thought into: OK, there are some issues here that might pertain to society that are some things that I haven't thought of, but by and large, I find and I found either with particular age groups that younger people had less of an issue with this thing. In fact one funny story was as follows. Before I sent out the press release on this CityWatcher project, it was a few months prior – I contacted a local newspaper that was a business journal and I talked to them about seeing if I could put in just some sort of like free announcement, you know, something to kind of test the waters, so they said, well here why don't you talk to one of our reporters and I talked to a reporter because they said there might be a story and I'm talking to this girl and she sounded pretty young and she said "Well what's this?" – she was their technology reporter so I said, "Well it has to do with microchips and implanting microchips and I'm explaining the whole process to her and I'm explaining what we're doing and interestingly enough we get to the end and she says "Well, I'm not… I don't think I'll do a real interview with you because I don't think there's a story there" …

Katina Michael: (laughing!)

Gary Retherford: And I'm like… and even I'm a little… I was a little…

Katina Michael: Taken a back…

Gary Retherford: Wait a minute… I've been sweating on this thing and you say there's not a story here!

Katina Michael: I know the feeling…

Gary Retherford: So she didn't do anything with it. She just said, "I don't think that there's anything here that you know, my readers would be interested in" and I'm like "Do you understand implantable? It goes in …" She goes "Yeah I got it". So I realised that okay, there's really kind of an age thing here. So the day that I did this and found the press release and it was a couple of days later, I'm standing in my bank and the tellers are all you know college kids or right out of high school and I told them what I was doing and here I'm thinking "Okay, lets see what kind of reaction I'm going to get…" I got a ho-hum reaction – it's just like, "Oh really that's cool. Hey, you know maybe we could do that."

Katina Michael: Yes.

Gary Retherford: So I'm thinking wow, there's some real dynamics going on here… likewise I've tested this in LinkedIn, on some of the security groups out there like "As Is" and some of the others. I put out there a few weeks ago: "Do you see a day where implantable microchips would be an alternative to access control cards?" Well I had five or six really hyped up on this thing and I mean they were just pounding away at it and they were sending back all their negative comments but relatively speaking to the size of the crowd, that was less than 3% probably. So the vast majority had no – either they didn't read it, or they just absolutely had no desire to make an influence.

Katina Michael: So, we've had different kinds of statistics for different markets … There are markets that have no problem with this kind of thing. My Chinese students for instance have lived all their life with ID cards, and I'd say with an ad-hoc show of hands in my class about 70% of them don't see the big deal.

Gary Retherford: Yes.

Katina Michael: Australians generally are traditionally less willing to adopt these kinds of new technologies. We've been renowned for squashing things like ID cards back in 1987 and 2007 … massive campaigns and primarily from the perspective of privacy… I look at politicians over there and I know some of them have strong leanings towards particular ideologies.…For example, Tommy Thompson, who was on the Verichip board, demonstrates that there are politicians who are proponents of this technology and are quite supportive. Then there are others who lead the anti-chipping law campaigns and we sort of tried to do a Republican versus Democrats breakdown by states of the U.S. two years ago with one of my honours students to see what laws had been passed and who was driving that.

To be honest Gary, I find that the majority of people just go with the flow. That's what I find. We're quite much the sheep mentality – if there are a few key leaders it's always got to do with word of mouth when you study what makes innovations happen and what makes them successful – yes user friendless is important and all these things are important but down to the fundamental, the lowest common denominator- Is it useful? Is there value in it? Can someone see value in it? And the second thing is: has a lead group taken it and implemented it in a particular context, therefore, their opinion followers will also adopt it in due course. But there was a study done … I don't know if you know the Wolk and Perakslis study back in 2005 which was a US-based survey that was looking at implants and biometrics for national security applications and the vast majority, but again, I don't know what the sample size was from memory, the vast majority said they would not adopt in this given context… sort of like a citizen ID number or something like that.

I'm continuously fascinated by the increase in employee monitoring using GPS and it's not really just the employees, it's the vehicle that they drive in most cases … but there are a number of, you know, I get emails from people that represent – that are senior counsels, representing unions that don't want their trucks to be tracked. Okay … So the cases are flying thick now. The US government is giving mobile phones with GPS devices in them to employees not really stipulating clearly or vocally that the device can actually track their location, circa certain proximity and accuracy. There was a case in October 2007 where somebody sued the government based on a contract, for hours that they proved he didn't work and he was dismissed on the spot. And he said "Well how are you dismissing me?" and they said "We've got you tracked" and he said "Well why didn't you tell me?" and they said "It was in the documentation you signed. You should have read it properly." And I'm not saying this … you know – that's Six Sigma, theoretically at it's best – in the sense … look at all the money you save busting people that weren't, you know, that were charging the government for work they didn't do but at the same token one could say, you know, hypothetically "Well maybe that guy didn't work those hours at that time. Maybe he forgot when he worked them exactly when he was filling out his time sheet. Maybe he didn't log his time properly. Maybe he worked for four hours in the middle of the night as many of us do sometimes" … So from an efficiency perspective, I can see where this thing is headed: major efficiencies. But from the social side, you know Gary, as humans we need a bit of slack, right? We need a bit of flexibility in the way we do things and we're often good at doing things routinely, within particular time limits, but I don't, you know, you mentioned the mining company –but I'm not sure, I'm really not sure, deep down that we can work like robots. Do you know what I'm saying?

Gary Retherford: Hmmm.

Katina Michael: I can see that these efficiencies will be gained using technologies and using newer emerging technologies, not just implants but location services and other things. I'm not sure that's going to give us … you know … are we entering an Orwell type of space where we have that monitoring happening and if an exception happens you're worried because oops, you know, "I didn't follow that procedure or I didn't do that right". Are we going somewhere perhaps or venturing towards a space that maybe wasn't meant for us as humans … I don't know … You can see how …

Gary Retherford: I think …

Katina Michael: Go ahead-

Gary Retherford: Well I think what you're saying makes sense in the sense that I've spent a lot of my time in the last few years working either independently for myself and I've worked for other companies, I've worked for large companies – I worked for Siemens for a while and in working I had cars that may have had GPS or … we had not let the GPS on them, but I understand what your saying – in the mind, it does make you wonder, "Okay, can I really work like this?", because you know if I want to veer off and stop off somewhere that's not typically where I would go next because I want to stop and get something to drink, a Coke or something, I mean, is that going to be in the back of my mind that I'm now outside of that control? And I do believe that that is certainly a risk that is coming at us as a society. That can we operate under those types of conditions? Unfortunately, I think on the other side of the fence to that, which is something that I found to be quite interesting when I was talking to people that could be potentially implanted with a microchip, and the question was, "What would be the driving force behind someone being implanted?", I realised that we also live in a society where … I found that there's always a question that always gets answered the same way and the question is: "Does my right to security supersede my neighbours right to privacy?", and what I found is, that even in, you could almost say in married couples, I mean, you know, "My security supersedes my wife's right to privacy". Because the question then comes is that – "Do I want to be chipped? No. But I don't really know my neighbour that well, so maybe Congress should pass a law that chips my neighbour." So I think it becomes a trade off because that could … see because what I'm wondering is what's going to be the driver? See, I think that when you said – earlier you talked about the government not interested in pursuing that track, however, what I have found is that …

Katina Michael: No the citizens ... the citizens didn't want to be chipped in the US survey. The government may want to pursue this track, I do not know but the citizens were saying they would not take that chip implant for a government-to-citizen context.

Gary Retherford: Oh okay. But I think that what I have found is that they said: "Definitely it's ok to chip prisoners".

Katina Michael: On occasion, this has been claimed, yes. Depending on where you are in the world.

Gary Retherford: Well, see what I think is, you will have a certain group of citizenry who will if you have a group within the citizenry that is of the "less desirable" ... remember that "less desirable" is the rest of the citizenry will say, "Okay, on that group, it's okay".

Katina Michael: Which, to be honest with you, I find to be quite dangerous and naive in the sense that: "What next?", you know ... Renowned in trials of smart cards, renowned in trials of biometrics, you know, have been minority groups... And it's funny, because ethical codes of conduct tell us you shouldn't approach people that are dependable on something like a patient in a hospital or a prisoner to a prison and where I find the danger in that is: No matter what technology – you just have to reflect back on history – no matter what innovation it's been either automatic or manual, its always found its face initially as a testing ground in prisons, in the mentally ill or cognitively slow individuals and patients is another one, in minors etc. If we go back to the holocaust, a typical one was the segregation of the Jews. We've always found that a technology is rolled out to a small minority, tested and then what's to say: "Next, let's implant all the schizophrenics", "let's implant all the people that look depressed to us", "let's implant all children because they're under the guardianship of the parents until the age of 17 or 18". Do you see that as a potential? Do you see what I'm getting at?

Gary Retherford: Absolutely. I think that's the way it's going to happen. I think that that is exactly – because I've already had that conversation – I've had the conversation with people who run gaols who wanted to talk seriously and I don't mean just in hypotheticals – that wanted to talk about putting readers in gaols. The reason was, that in the course of running a gaol where they had large numbers of prisoners that were coming in every year, maybe in the vicinity of 50,000 a year, but – that's 50,000 total processes, but it was 90% of that 50,000 that were repeat offenders. But the cost of processing was significant ... One of their large costs of processing was due to all of the files that had to be kept on them and their identity and it had been known that there had been prisoners that were let out because there was an identity mistake between a couple of prisoners. They were supposed to wear a band, they were supposed to wear ankle bracelets, whatever – but they would cut them off – they'd find a way to get out of them. So then it was the prisoner for all practical purposes... of course the ACLU is not sitting at the table during the discussion... but the prisoner for all practical purposes in the minds of the gaol officials ... they did not think of them any different as a dog. I mean ...

Katina Michael: Yes.

Gary Retherford: They really, absolutely did not think of them as humans.

Katina Michael: Can I ask you on that point because this is something I have been looking at recently, did they also think about the issue of visitors? So you mentioned to me the prisoners, but were they concerned at all about how to track visitors to the prison cells?

Gary Retherford: Umm.

Katina Michael: Or they didn't mention that to you?

Gary Retherford: No, we really didn't get into that.

Katina Michael: Because that's actually a biggie – knowing where visitors are at any one point in time and I've seen an Australian, Queensland state example, where they've got visitor tracking capabilities based on biometrics – iris code.

Gary Retherford: Okay.

Katina Michael: But the prisoners are more intrinsic – they're like the core competency so to speak to the gaols, but the visitors are another data flow and it's a difficult one to handle, that needs to be tracked for diverse reasons.

Gary Retherford: But going also on the same line of a certain group that is targeted …

Katina Michael: Yes-

Gary Retherford: I had a conversation with an Ohio State Senator – not a US Senator, but he was at the State level – and he asked me if we could invent a product that if you sent out, say a signal from an antenna that was on a school and if a chipped sex offender went within 1000 feet of a school, that they could set off an alarm.

Katina Michael: Yes.

Gary Retherford: And I said, "Well, you know, you can invent almost anything I guess if you really want to put your mind to it". But I mean, he basically gave me an open invitation and if we could invent it, he'd see that the legislation got passed to support it – to actually require it on the schools. Now, there again, I cannot find any mother …

Katina Michael: I know that …

Gary Retherford: … of a kid who would not support that legislation. But it does go back to what you were saying that "Yeah once you start with group A, then eventually it's going to be group B and before you know it it's going to be anybody that gets within 1000 feet … an alarm is going to go off".

Katina Michael: I'm intrigued Gary, when did you first conceive of microchip implants in humans? Was it that you heard about Kevin Warwick implanting himself? Did you hear about anyone or was it just something you came across by accident via that guy you were talking to me about when you started at Six Sigma?

Gary Retherford: I'll go ahead and tell you how it started. I was working as a security system integrator. I was in sales and business development and I was working for a company here in town and I had been working with an Art Museum who was interested in looking at some asset tracking for some high valued art and I began to do some research on their behalf and in the process came along this company called Verichip and up to that point I had not had any notion or concept ideas at all about implantable microchips – it never even entered my mind.

Katina Michael: Yes.

Gary Retherford: When I saw the product that they had I became fascinated with it. Even more fascinated when I realised after, I guess, maybe through the Google searches that I had done, and I came across Verichip – and so many hits associated with some of the evangelical slants in the market … and I think that kind of intrigued me, because I'm thinking: "Wow, where's this coming from?" So I actually had reached out to Verichip to find out about their asset tracking and simultaneously was going to ask about their implantable microchip product for access control because I was in the access control business. Interesting enough though, at the very same time and I'm talking almost down to the minute, I was getting ready to have lunch with the owner of the company called Citywatcher and I reached out to them because they were offering this service of doing video surveillance on servers and they were doing some work in the city of Cincinnati. So as I literally had my phone in my left hand getting ready to introduce myself to a contact at Verichip, I was reaching out with my right hand to shake the hand of the president and little did I know at that point that roughly a year later, what was going to eventually end up happening. So then I began talking to Verichip. We talked about their asset tracking component for the art, but I also started to ask them about their access control system and when I was beginning talking to them and their sales people/person that they had in charge, I realised that they had a little bit of a flaw, in my opinion, in the way they were trying to market their product. What they were trying to do was create a whole access control system and sell it as an entire system and I said, "Well, I have a suggestion for you. My suggestion is that in doing that you just take the reader and they can integrate with everybody else's access control."

Katina Michael: Exactly.

Gary Retherford: And I said "Now you're going to sell more readers and consequently, you'll sell more chips, because you're in the chip business, you're really not in the security business, per se.

Katina Michael: That's right.

Gary Retherford: So we talked about some technology and I talked to their technical person and they said "Okay, we get this, we understand and let's go to work on this." So in the meantime when I'm working with Citywatcher outfitting their server room, I actually sold them an access control system as well.

The owner kept losing his card, leaving his access control card everywhere – he was kind of an absent minded guy and it was also cutting edge, so I said to him one day—I said, you know – it's nothing more dramatic and this is what it was. I said to him: "You know, if you keep losing your card… you're not really keeping control of the server room". I said "I have an option for you." I said "How would you like to, instead of using a card, have a chip implanted?", and he jumped on it. He said "Absolutely." He said "I'll do that". So I called back to Verichip and I said "You know, you get this reader created and invented if you would". And I said "Send it to me, and I've got an application". And they, at that point, it was agreed on the phone that when they got it built, they would send it to me. My interest in this initially was only from the standpoint of how it was technologically speaking, so because we were going to integrate some technologies here and we wanted to see how those technologies were going to work together. But the more I started to really understand and read about some of these other social and ethical issues, I began to realise that there were a whole lot more here and I became fascinated by how it would be … How this product and how this concept enters into society. So therefore it became much bigger than strictly the integration. It was never done for any purposes that would be, you know, publicity or anything like that, it wasn't like a publicity stunt if you would …

Katina Michael: Yes.

Gary Retherford: I really wasn't looking for attention because interesting enough I never put my name out anywhere and nobody knew who I was.

Katina Michael: Yes.

Gary Retherford: And there were reasons for that, but I will say that when I got down to the day in sending out the press release – because I sent out the press release in the company at the time which was Six Sigma Security, because I was still doing the Six Sigma stuff too – I had a lot of thoughts regarding "Okay, think about the impact that this could have" because I realised that nobody as far as I knew had initiated and actually done this anywhere in the world. That it was going to be the first time that any place of employment, was having employees implanted with a microchip.

Katina Michael: Yes.

Gary Retherford: And I started to realise that this is going to really set in motion a lot of things and I have no idea of what all those things are but I wanted to see – and this is a little strange to say this – but I wanted to see what the world was made of. I wanted to see really what the human race was going to … how the human race would react and what the world was made of to deal with this concept and so when I sent out ... I remember when I clicked the little "send" button and I sent it to the company that was sending out the press release, I remember I had kind of a pause and a thought and a deep breath. Because I knew there was no coming back once I hit the send.

Katina Michael: That's right.

Gary Retherford: And that really set in motion some things that I had no fathom was going to happen over the next two weeks and then even beyond – what the impact was going to be. I underestimated it. So that's kind of how I got into it. There was only one interview that I ever did and it was really by accident, because somebody at Citywatcher gave my cell phone to … McIntyre.

Katina Michael: Oh okay, you've done one for Albrecht and McIntyre?

Gary Retherford: Well, she called me on my cell phone and caught me off guard at a weak moment and otherwise because I had committed … I had told myself I was not going to do any interviews.

Katina Michael: Yes.

Gary Retherford: And I had it set up because my name was never anywhere. That there was no record of who I was—

Katina Michael: Was it leaked through them, or they still didn't know who you were?

Gary Retherford: Well … What happened was is that one of the guys who worked there who decided not to have the chip implanted was a kid by the name of Corey Williams – wasn't really a kid – and Corey had just, I guess he got asked from Liz McIntyre, you know, "Who was the company that did this? Verichip or whatever?" And he said "Well, it's Gary and here's his cell phone." I guess I didn't think about them giving it out …

Katina Michael: The human factor.

Gary Retherford: So the next … I start getting this phone call which was totally unexpected and it was her. I had no idea who she was and she started asking me a couple of questions and I said "Well, I'll answer a couple of questions." But it was from that interview that my name ended up getting on the Internet.

Katina Michael: Okay.

Gary Retherford: Because, had I not talked to her I still could have probably stayed incognito. But I never did any of the interviews … The guys who owned Citywatcher or who worked there, I think they did fourteen radio interviews in two days. I know they were on CNN and FOX and everybody else, but I never appeared on any of those things. But I was actually the guy that was …

Katina Michael: Behind it.

Gary Retherford: Yeah.

Katina Michael: So in that sense, do you consider yourself an innovator?

Gary Retherford: I do because, well … Let me kind of define that. I'm not a technical person… I'm not a strong person in the area of technology, my background has been in areas of management, sales, business development, things of that nature. But I think that because of both my sales and business development background, I'm always looking at needs assessments and its kind of a natural thing for me to look at – not necessarily fulfilling the need now but where I think its going to go into the future and I ended up over the next few years, not only being involved in that, but I did … I was involved in a potential patent which I decided to drop, but it was the integration of geographical information systems and live video surveillance. And I was pushing the envelope. So I do kind of push the envelope and I guess if you would say that that's an innovator – I'm an innovator in concepts and ideas but I'm not the person to actually sit down and make the widgets and stuff. You know, I can't … technologically I can't. I'm not the techy guy.

Katina Michael: And so, can I ask for the real reason or the main reason, is it personal with regards with why you didn't want your name out there or why you didn't go forward with that patent? Or, can you …

Gary Retherford: Well the reason that I didn't put my name out there when we did the press release was: I think I knew enough at that time that there could be a potential for a pretty radical kick back …

Katina Michael: Yeah.

Gary Retherford: … but there was a guy who was at Citywatcher – he loved all that, in fact he had his own radio station.

Katina Michael: (laughing). Okay!

Gary Retherford: So for him being on the radio is like … he did one interview on radio – believe this or not – he did one interview on radio with I dunno who this was, it was the BBC or somebody – he did it while he was coaching his kid's basketball game because he was so comfortable with being on the air and had no qualms about it and I just wasn't sure that I knew yet what was going to be the kickback if you will or whatever the right term is – from some of the whackos out there, who I didn't really understand where they were going to come from quite at that point and I just didn't feel like I needed the aggravation so I just didn't. I had no interest on being on television I had no interest in being on radio so for me it wasn't that big of a deal not to get myself out there in the public eye – I was more interested in seeing what was going to happen in the world for my purposes of understanding it rather than whether it was for my purposes of going out there and trying to make a position.

Katina Michael: I was curious for the main reason that as this is our main research area, we've had that question but, I mean, as academics if we don't put our name to the article it's not published. But we've actually had those wacks on the head from various locations, for and against. So I know what you're saying. The aggravation is there. I do hope we haven't … my family hasn't been put in a situation whereby our heads are on the chopping block from either side but I do feel it gets personal at one stage and if your not mature to be able to discuss the subject matter just from a perspective of research then you will … You know, anyone getting close to this is often hurt to some degree. I don't know how to say it to you. It's how I began the interview with you telling you – basically that there are people that will read whatever they want to read and make whatever assumption they want about you and they'll let you know what assumptions they've made and it's not always pleasant. But if I was to ask going to the case study questions, I've got about half an hour before I get into class and I'm weary that I've kept you on the phone for maybe an hour and half now … How you feeling?

Gary Retherford: Keep asking your questions, we'll see how far we can get now.

Katina Michael: OK. You said a little bit about the Citywatcher trial. You mentioned that there were seven people that were implanted? Could you elaborate?

Gary Retherford: Well there were seven people implanted in total in Cincinnati that were at Citywatcher. There were just two employees and one contractor who were working at the company. So technically, only three people were implanted in this organisation. It was a small company with only six employees.

Katina Michael: So the name of the trial was?

Gary Retherford: Well, okay … That's where I guess it was not academic in nature. We never really thought of it as a trial. We were doing something in terms of testing some technology, but I never had an actual name for it.

Katina Michael: Okay, so you were dabbling … testing what worked?

Gary Retherford: Yes.

Katina Michael: And it was testing ID implants for access control? Was it access to the physical premise?

Gary Retherford: Yes. There were four doors. Two perimeter doors and two interior doors that were on access control using a swipe card and then one of those doors which was a server room that had a higher level of security to it which was only really accessed by three of the six employees. That server room access entry was altered. The old reader was taken off and the Verichip reader was put on. So you could enter the building if you had a card, but you could only enter the server room if you had either the chip implanted or the chip on a key ring.

Katina Michael: Okay.

Gary Retherford: And so one of the employees that was actually the IT manager, if you would, who ran the server, well he chose not to be implanted so he was given a key ring with a chip inside instead.

Katina Michael: So you basically asked the question and people volunteered to be implanted?

Gary Retherford: Yes.

Katina Michael: That IT manager, I am curious, why didn't he participate do you think?

Gary Retherford: For him he could just not get past the uneasiness of having it implanted. There was no … it was not for religious reasons. It wasn't a tracking reason. It wasn't a conspiracy for him. It was simply a reason that he just did not like the idea of having this item implanted inside of him.

Katina Michael: Did the other employees feel different to him do you think, did he feel left out?

Gary Retherford: No, I don't think so because there were still two other employees or three other employees who had decided not to have it done at all. I mean, they could have if they wanted to but it was kind of one of these things that was, you know, put out there as an option because at this point we were kind of more interested in seeing what the feedback was going to be like I guess. For example, there was one young man that worked there. His decision was "no" because he was influenced by his parents who felt that it was too much of an intrusion in his personal life – that it wasn't so much the "creepiness", as the other guy had pronounced– but it was more of a feeling of "Big Brother" tracking – even though you couldn't track. How do you get people convinced that you're not being tracked? So in their minds it was too much of a Big Brother type of an idea. So the problems were more privacy related and liberty related for him. And there was one girl who had originally said "no" and then she turned round and changed her mind. Which kind of led me to – and I'll elaborate a little bit because there was actually another project that was supposed to happen shortly after this one which did not happen because the reader that Verichip sent we could never get working properly. One of the problems is Verichip never really made good readers for access control applications.

At the time, I was talking to different companies, because there were about two or three companies who had an interest and were saying: "Yeah we can do that here, not throughout the whole company but maybe on a particular door here and there". But what was interesting was when I found out, is that while you might have say, for example ten people who might be asked if they would be interested in being chipped, then you might have two or three that would immediately say yes. I came to the conclusion after talking to these various company employees and just getting general feedback, was that once it started and you started talking about it, you would slowly get people who said "no" initially to say "yes". So what I realised was that when they saw that the other guys' arm didn't fall off, (laughing), okay, then … they realised that well if I do this, I'm not going to end up like a drone…

Katina Michael: (laughing)

Gary Retherford: They started to then say "well okay, I guess I will do it" because they saw that it was convenient and couldn't help but feel relieved … "Well, I like the fact that I don't have to worry about that stupid card anymore". So they would just automatically say "yes". So I came to the theory that, and this was strictly a theory, was that if you had 20% of the people say "yes" and 80% of the people say "no", that within a few months it would be flip flopped and you would have 80% that would end up saying "yes" to implants for access control and 20% who would probably still maintain their "no".

Katina Michael: Yes. And did those guys who were the early adopters, did they feel like they were breaking new ground? Was there a buzz in the office? Did they joke about it? And how long did it take before the novelty wore off?

Gary Retherford: Well, I think they thought that it was breaking new ground in the beginning but I don't think they understood the real impact it would have until after the phone was ringing off the hook for all the interviews. And afterwards CNBC came in and did a special. They did a special that was called "Big Brother, Big Business", and it's still shown periodically here in the States two or three times a year. And the last ten minutes of that two hour special was on this particular product, the implantable chip with respect to Citywatcher. Now, to answer your question, after it was probably two or three months … no maybe just one month, the interviews slowed down – there was no more of that. It never really felt like there was any regret of having done it.

Katina Michael: Yes.

Gary Retherford: The feeling the employees had was that they were glad that they did do it and it just became no different than any other aspect of their job. It was no longer considered novel after some time, but it wasn't like that was a bad thing. In fact, it was felt like "I'm glad I did this".

Katina Michael: And is Citywatcher still going Gary?

Gary Retherford: No, they went out of business.

Katina Michael: Okay.

Gary Retherford: No but it had nothing to do with the chip itself. Would you believe I still have the reader that we used?

Katina Michael: So how did you guys conduct the injections? One guy I think you said did them all?

Gary Retherford: Yes. There was a guy here in town, his name is Dr. Jim Scott and he works for a practice that's called Doctors Urgent Care. They're one of these extended hour urgent care type of places and they have eight offices and Dr. Scott is their lead physician.

Katina Michael: And did he have to do any extra research? Or for him, it was just like doing a normal local anaesthetic?

Gary Retherford: Yeah for him it was nothing.

Katina Michael: Yes.

Gary Retherford: I mean, he did converse with Verichip, I believe, just to get their input, but for a doctor, what I found out is that doctors are used to sticking people with needles all the time so for him it was nothing new. And going back to the seven people, he's done all seven people in Cincinnati.

Katina Michael: And it's in your hand or it's in your arm?

Gary Retherford: It's in the right arm and that's an interesting story too that really people do not know what happened there. This is really kind of interesting. Verichip's official direction was the upper arm in the back, on the back side in that fatty tissue area.

Katina Michael: Okay.

Gary Retherford: Okay, but here's what happened. When we did follow their directions … and we had to put on the readers. Well, we mounted the readers at Citywatcher, where we had our original readers. They were located there because they were used where you would have a card. Typically this is about the height of your belt buckle. So when we went to have everybody chipped, we said "where should we put the chip?" And it became obvious that you had to put it in something that was going to make sense relative to where the reader was. So the chip was too big to put in my hand so we had to find the closest fatty part of the arm and typically in the right arm where you could still put the chip in and not see it. So that ended up becoming in the front part of the forearm below the elbow… you know like down in the front part of the arm?

Katina Michael: Yes.

Gary Retherford: So we did that and then it ended up on television. And Verichip called me and wanted to know: "Why did you put it there?" And we said well we put it there because …

Katina Michael: It made sense …

Gary Retherford: … you would have to do it to use the reader. And they said that "We tell everybody that you have to put it in the upper part of the arm in the back". Well, they said, "Why didn't you just put the reader up there … you'll then be able to use the reader in such a way that you could use the back of the arm?" Well did you know why we didn't put the reader there? This is a question for you: Do you know why that reader is … See, if you look at any tape of Verichip, the reader that they have in their office in Florida sits up about the height about where your shoulder is and to get into a room, they turn around …

Katina Michael: And your reader's down … lower down?

Gary Retherford: Yes. Our reader's down. Here's the question and you may not know the answer, the question is why was our reader down lower? Well, because in the United States in the 1980s, George Bush Snr signed a law called the Americans with Disabilities Act…

Katina Michael: Yes exactly … wheelchair access or those with stunted height …

Gary Retherford: Yes wheelchair. It has to be wheelchair. So Verichip – they're a little miffed at me because I didn't put it …

Katina Michael: Yes I understand that …

Gary Retherford: ... the thing is not even allowed to be where it's at. I said technically you're supposed to move that reader. So that's how it ended up being in the lower right arm. So some people might think that was some other type of reason for it ... you know, "right hand" – all that kind of stuff ...

Katina Michael: No no ...

Gary Retherford: But it's actually because we had to follow the ADA compliance laws.

Katina Michael: Okay. You know, they always ... whenever you see the Verichip readers in media snippets, they're always the mobile ones ... they're the ones that you can pick up and move around to find the location of the chip, but that's another ergonomic question. Do you know Amal Graafstra at all Gary?

Gary Retherford: No.

Katina Michael: Okay. Well, Amal Graafstra is a hobbyist implantee and there's a whole underground group – possibly even thousands of these guys. They call themselves "the tagged." We did a full length interview with Amal in a book that we just released titled: *Innovative Auto ID and Location-Based Services: From Bar Codes to Chip Implants*. Anyway, Amal has got the implants in the webbing of his thumb and forefinger and the reason he said he did that is for usability and easy removal. He does not like commercial chips and he doesn't like even the remote idea of a government chip, so he said "I'm doing this because its my hobby". He's written a book called *RFID Toys* and this is where he discusses all his fun innovative software where you can use the ID number in the implants to unlock your front door, unlock your car and things like that. He was adamant about not having the implants in his arms because he did not like the idea of crouching down everywhere he went. He believes the hand is the best place, granting him maximum freedom and usability. But Verichip's problem Gary is that all the specifications, the original patent, the Food and Drug Administration approval, the anti-migration coating they have on the chip itself, it is all geared towards the Emergency Services market. Now, I had some very brief contact with the marketing manager back in 2003 when I was finishing my thesis but the timing was wrong for further contact. Gary was it a transponder or a tag that was implanted in the Citywatcher trial? Was it a glass cylinder or it was like a square?

Gary Retherford: Glass cylinder.

Katina Michael: And you mentioned it was big. I'm surprised. Like bigger than your thumb's fingernail?

Gary Retherford: Well they usually use the expression that it's about the size of a grain of rice. I think officially it's 12 millimetres long.

Katina Michael: And did you find through your experience at Citywatcher that the transponder worked effectively? That the device didn't fail – that it worked all the time?

Gary Retherford: The device worked great and when tested with a handheld device that would be used in say an emergency room or a medical application, it read every time like it was supposed to. If Verichip made a better reader, an access control reader, we would be selling them today. But Verichip did not have their own capabilities. They were not willing to make a commitment to invest in a better fixed reader – to make it smaller and more cost-effective. The reader that they manufactured for access control was approximately eleven inches high by five inches wide. So that's a big monster thing to put on somebody's door, when most readers today using a card are significantly smaller in dimensions.

Katina Michael: That's right.

Gary Retherford: … and today you can also get fixed readers in a variety of colours. So what happened was that Verichip made the reader too large and they made it to where they wanted someone to pay them on a cost price of about $500 a piece but you can buy an access control reader for as little as $50 to $75 cost and up to maybe $150. So the reason the reader was too large was because most of what's inside that reader is an antenna and the antenna is wrapped around in a way in order to make it so that it can read the chip from a distance of say three inches. What I know of it from having talked to some people is that you could make the reader smaller but now you have problems associated with how much heat it puts out in order to read the chip. For instance, too much heat could actually burn a person who's got the chip inside them. So I had talked to some people about a redevelopment of the reader and there was a little interest out there, but for the most part nobody wanted to make the investment.

Katina Michael: Why was that do you think?

Gary Retherford: The return on that investment was just too risky as you were still competing. Conceptually, if you would make a reader that could read both a card and a chip and you could make the reader you know, say, three or four inches by an inch or an inch and a half wide, you would have a marketable product. That's where the thing got hung up. It wasn't that anything necessarily did not work. It was just really because the reader needed to be redesigned and made more marketable.

Katina Michael: Did you ever see the Nokia 2005 design I think it was, where they had a very tiny reader at the bottom of the handset? I'll try and source a photo for you if you like.

Gary Retherford: Yes.

Katina Michael: So, they had a mobile phone which had a GPS device in it and it also had a reader at the bottom – and in that case you can track, so that an object could be identified using that reader, the ID number is stored somewhere either on the SIM card or sent back to base remotely using the GSM network and that GPS can tell you, at that timestamp, that's where the individual was, if they're outside of course. And through Wi-Fi you could even tell what cell they were in. So, this is when the hybrid devices become fascinating to me.

Gary Retherford: I'm always interested in talking to companies out there because I had reached out to the Malaysian companies who have some things going on in microchip readers and technologies.

Katina Michael: Gary, how did your attitude change toward microchip implants after you got your own implant?

Gary Retherford: It went from being I guess curious and had no opinion one way or the other to being in favour of and a supporter of. So I'm currently a proponent and unlike before when I was staying pretty quiet and incognito, I'm starting to change that to become more vocal.

Katina Michael: Okay. How does it make you feel to bear one at the moment while it's dormant, it's not being used?

Gary Retherford: Well, one of the issues that I have is that I think we have to have more of an infrastructure for it. I wish that we did have. I'm hoping that with things such as electronic medical records and I don't really work in an environment where I'd need an access control card anymore but I think that there's an application for it but unfortunately I need companies, I need hospitals, I need doctor's offices to be on board with it, so in that regard I can still get on it with my medical information and do things like that, but without there being an actual infrastructure there it remains just another idea. I have looked at it from the standpoint of it just being for personal identity purposes, but I think that if people have the opportunity to have a chip simply for the purposes of that they will also be able to have their information online and family members should have access in instances of a sudden death for instance. If there are directives in terms of last wills and things like that, that would be of great benefit. People might say, "I've lost x or y and I can't find the directives of what my relative wanted" but now if you have it scanned and put online there are some potential identity and informational benefits there but by and large there has to be an infrastructure to support the applications. So for my purposes, I mean, you don't even remember that you got it. It's not something that you walk around with thinking about. You forget about it. But from time to time, I wish the system was already in place where it would be functional and beneficial. I'm hoping that will change quickly.

Katina Michael: Do you think your implant is now removable? If you wanted to get rid of it one day … or you wanted to upgrade it, do you think you could remove it? Or it's in there enmeshed with the tissue?

Gary Retherford: The doctor has told me that it is removable but I don't know that it is something that is not. If it's not that doesn't bother me.

Katina Michael: and with multiple implants … how would you feel if you had multiple implants for different applications or functions?

Gary Retherford: Well I think that that's okay but I don't think that's the future. I mean, I think that might be the case in the short term -- there might be multiples and that's okay too to some degree. Verichip just announced that they have just come out by shortening the twelve millimetres transponder down to eight so they're going to continue to make it smaller so as long as there's some place to put it, I don't see a problem with that. The difference is that is with the Verichip type of implant, its only good for having a 12 or 16 digit number and it doesn't really hold any of the data whereas if you take the Malaysian microchip which is about the size of a head of a pin and can hold vast amounts of data, I think if anybody would ever get to the point that you could encapsulate that into the glass cylindrical tranponder so that it could become implantable, then I think that would make huge strides.

PART TWO

Katina Michael: Gary, I have to say that throughout the whole day I was in class I could not help but to go back to what we were talking about yesterday. I think we're coming at it from very different angles. I guess our backgrounds in business are the same, I used to work for Nortel Networks in presales engineering– so technical bid support- and it's very interesting you're in business development role and still sort of working in sales and I was thinking about your technology know-how and how I think it doesn't really matter whether someone knows the ins and outs of how technology works exactly – you may have a good business idea and that's really what it's all about in this arena. However, in the case of RFID, I was considering whether you had any background in its security. So you're in the area of security in Six Sigma, but have you done any research with respect to the security of RFID systems?

Gary Retherford: Any research?

Katina Michael: Yes. Like whether it's a robust technology or what some of its flaws are or its pitfalls or how it may be intercepted or cloned, or how people can remotely kill an RFID device even if it's within the body and render it dysfunctional? I mean have you looked at some of the limitations of the technology itself?

Gary Retherford: Not to that degree, no. The only security issues I came across were identified after the Citywatcher project and then when I was in the conversation with Liz McIntyre she asked me if I knew that RIFD tags had been hacked. But the thing of it is that— I never did take the time to do any of the research on RFID from the perspective that you're talking about. I have been involved in … when I was working for other companies and for myself … to look at selling … companies that were looking at using RFID for tracking of whether it be pallets or something inside the distribution centre but I never went into looking at the hacking part of it from a technical standpoint, no. But after considering what Liz McIntyre told me, honestly, I didn't consider it to be too much of a concern. So I was not really compelled and wanted to research it because I didn't see the concern in it.

Katina Michael: Okay.

Gary Retherford: And here is the reason why: the reason why is that on the one hand, people can say that they're concerned about what's wrong with a particular technology but the alternative is of equal risk. For example, on the issue of cards those are very easily stolen. I just read the other day that a stolen credit card with pin is only worth 64c on the black market because there's so many of them being sold every day. I'll give you a personal example. I went to a restaurant about a year ago and used my credit card. At the time I didn't think anything of it, but it was rather odd that the person that had my card seemed to have run it through some other type of machine just real quickly and I guess I wasn't thinking and I didn't realise what they were doing exactly. I got a phone call from my bank about two or three days later that my credit card was being used at a Wal-Mart store in Los Angeles and what I found out was that the credit cards are regularly being re-swiped by a little box about the size of a matchbox and then they're transferred. The information/the data is transferred into Mexico where it is converted back into another card and then brought back across the border and used in places along the Southern US/

Mexican border. The point of this is that I think that it's our problem with RFID in terms of some type of potential hacking. The whole thing being equal relative to whatever else is on the market, not to say that that's a good thing, but it's the other areas inside the realm of security that are perhaps equally or even more unsafe but you have to put it all into perspective – what's relative one to the other.

Katina Michael: Yes. What do think about biometrics?

Gary Retherford: I think biometrics are good. I have talked to and did a little bit of research in that area only to come away with that it's not 100% – that there are some possibilities always that they're not giving you the same read. I've had this conversation with people here just even recently within the last six months that have indicated that they wish that there was a better consistency to either iris scans or generally biometrics. What I foresee happening in the future – and I've had this conversation with some of the technical people, is that they see a world where you're combining two things. In other words, it may not be always the case where it's implantable chip versus biometrics or biometrics versus something else, but it would be biometrics along with an implantable chip so that maybe in some certain secure areas you would have either fingerprint, hand, eye – which is an identifier with no data along with an implantable type microchip which *is* an identifier which could include data and that you would need both of them in order to gain access either to a computer or to a room or something of that nature.

Katina Michael: So similar to what some state departments of social security in the US have right now, though instead of a card and biometrics, it is an implant and biometrics.

Gary Retherford: So I don't see a world necessarily where you have implantable microchips and nothing else. I see it as a world where there is a combination of things.

Katina Michael: In fact, the technical term in the innovation literature is selection environment to describe all those different product devices that might be out on the market and that's exactly the outcome of my PhD thesis – it's basically that we'll have several technologies being used for different aspects of whatever it is that we're looking at and there's a best fit for a particular application, whether it's transport, telecommunications, e-health or whatever it is – that depicts the design … and instead of migration of technology we are talking more integration, convergence and co-existence.

Gary Retherford: I want to go back to one thing that you said too about the hacking of the RFID or if RFID can be read. There's two parts to this. One is: my proposal from what … well, from my perspective the way that I'm approaching it is that I see ultimately sooner probably as opposed to later, the use of the implantable chip as being the primary– at least a positive way to go. The technology – the "scientists" if you would – they can always come up with a way to improve something. So, if it's put out there and we say: "[o]kay, we're going to do this" whatever *this* is, and there is enough of a reason to develop the technology then it will go ahead. Just like the example I gave of the highly encrypted Malaysian national ID card. They developed a chip that is still highly regarded as the gold standard. Well, that's because there was a need for it. So, my purpose is to push the envelope to see it implemented and then when they say "well"… I will respond "Okay well, here are the problem(s), go ahead and fix the problems". I'm not going to say "[w]ait to fix the problems before you start implementing."

I think the biggest trade off between cards and chips is identity theft. And people are not who they say they are and as long as I can get your purse away from you or your wallet away from you, I'm you – and that's easy to do. That's not hard. So for say a low-end thief who probably is not going to know how to steal an RFID because of lack of technology know-how you are limiting theft. Certainly when you walk away from your car and if you leave your door unlocked, now somebody else is now you – and that's what I see as being a bigger problem. So in the world of the trade-off between the problems – what's the lesser of the two evils – what's the easiest problem to solve first, the problem to solve is going to be making sure that you're you. That's the biggest problem with cards right there. I see it all the time. It's typical in a corporation for people to hand other people a card and say "Here just go ahead and use my card to get in" and it's not uncommon for people to swap cards, or for people to just easily take a card or grab a wallet. That to me is where the real issue is. While people are focused in one direction of saying "Well here is the problem with RFID" well, again, like I say it was just announced the other day – it is so easy to steal a card and a pin now it's only worth sixty four cents on the black market. There's just so many of them. So we're going into a world where I think that you're going to have actual people saying "Look, I need my identity back and I need to have some way to secure identity". So how is that? And someone says "Instead of having a card, here's an implantable chip" and that can secure my identity, because if it was a chip there's no way I could have been taken by a two bit thief. There's no way that I would have ended up buying something in a Wal-Mart store in Southern California. Now to a professional maybe, they could have found a way to read these, but I think technology will quickly overcome that problem.

Katina Michael: And what of the fundamental problem that you could just zap to oblivion any chip just by walking past it with a particular device, say a microwave device at a higher frequency. I mean we already have at trade fairs for example, people with Cochlear implants or pacemakers are warned to stay away from particular displays because of the frequency and the issues with particular emissions. You know, there was a demonstration made in the US by somebody who was remotely able to detonate an explosive using an RFID device and just doing nasty things to it, so it was able to detonate a nearby bomb for example. We noted this example in an *IBM* paper we wrote. We've also sourced an article from the *IEEE Security and Privacy Magazine* from 2007 when someone demonstrated that via holding a small on-board carry bag that e-passports could be both skimmed and killed if they were not covered by a basic Faraday Cage. Now I totally understand what you're saying and the pluses and benefits of an implantable chip as opposed to one that is outside the body, but what of the other problems you would be unleashing by such a system? For one, people could have a field day just "killing tags" rendering them unusable. The second thing is that they're downloading viruses onto the tag because there is a wireless capability to do that depending on the type of tag and it doesn't necessarily have to do with encryption at that point. Some aspect of the tag may be attacked. The third thing is to do with the more mundane issue of brand. You know, we all understand brand in the credit card world even though EMV Europay MasterCard, Visa have got this global agreement, but what if tomorrow we've got issues with branding of the device and who to subscribe to. I tried to get to this question previously but I don't think I was very clear this morning when I was saying: "Who's going to be the owner of that tag? Do you perceive that it will be a Verichip?" I mean, I don't. I can see a big loyalty thing happening whereby we do have potentially several chips in our bodies even though it is not required. Others will be just dysfunctional and when I asked you about the removal – you were told that it could be removed ... but I know that very well from the people that I've spoken to: Kevin Warwick the Professor of Cybernetics at the University of Reading, Amal Graafstra and even the American Medical Veterinary Association (AMVA) that it is not a straightforward procedure depending on the implantation zone and whether or not you have the right tools for removal. When the recent who-haa happened about the implants and cancer which was linked back to a Verichip trial, something came out in the Associated Press. Lewan reported about issues relating to tumours… But there was a reaction by the AMVA with bold writing on their official website stating "Do not try to remove the implant from your dog or cat. You'll kill it" basically. "Don't try to do it yourself if you think it's going to cause it cancer". So it said "Stop press," basically "We're going to investigate this further" and the next thing they said – so they didn't actually just say forget about it, it's not an issue, we don't have to investigate – they said we're investigating it. And the second thing they said is "Don't try to remove the implant because you'll destroy the subject" – and that was the word they used: destroy the animal. So, it's not as easy as you think Gary to remove one of these things. The body accepts it like it is part of everything in there and it has to be a surgical removal. I think it's very easy to implant. It's very hard to remove. But brand to me is going to be a massive thing. We can switch now. We can go from one supplier to the next. I don't know how that would be organised when it comes to storage of information or unique IDs or who's going to be the major player there. That's where my questions were. I've got the state of play on the value chain. It hasn't even really been figured out very much. You know, who are the stakeholders in this space? We know integration is a big thing, we know that the supply of the transponder is a big thing, but who else is running the show? That's what I'm trying to grapple with at the moment.

Gary Retherford: Well, one thing that I see that kind of plays to that question is in the area of electronic medical records because in my Six Sigma world, one of the things that I have been doing is working with some other Six Sigma guys who are interested in, as I have interest in, the medical field because Six Sigma is really coming into the medical field and particularly long-term care, group practices, things of that nature – hospitals certainly. It's probably leading the way. One of the guys that I network with is a former medical doctor in the Army. He was an army doctor and he's also a black belt in Six Sigma and between him – he's been involved in electronic medical records implementations and a couple of other guys have more experience than I have. What I've learned is that: that's one of the things that they're grappling with inside electronic medical records industry is standardisation. Because there's a plethora of players in the market and what they're working towards is some type of standardisation so that not necessarily just because of say the type of device, whether it's a card, tag, whatever the case may be, however they identify a particular individual, but also in terms of the format, the different type of coating that is used to make one type of system integrate with other systems so for example multiple hospitals, multiple doctors offices with multiple hospitals. How can it all be integrated? There's a company here in Cincinnati called Health Bridge which as I understand it and I don't know that much about them, but I understand that that's what they do: they take the integration between multiple medical facilities and make it so that if you go into one hospital that hospital can access the database of you from other locations who are all part of that Health Bridge system. But what this was leading to was: What would be …. there has to be some kind of standardised system developed so that this communication can take place. Now what I see happening is that through organisations, and I don't remember who the organisations were, but through various organisations that would come together and there would be a governing international body.

Katina Michael: Like the ISO …

Gary Retherford: … say the international standard and eventually it all works its way through an international standard body that comes up and says "[t]his is the standard that now we're going use". I think that we're still in the early stages of that obviously, but it appears as though there is a process albeit, maybe it doesn't look real pretty, but there is a process in place that eventually works all that out. So what I see as being multiple manufacturers – there could be. But multiple manufacturers working on one standard that says "Okay, its either going to be a 16 digit number 15 digit number or it's going to hold this information, it's going to hold that information, and these are the formats that have to be used if it's integrated into some type of networked IT system." But I think that all that would be worked out – I just don't know who all those players are, but I've heard their names mentioned.

Katina Michael: So back to that very notion linking what we were talking about before about ID theft. What if IDs were leaked for example? Or what if devices were tampered with externally and viruses downloaded onto these devices rendering them unusable? I know you're saying that lets deploy and then let's fix, but I think that's one of the big problems with e-passports at the moment – we've deployed very quickly, extremely rapidly and nobody has thought about it and now we have people actually skimming left right and centre all over the world – it's been proven: your passport ID number, your name, your date of birth, your place of birth, your citizen and nationality and we haven't seen yet how this information has been used or applied, but I know that in many countries e-passports or passports themselves create points to get certain certificates that are government organised. How do you see that? If I was to tell you that they could be zapped and rendered unusable, what does it that say to your business case then? How do you interpret that?

Gary Retherford: How do I interpret that they can be rendered unusable?

Katina Michael: Yes, remotely by anyone who's got access to a smart phone … you know what the script kiddies are doing? I know of them because I suspect I have taught many of them dabbling in the hacker realm. These are kids just out of high school who do not know much about mainstream Information Technology but think it's cool to go around hacking sites. They download a script, they don't know what it does exactly but they know it's fun because they watch their peers accept notifications on their smart phones or PDAs using Bluetooth for example and then they just create havoc. It's like victims think "Oh my god, my PDA is just looping non-stop, starting up and I can't control it" and people in panic while the hackers are just getting maximum laughs from observing their victims. Can you see that this is highly possible in that implant environment? Everyone's carrying one so therefore everybody's got a chance at doing harm to others if they so choose to. And we could institute laws to prevent that from happening, but then go and prove it basically – it's over a wireless infrastructure so your guess is as good as mine as to where the attack came from. I've seen how easy it is to render someone's PDA dysfunctional and to me it's no different whether someone is wearing it or carrying it or has it embedded. But I do take your point that … in the perceived sense, that if it's embedded, people perceive they can't get to it as easily … but there are other pitfalls …

Gary Retherford: Well, here's kind of how I see this. And I'm not … I don't want to sound like … it appears to be over cavalier about the fact that there are not problems that could and probably do exist, however, I think that there's a trade-off between how far you go. Vis a vis– take a look at the Internet. I mean, I think in some respects there could be the argument made and again I am not trying to sound too flippant about this, but had somebody mentioned to Bill Gates in 1980 "Let's not work on creating machines that can talk to one other across something called the World Wide Web because we haven't foolproofed the method at which those machines cannot be hacked by another source". Now the problems we have today with viruses and people that are deliberately trying to steal identities over the Internet, I mean it's unbelievable, but it did not stop the intent to move forward and I don't know that … you know, I don't want to put myself in Bill Gates' head … but I don't know that the people that were in the early stages of computing – Apple Computer and Microsoft – were they going to stop doing their progress – if you call it progress – of moving forward with those technologies because someone came along and said "Oh, you better not do this because somebody could end up causing you a problem on your computer?" Well, no, they moved forward and I think that's the trade-off here – is that what's the alternative. I see the alternative being – because it's always the case of one of alternative versus the other – and the alternative I see right now is because I think that there are more and more people getting fed up with identity theft and if somebody comes along and says "Well, here's an alternative now you have this risk", but now what's my risk relative to what? Is my risk relative to: somebody could potentially end up causing this to be useless – which is a risk – but people might look at that and say "Okay, but what is the probabilities of it being useless versus what is the situation as it pertains to my other credentials becoming lost or stolen or whatever the case may be?" So what I see is that we live in a society … it's human nature – and this is why I've always said that when people say, "Well the chip will never happen" and I say "Yeah it will and the reason it will is because human nature says that it has to" …

Katina Michael: I agree with you regarding the human nature aspect- I guess the famous adage "you can't stop science" is also equally true.

Gary Retherford: … human nature demands that we have – not me or somebody – but that it is human nature that we go into this technology and that's because we've been told you can't go there… and as long as there's the possibility – you know, it's like "Why did he climb Mt. Everest? Because it was there". Why do we push the envelopes? I know that in this country, we're pushing the envelopes on everything from same-sex marriage to we're still grappling with the pro-choice abortion issues, we're grappling with all kinds of issues and if you look at this over time, we have to continue to push the envelope. So, will we have same-sex marriage in this country? Absolutely. There's no question about it. That's because it's human nature – we have to. We can't go in reverse, we can't stop where we are, we always have to go to the next level.

Katina Michael: Yes.

Gary Retherford: … and the same thing goes here is that we will go there. So, that while you raise the point – which is a good point to raise that there's something that has to be done – that's not going to be a reason to say don't go forward.

Katina Michael: And for you Gary, going back to the Microsoft example or Apple example – you don't see a distinction between a technology which is outside the body and one that is embedded? For example, the frustration that most global corporations and professionals face today is the so-called Blue Screen of Death … you know the Microsoft operating system when it crashes or doesn't work as it should … and I don't wish to knock Microsoft – it's enabled so many wonderful things, but at the same time, can you imagine patch upon patch upon patch upon patch – actually having to maintain your chip … you know, your system breaking down or not being available or the infrastructure not being up – and we're making one basic premise here and one basic assumption is that electricity is up 24/7 and I don't know about the US, although I can tell you about a few case studies in Los Angeles about power grid failures but definitely in Australia of late we've had some massive problems so one problem is the actual infrastructure and the other is it relies on electricity and it relies on other things – there is this whole value chain analysis that you could conduct on dependability and availability of systems. Now what of systems that the chip might rely, could they be likened to the Microsoft problems? You know, our computer engineers and developers are getting better with each iteration of new software, but we've gone through a lot of teething problems with regards to that. How would you feel if you had a chip inside you that functioned as if the Microsoft operating system did say back in the 1990s?

Gary Retherford: Well okay, I think that … while, not to say that would not be a frustration, I think that we're talking about two different things and here's the reason why. What I mean by this is that I think the psychology of having it implanted is not going to be the hurdle. I think people will become accustomed and when I say people, I'm going to now kind of shift the tide back to the younger generation so people that are today in their 20s say for example versus someone in their 60s– but I think that they will just view that as "Oh my iPod doesn't work or my cell phone doesn't work and my chip doesn't work so therefore I need to go get some maintenance done". In other words, the fact that it's a chip and it's inside you or inside the person, I think is a different issue as opposed to – and when I say different issue, I mean it's not going to be any more of a negative impact on an individual just because it happens to be a chip that's inside them. I think it's going to go with the territory because I think that it still stands that there's a lot of people out there particularly – and I don't … I always get proven wrong every time I say this but I think generally speaking, overall I'm correct – is that people under the age of 30 and under 25 that are part of this whole technology world are more and more acceptable to an implantable microchip and from that group that is here today that is 20 and for the ones that are just being born today, they're going to look at this and they're going to wonder why we ever had this discussion because they're going to say "Well, I have the chip and somebody did knock my frequency out or they did render it a problem so now I need to go in to my doctor or to some place – the clinic or whatever it is – a chipping centre if you would, I need to go in and get my chip reprogrammed or hook me up to a scanner and program me or something like that. I really do see that that's going to be the way they're going to think.

Katina Michael: And you …

Gary Retherford: So I see problems associated with people that are trying to destroy these things, I also see that it's going to become no different from people that are trying to destroy a cell phone and somebody says, "Well I need to go get my cell phone fixed".

Katina Michael: And you don't see however the fact that you … Gary, you have full control over your PDA. You can't touch your implant in the same way. One day we might have these remote applications which you could do self maintenance or run diagnostic tests on the implant device but you're talking to me about going to a third party to fix it during times of outage. I see that as a massive problem regarding control and control over one's body, oneself, one's identity, one's personhood so to speak. Having to go to someone else to fix it is where the problems begin I think and it's not like we don't take our computer elsewhere to get fixed, but this is something within you and so I can always throw out my cell phone if I want. I can always delete the applications on it. I can always opt-out of specific services. I don't know how easy it would be to opt out of an implant application. We are making the assumption Gary that these devices will remain passive, that they will remain just a dumb sort of ID number device, but you did allude to the fact that you can see additional applications will be built in and more storage will be made available on these devices, but let me go further and can say to you: okay, we have … let's say it's 2015. We've got active tags and we've got tags that can be likened to computers within our body and these computers can emit signals to our brain for example and I'm talking now about deep brain stimulation which has been demonstrated to help people with Tourette's syndrome and Parkinson's from shaking unrelentlessly. Can you see that once somebody has, potentially, control over certain aspects – depending on what type of person they are, so for example you are diagnosed as being depressed: "Oh, okay you require this upgrade to your implant. We'll make sure that whenever we feel that you're getting dark or melancholy, we'll emit certain signals so that you can feel better" – and this is perhaps a new generation of drug delivery or stimulants that we haven't considered in the past but is being written about widely at the moment in the area of brain computer interfaces and deep brain stimulation so I'm not talking hocus-pocus now this is happening in a big way.

So, medical is pushing this at the moment and we're talking about obvious applications in security potentially banking, potentially e-health, potentially a lot of other vertical sectors, but it's going to happen the other way around in the sense that the biomedical fraternity are already doing cochlear implants, brain pacemakers, heart pacemakers – they have been since the 1950s and 1960s in the heart pacemaker area but in the cochlear implant area sort of late 1990s. Can you see that these things, these future implants are not just going to be passive? They're not going to be just linking your ID to your database, it's going to be a much more powerful machinery within and that's where, for me, I draw the line. When somebody else has control over drug delivery for example or monitoring certain perhaps emotions or corrective kinds of things.

I must be clear here, that I am not talking about implants for prosthesis- I am a strong supporter of corrective technologies to help the blind see, the deaf hear or otherwise. But I'm talking about those other sorts of grey areas. You're talking about a passive device, I think most of the time, which are simple things, but what if the move and as you say you can't stop science, but what if the progress was to go more than just a passive implant, it was going to be: hold that computer within the body … as a central CPU controlling things in the body and amplifying things with a greater sensation of touch, smell, sight and hearing. What if you could amplify or enhance humans. Then what would you think?

Gary Retherford: Well, I think that the stage is already set for it. Let me kind of throw another part in that I can tell you why I believe that it will end up there. We have, for example … in… I don't know … do you guys in Australia, do you have national healthcare?

Katina Michael: Yes. It is called Medicare.

Gary Retherford: Medicare. Okay, so you don't have it at all ages … It's only …

Katina Michael: No, everyone …

Gary Retherford: … like our Medicare. Medicare here's only up to 62.

Katina Michael: No. No, we have a blanket coverage safety-net Medicare for everyone from birth and then we have additional private health insurance.

Gary Retherford: Okay. Well, here I know, of course, you know we're now fighting over and moving into this national universal healthcare thing and one of the things that came out of that the one side has kind of thrown up as being a potential problem is that it implies in some of the language that the government makes the final decision on what types of care you can get. In other words, an actual decision on a particular cure if you would, to a problem – which may or may not be the case, but my point is that what we're seeing of it – and you have a lot more history of it there – but what we're seeing of it is that you already have an interest industry that for the most part have the major control over which drugs you get and how certain things are handled in terms of your care. Now, if you take for example the drug issue and it's being dispensed somehow through a microchip that's controlling you, if you step back and look at that just for a second, to some degree you already have that issue, it's just that is not internal. So if you need a particular drug to handle depression, you can't get the drug or there are stipulations that you can't do things unless you take the drug. So the fact that it's either internal through an implant or external I think at some point becomes moot because someone else still has the control and now you go back to what's the trade-off in terms of cost. So when the argument gets placed at: well, it's more cost effective to do it through an implant than it is for a person to go through these other steps, other processes in order to get the drug say, for example if they do an analysis and they come away and it says: well it cost four times more in taxpayer dollars … or just anybody's dollars to get this drug into your system and maintain you as opposed to doing it through an implantable chip, well now you've got the push on … towards using it as the chip.

Now, you asked me what do I feel about all of that. Well, I'm one of these people that I do not make necessarily a decision on what I like or what I don't like. I make the decision or I look at things based upon where I see it going and understand and now you have to work with these parameters. So, I'm totally a proponent that we should move forward with the chip given even the comments that you made about "well what if this, what if this and what if this", because ultimately that's where the argument is going to end up anyway – not should we do this, but how do we implement it. I think we're already at the stage of figuring out: okay how do we … what laws are going to be passed for example, to how far a medical facility or a decision can be made on somebody with a chip – not should we use the chip or not use the chip – I think that discussion has already been done. I think that discussion is over with. Because, the fact that it is within the sights of the technological world to achieve it, you no longer make the argument on should it or shouldn't be, now you go into the realm of: okay, how do we discuss and maybe legislate on the trade-offs between how much control and how much not control, because we're only there for all practical purposes we're there now even if you don't look at it is an implantable chip. We're there…

Katina Michael: That is certainly one school of thought.

Gary Retherford: So anyway, I guess that's kind of where I'm coming from. I'm not viewing the question that you proposed to me in terms of: if we believe this argument then we should shut this down? No. It's an argument that says we're already there really, even though it may not be. I guess in some respects you could apply it to almost anything. I'm kind of past the argument of saying "Well, this is implanted inside you". I think that's a moot point. I think there is enough acceptance in the general population and there are a lot of people that are just waiting for it. In almost all the people I've talked to … and I've talked to literally probably hundreds – between 2005 and 2007 when I was really doing a lot of the chipping, when I was just bringing it up into groups where I wasn't even going to be looking at these people as being potential users or receivers of the implant. Every single time, I mean, it was literally 99% of the time there was agreement. Even when people said "Well, I do like it, I don't like it", they all said: "Oh but I know it's coming". So, society has accepted it. I mean, they've basically said flat-out: "I know its coming". So, its like you're waiting for the train – you know its coming. So it was really interesting that they just resided to the whole idea, and that I probably am not going to have a choice and so therefore I have mentally accepted the whole notion that at some point in my lifetime, you know, relative I guess to how old one is and how long one thinks they're going to live, but relatively at some point in one's lifetime they'll be receiving an implantable microchip for a variety of potential reasons, typically medical or security. So in some respects I think that we become a society that becomes more tolerant of government intervention or of corporate intervention into our lives and the notion that we have people out there that are fighting this – well, yes that's true, then that's when I go back to: it's a vocal minority.

Katina Michael: So what will drive this implantable regime?

Gary Retherford: I think that it will be driven by commerce. I think it's going to happen in small incremental ways, in efforts that are first imposed on smaller groups. i.e., for example: Verichip had started to do this thing in Florida back in 2007 where they were going to implant microchips in an Alzheimer's facility in Florida. The way I think I see it happening is it's going to be an Alzheimer's facility here or a nursing home there, or it's going to be some prisoners here or it's going to be … you know a corporation that does it with an access control system – but they have to get those readers working better. I mean, I can tell you right now if they would have given me a reader that worked …

Katina Michael: A mobile reader?

Gary Retherford: No. A reader for an access control door.

Katina Michael: Okay a fixed reader.

Gary Retherford: If they would have given me what I had asked for- you know how many companies I had already lined up to begin putting readers in? Several!

Katina Michael: Right.

Gary Retherford: But it wasn't the problem that I had people telling me: " no." My problem was that I didn't have the right device. I had a company that was ready to go with the implantables, we had picked out the door, we started to go to the employees to talk about getting chipped, but we could never get the reader to work right consistently and at the end of the day I don't want to chip for the sake of chipping – we had to have an actual functioning system. But we had already drilled holes and everything else. The guy had even run the wires. We were already there. I had no problem. In fact, I had another company even after I had given up – I was just in conversation with another company who said they wanted to do it. So my point is there's no problem with having people doing it. I mean, that wasn't the hard part at all. Now, inside any given company that I had this conversation with, of course, I did not have 100% of the people say that they wanted to have a chip, but I did have enough people who said they wanted to have a chip as opposed to carrying a card and there was not a problem with people saying that "I don't want to do it". In fact, sometimes they were saying they wanted to do it and there was really no logical reason why. They just said, "Well, oh, I just want to get chipped". So, I guess I don't see it coming as a massive wave because I think that would cause a revolt. But even with that I think it would be short lived revolt because I still think that it goes back to the comments that I had made to me from people that said "It's coming. I know it's coming. It's just a matter of you know, time". So I could be wrong, it could be that there's going to be a massive rollout of it – what I could also see happening is that in my world – because I'm still looking for opportunities where it can be done, the only thing that stops me right now is Verichip because they're not selling any more chips. They've basically shut down their human chip business for access control ... I mean even if I had a guy today that said "I want a chip", I couldn't sell them one because Verichip is not letting them out ...

Katina Michael: But what about free will? I'm challenged by the very notion of human rights and the question that I'm getting at is: Do you think we will need to reassess what human rights means to people? You know, we have this traditional view of what human rights mean and you look back at conventions that were formed post-World War II, but do you think they are irrelevant now? Do think we are entering a new stage where the very notion of human rights that we conceived of, back in, you know, forty, fifty years ago – is it completely different today? Do we need to reassess what human rights means today?

Gary Retherford: Well, yes. I think that that's a valid argument or a valid statement. One of the things that I keep looking at is that there's a natural tendency to make the statement that you're making, you know, if you're in this world view that you're in where you see these types of things, but I also take it a little bit further in the sense that what I see it is a continuum of things that are not necessarily based upon a point of time, but it's a continually evolving thing. I guess the reason I point to that is that-we all become a product of our environment and we all become accustomed to what we see and what we're used to and I go back to various age groups and where they're at within our society. So, if you take a person who is 40, for example, and they have a particular notion in their heads, they can sit down and articulate maybe to some degree, what they see as human rights, okay. Now you take a 20-year-old and you have them sit down and do the same exercise and they won't look anything alike. And where I'm going with that is that: while I agree with your statement is that do they need to be rewritten, it's almost like they need to be rewritten every year. Because we're constantly, constantly moving quicker now ... faster... but I think we're already behind having rewritten these rules probably twenty times. In other words, technology is moving us – not just technology. But a similar example is my daughter you know, who watches … she's 17 years old and she's been watching TV like everybody does, they watch the TV sometimes, and she's been watching it for the last five years and one of the shows that we watch you know just occasionally is the show *Friends*, okay.

Katina Michael: Yes.

Gary Retherford: Well, we did not realise … I'm a product of the 70s and I remember what was considered the social acceptance at that time and now we look on TV and what is the social acceptance of today. So my daughter is making statements one day about what is socially acceptable and my wife and I are, you know, our mouths drop. That's because we have been subliminally impacted by a change in mores and over the last thirty or forty years because of what is in our environment. So what I see is that while what you say is true, it's almost moving at such a fast pace now that I'm not so sure that a 20-year-old would even think that whatever you come up with or what I come up with as these new rules, they would still look at it as saying "I don't even see where those apply." I don't think there's anymore a twenty year or say forty year age difference between a 60-year-old and a 20-year-old, I think that between 60-year-olds and 20-year-olds today, the age difference is about two hundred years. So I guess it was kind of a long answer to a short question, but I think it's just to the point where we're just not catching up to the changes that we are going to need to be able to make to address what you just talked about.

Katina Michael: And do you see any risks associated with the pace of change?

Gary Retherford: Do I what?

Katina Michael: Do you see any risks associated with the pace of change we are living in at the moment?

Gary Retherford: Well see, there again, I think it's all relative. It's relative to who you're asking that question to. I don't think … I mean I can easily answer "No, there's no risk at all", but I could also easily say that there's tremendous risk. Why? Because I can tell you that I think relative to what I know of my past of living in the 1970s, I think that where we're going is horrible, okay. But I can also tell you if somebody who was born in the 1990s, they would look at me and say – because they can't conceptualise what I did in the 1970s. It's unfathomable for them to ever think that I was watching a television with only three channels and I did not have a cell phone and I did not have a computer and, you know, so the differences are so vast that I think it's … it's … the question that you pose is a legitimate question, but it can be answered multiple ways. So I don't get … I guess if I had to pick an answer I'd say "No, I don't think there are risks because it's a moot point to say that there are."

Katina Michael: Yes. Meaning, so what? There are risks in everything?

Gary Retherford: Right. Yes well, you deal with the new risk. So all you have to do – there's new legislation, new law, they battle it out between various you know lobby groups that have their opinions, but ultimately we continue moving forward in the same direction. It's just like in this country, we have the problem where we talk about the Democrats and the Republicans and we talk about the Liberals versus the Conservatives, so everybody says "Well, you know a politician that wants to get elected moves to the centre". Well, what they now realise and have probably realised is that the centre keeps moving to the left. So therefore it's like the Republican Party is struggling today in this country because many of them want to maintain their original ideology and their core values. Well those core values that they're choosing are core values from the Reagan years. Well, those core values are no longer applicable. So the core values are changing. So now in order to be central and to be immoderate, you know, I mean a sort of conservative today is 20 years ago the moderate. In other words, it keeps moving to the left – the centre keeps moving.

Katina Michael: So is it the same for you in the case of technology innovation?

Gary Retherford: Yes. So, I think that's the same thing as we're talking about in this technology, I think it's a moot point. I don't think it's even an argument anymore. I think the arguments can be thrown out there, but at the end of the day it's going to be found to be … you know, the new argument is: okay, it's going to happen so now how do we do it. Let's address the issues that have to be based on "fact" (quote, unquote), that it's going to happen anyway. I mean, the things you're telling me are really hard for me to comprehend. I mean, I comprehend them because I know they're going on – my involvement with GIS and the tracking and data mining and all that – I understand all that, but I probably understand a lot better than most people because the world I'm in is just like yours. But, I've kind of accepted the fact for my own purposes that you know, I'm no longer going to make the argument that it's a bad idea because it's not relevant to me to make that argument.

Katina Michael: Yes. I understand. I understand. You know how you mentioned to me and I find this intriguing, you said to me just a second ago, "I know where we're going. To me, the 1970s me, sort of thing is horrible" and I want to stress that point because many people that I've spoken to have said that sort of … it's an anomaly really, it's an oxymoron. We're going to a place that we know … there's a trajectory that we're on … we're riding the trajectory at the moment – it's like riding a wave, you know you're going to get to the shore. But you can see it's horrible and I'm assuming that you're talking about social structures perhaps … what may happen to social structure… I'm not sure what you were referring to as potentially horrible, but can I say to you that's exactly what we're thinking. We know it's going to happen. So we're on this trajectory, it's going to happen, okay we know … again, the risks, who cares – that's not relevant anymore, to many people, I agree with you. But what's going to be the aftermath? What's going to be the result? What will it mean for social structures, for the way people interrelate with one another and depend on one another, to families – if family is going to be even a concept we can understand in the future, I'm not sure. And I'm not saying it's the correct thing or the right way … I'm going to a conference in Spain next week which is actually about nanotechnology. We're talking about devices that are the size of one hundredth of a hair … things that are so small they can be ingested. You won't even know that you've ingested it or you're wearing it or you could be covered in cloth that is made out of this polymer or you know like GPS devices under the skin created using special polymers. What do you think these technologies will do to society? Or is that an irrelevant question too?

Gary Retherford: Well, I think that there's other things that are … when you talk about society, there's so many impacts on that. From what I was describing as the thing with watching on television to now, you know, this is a strange one. Now they've got to the point where: okay, we're battling in this country with the question of same-sex marriage. Well, now the next thing to come on the scene which I just heard about two weeks ago is called the triad marriage, okay. Have you heard of that term?

Katina Michael: Yes. I've seen documentaries that have alluded to that notion. Similarly, what about the sex bots being developed in Japan which have been around for two or three years now – you know the machine mistresses. The notion that it's not going to even be human-to-human relationship any longer. Some of the Japanese feel very comfortable with sex machines or dolls and it is sort of acceptable in certain cultures. I'm referring here to extramarital affairs with machines which can be likened to human females. You know, it's not even human-to-human contact anymore. It's human-to-nonhuman, but then what's going to be classified as human in the future is yet to be defined properly.

Gary Retherford: Yes, so I guess in that respect, what I think when you say you know what's going to have the impact on society? Well if you probably look at it – I don't know – but if you look at it fifty years from now or one hundred years from now and somebody can look back – the implantable microchip might end up being the least of the bizarre …

Katina Michael: Yes, you are probably right.

Gary Retherford: When people look at it they might think, well you know, I mean what you just described in Japan to me is pretty bizarre. So you know, where on the scale of bizarre-ness does the implantable microchip fit? I think the implantable microchip is going to have its impact on the human … on society as it relates to what you refer to as control or as control and privacy issues, but then it becomes relative to security and that's where I go back to the whole idea – the first … one of the questions that I brought up earlier which is: you know, does my security supersede my neighbours right to privacy? And I think that always becomes the compelling question.

When I work with people that are in the security realm, Homeland Security people that I know, police chiefs for instance- their whole attitude is like "It's my job to protect you regardless of what it costs you". That's almost their philosophy. So I think that their intent is there, but it's almost as though the attitude comes through that- "I will protect you no matter how much of your privacy and your liberties I destroy, it's my job to make sure that there's no harm that comes to you". Well, I mean I've had that conversation – not in those exact words, but that's the attitude that you get. But I think the human nature attitude is that no-one wants to have it happen on their watch. Did George W. Bush go too far? You know, now we've got that question on water boarding. So, you know, well, we were protected, but to what cost to our social structure? I think 9/11 was really the big kicker although it's been coming along for many years, but 9/11 certainly had a big impact on pushing the envelope further and further.

PART THREE

Katina Michael: Gary. I'm wondering what your main motivation is for entering the chip industry.

Gary Retherford: I guess there's two ways to answer this. First, I see an alternative to the system that we have right now that is not very secure and I think that coming at it primarily from the security slant but also from the medical slant as well, I think it makes too much sense to not go down the RFID implant path. I mean, it makes too much sense to not be into chipping. In other words, I think it's what's really going to propel us into a world of being more secure relative to the people around us. In other words, our individual security is going to depend upon this particular technology, not that it doesn't have it's faults at this point, but I still go back to that I think it needs a proponent and I'm a proponent of it and so I believe that it's not only going to be an opportunity in the future but I also think that it's needed and I think it's something that we need to continue to push for as a global society.

Katina Michael: And the second part, you said that there were two motivations there …

Gary Retherford: Well, the other … I guess the other one is that I to some degree get a little frustrated that there's too much negativity for what I think is irrational paranoia or whatever the case may be and that I think that the arguments that some people might make – because I, quite frankly, I don't think it's as big of a deal that some people are making it out to be. I think there is …there's certainly those people who either … whether it's the conspiracy theory thing or if it's, you know, the Big Brother thing or whatever the case may be, I think it's a minority – vocal minority, but I think by and large, I think society is ready for it … I'm one of these people that if I know something is coming, I don't try to not embrace it, I go ahead and embrace it and look for the positive sides of it and on this it was pretty obvious to me that there was a positive side. So I guess that's the second half of it. It's almost like I'm doing it also because I feel like there needs to be this side of us who believe – I don't know who else is out there, but I'm on the side that says we should go forward with this and not be trying to … not be trying to stop it.

Katina Michael: Okay on that point, I have a question which is about how we can obtain or measure community opinions about microchip implants? For example, you've mentioned to me you know the sort of informal surveys that you've conducted with the community stakeholders you've been in contact with. I have a very different view in Australia from the Australian individuals I've had contact with. In fact, we ran some classes, focus groups earlier this session and we asked questions about implants and location-based services and it was the vast minority who said they would adopt for convenience mostly – that was their reason toward adoption, not really medical etc. But what kind of instrument can we use to gauge community opinions in various states and jurisdictions?

Gary Retherford: One of the things of course was my informal stuff, but I did engage one person at [UNIVERSITY NAME WITHELD] who continued to do some informal and again call it informal because it did not follow any properly set out focus group policy or procedures but he did go around and did some of his focus groups to various groups on the campus back a few years ago and he reported back to me that overwhelmingly the vast majority of the you know, 18 to 22 year age group that was at the campus that he interacted with (and this was a guy that was in marketing so he was accustomed to concepts of focus groups), and he came back to me and reported that the vast majority of the people that he talked to in that age group saw absolutely nothing against it, nothing wrong with it, and if given the opportunity, they would do it. So I think … I don't know whether I want to necessarily go around and say that it is an age issue, but I think when you start to look at any type of group that is out there and you're right, the question is how you get them … you know, who do you get engaged into these conversations? But you have to be careful that the only ones that show up are not the ones that already have a predisposition against the chip.

Katina Michael: Yes.

Gary Retherford: ... and that somehow you're getting a cross-section of society. I think the only thing you do, maybe it's some type of random sampling, phone calls, something like that, but here again you have to look at what age groups you're looking at – perhaps age groups and maybe other demographics, to try to keep it fair and a quality study. I believe there was a study done and here again I don't know who did this study, but there was a study done a few years ago as I understand, where the number was up closer to over 50% of the population. I believe this was around 2007 and over 50% of the population at that time would accept the chip. I think the issues regarding the chip might be less in terms of the religious side or the Big Brother side and are probably just more connected to what do I get out of this in return? So, if you're saying "Well, you know, it's improving of your healthcare and here's the reason why or it's improving of you personal identity and your personal security and here's the reason why". I think people can comprehend and make their own judgement based upon the risk return trails to say, "Well, relative to what you're giving me as a benefit, then I would say 'yes'". Those are the discussions that it should be about. For people to come in and start engaging in the discussion and say "Well, it's Big Brother", well – or whatever your other paranoias are – then you have to start looking at: 'Okay, well then why do you have a cell phone? Or why do you use the Internet? Or why are you all these other things, because those same things could have been argued 25 years ago.'" So now you know, you have to consider the arguments of the people that are against it and really come down to a rational reason why people would argue either for or against either way and I don't consider paranoia or just someone of a radical viewpoint to be a rational reason to either include not include something.

Katina Michael: So just to clarify there, because we've asked this question of all our stakeholders on the commercial side and academic side, so for you for instance religious conjectures are radical or irrational thoughts or ideas regarding RFID implants? Would I be correct in assuming that these notions or these views you don't pay much attention to because they are not important from your innovation perspective?

Gary Retherford: No, because I think they're a vocal minority. I think they're people that are going to vent over something. It doesn't matter what it is, there's going to be something that they want ... if it wasn't the chip, they would find some other thing to bring up their frustration with ... the chip just gives them a convenient opportunity to express their opinion on something.

Katina Michael: Yes.

Gary Retherford: So, if you didn't have the chip ... probably in the 1930s it was the Social Security Number in the United States. You know and then later on it was going to be something else. There's always a group that's always going to find a reason to vocalise their resistance against any kind of new technology or what they consider to be something that's going to be an infringement. But they will always find that. So right now, the chip ... if you can go to some websites and they're going to find just about anything they can to have their arguments against, so what I do is, I mean for me personally, I have to look at these groups ... and essentially for my sense of purpose I ignore them. After the Citywatcher project back in 2006, Verichip had offered me, and for a while I was the person who was supposed to be the national rat for the access control product in the United States, I wasn't the only one who was offered it, there were others ... A couple of the other people that I know that got hung up on the idea that they didn't want to really address going out and doing the implantable microchip in an access control environment until they had addressed the issues of say in some cases, the religious groups and my position was "No, you ignore them", because, you know what: you're wasting your time in that debate because that's what they want to do, they want a debate and that's not my position. My position is to move forward and not pay attention to that particular fringe.

Katina Michael: And Gary on that thought process, how do you view critically the US anti-chipping laws, do you ignore them as well? Is that part of the debate in the sense that …

Gary Retherford: Well, it is interesting because I actually was not even aware of the anti-chipping laws when we did it and what I find out was that back in Ohio where we did this, the anti-chipping law that they have here had not passed. It was not actually even put on the table. There was not a committee on it until after we did what we did with Citywatcher project …

Katina Michael: Yes that's right.

Gary Retherford: And I believe looking at the dates that Wisconsin – I'm not sure about Wisconsin – California, Ohio, and I believe it was one of the Dakotas …

Katina Michael: North.

Gary Retherford: I believe this took place after we did what we did. So certainly what we did … you know must have caused some rumblings in some state houses, but like Ohio it did not pass ... They did not have a law at the time as I understand it… But there again you have a minority of states, I believe at the last count there might have been six that had something that they were looking at in this regard. I'm not saying that there aren't some laws that are not needed. I don't want to say that there are not some conditions in which some rules perhaps, some legislation should apply. But the anti-chipping law the way I saw it was not against the use of the chip, it was just against being an enforced implantation …

Katina Michael: That's correct.

Gary Retherford: I don't see chip implants ever being forced. What I see happening is it's going to be a situation where there are people that are going to take it but they will be given an option. I think that the options that people look at might be "Well, given the possibilities I think I would rather have it than not have it", but we already live in that today. I found it interesting enough that it is not required by law that you have to have a Social Security number in the United States.

Katina Michael: That's right.

Gary Retherford: But that it is an option. Most people don't know that anymore – that you can go through life without a Social Security number but it certainly makes it difficult, because if you want to do anything else you've got to have one. So the chipping law may always be the case where it may be an option, but the question is: what do you give up by not using it?

Katina Michael: Gary, could I ask you why you got in touch with us? What was the intent behind this communication? I know it was after the article that appeared in *PerAda Magazine* but what did you hope to achieve from calling us and getting in touch with us?

Gary Retherford: Well, what I'm doing – here's a little bit of background now – when I had done the Citywatcher thing and I had said earlier that I was basically wanting to see what the world was made of, there was also a part of me that felt that there was nothing wrong with this, in the fact I felt and still do, that society should go forward with this. Part of the reason I also put it out there was because I know that as humans we begin to do things more and more as we get used to it. So there's the initial shock, but once you get over it, whatever "it" is, we tend to absorb it, we bring it into our minds, we're able to wrap arms around it and we continue to move forward. Now sometimes it takes a little bit longer for some than for others, depending on the situation or what it is. Part of the reason I sent that out there at the time was because I wanted to get it talked about, I wanted to get it out there in society, get people thinking about it. One of the things I knew was that as that happened, it would be more likely– and I was confident in this– that it would be actually driving towards an acceptable issue. Because once people got over the shock– and I'll give you an example of this– is that when I first did this, the first one, we had all the TV and the radio and everything else, they were showing up for Citywatcher. Then when we proceeded on over the next couple of years I did a couple more chippings. I finally got to the point where nobody wanted to show up. We had no TV people show up, no radio people, it was no longer a story. The last time I did this, we did it in 2007 and there was a guy who was chipped here in Cincinnati and all the TV stations knew about it, you know we put some word out there, but when nobody was showing up, except one TV station did and then it was not the main story, I said to somebody, I said "Okay, it's here, because once you get to the point where it's not the story then it is now acceptable."

Katina Michael: Yes.

Gary Retherford: … because nobody turned up, it was no longer of any shock value. Now what I did at the time, to get back to your question, what I did at the time I had still stayed for the most part in the background. I have recently changed my position in the sense that I'm no longer anonymous. I'm still a proponent of and will still continue to push for the use of this product, but the difference is now I'm no longer going to be sitting in the background. I've taken a different stance in the sense that I'm coming out, I'm not behind-the-scenes, but rather to say that I am for it openly, and I believe that we should continue to work towards embracing it, deal with the issues that we may have as regards to any of the other concerns that people may have, not be afraid to go into legislation and have State or Federal people, you know, politicians look at, you know, what are the factors here and there. But not to say that we should just kill it outright, because I don't think we should kill it outright, I think it needs to go forward but it's just now I'm willing to come out and when I saw your article I thought well, I would contact you and you know wanted to start a dialogue because I'm not sitting back anymore being quiet about it.

Katina Michael: Gary would you like to be named in this transcription, yes?

Gary Retherford: Sure.

Katina Michael: And we can mention the name of Citywatcher?

Gary Retherford: Yes, you can use the name I guess because they are no longer in operation.

Michael Michael: Katina, can you ask Gary is he aware of Professor Kevin Warwick's research?

Gary Retherford: Was I aware of what research?

Katina Michael: Professor Kevin Warwick at the University of Reading ... the guy who did cyborg 1.0 project ...

Gary Retherford: No I wasn't. Was that something that you mentioned to me earlier?

Katina Michael: Yes.

Gary Retherford: Okay, I think that I have that written down.

Katina Michael: I'll send you his details and also you could find all these interviews in our book that was published in March of this year. There are six or seven full-length interviews, three of which were with implantees and one which was with the director of the Biomedical Institute at Imperial College, but I'll send you a link to that book in case you don't have it yet, but I think you'd be very interested in reading what Kevin Warwick had to say and what Amal Graafstra had to say.

Gary Retherford: Sure.

Michael Michael: Gary, I know Katina asked you before, but she may not have asked it in this way. Do you agree or do you believe that eventually if this happens, there's going to be quite a massive abuse of this system? Or are you more idealistic about its application or its use by the government or by private enterprise? Or are you suspicious that one way or another the system will be abused and to what extent do you think that it will be abused and do you think it's going to level out?

Gary Retherford: I think that it will be ... first of all I feel it's definitely going to end up coming. So it's not like "Yes" in my opinion, it's just a matter of when. I think it will be incrementally over time. I don't think it's all the sudden being tomorrow or one day they wake up and say "Okay right, the world must do this". I think it will be incrementally brought into society. I think that there will be attempts to maybe abuse it by those people who are already trying to attempt to do these things that are either external to the government or you know, say other organisations that we interact with. But I don't think it's going to be any different from what we see today. For example, I just saw on the news today in the United States they reported that from 2005 to 2007, the number of cyber attacks in this country had went from somewhere in the vicinity of 6 or 10,000 or whatever it was in 2005 to 72,000 this year. Well, that's obviously something that we live with as a society. We live with the fact that people are trying to steal our identities through credit cards and our pin numbers– that has increased. So I think we will look at the trade-offs between what we have as the benefit versus the cost. So for an argument to be made, well there will be an abuse of this, well can we say that there's no abuse today? There is. So what's the return? Well the risk return is still that it's better to have than not to have and I think that it's actually going to give us better control over these abuses because one of the abuses, one of the places that we're vulnerable today is that our identity is so easily taken away from us and there's nothing that prevents somebody else to be us okay, because our system is too weak. Right now our system allows the somebody to take my identity go to another part of the country and completely start all over or start using me, but if I'm not... if I was implanted for example, I would have to be there physically for that to be possible. In other words, there is going to be abuse, but it's going to actually, I think, make the abuses much tougher by having this particular product than the abuses we have today. I think we'll look back at this 100 years from now and say wow, you know, they should have been onto that chip thing much earlier than they were because you can see how abused the system was before the chip and how much better it was after the chip. So I think it all has to be taken in the context of looking at what the alternative is and I think the alternative is to us in this society, today, is that the alternative to not having the chip is much worse than having it.

Michael Michael: Gary, just one last question before I let you go. Okay, there is so much we can say here, but I want to get another opinion. Given that this will be ... this chip will be so intrusive and of course that opens up other questions, but I want to get to the point of, the question of autonomy. Don't you think that it's going to have an effect on the way we act? In other words it's going to affect our choices and our decisions and maybe a great part of our life will be acted out in theatre. We may lose our spontaneity and we may act the way we think we have to act. I'm thinking in terms of identity, personality, consciousness. Do you think of these issues and are you concerned about them or do you think they're philosophical concerns are not really touching reality? Or do you think they are real concerns?

Gary Retherford: I think they're concerns if you knew the alternative and I'll give you an example. I myself am 50. I was working here recently about a year ago with a group of guys who were around 23 and they were very high-tech guys and we got into the discussion of me remembering the 1970s and them not even being born until they were in you know the late 1980s. So what I realised was that they could not comprehend at all in their minds what the 1970s were like. So what I realised is that when we put up as a concept of what you just said, you know, is this going to have an effect or not, I think it all becomes relative to where we start and the reality of it is as people we don't live forever, we die. Now, not to sound morbid but what that means is that as we move forward everything becomes relative to what you're accustomed to. So I don't think that over time the question that you pose is ... Well, I think it's legitimate to throw it out there, but I don't think it's going to have legitimacy as we move down the path of time. Because as we move down the path of time, yes we're certainly going to be different today than they were in the year 1930 and in the 1930s they were different than they were in the year 1850. So every generation that comes along is different. So the question is a legitimate question, but the question is just as legitimate to be asked in 1930, as it is to be asked in 2009. But will that change anything? No, it won't change anything at all. It just means that as time marches on, everybody adapts to their environment. So the humans, those people that are not yet born, the ones that will be born next week– those people will adapt to this ... whatever new technology surrounds them in the year 2025 we'll say, then they will be adapted to that and that will set the their benchmark at that point.

Katina Michael: Thank you ...

Gary Retherford: So I don't see... I think that's why I say it evolves. I say it evolves because we do not live forever. We have a limited timeline whereas technology does not. Technology's timeline is infinite.

Katina Michael: Thanks Gary for that reflection. I'm conscious that it's coming up to just over 30 minutes and I said five questions and we've asked about nine. I just wanted to close Gary by thanking you for all your time, I think to be honest with you, I feel something special happened here over the last few days. This is a groundbreaking interview.

Gary Retherford: Well this has been really good for me, I've really enjoyed this and being able to have this discussion because you know, it's made me think.

Katina Michael: Yes, us too.

Gary Retherford: But I appreciate it. It's been very good. Thank you.

Katina Michael: Okay, thanks for all your time, and we look forward to further contact in the future.

KEY TERMS AND DEFINITIONS

Autonomy: Independence or freedom, as of the will or one's actions; the autonomy of the individual.

Big Brother: The most common sense of "Orwellian" is that of the all-controlling "Big Brother" state, used to negatively describe a situation in which a Big Brother authority figure — in concert with "thought police" — constantly monitors the population to detect betrayal via "improper" thoughts.

Choice: The freedom to choose between several options, such as mainstream or alternative options.

Citywatcher.com: A video surveillance company in Cincinnati, Ohio, USA.

Free Will: Is the ability of agents to make choices unconstrained by certain factors.

GIS: Geographic information systems are designed to capture, store, manipulate, analyze, manage, and present all types of geographical data.

Human Rights: Are universal and apply to all people. These rights may exist as natural rights or as legal rights, in local, regional, national, and international law. See for example, the European Court of Human Rights.

Microchip Implants: Are those implanted into the human body for therapeutic and non-therapeutic purposes. Usually the implantation zone is in the lower half of the arm, for mobility and interaction for various reader heights and ease of use.

Risk Return: Principle is that potential return rises with an increase in risk. Low levels of uncertainty (low-risk) are associated with low potential returns, whereas high levels of uncertainty (high-risk) are associated with high potential returns.

Rules: A rule pertaining to the structure or behavior internal to a business toward decision-making.

Six Sigma: Is a set of tools and techniques/strategies for process improvement originally developed by Motorola in 1985. Six Sigma became well-known after Jack Welch made it a central focus of his business strategy at General Electric in 1995.

Technological Evolution: Is the name of a science and technology studies (STS) theory describing technology development. It has been applied to other theoretical frameworks with respect to incremental innovation, or changes to technology over time.

Presentation 4.3
DangerousThings.com
Mr. Amal Graafstra
Presentation presented at ISTAS10 on 7 June 2010, University of Wollongong, Australia

RFID PASSIVE AND ACTIVE SYSTEMS

The first thing I want to talk about is that radio-frequency identification (RFID) is very diverse. There are a lot of different technologies involved in making RFID work, and not a lot of people are aware of just how big that diversity is. So in very basic terms, there are two types of systems.

Passive RFID systems are those that induct and modulate magnetic fields to derive power from readers and communicate with those readers. And they come in three basic frequencies:

1. Low frequency RFID which is the type of tags that are used in implants, pet chips. The reason for that is that the data can communicate through flesh and not be absorbed; that is, the signal is not absorbed.
2. High-frequency RFID which is typically used for access key cards and the like; and
3. Ultra-high frequency RFID which is used for other things.

Each type of RFID has different advantages and disadvantages.

Active RFID systems have a battery and a power source. They transmit much like in the same way your mobile phone or Wi-Fi network or radio beacon works. And in essence, a mobile phone, Wi-Fi network, and a radio beacon all have unique identifiers; so in effect they are an active RFID system of some sort.

PROTOCOLS

Data protocols vary with respect to RFID: the air interface, the way that RFID tags communicate, how data moves from the reader to the tag and then goes back. There are some ISO standards, but many are still proprietary with respect to RFID. Encryption standards are almost nonexistent; a very high percentage is proprietary encryption, which exchange effectiveness for speed and convenience, and that's mostly due to limitations in power and processing in the tag. Normal RSA (Ron **R**ivest, Adi **S**hamir and Leonard **A**dleman) security banking style encryption would take a very long time for this low power processor to set up a secure channel.

IMPLANTS

So I'm diving right into DIY (do-it-yourselfer) RFID implants. This is some of the stuff that I kind of toyed with on my desk. There's a standard reader here for $30, it has a TTL (Transistor-to-Transistor Logic) interface with a microprocessor. You can buy it and incorporate it into your own DIY projects. Some example glass tags I've worked with include:

1. The large sized RFID which are used for cattle and large animal use;
2. The middle sized RFID which is about a 3 millimeter by 13 millimeter cylinder, which I have in my left hand;
3. A small sized RFID which is the size of standard pet chips, and is the size of the tag I have in my right hand, which can be injected. That is, it's small enough to fit into an injector kit.

Left Hand Implant

In my hands I have the EM4102 chip which is the type of chip in the tag in my left hand. It's a 125 kilohertz low frequency tag. My right hand has a Philips HITAG which has more capability. Here's an x-ray image of my hands, that actually took quite a bit of effort to get; I can't believe how convoluted the healthcare system is in the US. But the reason why I chose this area in the hand is illustrated here. You have two major nerve bundles in each hand, but then in between it is fairly devoid of major nerve fibers and it is kind of a squishy zone between the thumb and forefinger. There's a lot of padding there to absorb shock and things and to protect the tag.

So there's some detail on the left hand (see http://www.youtube.com/watch?v=kraWt1adY3k). That's the tag before it went in. And there are the tools that were used to put it the RFID tag. And that's just immediately after the injection. That's me with the reader, doing my first access control project, and that's the scar that was left about a week later, and that's the scar that's there today. It is just a little, little thing, that you can see right there.

Right Hand Implant

In my right hand I have a Phillips High Tag, and this is a little more interesting. It has some crypto-security features – not a lot, about 40 bits – but it's enough to ward off kind of momentary attackers. It also has 2,048 bits of read/write memory, which is kind of cool. You can store data on it, you can change it, and as Mark Gasson alluded to earlier, you can even put viruses on it, apparently. So there's the tag before implantation. And there's the gear used to put it in. There's my doctor, just your every day family doctor. We used a pet implant kit: we took the injector, took the pet tag out, threw it away, put the hitag in, and after a simple sterilization process, it was only then a two-second deal. There we go. So that's immediately after, and that's today. I use this tag daily, so I would liken it to an enhancement. I can get into my front door and tap the little deadbolt system there with the tag, and then add on RFID as an authentication method. I can still use the key, and I can still use the PIN code but I just wanted to add the ability to use an RFID tag as well. I could also use a key card or the implant, either one.

This fire safe was modified to allow PIN code access or RFID tag access. That's me getting into my car. This is a different application- I only wanted to get into the car, I didn't actually want to use the tag to start it, because there are security issues of course if somebody were to get my tag ID and somehow emulate it. So getting into the car is great, because then there's a hidden key, and I can use my knowledge to get the key and start the car.

I ripped apart a keyboard, modified a reader and put it in there, used that to log in, and I have that set up here if anybody's interested, you can see that happen. I can also start my motorcycle, and this is a little different. I just went ahead and said, "Yeah, let me start the motorcycle," because it normally stays in my garage and I'm not too scared of somebody wanting to break into my garage- they would probably take the bike regardless.

So I actually use active RFID in my daily life too, after having had my laptop stolen. The police said they could not do anything about it, even though I suspected a neighbour. So I got this locator device – it's just a standard locator active RFID system, it has a reader with a directional antenna and two tags it came with. I ripped the tag apart and put the three-volt regulator on it, dipped it in some plastic, took my laptop, took it apart and embedded it there in the laptop. So now I can find my laptop if it's stolen. The range isn't great, but you know, it's enough to tell the police, "Hey, it's in there – go get 'em."

The other thing about this which is of interest is that in active RFID, typically the transmitters constantly transmit, which can also be considered a security or privacy issue. But this system is set up specifically so the tags do not transmit until the appropriately paired receiver tells them to. So the receiver has to say, "hey, I'm looking for you," then the tag responds. So it's kind of a neat setup.

PUBLIC REACTION

Public reaction – angry. I get a lot of angry emails, calls, and things like that. There are some people that wish I'd just go away, and there are others claiming that I am somehow helping "the conspiracy". This is just kind of a little thing that I thought up, about the cycle of fear that I've noticed when talking to people. So when people come to me and they're angry about things, I try to engage them in conversation but usually they're afraid of misconceptions about the technology. They think that somehow the GPS satellites are communicating with this tag – which really only has a three-inch read range – and somehow reporting my location, "Can't *they* track you?" … the elusive "they".

So you know, they're afraid of something they're not sure of and they take action because they're afraid. Then people that know about it respond, usually poorly. This interaction reveals to the angry people that they really don't know what it is they're talking about. And what's interesting is that they have a new fear then, and that fear causes them not to want to learn about the technology. They don't want to engage, because they somehow feel that if they learn about it, maybe their fears are unfounded or whatever. But it's a cycle that repeats quite often. So the concept is that, you know, somehow now your body is up for sale, and companies and governments are vying for it.

Here is a picture of my x-ray image misappropriated. It is used all over the place, but I think it's kind of interesting to do a Google search on it every once in a while and see where it's been used. The *mark of the beast*, of course, has to come up. This is an email that I got, I think the second day after injecting the RFID tag. The first email that I got actually was, "You're the Devil's mouthpiece," and I thought that was kind of interesting. So what I notice is a fear- that somehow- this is going to be compulsory. "I'll never take that stupid chip in my hand", taking it meaning they're just going to have to take it. Just very interesting emails And I get a lot of them. But I'm just kind of showing the reaction that I've been getting from certain segments of the population.

SECURITY

Many RFID tags used today are not designed with security in mind. The IDs can be easily read by a standard reader and this leaves those systems open to attack. But in a lot of contexts, that's usually irrelevant but in some contexts it's not. The RFID tags that are designed with crypto-security features, most have been cracked or otherwise defeated. Entire systems need to be designed with comprehensive security features and not to just rely on the RFID tag's encryption mechanism to secure the system.

An example of that is the Texas Instruments DST (digital signature transponder) tag. It is used in this key fob, which is ExxonMobil's speed pass in the US. It's used to buy gasoline, and I think you can buy fries at McDonalds with it as well. And then the same DST tags are used in some automobile keys to immobilize the system. It's a 40 bit cryptographic key, it emits a factory set 24 bit identifier and authenticates itself through a challenge response mechanism. So there's the DST tag there being used for the speed pass.

These guys at John Hopkins were able to use 16 parallel FPGAs (field-programmable gate array) and basically crack the algorithm, in about two hours. So they broke the algorithm for five different DST tags and created a common algorithm out of that, and they just set up a simple system that can randomly attack anybody with a DST tag. So they were out stealing cars and stealing gas. Kind of interesting… That site is no longer available, but I think you can go to a cached version somewhere on the Internet.

FAILURE

Basically security mechanisms can fail, and the one in the previous example did so. And the problem is, once they gained access to the algorithm, they could use it on anybody that had one of those tags. The possible remedies are to use stronger encryption based on open standards, not proprietary standards. You can also rotate one-time use keys that are written to read/write memory blocks. You can overcome power processing limitations to merge Smart Card technology, which is actually a small processor, with a more powerful capability using the contactless features of RFID. Or you can get these Faraday cage pants made by a friend of mine, Mikey Sklar. He keeps his keys and everything in those pants, and nothing can be read through those Faraday cages.

COMMON SYSTEMS PRONE TO ATTACK

The severity of security issue depends on the context. So this is kind of where we look at the different uses, a business use where there is typically a high risk involved such as payment systems, high security access, medical records etc. Those types are all risky business. The attacks can be random because the systems are common. So you've got a VeriChip system or PositiveID (or whoever they are now) to access medical records. Well, there are other mechanisms that can be used to secure that system, but the tag itself is completely unsecured. And the system is common. So in a payment solution, let's say you have your credit card – well that credit card now is RFID-enabled. It's a common system. So any attacker can, once they figure out the system, they can attack anyone with a credit card. The attacker does not have to know that individual person- it could be a purely random attack. Then they know exactly what to do with that data – they can go anywhere and use the data and buy things.

So the common design makes it easy to attack and expensive to modify. It's been deployed, so to modify that system you have to replace millions of readers, millions of tags. Personal use is quite low-risk, even though it is your front door access which is in question. This seems risky but it's pretty low-risk because the attacks have to be extremely targeted. With the DIY context, it is a random design, so there's not a common system- you're not quite sure what you're walking into as an attacker.

Other things that come into play are things like the reality of attacks. So my car has RFID in it, and after giving a talk one time, I came out to find my car burgled. So, I almost wished at the $500 price tag of that smashed window I had to repair thereafter, that they had used my RFID tag to get in and just take the $5 worth of change that they did end up taking.

PRIVACY ISSUES

So RFID implants like mine have a three inch read range. Reading the tag is deliberate and consensual. Logs, if kept, are mine and mine alone. Active tags and other types of tags can be read at a greater range. These could be used for "tracking", which I put in quotes, which is really just logging, possibly without consent.

And the thing that I want to make clear here, is that tracking and logging are things that we do every day. For example, every time we make a phone call or use a credit card, it's not locating. And I reiterate that here. RFID is an identification technology, not a locating technology like GPS, RF beaconing, or mobile phone triangulation. Logging, where a person was and when, is truly standard practice in today's society, and we find this practice in loyalty card schemes, computer logins, credit cards, mobile phones, traffic toll tags – they all keep data of this type.

So how do we proceed? Focusing on RFID or any single technology is a waste of energy, in my opinion, particularly when you're trying to somehow eradicate the technology or stop it altogether. Instead, I think that intelligent legislation needs to be created to broadly address the real issues behind those concerns, and not control technologies that enable the issues to arise in the first place.

With regards to the technology, be it RFID, biometric scanning, credit card purchases, mobile phone location- and this is just something I'm putting out for discussion- there should be awareness, consent, control, and licensing. Let me elaborate briefly what those mean.

AWARENESS

A person, user, employee, customer, must be made aware that these technologies are in use, and who to contact with questions and concerns about them or the business process they are used in. Also public awareness in general needs to be raised, which slowly I think is happening but unfortunately RFID I think is getting the bulk of the attention, while there are facial recognition systems, biometric systems, systems that enrol you just simply by walking through an area and you don't even know about it. At least with RFID, you can opt out by leaving the RFID card at home, shielding it, or otherwise not using it.

CONSENT

Systems should be designed around the idea of consent, difficult as it may seem in some situations. RFID cards could be designed with momentary switches that only enable the antenna when it's intended to be used. An example would be like an access card for work. You have a badge, it's got a card in it, and if you need to go through a door, you just give it a little squeeze so it connects the antenna, and only at that time can the card be read. At every other time it's inactive, you can't read it. So that's a consent process.

Another example is the US Government issued me an access card which allows me to travel quickly to the Canadian border. They issue that card with a copper film sleeve that blocks the card from being read, and instructions on how to keep the card in the sleeve when not in use at the border, and that's a consent process. So you take the card out, you're saying, "I'm allowing this to be read now."

Authorization must be given for each application the collected data is used for, as well as when data is shared with or sold to another party. So this idea is kind of going into the concept of legislating these mandates, where if you're involved and are enrolled in the system.

A few years ago I went to Disneyworld, and after buying the tickets – which were very expensive – and travel, and hotels, and everything, I walk up to the front gate and there's a fingerprint reader. And you have to match your fingerprint to the ticket. And I thought, "Well, there's no notification before I went spending all this money that I had to do that- I had to give up that biometric information to the Disney Corporation."

Even worse, by giving it up, I can't really change my fingerprint, whereas if I had an RFID card issued to me or whatever, I could choose to shield it or use it to get through. But that biometric data, once given up, it's hard to opt out – you really have to depend on that third party to opt out. Not to mention I have no idea what the licence agreement is to use that information – how long do they keep it? Nothing like that. After asking the Head of Security there at the gate, disturbingly, nobody else had asked or even wondered if there was another option, which there was not. So I gave in and gave them my fingerprint, so I'm kind of upset about that.

CONTROL

Participants should have the option to opt out and/or remove their identifying data from systems in question. This can only really be mandated through legislation I think, and basically something that says that if a company or a corporation or even Government is going to collect this information about you, if you choose to opt out of their system or service or product or whatever it is, that you should also have control over your data, be it biometric or RFID or otherwise. Citizens should always have manual or anonymous options offered to them as an alternative.

The previous presentation today about the Japanese universities where the entire university had payment systems set up for RF cards, they put cash into a terminal, it charged up their card, and then they used their card everywhere. Well, their card identifies them as buying a can of soft drink or whatever it is that they're buying, and they should have the option to have an anonymous purchase at that machine. So this is something to think about, because by mandating that we can only use an identifying payment technology, that's kind of against the rules in my opinion.

And this is kind of an odd concept, but licensing – licensing your data. If, you know, quote, unquote, "free society" – commercial interest usually spurs the efforts to collect this type of data. Citizens should realize that their mundane activities have value, and companies and even governments should pay to collect and license that data from you. And that license term should favor the data supplier, you, and integrate the previous points of consent and control.

KEY TERMS AND DEFINITIONS

Authorisation: The function of specifying access-control rights to resources as related to RFID security and information security.

Control: The act of having authority and power over one's own personal information and the ability to limit its distribution.

DIY: Do-It-Yourselfer approach is a method of design and construction of a computer system, without the aid of experts or professionals. DIYers usually like to deviate from commonly used computer systems, customizing systems to their own personal needs.

DST: Digital Signature Transponder is a cryptographically enabled radio-frequency identification (RFID) device used in many every day systems. It was developed by Texas Instruments.

FPGA: Field-Programmable Gate Array is an integrated circuit (IC) designed to be configured by a customer.

HITAG: Is a well-established brand in the low frequency (LF) RFID market. It is particularly sturdy in harsh environments.

Immobilise: The ability to withdraw a vehicle or object from circulation or theft.

License Agreement: A mutual understanding between one party (licensor) and another party (licensee) that is legally binding.

RF: Radio-frequency is a rate of oscillation in the range of about 3 kHz to 300 GHz, which corresponds to the frequency of radio waves.

RSA: Is an Internet encryption and authentication system that uses an algorithm developed in 1977 by Ron Rivest, Adi Shamir, and Leonard Adleman.

Sterilization: The elimination of microbiological organisms to achieve asepsis, a sterile microbial environment.

TTL: Transistor–transistor logic is a class of digital circuits built from bipolar junction transistors (BJT) and resistors.

VERICHIP: The VeriChip, which then became known as PositiveID, was the only Food and Drug Administration (FDA)-approved human-implantable microchip.

Chapter 5
Knowledge Recovery:
Applications of Technology and Memory

Maria Burke
University of Salford, UK

Chris Speed
University of Edinburgh, UK

ABSTRACT

The ability to "write" data to the Internet via tags and barcodes offers a context in which objects will increasingly become a natural extension of the Web, and as ready as the public was to adopt cloud-based services to store address books, documents, photos, and videos, it is likely that we will begin associating data with objects. Leaving messages for loved ones on a tea cup, listening to a story left on a family heirloom, or associating a message with an object to be passed on to a stranger. Using objects as tangible links to data and content on the Internet is predicted to become a significant means of how we interact with the interface of things, places, and people. This chapter explores this potential and focuses upon three contexts in which the technology is already operating in order to reflect upon the impact that the technology process may have upon social processes. These social processes are knowledge browsing, knowledge recovery, and knowledge sharing.

INTRODUCTION

This chapter is concerned with the implications upon the processes of storing, recalling and passing on memories as emerging digital technologies offer people the ability to associate data with physical artefacts. The network society has grown up using screens as the familiar interface with which they are able to access digital networks. Televisions, computer screens and mobile phones have all manifested digital data behind a glass screen. As the reach of ubiquitous computing extends from urban contexts into the rural we are beginning to experience places which are always in reach of the internet, this coupled with the ever increasing range of devices that are able to access it, offers a

DOI: 10.4018/978-1-4666-4582-0.ch005

context in which we may no longer need screens to interact with the internet. Described as an Internet of Things (coined by Kevin Ashton at the Auto-ID research group at MIT in 1999 (Ashton 2009)), many new manufactured objects have barcodes or Radio Frequency Identifying (RFID) tags attached to them to allow them to be scanned and identified. In range of an RFID reader, these artefacts become part of the internet and access points to data that is associated with the object.

The emerging tendency to tag objects with RFID and barcodes that can also link to data, is accompanied with a proliferation of tag readers that are appearing as hardware and software applications on smart phones. In the hands of the public who can read these tags, objects are beginning to become interfaces to the internet.

We are interested in exploring a technology that for many years has been in the hands of check out staff of supermarkets, but one that is now available to anyone with a smart phone. Used generally to recall logistical information on a read only basis, the public have rarely understood how artefacts with barcodes were part of the internet because until now they didn't have the technology to connect their packet of breakfast cereals to the web. In addition the public presume that the only data that would be available from a barcode is likely to be logistical: name, price and weight. However, recently a series of web technologies have become available that link logistical data to identify an object with social data. Since each barcode is a signature for a product in an internationally available database, entries can also be associated with other media such as advertising media or special offers. Whilst this extends the potential for using barcodes and tags to "read" media from the internet, what is of special interest to the authors is the introduction of some systems that allow the public to "write" information themselves to a particular tag.

The ability to "write" data to the internet via tags and barcodes offers a context in which objects will increasingly become a natural extension of the web. And as easy as the public was to adopt cloud based services to store address books, documents, photos and videos, it is likely that we will begin associating data with objects. Leaving messages for loved ones on a tea cup, listening to a story left on a family heirloom, or associating a message with an object to be passed on to a stranger. Using objects as tangible links to data and content on the internet is predicted to become a significant means of how we interact with the interface of things, places and people.

This chapter aims to introduce the varied applications and relationships of technology and memory, where knowledge is the key which links the two areas.

This is achieved in three ways. First by the introduction of the different aspects of knowledge management – that of knowledge browsing; knowledge recovery and knowledge sharing; second by a discussion of the research project "Tales of Things Electronic Memory" (TOTeM); and third to present findings of relevant case studies.

KNOWLEDGE

In its simplest form, knowledge can be categorised as explicit or tacit knowledge. Explicit knowledge can be defined as documented knowledge whilst tacit knowledge in general is that which has not been recorded. (Ali & Ahmad, 2006; Brooking, 1996; Jain et al., 2007; Selamat & Choudrie, 2007; Zheng, 2005; Song, 2002; Kim & Lee, 2006; Brent & Vittal, 2007).

Knowledge is produced from raw information by members of a society. Society in general is organized into many different systems, (organisations), which are often controlled by technology. Within organizations knowledge systems utilize the available technology in order to undertake particular parts of the information management process – including careful planning of the way in which the information flows within the organization structure – resulting in overall improved control of the way in which the knowledge is managed. Due to the current, continuous nature of change in organizations today, it is critical that managers are able to respond and take prompt good quality decisions. The new tagging technology we refer to in this chapter will provide a means of improving management decisions by offering a very new way of browsing, recovering and sharing information. For example, an estate agent will be able to measure the performance of property adverts in local papers by offering clients the facility to "read additional information" on a property through QR codes placed beside the property photo. The recruitment manager may be able to dramatically affect the induction process by setting up a process for leaving "hidden" memories in the form of messages embedded in QR Codes, on items in offices, such as printers, keyboards; desks; walls and on architecture, in order to speed the levels of efficiency and effectiveness, and again improve performance of the organization. The charity store manager may have the facility to personalize each donated gift through a technological facility where customers can listen to powerful memories associated with object by "reading" the QR code.

In these examples, knowledge that was tacit becomes available – available for others to browse; available as a form of recovered (known but never written) almost a type of mythological knowledge; and knowledge which is available to be shared using new forms of technology. All of these aspects are part of a broader discipline of knowledge management which can be defined as the process of locating, organising, transferring and using the information and expertise within an organisation.

There is no one formal accepted definition of KM but all definitions are concerned with information, management and some kind of system. Other standard definitions include:

Knowledge Management is the attempt to recognise what is essentially a human asset buried in the minds of individuals, and leverage it into an organisational asset that can be accessed and used by a broader set of individuals on whose decisions the firm depends. -Prusak and Cohen (1997)

Any organisation that wants to excel at managing knowledge will need to do three knowledge management processes well i.e. generation, codification and transfer of knowledge. -Davenport and Prusak (2000)

Knowledge management is the process through which organisations generate value from their intellectual and knowledge based assets and will highlight issues concerning companies in emerging economies which could have the potential to feed into government strategy and policy. -Brown and Duguid (2000)

All these definitions reflect societal and managerial trends at the time of writing, yet still have value today. In fact what does not change is that the knowledge management process is always supported by the four key enablers: leadership, culture, technology and measurement. Leadership in terms of managing the people within organisations; culture in terms of communicating organisational values, beliefs and assumptions; technology in terms of cost efficiency and finally process improvement and measurement in terms of increasing performance in order to successfully compete in the market place.

More formally, knowledge management can be defined as "the generation, representation, storage, transfer, transformation, application, embedding and protecting of organisational knowledge" (Schultz & Leidner, 2002) and this is the one that we consider best represents the work outlined in this chapter. The old adage that the overall success of the organisation, however, rests on one aspect, that of sharing information is still true – but with the onset of social media and newer more accessible technologies the ways of dealing with knowledge is changing. Now it is easy to share and indeed sometimes it is difficult not to share. What has become important and what we want to introduce next is the ability to both "browse knowledge and to recover knowledge" and to show how tagging technologies can be applied in these areas.

KNOWLEDGE BROWSING

The confidence to browse suggests that an individual or organisation are comfortable in a context to afford them the time to survey products, services and perhaps people with whom they would like to connect. The act of browsing also suggests an open-minded disposition that is receptive to new modes of practice and interested less in finding answers to specific questions, but to understanding novel solutions, or even opportunities of which they were previously unaware.

What is important in this technological context is that aspect of browsing which we can define as "uncertainty". We will deal here with organisational uncertainty. Uncertainty can be viewed from two areas, that of "relational uncertainty" (according to Berger, 1975) where it is difficult for employees to predict the beliefs and behaviour of colleagues and the that of "informational uncertainty" where the accuracy of the actual information is called into question - as addressed in the information seeking literature (Burke, 2003; 2006; 2007; Choo 2001; Kuhlthau, 1993;). Both these areas are concerned with three issues. The first is the fear and trepidation experienced by organizational members about levels of accuracy and quantity of information; the second issue is about both trusting the source of the information and a willingness to trust co workers enough to share information whilst the third issue is about having sufficient relevant knowledge to make quality decisions.

Whilst tagging technologies probably at the moment cannot solve relational uncertainties, it can certainly be employed in order to assist with informational uncertainties. Let us consider the three issues previously outlined, information accuracy; trust and relevance.

Information accuracy and fears surrounding this are affected by organisational size, by culture, by industry sector and by any kind of punitive "punishment" for "getting it wrong". If we can use tagging to enhance the process of accuracy by looking at some kind of new business process which would encapsulate knowledge – from creation, make it available for browsing for a set period of time (tagged with a quality mark) then all users would know – and trust the accuracy of the information However, of course in doing this the danger could be the great loss of creativity and freedom to innovate ideas which had not somehow been "approved". This of course would have to be tempered and carefully implemented as if one follows this route there is a danger of it being seen as a communist, rather than managerial ideologist.

Trust, information and organisations have always been problematic. Using social media which allow "messages" to be added and commented on, post after post, blog after blog do initiate a stronger level of comradeship and of communication. However, they are not attached to singular objects but to individual people. An organisation needs continuity and cannot change each time an employee arrives and departs. The secret success of tagging is that it is object orientated – literally. QR Codes are permanent and give out information – employees can "browse" the knowledge, at whatever time they choose.

The third issue of uncertainty is that of "relevance" - issues about how we decide if a piece of knowledge is relevant to the current task (Basden and Burke, 2004). Usually it is based on our own expertise, our own experience and whether we trust the source. By having the ability to "browse" the tales of objects and of documents, we can leave messages and build up a database of categories of relevant information assigned to various organisational processes.

So, knowledge browsing can be enhanced by tagging technologies – it opens new avenues and allows new methods of verifying, validating and storing information.

Case Study One: A Contemporary Example of a Smart Phone Application that Supports the Reading of Barcodes

Property podcast (http://www.propertypodcast.co.uk) is an example of a web based service for Estate Agents that links QR (Quick Response barcodes) to videos and PDF documents of houses for sale. Users may use any barcode reader application than runs on a smart phone. Upon reading a tag that may be located on a sale board outside a house or in the window of an estate agent, an internet browser will launch and play a short movie that describes the interior of the house through photographs and accompanying voice over. Like many software systems of its kind, Property Podcast allows the general public to pursue search enquiries on their terms in their own space and time without the need to go through an actual human agent. However the "knowledge" received is often an extended form of marketing and travels one way under the control of the publisher.

KNOWLEDGE RECOVERY: THE CONTEXT OF MEMORY

What then is knowledge recovery – this is a new term and one that can be used to discover and recover information – to find out about memories and about identities of artefacts, to engage almost with history. This kind of knowledge is embedded personally in an individual experience and depends on other factors such as personal belief,

perspective and the value system Gourlay (2002) discovers that tacit knowledge has the identical phrase and defines it as practical know-how. It is informal rather than formal among professional groups including managers. What is particularly interesting is that new forms of digital technology are being used to enhance this process. For example, the web site talesofthings.com which allows users to record a "tale" about any object and to upload to a database is a form of both knowledge sharing and knowledge recovery.

As individuals we are able to share with relative ease – however this becomes more problematic for us as we spend most of our lives dealing with or as part of organisations which operate within an ever changing external environment How then, can knowledge recovery – both implicit and explicit be enhanced through digital technology? We may start to approach this problem by analysing types of societies. This may be helpful as it allows us to consider the aspect of sharing information and the management of knowledge from quite different perspectives than technology and sociology. For example, Van der Rijta's (2007) work was concerned with the two concepts of societies which displayed characteristics associated with individualism and collectism. These types of societies are important and means of charting differences in the concept of sharing (Chen et al., 2001; Hofstede & Hofstede, 2005).

Case Study Two: An Example of the use of Barcode Scanning that Offer Multi-Dimensional Enquiry and the User More Choice

Snap Tell smart phone application: Snap Tell is an example of one of many smart mobile phone applications that integrate a series of digital technologies to allow users speedy access to best price options and geographic knowledge about books, music CD's and DVD's. The Snap Tell application allows the user to take a photograph of a book cover (for example) using the built in camera of a smart phone. The application then uses image recognition technology to find matches between the photograph and products by communicating with online databases. Once a match is identified, the software presents prices of where the book may be bought at online stores such as Amazon and Overstock, as well as offering prices at actual shops in the local vicinity. The location function is limited to the United States however the integration of photographic technology, image recognition software, database interrogation and geographic services transforms the traditional relationship between knowledge management and shoppers as they are presented with this new data. No longer are consumers restricted to acquiring knowledge from "experts" who are in the immediate vicinity, the tag (in this case the book, CD or DVD cover) is a conduit to an internet of data that, if organised well, can offer multiple access routes to information and knowledge.

KNOWLEDGE SHARING

Sharing generally happens within the context of an information system or a knowledge management system. Yet the popularity and availability of social media sites has made "sharing" a much more social activity.

Sharing in organisations only takes place where there is trust and where there is a shared feeling of ownership of goals. The reasons behind the tendency to share are based on the kind of interpersonal relations between co- workers inherent within the organisation and the effects of social relationships on organisational teams. Strengthening the social relationships between individuals in the team is crucial in motivating team members to share knowledge.

The current thinking in the research community about knowledge sharing within organisations is that barriers to knowledge sharing can be classified into individual barriers, organisational barriers and technology barriers. The UK has a rich array of examples where attention has been paid to knowledge management initiatives in order to set up major knowledge management systems, e.g. the Health Service and Banking sectors. Although these have not always been wholly successful, UK Companies have taken up the ideas of knowledge management and have endeavoured to identify and overcome barriers to sharing. (Wong & Aspinall, 2005). Of particular interest is the work by Elenurm, T. (2004) who looked at knowledge sharing capabilities and knowledge development needs in the context of East-West technology.

However, in order for even the most basic KM system to work effectively, as we have seen previously, there must be a sense of trust in the organization and this trust is crucial to the open sharing of information. Sharing only takes place where there is trust and where there is a shared feeling of ownership of goals. Within a business, this is often done through a framework of knowledge sharing networks. For example, Dyer and Nobeokai's (2000) work on the Toyota's network can be seen is a purely classical way as having solved, "three fundamental dilemmas with regard to knowledge sharing by devising methods to (1) motivate members to participate and openly share valuable knowledge (while preventing undesirable spillovers to competitors), (2) prevent free riders, and (3) reduce the costs associated with finding and accessing different types of valuable knowledge. Toyota has done this by creating a strong network identity with rules for participation and entry into the network. Most importantly, production knowledge is viewed as the property of the network."

Yet knowledge sharing in business is also about social relationships and tagging technologies can enhance social relationships by accentuating that relationship rather than, or as well as, the business relationship.

Case Study Three: An Example of a Scanning Platform that Allows the user to Not Only Read Information from a Source but also to "Write Back" and Contribute to Database

Tales of Things self tagging service: Tales of Things (http://www.talesofthings.com) is a web-based application that is able to associate different media types to a unique two-dimensional barcode. Members of the "free to use" system are able to submit an object to the online database with a photograph and other information: name/title, keywords, location and most importantly a story that the object evokes for them. The interface also requests for other media to be associated with the artefact: sound and video clips of the owner telling a story about the object that are stored on services such as YouTube and AudioBoo. Once submitted, the Tales of Things system creates a unique two-dimensional barcode (QR Code) for the artefact which can also be printed out and attached to the object.

Tales of Things also provides mobile applications for Android and iPhone platforms that allow the user or more importantly, anyone who comes across an object tagged with a Tales of Things code, the opportunity to scan and retrieve stories, video and audio clips. The same phone applications also allow people to add additional stories to the artefact using text and video and thus contribute to the objects history. Tales of Things offers a unique social and storytelling focus for both the browsing and sharing of knowledge. This is in stark contrast to the current deployment of tagging technology that often focuses upon providing consumers with marketing material about products and offers no portal through which they can feedback.

CONCLUSION

In many ways the order that we have used to describe the three opportunities for knowledge that is associated with objects, follows the order in which individuals have learnt to interact with the internet:

1. Recalling data is akin to basic web searches,
2. Knowledge browsing represents a confidence of the user to look through web based media at their leisure, and
3. Knowledge sharing implies a further confidence to post and distribute media for others to search and browse.

The three case studies demonstrate the multi-dimensional properties of acquiring and sharing knowledge through the relatively new technology of public bar code scanning:

1. **One-Way, Closed Media Channel:** QR codes are becoming a popular interface to recall marketing material however this is limited to a specific service and is edited by the provider.

Figure 1. Burke and Speed's Tagging Media Matrix (2011)

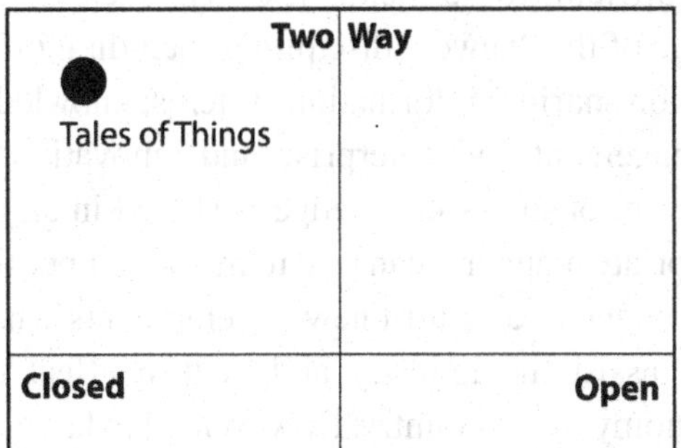

2. **One-Way, Open to More Media Channels:** QR codes and other forms of tags are associated with one product but an intermediary service offers the user multiple choices about where the product may be bought and at different prices.

3. **Two-Way, Closed Media Channel:** An emerging characteristic of tagging may be the ability to "write-back" to the database that a tag is associated with. Whilst this service is extremely limited at present for tags, just like the emergence of Web 2.0 technologies the public are now familiar with the ability to comment and contribute to the internet. However due to the nature of the research project to record the memories associated with a single item, the database is closed and only includes items that are within its own database.

The Tagging Media Matrix (Figure 1) presents the difference between the three case studies and acknowledges a fourth space in which a technical platform may offer an open and two-way platform in which the public are able to explore multiple knowledge sets through a tag, but are also able to contribute to the knowledge. This is the initial introduction of this model – further work will follow on this in later publications.

The concept of knowledge sharing is inevitably difficult to define, as it covers such a wide range of the "newer" disciplines including information sharing; information systems; knowledge management and enterprise and innovation. If relevant business knowledge is shared in an appropriate manner it can lead to major competitive advantage and in turn new developments which will assist the industry and in turn affect the economy of the country. However, what has been obvious so far in this research is the clear energy, passion and commitment to bringing the latest ideas to their enterprise, regions and ultimately their countries. It is interesting to consider the different perspectives which are taken when sharing information is a new factor. New models and frameworks need to be devised in order to incorporate changed societal and organizational culture. Whether the future for the development of Knowledge Management is sustainable is still to be seen, but from the evidence there is certainly both growth and hope in the area. No doubt the final way forward will depend on two factors - the cooperation of relevant bodies and the appropriate resources being made available, and the take up of the new technology by business organisations and the wider society. It is hoped that now, today in the freedom of the 21st century both these factors can be given reasonable consideration and a positive response.

ACKNOWLEDGEMENT

The Tales of Things project is supported by a Digital Economy, Research Councils UK grant, and made "real" by our team: Barthel, R., Blundell, B., Burke, M.E., De Jode, M., Hudson-Smith, A., Leder, K., Karpovich, A., Manohar, A., Lee, C., Macdonald, J., O'Callaghan, S., Quigley, M., Rogers, J., Shingleton, D. and Speed, C.

REFERENCES

Ali, H. M., & Ahmad, N. H. (2006). Knowledge management in Malaysian banks: A new paradigm. *Journal of Knowledge Management Practice, 7*(3).

Ashton, K. (2009). That 'internet of things' thing. *RFID Journal, 22*.

Barclay, R. O., & Murray, P. C. (1997). *What is knowledge management*. Retrieved from www.media-access.com/whatis.html

Basden, A., & Burke, M. (2004). Towards a philosophical understanding of documentation: A Dooyeweerdian framework. *The Journal of Documentation, 60*(4), 352–370. doi:10.1108/00220410410548135

Berger, C. R. (1975). Beyond initial interaction: Uncertainty, understanding and the development of interpersonal relationships. In H. Giles, & R. St Clair (Eds.), *Language and social psychology* (pp. 122–145). Oxford, UK: Blackwell.

Brent, M. H., & Vitall, S. A. (2007). Knowledge sharing in large IT organisations: A case study. *VINE: The Journal of Information and Knowledge Management Systems, 37*(4), 421–439.

Brooking, A. (1996). *Intellectual capital.* London: International Thomson Business Press.

Brown, J. S., & Duguid, P. (2000). *The social life of information.* Cambridge, MA: Harvard Business Press.

Burke, M. (2003). Philosophical and theoretical perspectives of organization structures as information processing systems. *The Journal of Documentation, 59*(2), 131–142. doi:10.1108/00220410310463482

Burke, M. (2006). Achieving information fulfilment in the networked society: Part 1: Introducing new concepts. *New Library World, 107*(9/10), 21–26. doi:10.1108/03074800610702624

Burke, M. (2007). Cultural issues, organizational hierarchy and information fulfilment: An exploration of relationships. *Library Review, 56*(8), 236–245. doi:10.1108/00242530710818018

Chen, C. et al. (1998). How can cooperation be fostered? The cultural effects of individualism-collectivism. *Academy of Management Review, 23*(2), 285–304.

Choo, C., et al. (2001). Environmental scanning as information seeking and organizational learning. *Information Research, 7*(1).

Davenport, T., & Prusak, L. (2000). *Working knowledge: How organisations manage what they know.* Boston: Harvard Business School Press.

Davenport, T. H. (1993). *Process innovation: Re-engineering work through information technology.* Cambridge, MA: Harvard Business School Press.

Dyer, J. H., & Nobeoka, K. (2000). Creating and managing a high-performance knowledge-sharing network: The Toyota case. *Strategic Management Journal, 21*, 345–367. doi:10.1002/(SICI)1097-0266(200003)21:3<345::AID-SMJ96>3.0.CO;2-N

Elenurm, T. (2007). *Entrepreneurial knowledge sharing about business opportunities in virtual networks.* Paper presented at the 8th European Conference on Knowledge Management. Barcelona, Spain.

Forrester, J. (1965). *Industrial dynamics.* Cambridge, MA: MIT Press.

Gourlay, S. (2001). Knowledge management and HRD. *Human Resource Development International, 4*(1), 27–46. doi:10.1080/13678860121778

Hofstede, G., & Hofstede, G. J. (2005). *Cultures and organisations: Software of the mind.* London: McGraw Hill.

Jain, K. K., Manjit, S. S., & Gurvinder, K. S. (2007). Knowledge sharing among academic staff: A case study of business schools in Klang Valley, Malaysia. *Journal of the Advancement of Science and Arts, 2*, 23–29.

Kim, S., & Lee, H. (2006, May/June). The impact of organizational context and information technology on employee knowledge-sharing capabilities. *Public Administration Review*, 370–385. doi:10.1111/j.1540-6210.2006.00595.x

Kulthau, C. (1993). A principle of uncertainty for information seeking. *The Journal of Documentation, 49*(4), 39–55.

Prusak, L., & Cohen, D. (1997). Knowledge buyers, sellers and brokers: The political economy of knowledge. In *The economic impact of knowledge*. New York: Butterworth-Heinemann.

Schultze, U., & Leidner, D. (2002). Studying KM in IS research: Discourses and theoretical assumptions. *Management Information Systems Quarterly, 26*(3), 213–242. doi:10.2307/4132331

Selamat, M. H., & Choudrie, J. (2007). Using meta-abilities and tacit knowledge for developing learning based systems: A case study approach. *The Learning Organization, 14*(4), 321–344. doi:10.1108/09696470710749263

Song, S. (2002, Spring). An internet knowledge sharing system. *Journal of Computer Information Systems*, 25–30.

Storey, J., & Barnett, E. (2000). Knowledge management initiatives: Learning from failure. *Journal of Knowledge Management, 4*, 145–156. doi:10.1108/13673270010372279

Van der Rijta, P. (2007). *Precious knowledge: Virtualness and the willingness to share knowledge in organisational teams*. University van Amsterdam.

Wong, K. Y., & Aspinwall, E. (2005). An empirical study of the important factors for knowledge management adoption in the SME sector. *Journal of Knowledge Management, 9*(3), 64–82. doi:10.1108/13673270510602773

Zheng, W. (2005). A conceptualisation of the relationships between organisational culture and knowledge management. *Journal of Information and Knowledge Management, 4*(2), 113–124. doi:10.1142/S0219649205001110

KEY TERMS AND DEFINITIONS

Knowledge: Can be implicit or explicit and pertains to familiarity with someone or something. These include: facts, information, descriptions, or skills acquired through experience or education. Knowledge is different to both data and information.

Knowledge Management: Is a system of strategies and practices typically used in an organisation to identify, create, represent, share, and enable adoption of insights and experiences.

Knowledge Sharing: Is an activity through which knowledge (i.e., information, skills, or expertise) is exchanged among people, communities, or organizations.

Memory: In psychology, memory is the three-step process in which information is encoded, stored, and retrieved.

Organisation Structure: Consists of activities such as task allocation, coordination and supervision, which are directed towards the achievement of organizational aims. Some organisational structures are flat, while others are hierarchal such as in bureaucracies.

QR Code: Quick Response Code is the trademark for a type of matrix barcode (or two-dimensional barcode) first designed for the automotive industry in Japan.

Storage: In essence our memory, a psychological and physiological process that takes place in the human brain. In computer hardware we refer to data storage devices.

Tag: A tag usually is used to identify an object or a subject using a unique identifier. For example the dog tag is used to identify military personnel, and ear tags are used to identify animals and pets.

Trust: Reliance on another person or entity.

Section 3
Adoption of RFID Implants for Humans

Chapter 6
Willingness to Adopt RFID Implants:
Do Personality Factors Play a Role in the Acceptance of Uberveillance?

Christine Perakslis
Johnson and Wales University, USA

ABSTRACT

This chapter presents the results of research designed to investigate differences between and among personality dimensions as defined by Typology Theory using the Myers-Briggs Type Indicator (MBTI). The study took into account levels of willingness toward implanting an RFID (Radio Frequency Identification) chip in the body (uberveillance) for various reasons including the following: to reduce identity theft, as a lifesaving device, for trackability in case of emergency, as a method to increase safety and security, and to speed up the process at airport checkpoints. The study was conducted with students at two colleges in the Northeast of the United States. The author presents a brief literature review, key findings from the study relative to personality dimensions (extroversion vs. introversion dimensions, and sensing vs. intuition dimensions), a discussion on possible implications of the findings when considered against the framework of Rogers' (1983; 2003) Diffusion of Innovation Theory (DoI), and recommendations for future research. A secondary, resultant finding reveals frequency changes between 2005 and 2010 relative to the willingness of college students to implant an RFID chip in the body. Professionals working in the field of emerging technologies could use these findings to better understand personality dimensions based on MBTI and the possible affect such personality dimensions might have on the process of adoption of such technologies as uberveillance.

DOI: 10.4018/978-1-4666-4582-0.ch006

1 INTRODUCTION

The purpose of this study was to investigate differences between and among personality dimensions and levels of willingness toward implanting an RFID chip in the human body (uberveillance). Specifically, the researcher examined levels of willingness toward uberveillance taking into account whether participants were categorized as an extrovert or introvert (where an individual primarily directs his or her energy), and whether participants were categorized as sensor or intuitive (how an individual prefers to process information) as defined by a personality assessment based on Typology Theory and known as the Myers Briggs Type Indicator (MBTI).

This quantitative, descriptive study employed two instruments: one attitudinal questionnaire measuring willingness toward uberveillance; the second measuring personality dimensions utilizing the MBTI. Descriptive statistics, including measures of central tendency, measures of variability, and frequency counts were run and t-tests were used to determine if there were significant differences in levels of willingness toward uberveillance based on personality dimensions of participants. The findings are presented and interpreted taking into consideration the reported willingness of participants based on Typology Theory, Concern for Privacy (CFP), and Diffusion of Innovation Theory (DoI). The objective of this chapter is to provide professionals working in the fields of emerging technologies with findings to better understand personality dimensions that might influence the adoption of technology such as uberveillance.

2 BACKGROUND

2.1 Uberveillance

RFID implants, also known as uberveillance, are defined as an omnipresent electronic surveillance, which utilize technology that makes it possible to implant devices into the human body to track the who, what, where, when, and how of human life (Michael & Michael, 2009). In 2004, the FDA (Food and Drug Administration) of the United States approved an implantable chip for use in humans in the United States. The tiny RFID chip, which is implanted in the body, can be smaller than the size of a grain of sand. The implanted chip is being marketed as a potential method to detect and treat diseases, as well as a potential lifesaving device. If a person was brought to an emergency room unconscious, a scanner in the hospital doorway could read the person's unique ID on the implanted chip. The ID would then be used to unlock the medical records of the patient from a database. Authorized health professionals would then have access to all pertinent medical information of that individual in a database including medical history, previous surgeries, allergies, heart condition, blood type, and diabetes, to care for the patient appositely.

Technological developments are reaching new levels with the integration of silicon and biology; implanted devices in humans can now interact directly with the brain (Gasson, 2008). Implantable devices in humans for medical purposes are often believed to be highly beneficial in restoring functions that were lost. Such current medical implants include cardiovascular pacers, cochlear and brainstem implants for patients with hearing disorders, implantable drug delivery pumps, implantable neurostimulation devices for patients with urinary incontinence, chronic pain, or epilepsy, deep brain stimulation for patients with Parkinson's, and artificial chip-controlled legs (Capurro, 2010).

2.2 Social Concerns

Social concerns plague this technology (Masters and Michael, 2006). In the United States, many states are crafting legislation to balance the potential benefits of uberveillance with the disadvantages associated with the technology;

privacy and security concerns abound around data protection and such potential misuse as surveillance of individuals. California, Georgia, Missouri, North Dakota, and Wisconsin are among states in the United States which have passed legislation to prohibit forced implantation of RFID in humans. Under the Microchip Consent Act of 2010, which became effective on July 1, 2010 in the state of Georgia in the United States, no person may be required to be implanted with a microchip (regardless of a state of emergency), and voluntary implantation of any microchip may only be performed by a physician under the authority of the Georgia Composite Medical Board (Georgia General Assembly, 2010).

Through the work of Rodata and Capurro (2005), the European Group on Ethics in Science and New Technologies to the European Commission, issued an opinion in 2005. The objective of the opinion was primarily to raise awareness and dialogue concerning the dilemmas created by both medical and non-medical implants in humans which affect the intimate relationship between bodily and psychic functions basic to our personal identity. The opinion stated that implants (referred to as ICT implants or Information & Communications Technology implants), should not be used to manipulate mental functions or to change a personal identity. Additionally, the opinion stated that principles of data protection must be applied to protect personal data embedded in implants. The implants were identified in the opinion as a threat to human dignity when used for surveillance purposes, although the opinion stated that this might be justifiable in some cases for security and/or safety reasons (Rodata & Capurro, 2005).

Researchers continue to investigate social acceptance of the implantation of this technology into human bodies. A 2010 survey by BITKOM, a German information technology industry lobby group, reported 23% of 1000 respondents in technology fields would be prepared to have a chip inserted under their skin for certain benefits; 72% of respondents, however, reported they would not allow implantation of a chip under any circumstances. Sixteen percent (16%) of respondents reported they would accept an implant to allow emergency services to rescue them more quickly in the event of a fire or accident (BITKOM, 2010). Perakslis and Wolk (2006) reported higher levels of willingness relative to implantation of an RFID chip (uberveillance) when college students perceived benefits from this technology.

2.3 Predicting the Acceptance of Technology: Theories

Many models have been developed and utilized to understand factors that affect the acceptance of technology such as: The Moguls Model of Computing (Ndubisi, Gupta and Ndubisi, 2005), Diffusion of Innovation Theory (Rogers, 1983; 2003); Theory of Planned Behavior (Ajzen, 1991); The Model of PC Utilization (Thompson, Higgins, & Howell, 1991), Protection Motivation Theory (PMT) (Rogers, 1985), and the Theory of Reasoned Action (Fischbein & Ajzen, 1975, 1980). Diffusion of Innovation Theory categorizes members of the population based on traits as exhibited through behavior; Typology Theory also assumes behaviors of members of the population are predictable and classifiable (Jung, 1971).

2.3.1 Diffusion of Innovation Theory (DoI)

Everette Rogers' (1983; 2003) Diffusion of Innovation Theory, which has been used as a theoretical basis for adoption of technology (innovation), outlines stages through which technological innovation progresses toward acceptance (knowledge, persuasion, decision, implementation, and confirmation), key characteristics of an innovation which are believed to affect adoption rates (relative advantage, compatibility, complexity, trialability, and observability), and adopter categories which group members of a social system based on common traits which are believed to affect the rate of adoption. Rates of adoption are defined as the relative speed with which members of a social system adopt an innovation (i.e. RFID implants, in this study); rates of adoption are measured by the length of time required for a member to adopt the innovation. The types of adopters as defined by Rogers are as follows.

- **Innovators:** Are believed to be those members of a social system who are the fastest adopters of technology; they are believed to be venturesome and enjoy being on the cutting edge of advancements. These opinion leaders, who are believed to possess higher levels of social influence, would openly communicate the benefits or disadvantages of the innovation thereby affecting other members of the social system. Research has revealed that the percentage of members believed to be associated with this category is 2.5% of a social system (Rogers).
- **Early Adopters:** Are believed to be those members of a social system who use the data provided by the implementation of innovators to inform their own decisions about adoption. If the innovation has been effective for the innovators, early adopters will be encouraged to accept the innovation. Research has revealed that the percentage of members believed to be associated with this category is 13.5% of a social system (Rogers).
- **Early Majority:** Are believed to be those members of a social system who are more cautious than early adopters; these members are likely to stall adoption until social and/or economic benefits are perceived. Contextual pressure can motivate adoption. Research has revealed the percentage of members believed to be associated with

this category is 34.5% of a social system (Rogers).
- **Late Majority:** Are those members of a social system who are assumed to rely on trusted opinion leaders; they are believed to be more cautious and suspicious of innovation. Research has revealed the percentage of members believed to be associated with this category is 34% of a social system (Rogers).
- **Laggards:** (The slowest adopters) Are those members of a social system who are the slowest adopters; they are often very traditional and can be isolated in their social system. Research has revealed the percentage of members believed to be associated with this category is 16% of a social system (Rogers).

Rogers' DoI (Rogers, 1983; 2003) includes five characteristics of innovation which are believed to affect adoption rates of new technologies. Innovation decisions can be optional, collective, or authority-based. Optional decisions occur when the individual has the choice to adopt or reject the innovation; collective when the decision to adopt is reached by consensus among members of a system, and authority-based are those decisions which are imposed by another person or authority. In this study, it was assumed that participants would be answering questions exercising optional decision-making. The five characteristics of innovation are defined by Rogers as follows:

- **Relative Advantage:** Is "the degree to which an innovation is perceived to be better than what it supersedes. The degree of relative advantage may be measured in economic terms, but social prestige, convenience, and satisfaction are also important factors. The greater the perceived relative advantage of an innovation, the more rapid its rate of adoption" (Rogers, 1983, p15; 2003).

- **Compatibility:** Is "the degree to which an innovation is perceived as being consistent with existing values, past experiences, and needs of potential adopters" (Rogers, 1983, p 15-16; 2003).
- **Complexity:** Is the degree to which the innovation is perceived as difficult to utilize/understand.
- **Trialability:** Is the degree to which an innovation may be experimented with on a limited basis.
- **Observability:** Is the degree to which results of an innovation are visible to others.

2.3.2 Typology Theory: Personality

Throughout the 20th and 21st centuries, researchers have utilized a variety of personality theories to predict attitudes and behaviors of individuals relative to marketing, work environments, management, group effectiveness, and teams (Amato & Amato, 2005; Halfhill et al., 2005; Tieger & Barron-Tieger, 1998). Research has revealed that personality dimensions can be useful predictors of the attitudes and beliefs of individuals relative to the acceptance and use of technology (Devaraj, Easley, & Crant, 2008; Nov & Chen, 2008). Individual differences, as categorized by personality traits, have been found to exert a noteworthy force in determining success of adoption of technology (Devaraj et al.; Nov & Chen).

Gingras (1977, as cited in Zmud, 1979) and Zmud (1979) identified the following four primary psychological variables that can help classify those who would accept, or reject, technologies: user-situational variables, demographics, cognitive style, and personality. Personality, in this context, refers to "the cognitive and affective structures maintained by individuals to facilitate their adjustments to events, people, and situations encountered in life" (Gough, 1976 as cited in Zmud, 1979, p. 967). When considering person-

ality as a primary psychological variable, eight traits of personality were identified as key predictors of implementation success of technology: 1. degree of defensiveness; 2. locus of control; 3. risk taking propensity; 4. need for achievement; 5. dogmatism; 6. ambiguity tolerance; 7. extroversion/introversion; and 8. anxiety level (Klauss and Jewett, 1974, as cited in Zmud, 1979). Research revealed that information search activity was greater for those individuals with internal locus of control (Lefcourt, 1972; Phares, 1976, as cited in Zmud 1979), low levels of dogmatism (Lambert & Durand, 1977; Long & Ziller, 1965, as cited in Zmud, 1979), and a propensity for risk-taking (Dunnette, 1974 as cited in Zmud, 1979). Individuals with low levels of dogmatism were reported as more deliberate (Long & Ziller, 1965, as cited in Zmud, 1979) and less confident decision makers (Dunnette, 1974 as cited in Zmud, 1979).

Only recently, researchers have begun investigating the effect of personality on adoption of such technologies as Uberveillance, which includes the implantation of RFID in humans.

Most personality theorists believe that the whole personality of an individual can be defined as the various traits, or characteristics, that an individual possesses, in addition to the way in which the traits are related to and interact with one another. Personality traits are described as characteristics that are stable over time, motivated by needs that drive the behavior of a person, and are psychological in nature. Personality is believed to have energy that is fueled by such need systems as the necessity to maintain an adjustment of the self in relation to the world. Need systems, which create motivational drives, exist within the personality of an individual and these produce energy for the individual to behave in certain ways. The amount of energy in need systems varies because some needs are more important to an individual than others (Argyris, 1957; Bales, 1950).

Carl Jung, a psychoanalyst, was among the first to discuss Typology, or Type, Theory describing human behavior as predictable and classifiable with individuals born predisposed to certain personality preferences (Jung, 1971; Evans et al., 1998). Typology theory examines "individual differences in how people view and relate to the world" (Jung; Evans, 1996, p. 179 as cited in Evans et al.). Typologies are not developmental in the psychosocial or cognitive structural modes, but rather are intrinsic differences of the mental processing of individuals. Jung maintained that individuals have innate ways to organize experiences such as how one learns and what interests him or her. Personality theory focuses on the perception and judgment of individuals, which is thought to govern much of the outer behavior of the individual. An individual demonstrates preferences when dealing with the external environment, with perception determining what an individual sees in a situation; and judgment determining what an individual decides to do about it (Jung, 1971).

Jung (1971) contended that type development is an ongoing process in which individuals can learn to access any of the functions despite having preferred methods of interfacing with the environment. He asserted that the environment influences the development of type and that individuals are capable of utilizing each of the functions despite having a natural preference for one over another. During youth, individuals develop preferred functions; later in life individuals can achieve adequate competency in even their least preferred functions (Jung).

2.3.2.1 Myers-Brigg Type Indicator (MBTI)

In the 1940s, Isabel Myers and Katherine Briggs designed an instrument to operationalize Carl Jung's theory of psychological types. This personality assessment, known as Myers-Briggs Type Indicator (MBTI), one of the most widely used to date, classifies individuals into one of sixteen personality types. Each personality type is comprised of four of the eight total primary dimensions (Dimension 1: extroversion vs. introversion; Dimension 2: sensing verses intuitive; Dimension 3: thinking vs. feeling; Dimensions 4: judgment vs. perception). Results of the MBTI assessment identify such preferred functions as how an individual is energized (extraversion vs. introversion dimension), the kind of information to which an individual naturally pays attention (sensing vs. intuition dimension), how an individual makes decisions (thinking vs. feeling dimension), and how an individual likes to organize his or her world (judging vs. perceiving dimension) (Myers and McCaulley 1985; Tieger and Barron-Tieger, 1998). The MBTI was chosen for this study because this forced-choice instrument is most widely used with high school and college students (Evans et al., 1998).

3 A PRELIMINARY STUDY: PERSONALITY DIMENSIONS AND WILLINGNESS TO IMPLANT

3.1 Instrumentation

This preliminary study was undertaken in 2010 in the northeast of the United States. There were two instruments utilized in this study, with a combined total of 86 individual questions. The personality assessment instrument utilized in this study, the MBTI, is an intact instrument, commercially-prepared and quantitative in design. The attitudinal instrument utilized was adapted from an existing survey (Perakslis & Wolk, 2006) and content questions were further developed based on the literature. Experts were asked to review questions and format of the attitudinal instrument. Individual questions and dimension creation for the instrument were based on constructs gleaned from the literature relative to usages of uberveillance. Such usages included RFID implants in humans to reduce identity theft, as a lifesaving device, for trackability in case of emergency, as a method to increase safety and security, and to speed up the process at airport checkpoints. The attitudinal survey used a 5-point Likert-type scale ranging from "Strongly unwilling" (1) to "Somewhat unwilling" (2) to "Neutral/no opinion" (3) to "Somewhat willing" (4) to "Strongly willing"(5).

3.2 Participants

Intact groups were chosen as participants ($N = 97$) in this study. The majority of participants were not randomly assigned but rather units were chosen by means of administrator selection (Shadish, Cook, & Campbell, 2002). No special screening criterion were set up by the researcher to ascertain that participants possessed certain characteristics, with the exception that each participant was a student currently enrolled in an institution of higher education. The majority of participants, or 55%, were enrolled at institutions of education in the northeast of the United States at either the junior or senior level of college. Of participants, 76% were between the ages of 18 and 24; 24% of participants were 25 years of age or older.

3.3 Findings

The purpose of this study was to investigate differences between and among personality dimensions and levels of willingness toward implanting an RFID chip in the body (uberveillance). The findings of this study suggest there were statistically significant differences, with medium to high effect sizes, in levels of willingness toward uberveillance when taking into account whether participants were categorized as an extrovert vs. introvert (where an individual primarily directs his or her energy), and whether participants were categorized as sensor vs. intuitive (how an individual prefers to process information) as defined by the MBTI.

3.3.1 Comparison between 2005 and 2010 study: Greater Willingness to Adopt RFID Implant

Utilizing three questions previously asked of college students by Perakslis and Wolk (2006) in 2005, this current study in 2010 asked of college students the same three questions. The researcher then compared frequency counts between 2005 and 2010 relative to the willingness of college students to implant an RFID chip (uberveillance). In both studies, students were asked the same three questions: "How willing would you be to implant an RFID chip in your body as a method … (1. to reduce identity theft, 2. as a potential lifesaving device, 3. to increase national security)". A 5-point Likert-type scale was utilized varying from "Strongly Unwilling" to "Strongly Willing".

Comparisons of the results of the study conducted with college students at private and public institutions of higher education in 2005 to the results of the current research with college students of private and public institutions of higher education in 2010 showed shifts from unwillingness toward either neutrality or willingness to implant a chip in their body to reduce identity theft, as a potential lifesaving device, and to increase national security; unwillingness decreased for all these areas and willingness clearly increased.

3.3.1.1 Less Unwillingness in 2010

As depicted in Table 1, between 2005 and 2010, the unwillingness ("Strongly unwilling" and "Somewhat unwilling") of college students to implant an RFID chip into their bodies decreased by 22.4% when considering RFID implants as a method to reduce identity theft, decreased by 19.9% when considering RFID implants as a potential lifesaving device, and decreased by 16.3% when considering RFID implants to increase national security. Between 2005 and 2010, the willingness ("strongly willing" and "somewhat willing") of college students to implant an RFID chip into their bodies increased by 9.2% when considering RFID implants as a method to reduce identity theft, increased 24.4% when considering RFID implants as a potential lifesaving device, and increased 10.1% when considering RFID implants to increase national security.

3.3.1.2 Greater Willingness in 2010

The most dramatic shift in willingness with college students appears to be relative to implanting RFID chips for use as a potential lifesaving device. As depicted in Table 1, the willingness of college students in 2010 increased from 2005 by 24.4%, shifting from less unwillingness (-19.9%), and less neutrality as well (-4.5%). Perakslis and Wolk (2006) reported more willingness among college students when benefits from this technology were perceived. The chip is being primarily marketed as a potential method to detect and treat diseases, as well as a potential lifesaving device; however, an analysis of current marketing efforts of this technology and/or the exposure of participants to the marketing of this technology was outside the scope of this study. Fertile ground for future research may, therefore, exist relative to changing perceptions of the benefits of this technology.

Although key aspects of personality variables are considered to be innate, or inherited, external influences are believed to play a significant role in the manifestation or expression (behaviors) of these innate attributes (Costa & McCrae,1994, Jung, 1971). The broad sociocultural context of generations continues to be proved to affect personality development over time (Smits et al, 2011). In 2012, Perakslis & Michael investigated generational differences relative to openness to implantation of RFID chips in humans and found very statistically significant differences of opinion. In the findings of the study using chi-square analysis, Millennials (those born 1981-2000) perceived the use of microchip implants in humans as a more secure method for employee identification and access in organizations more than expected (31 vs. 16.5, adjusted residual = 4.4) ; Baby Boomers (those born 1946-1964) perceived this as a more secure method less than expected (31 vs. 16.5, adjusted residual = 4.4). Differences of opinion were clearly found when comparing generations ($\chi2 = 29.11$, $df = 2$, $p = .000$). One may conclude that the generation to which one belongs affects the expression of personality. Therefore, it may be prudent to give consideration to the generation to which the participants belong in the study conducted in 2010 because all participants were

Table 1. Willingness to implant an RFID Chip (US): Research in 2005 compared to Research in 2010

Willingness to implant an RFID Chip (U.S.): Research in 2005 compared to Research in 2010		Strongly & Somewhat unwilling	Neutral/no opinion	Strongly & Somewhat willing
IDENTITY THEFT: Willingness to implant a chip to reduce identity theft	2005: Research (Perakslis & Wolk; 2006)	55.0%	11.0%	34.0%
	2010: Research (Perakslis, 2010)	32.6%	24.2%	43.2%
	% change	*-22.4%*	*13.2%*	*9.2%*
POTENTIAL LIFESAVING DEVICE: Willingness to implant a chip as potential lifesaving device	2005: Research (Perakslis & Wolk; 2006)	42.0%	14.0%	44.0%
	2010: Research (Perakslis, 2010)	22.1%	9.5%	68.4%
	% change	*-19.9%*	*-4.5%*	*24.4%*
NATIONAL SECURITY: Willingness to implant a chip to increase national security	2005: Research (Perakslis & Wolk; 2006)	50.0%	18.0%	32.0%
	2010: Research (Perakslis, 2010)	33.7%	24.2%	42.1%
	% change	*-16.3%*	*6.2%*	*10.1%*

categorized as Millennials (those born 1980-2000). To provide the sociocultural context of the Millennials, the following section will present a synthesis of external influences on, and inherent generational traits of, Millennials that may significantly affect the adoption of technology.

One of the most significant aspects of Millennials is that they are accustomed to having gadgets that allow them to be the always-connected generation. Speed and access are keys to engage these individuals, which is often accomplished for them through technology (Pew Research Center, 2010). Researchers (Curtin et al., 2011) report that technology created information-laden formative years for the Millennial. Having grown up in this information-rich environment, Millennials are reported to value a free flow of information with much transparency. Fertile ground exists to explore if increased willingness is affected because this generation perceives value in ubiquitous living.

In addition to a wide and open acceptance of technology, this generation is believed to possess key inherent attributes that may further affect openness to technologies. Millennials are team-oriented and socialize in groups, and are better adjusted to diversity in environments than other generations (Curtin et al., 2011; Howe & Strauss, 2000). They are diverse and accepting (The Futures Company, 2011). They easily blend their in-person interactions with virtual interactions (Lippincott, J., 2010). According to a review of the literature done by Meyers (2010), more than preceding generations, Millennials value teamwork and are accustomed to collaboration. Millennials report that working and interacting with other members of a team makes work more pleasurable. Millennial workers are likely to be actively involved, fully committed, and contribute their best efforts to the organization when their work is performed in a collaborative workgroup or team (Myers, 2010; Alsop, 2008). Perhaps the openness of Millennials will be affected by a perception that implants may allow for more interconnectedness and cooperative approaches across diverse groups.

Although Table 1 shows increasing openness of Millennials over time, fertile ground exists to explore if the Millennials may have less openness to this emerging technology when taking into account the reality of the social context within which this technology may eventually exist. In example, there may develop a digital divide, or the inequality between groups in terms of the equal access to, and/or the benefits derived for all of society through the use of this technology. If Millennials perceive a digital divide, this generation may perceive issues because all members of society are not able to choose, or take advantage of, this technology. This generation is known to be greatly attuned to social responsibility and consensus; they are known to make decisions based on these factors. If some members of society could afford the implantation for enhanced safety and security, and others could not, there is an outcome which is one of inequality, potentially leading to issues of fairness and equity.

3.3.2 Finding: Extrovert vs. Introvert and RFID Implants to Reduce Identity Theft

An individual is categorized by the MBTI as either extroverted (directing energy to the outer world of activity and spoken words) or introverted (directing energy toward the inner world of thoughts or emotions). Based on data compiled between 1972 and 2002, estimates of relative frequency of extroverts in the population (49.3%) is slightly lower (1.4%) than estimates of introverts in the population (50.7%) (Center for Applications of Psychological Type; CPP, Inc; Stanford Research Institute, as cited in The Myers Briggs Research Foundation, 2010). In line with population estimates, this study revealed slightly fewer extroverts than introverts. Extroverts accounted for 46.3% of participants; introverts accounted for 53.7% of participants in this study.

Results of t-tests conducted in this study revealed statistically significant differences in the willingness of participants to implant RFID chips in their bodies when taking into account whether participants scored as extrovert vs. introvert on the MBTI. Participants in this study who scored as introverted, ($n = 52$ or 53.7% of the participants in this study), reported less willingness to implant a chip in their body as a method to reduce identity theft ($t = 2.00$, $p < .05$, $d = .41$, $M = 2.77$, $SD = 1.15$) than those participants defined as extroverted ($n = 44$ or 46.3% of participants in the study, $M = 3.27$, $SD = 1.29$).

3.3.2.1 Discussion: Concern for Privacy and Diffusion of Innovation Theory

Introverts may be less willing to implant chips in their bodies to reduce identity theft because introverts are likely to have more of a concern for privacy (CFP) when compared to extroverts, based on the definition of CFP and the trait categorizations of Typology Theory. Research revealed that personality traits appear to affect CFP, and CFP can have a negative influence on the adoption of technology (Junglas, Johnson, and Spitzmüller, 2008). CFP can be defined as an anxious sense of interest within an individual, due to various threats to the individual's freedom from intrusion (Junglas, Johnson, & Spitzmüller, 2008). Privacy issues abound relative to implantation of chips into the human body, in large part due to society's perceptions of the surveillance that would be possible; significant amounts of personal information housed on a chip would be accessible to various parties (Tootell, 2007). If privacy is the right to be left alone (Warren & Brandeis, 1890), introverts are believed to value privacy intensely. When considered through the lens of Typology Theory, introverts are often defined as private because they are believed to be withholders of information, difficult to get to know, and often secretive. In contrast, extroverts are believed to be more comfortable with sharing their lives with the public. As self-contained and self-reliant individuals, introverts are believed to

desire solitude, and freedom from interruptions and/or the involvement of others (i.e. intrusions) more than extroverts. As opposed to being defined as private, extroverts are believed to have a strong need to be around, and involved with, others (less fear of intrusion than introverts). With less drive for interaction with the external world, an introvert is assumed to have little concern with how he or she relates to other people or things, but rather how things or people relate to him or her.

The differences between extroverts and introverts, as defined by personality dimensions based on Typology Theory, provide insight into statistically significant differences found in this study relative to willingness of participants to implant a chip in their body to reduce identity theft. Introverts, as defined by the aforementioned characteristics and behavior, are likely to have higher levels of concern for privacy than their counterparts, the extroverts. Therefore, The personality dimension of introversion may coincide with higher levels of CFP, and may account for the reported lesser willingness of participants categorized as introverts in this study to implant a chip into the body with personal information that can be accessed by others (intrusion).

In this study, extroverts reported statistically significant greater willingness than introverts to implant chips in their bodies to reduce identity theft. When considering this finding against the framework of Rogers' Diffusion of Innovation Theory (DoI) (Rogers, 1983; 2003), the traits of extroverts appear to be more closely associated with the earlier (faster) adopters (innovators and/or early adopters) when considering the five adopter categories as defined by Rogers (innovators, which are the fastest adopters, early adopters, early majority, late majority, and laggards, which are believed to be the slowest adopters).

The faster adopters, who are defined as innovators and/or early adopters, are defined with such traits as very social, higher degrees of opinion leadership (informal influence over the behavior of others), and positions of central communication (i.e. more personal communication networks). Similarly, Typology Theory describes extroverts as more socially forward, possessing a propensity to share their opinions and influence others, and desire to be "tuned in" to what is happening around them while sharing their lives with others. Conversely, the traits of introverts based on Typology Theory appear to be more closely related to such later adopters as early majority or late majority. Introverts are believed to have higher degrees of skepticism and less contact with others. The similarities between traits of early adopters as described by Rogers (1983; 2003) and personality dimensions as defined by MBTI may prove fertile ground for future research to further investigate potential relationships between adopter categories and personality dimensions.

3.3.3 Findings: Sensing vs. Intuition Personality Dimension and RFID Implants

Participants in the study categorized as sensors (those who prefer to process information in the form of known facts or familiar terms) reported statistically significant more willingness than participants categorized as intuitive to implant a chip in their body for all four purposes under study.

Sensors prefer to process information in the form of known facts and familiar terms. Intuitives prefer to process information in the form of possibilities or new potential. Based on data compiled between 1972 and 2002, estimates of relative frequency of sensors in the population (73.3%) is higher than estimates of intuitives in the population (26.7%) (Center for Applications of Psychological Type; CPP, Inc; Stanford Research Institute, as cited in The Myers Briggs Research Foundation, 2010). In line with estimates of relative frequency of these two personality dimensions in the population, in this study sensors accounted for 82.6% of participants; intuitives accounted for 17.4% of participants.

3.3.3.1 RFID Implants as Lifesaving Devices: Sensors vs. Intuitives

Results of t-tests conducted in this study revealed a large effect size with statistically significant differences in willingness of participants to implant RFID chips into their bodies when taking into account whether an individual scored as a sensor vs. an intuitive on the MBTI. Participants who scored as a sensor type, ($n = 87$ or 82.6% of the participants in this study), reported statistically significant more willingness to implant a chip in their body as a potential lifesaving device ($t = 2.32, p < .05, d = .72, M = 3.63$) than those participants defined as intuitive ($M = 2.63$).

3.3.3.2 RFID Implants for Trackability in Case of Emergency: Sensors vs. Intuitives

Results of t-tests conducted in this study revealed a large effect size with statistically significant differences in the willingness of participants to implant RFID chips into their bodies when taking into account whether an individual scored as a sensor vs. an intuitive on the MBTI. Participants who scored as a sensor, ($n = 87$ or 82.6% of the participants in this study), reported statistically significant more willingness to implant a chip in their body to be trackable in case of an emergency ($t = 2.34, p < .05, d = .89, M = 3.20$) than those participants defined as intuitive ($M = 2.13$).

3.3.3.3 RFID Implants to Increase Safety and Security: Sensors vs. Intuitives

Results of t-tests conducted in this study revealed a large effect size with statistically significant differences in the willingness of participants to implant RFID chips into their bodies when taking into account whether an individual scored as a sensor vs. an intuitive on the MBTI. Participants who scored as a sensor, ($n = 87$ or 82.6% of the participants in this study), reported statistically significant more willingness to implant a chip in their body to increase their own safety and security ($t = 2.16$, $p < .05$, $d = .76$, $M = 3.32$) than those participants defined as an intuitive ($M = 2.38$).

3.3.3.4 RFID Implants to Speed Up Airport Checkpoints: Sensors vs. Intuitives

Results of t-tests conducted in this study revealed a large effect size with statistically significant differences in the willingness of participants to implant RFID chips into their bodies when taking into account whether an individual scored as a sensor vs. an intuitive on the MBTI. Participants who scored as a sensor, ($n = 87$ or 82.6% of the participants in this study), reported statistically significant more willingness to implant a chip in their body to speed up the process at airport checkpoints ($t = 2.35$, $p < .05$, $d = .97$, $M = 2.91$) than those participants defined as intuitive ($M = 1.75$).

3.3.3.5 Discussion

Participants in this study who were categorized as sensors reported statistically significant greater willingness than intuitives to implant chips in their bodies relative to the four potential usages. The sensing and intuition scale (sensor vs. intuitive) is believed to represent the greatest potential for differences between people. The sensor and intuitive dimensions of personality categorize how an individual prefers to take and process information, which is believed to have a considerable influence on one's world view. The greatest potential for difference is assumed to be between these two personality dimensions. The substantial delineation that is assumed between these two dimensions of personality might inform the statistical significance in this study that was found to be consistent across four potential usages of uberveillance. The researcher opted to compare behavioral (personality) traits as defined by MBTI with traits defined by Rogers' DoI (Rogers 1983; 2003), which addresses adoption of technology.

When reviewing the personality dimension of sensors, the traits associated with these individuals may affect the perception of the innovation when considering one of Rogers' (1983, 2003) characteristics of innovation: relative advantage. Relative advantage is described as the degree to which an innovation is perceived as better than what it supersedes and the degree to which the innovation can be measured. Sensors are likely to perceive RFID implants with higher levels of relative advantage than intuitives because sensors have a high regard for practical solutions. RFID implants are likely to be viewed as a concrete, functional solution as a potential lifesaving device, for trackability in case of an emergency, to increase one's own safety and security, and to speed up the process at airport checkpoints. Conversely, intuitives are known to be more theoretical, focusing on the underlying causes of problems or the possibilities offered by new ideas. Intuitives might prefer to consider how else RFID could be applied in various settings as opposed to the solution presented. Sensors favor established actions that are designed to have an immediate effect; especially if the effect can be somehow measured (i.e. a lifesaving device or increasing one's own safety and security) and Rogers reports that measurability is directly related to higher degrees of perceived relative advantage. Intuitives are more likely to think about such future implications as: what is "between the lines" relative to this new technology; and what is the meaning of this new technology. Individuals who are categorized as sensors also place high value on setting up systems and following procedures; they are believed to need tools that will allow them to keep a logical, practical flow to their work (i.e. such measurable systems as airport security systems that ensure robust processes). Intuitives are known to prefer innovative and creative solutions, solutions not necessarily built upon established systems but rather upon what could be. As literal individuals who prefer facts and details to interpretations, sensors are also assumed to accept new applications (i.e. RFID implants) for what has already been invented or established (i.e. RFID as currently used in society). Although uberveillance is an emerging use of RFID technology, RFID is likely to be perceived by sensors as having long been established with various usages in everyday life.

4 RECOMMENDATIONS

Professionals working in the field of emerging technologies can utilize Typology Theory, as defined by such tools as MBTI, to better understand rates of adoption of new technologies, as well as how the characteristics of an innovation are likely to be perceived by individuals with certain personality dimensions. Millennials were highlighted as key participants who reported increasing openness to implants in 2010 when compared to the openness of participants in 2005. Personality dimensions could be taken into account to tailor information about innovation to address key internal motivational drives identifiable based on personality dimensions.

Concern for Privacy, which is likely to be high for introverts who had less willingness to implant chips in this study, must be addressed with vigor and in a manner that alleviates the potential for intrusion (i.e. violation of privacy). Extroverts, who are likely to be faster adopters when compared to introverts, are likely to influence the progress from creation/introduction to adoption as this technology is communicated among the members of a social system. Extroverts are believed to be opinion leaders (influencers) more so than introverts. These imperceptible forces of social influence within individuals and between individuals in a social system make people conform. Such forces as social proof and conformity explain how individuals are influenced by the actions of others (i.e. Rogers' opinion leaders) and even more so by the actions of many others (Bandura & Menlove, 1968; Napier & Gershenfeld, 2004). Overcoming the objections of extroverts at the outset of innovation is likely to affect rates of adoption because extroverts often are opinion leaders affecting later adopters.

Persuading sensors to perceive the relative advantage of technological innovation in a more enhanced manner, is likely to affect rates of adoption with this personality dimension, which are believed to account for approximately 73.3% of the population. Robustly addressing the future implications and "what's between the lines" of this technology is likely to resonate with intuitives.

5 FUTURE RESEARCH DIRECTIONS

The research and findings presented in the chapter were limited to a small group of participants enrolled in only two colleges in the United States, and in particular students who self-assigned to courses; this decreases likelihood that the findings could be generalized to the population. To expand the scope of the study, researchers might test the transferability of the results of the findings with students at alternative institutions of higher education (Brown, 2001; Creswell, 2003). Additionally, fertile ground exists for future research that would investigate potential correlations between Rogers' (1983; 2003) adoption categories and personality dimensions to inform adoption rates.

The need for qualitative data to inform the quantitative findings also serves as fertile ground for future research. Focus groups are likely to yield valuable data to better explain the role of personality dimensions relative to the perceptions of participants about such technological innovations as RFID implants in the human body.

6 CONCLUSION

The author presented key findings relative to the personality dimensions (extroversion vs. introversion dimensions, and sensing vs. intuition dimensions) and willingness to implant RFID chips into the human body (uberveillance). Extroverts reported statistically significant more willingness than introverts to implant RFID chips in the body to reduce identity theft. Sensors were more willing than intuitives to implant RFID chips in the body for four such potential usages: as a potential lifesaving device, to be trackable in case of an emergency, to increase one's own safety and security, and to speed up the process at airport checkpoints. The author concluded with recommendations based on Typology Theory and rates of adoption.

REFERENCES

Alsop R. (2008, November 2). Coddled kids hit corporate culture. *St. Petersburg Times*, p. F2.

Amato, C., & Amato, L. (2005). Enhancing student team effectiveness: Application of Myers-Briggs personality assessment in business courses. *Journal of Marketing Education*, 4(27), 41–51. doi:10.1177/0273475304273350

Argyris, C. (1962). *Interpersonal competence and organizational effectiveness*. Homewood, IL: Irwin.

Bales, R. (1950). *Interaction process analysis: A method for the study of small groups*. Reading, MA: Addison-Wesley.

Bandura, A., & Menlove, F. (1968). Factors determining vicarious extinction of avoidance behavior through symbolic. *Journal of Personality and Social Psychology*, 8(2), 99–108. doi:10.1037/h0025260 PMID:5644484

Brown, J. (2001). *Using surveys in language programs*. Cambridge, UK: Cambridge University.

Capurro, R. (2010). *Ethical aspects of ICT implants in the human body*. Paper presented at the Meeting of the IEEE Symposium on Technology and Society (ISTAS10). New South Wales, Australia.

Devaraj, S., Easley, R., & Crant, J. (2008). How does personality matter? Relating the five-factor model to technology acceptance and use. *Information Systems Research*, 19(1), 93–105. doi:10.1287/isre.1070.0153

Ethics Resource Center. (n.d.). Millennials, gen x and baby boomers: What do they think about ethics? In *2009 National Business Ethics Survey*. Retrieved from http://ethics.org/files/u5/Gen-Diff.pdf

Evans, N., Forney, D., & Guido-DiBrito, F. (1998). *Student development in college: Theory, research, and practice*. San Francisco: Jossey-Bass.

Futures Company. (n.d.). *Millenials ahead*. Retrieved from http://www.lifebenefits.com/lb/pdfs/Millennials_Ahead_Report.pdf

Gasson, M. (2008). ICT implants: The invasive future of identity? *Advances in Information and Communication Technology*, 262(2), 287–295.

Georgia General Assembly. (2010). *Senate bill 235*. Retrieved January 12, 2011, from http://www1.legis.ga.gov/legis/2009_10/versions/sb235_As_passed_Senate_5.htm

Goleman, D. (2006). *Social intelligence: A new science of human relationships*. New York: Bantam Dell.

Goleman, D., Boyatzis, R., & McKee, A. (2004). *Primal leadership: Learning to lead with emotional intelligence*. Boston: Harvard Business School Press.

Halfhill, T., Sundstrom, E., Lahner, J., Calderone, W., & Neilsen, T. (2005). Group personality composition and group effectiveness. *Small Group Research*, 36(1), 83–105. doi:10.1177/1046496404268538

Howe, N., & Strauss, W. (2000). *Millennials rising*. New York: Vintage Books.

Huff, L., Cooper, J., & Jones, W. (2002). The development and consequences of trust in student project groups. *Journal of Marketing Education*, *24*(1), 24–34. doi:10.1177/0273475302241004

Jung, C. (1923/1971). *Psychological types*. Princeton, NJ: Princeton University Press.

Junglas, I., Johnson, N., & Spitzmüller, C. (2008). Personality traits and privacy perceptions: An empirical study in the context of location-based services. *European Journal of Information Systems*, *17*(4), 387–402. doi:10.1057/ejis.2008.29

Lippincott, J. K. (2010). Information commons: Meeting millennials' needs. *Journal of Library Administration*, *50*(1), 27–37. doi:10.1080/01930820903422156

Masters, A., & Michaels, K. (2007). Lend me your arms: The use and implications of humancentric RFID. *Electronic Commerce and Applications*, *6*(1), 29–39. doi:10.1016/j.elerap.2006.04.008

McCorkle, D., Reardon, J., Alexander, J., Kling, N., Harris, R., & Iyer, R. (1999). Undergraduate marketing students, group projects, and teamwork: The good, the bad, and the ugly? *Journal of Marketing Education*, *21*(2), 106–117. doi:10.1177/0273475399212004

Michael, K., & Michael, M. (2004). The social, cultural, religious, and ethical implications of automatic identification. In *Proceedings of the Seventh International Conference in Electronic Commerce Research*, (pp. 433-450). IEEE.

Michael, K., & Michael, M. (2006). A note on uberveillance. In K. Michael, & M. Michael (Eds.), *From dataveillance to uberveillance and the realpolitik of the transparent society* (pp. 9–25). Wollongong, Australia: University of Wollongong.

Myers, I., & McCaulley, M. (1985). *Manual: A guide to the development and use of the Myers-Briggs type indicator*. Palo Alto, CA: Consulting Psychologists Press.

Myers, K. K., & Sadaghiani, K. (2010). Millennials in the workplace: A communication perspective on millennials' organizational relationships and performance. *Journal of Business and Psychology*, *25*(2), 225–238. doi:10.1007/s10869-010-9172-7 PMID:20502509

Myers Briggs Research Foundation. (2010). *How frequent is my type?* Retrieved January 3, 2011, from http://www.myersbriggs.org/my-mbti-personality-type/my-mbti-results/how-frequent-is-my-type.asp

Napier, R., & Gershenfeld, M. (2004). *Groups: Theory and experience* (7th ed.). Boston: Houghton-Mifflin.

Nov, O., & Chen, Y. (2008). Personality and technology acceptance: Personal innovativeness in IT, openness and resistance to change. In *Proceedings of the 41st Hawaii International Conference on System Sciences*, (pp. 433-450). IEEE.

Perakslis, C., & Michael, K. (2012). Indian millennials: Are microchip implants a more secure technology for identification and access control? In *Proceedings of IEEE International Symposium on Technology and Society* (ISTAS12). IEEE.

Perakslis, C., & Wolk, R. (2006). Social acceptance of RFID as a biometric security method. *IEEE Symposium on Technology and Society Magazine*, *25*(3), 34-42.

Pew Research Center for the People and the Press. (n.d.). *Millenials: A portrait of the generation next*. Retrieved from http://www.pewsocialtrends.org/files/2010/10/millennials-confident-connected-open-to-change.pdf

Pew Research Center for the People and the Press. (n.d.). *Millenials: A portrait of the generation next*. Retrieved from http://www.pewsocialtrends.org/files/2010/10/millennials-confident-connected-open-to-change.pdf

Rodota, S., & Capurro, R. (2005). *Opinion n°20-16/03/02005: Ethical aspects of ICT implants in the human body.* Retrieved December 12, 2010, from http://ec.europa.eu/european_group_ethics/docs/avis20_en.pdf

Rogers, E. (1983). *Diffusion of innovations.* New York: Free Press.

Rogers, E. (2003). *Diffusion of innovations* (5th ed.). New York, NY: Free Press.

Shadish, W., Cook, T., & Campbell, D. (2002). *Experimental and quasi-experimental designs for generalized causal inference.* Boston: Houghton Mifflin.

Smits, I., Dolan, C., Vorst, H., Wicherts, J., & Timmerman, M. (2011). Cohort differences in big five personality factors over a period of 25 years. *Journal of Personality and Social Psychology, 100*(6), 1124–1138. doi:10.1037/a0022874 PMID:21534699

Tieger, P., & Barron-Tieger, B. (1998). *The art of speedreading people: Harness the power of personality type and create what you want in business and in life.* Boston: Little, Brown and Company.

Tootell, H. (2006). A note on uberveillance. In K. Michael, & M. Michael (Eds.), *From dataveillance to uberveillance and the realpolitik of the transparent society* (pp. 9–25). Wollongong, Australia: University of Wollongong.

Warren, S., & Brandeis, L. (1890). The right to privacy. *Harvard Law Review, 4*(5). doi:10.2307/1321160

Zmud, R. (1979). Individual differences and MIS success: A review of the empirical literature. *Management Science, 25*(1), 966–979. doi:10.1287/mnsc.25.10.966

ADDITIONAL READING

Fischer-Hübner, S., Duquenoy, P., Zuccato, A., Martucci, L., & Gasson, M. (2008). ICT implants: The future of identity in the information society. *IFIP Advances in Information and Communication Technology, 262*(2), 287–295.

Michaels, K., McNamee, A., & Michaels, M. (2006, July). The emerging ethics of humancentric GPS tracking and monitoring. Proceedings of the International Conference on Mobile Business. Copenhagen, Denmark.

Michaels, K., & Michaels, M. (2006). A note on uberveillance. In K. Michael, & M. Michael (Eds.), *From dataveillance to uberveillance and the realpolitik of the transparent society* (pp. 9–25). Wollongong, Australia: University of Wollongong.

Schummer, J. (2005). Societal and Ethical Implications of Nanotechnology. *Meanings, Interest Groups, and Social Dynamics, 8*(2), 56.

KEY TERMS AND DEFINITIONS

Dataveillance: The tendency for governments and private businesses to use computerised data to make decisions affecting, or monitor populations and consumer behaviour. Frequently, decisions about public or private service provision are based on the data about a person, rather than a person (see surveillant assemblage).

Due Process: The series of rules, procedures and rights designed to protect the individual against the power of the state, constraining the activities of state agencies in dealing with citizens under the criminal law. Examples of rules that involve due process include the right to silence, the right to legal representation during police questioning, the obligation on police to have clear evidence of criminal behaviour before entering and searching private property.

e-Governance: The use of computerised methods of data collection and sorting to streamline government and public service delivery.

ID Scanner: An electronic device comprising various combinations of a portable camera, image scanner, biometric fingerprint reader and computer, designed to create a replica of a person's identity documents to ensure authorised entry into public or private premises. In the night-time economy, these devices enable a person's identity to be collected and stored in a licensed venue and/or a computer network to enable security personnel to prevent undesirable or banned patrons from gaining entry. These technologies are considered to minimise the prospect of disorder and violence occurring within hotels and nightclubs via deterrence, and promoted as enhancing the ability to identify offenders.

Night-Time Economy: The development of urban precincts to enable increased commercial trade, largely through entertainment, restaurants and licensed venues, that are specifically promoted to operate outside of daytime business hours.

Privacy: The series of legal protections governing the use of personal information for public and private service delivery.

Public Policing: Government agencies and agents with specific legal powers to help promote order and investigate crime for the public benefit.

Security: The range of human and technological measures designed to prevent losses to governmental agencies, private businesses and the community. Security is commonly linked to preventing crime, rather than prosecuting suspected offenders after a crime has been committed.

Surveillant Assemblage: The multiple agencies (state, non-state and hybrid) engaged in surveillance and the various processes, forms and purposes of surveillance that is brought together through information sharing networks or other means (from the sale of access to databases through to secret court orders demanding technology companies transfer megadata to state agencies) to create more intensive and comprehensive surveillance capacities. This data can then be used to inform policing practices to target or distinguish between 'desirable' and 'undesirable' in a range of settings, further intensified by the use of surveillance technologies to enforce such divisions.

Technological Determinism: The belief that automated technologies can solve complex social problems, including crime.

Interview 6.1
Conversation with a Minor
about Chips and Things

Minor, Male, 16 years of age, Campbelltown, NSW, Australia
Interview conducted by Katina Michael on 19 February 2009 at the University of Wollongong

Katina Michael: As I said, my area of expertise is in the social implications of technology. My first question is whether or not you were aware that people have implanted themselves with chips that allow for instance their front door to unlock?

Interviewee: No idea.

Katina Michael: And what are your immediate reactions to learning about this practice?

Interviewee: I am amazed, actually. It's interesting. It's different, but … yeah. I don't know. It's all come as a shock to me.

Katina Michael: Okay.

Interviewee: It's good to know more about it.

Katina Michael: Yeah. Would you ever consider getting a chip implant?

Interviewee: Yeah.

Katina Michael: You would?

Interviewee: Definitely.

Katina Michael: On what grounds?

Interviewee: To be different, and to do something, like, out of the ordinary. To have something that other people don't.

Katina Michael: Okay. And as a teenager, do you think you would require consent from your parents to get one, or do you think that should be your own choice?

Interviewee: I would tell them. I'd tell them beforehand, but whether they say "yes" or "no", depending on the outcomes and possible results, I'd still do it, if it interested me.

Katina Michael: Okay. Why do you think people might not get a chip? Just say tomorrow, a scenario was that everyone was to receive one, and it wouldn't make you different. For example, it wouldn't be unique, or it wouldn't be a small minority that were chipped. Do you think you'd still get a chip implant, or then you'd try and resist it if everyone had one?

Interviewee: Probably not, if everyone had one.

Katina Michael: Right- okay.

Interviewee: I'd ... again, depending on what it does. Like, if it performs something amazing, then yes, but if it's just ... like, for example, just like the ID thing, then no. If, it was something really unique and amazing, definitely, but other than that, no.

Katina Michael: Okay. Would you differentiate between a chip implant in the body and piercing, as in body piercing? Do you think there's a difference between the two?

Interviewee: Yeah, definitely. One's ... I mean, body piercing, if you get a ... what's it called ...

Katina Michael: The implant.

Interviewee: The implant, sorry, that'd be more for a purpose of something different, whereas piercings are more for just appearance to people. Like, you wouldn't get an implant to show off, sort of thing, whereas you get earrings to, like, look good.

Katina Michael: Okay. So you sort of think one is visible to the outsider ...

Interviewee: Yeah.

Katina Michael: ... the other is a hidden thing?

Interviewee: Yeah. Well, I mean you're going to still show people, but it's more for a purpose. Like, you're not getting it just to please everybody, pretty much.

Katina Michael: Okay. That's very interesting. Do you differentiate between different types of implants? For example, someone who needs a cochlear implant to hear, vs. an implant for convenience that would allow you to open your car door?

Interviewee: Definitely the hearing. Yeah. The car, it's just selfish.

Katina Michael: Okay. You said at the beginning that you might consider getting implanted because it is something cool or something different, something unique. But what kinds of applications do you think you would go for... What kind of application would convince you to go for it?

Interviewee: That's a tough one. Something that's … that will freak people out. Like, something that … like, I don't know. I can't think of anything on the spot, but, like, something that can possibly change things. I don't know, like … I can't explain this. Yeah, I don't know. But something different and unique that, like, can freak people out, pretty much. Like, you know, those freaks on TV and stuff. You could do something like that. You know, maybe make part of your body grow, and then shrink, or I don't know. It's just … something like that.

Katina Michael: Yeah. So not just like your everyday computing, but something pretty much that we've not seen before. Something …

Interviewee: Yeah. Something …

Katina Michael: Advanced?

Interviewee: Yeah.

Katina Michael: Interesting. Do you think that would be … well, like, if we call these technologies enhancement technologies, or amplification, there are some guys out there that are looking at, like, a second ear function, or hearing further away, or seeing further away.

Interviewee: Yeah, something like that.

Katina Michael: So … would that be acceptable to you, do you think?

Interviewee: Yeah. Like, if you could be able to hear someone. You know …

Katina Michael: Tune in.

Interviewee: If they're a long way away, and you can just do that, and you can hear what they're talking about. Yeah, that'd be amazing.

Katina Michael: Okay.

Interviewee: Even see further, something like that.

Katina Michael: Do you think humans should go that way?

Interviewee: Well, I believe that if that does happen people are going to start getting greedy, and then … defects will start happening. Like, wrong things will happen, and then people will just … I don't know, possibly just … things will happen computer-wise to the body, and then, you know, you … yeah, stuff like that. They might overtake the human body and, you know, God knows what will happen next.

Katina Michael: And tell me what you think … at that point, you mentioned the positive effects. Like, what are the social benefits, do you see, of this kind of implant technology?

Interviewee: Well, you'd be seen differently, obviously. Like, people would think either you're a freak, you're cool, or, or, yeah, you're just pretty much different. It just depends on what crowd you're actually targeting, or showing that to, or … like if you're doing it strictly for computing, or research, obviously it's important to them, but if you're just doing it for a laugh with your friends, or just to show off, sort of thing, like, seeing further away … people might see you as a freak, or they might think, "You know, he's cool, let's hang out with him," and stuff like that.

Katina Michael: And if I was to ask you about applications with the implant that you would never want to see implemented in society …

Interviewee: What do you mean?

Katina Michael: Do you believe that we should control which applications should be developed, or do you think people should be free to develop whatever applications they want?

Interviewee: Yeah, I mean, if you can do it, do it. Like, if it's possible, it would be amazing to, you know, like you said, accelerate human knowledge to science and technology. If you can, it would be amazing. But then again, you've got to think about the side effects and what possibly could happen. Always look at the bad things before the good, sort of thing.

Katina Michael: Okay. And do you see any social costs or risks with this kind of technology implementation in the future?

Interviewee: I'm guessing it's really expensive to do, and then you always have to find someone that's really keen to put themselves as a guinea pig, pretty much, like you can't just choose some average Joe on the street and say, "Hey, do you want to get shot" you know. You can't say, "Let's put it in there … let's put some piece of metal in you." So it's got to be someone that knows what they're doing, possibly someone that's really experienced with that technology, and take it from there.

Katina Michael: Okay. That's very interesting. Do you think there are any risks to the person, even if they're experienced?

Interviewee: Well, it depends what the implant is.

Katina Michael: Okay.

Interviewee: Well, I mean, there's … if you were trying to change the brain with sight, for example, hearing, I guess maybe possibly it could deafen them more, or blind them even more. You know, they could lose feeling in the nervous system to any part of the body.

Katina Michael: Yes.

Interviewee: … and cause them to start growing… anything is possible I guess but, it depends what the implant is.

Katina Michael: You mentioned before that you could be considered cool or a freak, depending on which group you were in. So do you see this as cultural issues? Like, perhaps, do you think the Australian culture would be different in perception of these technologies to maybe the Chinese culture?

Interviewee: No, not really culture. I'm just saying, like, as a teenager, like, I see a lot of people being accepted and not accepted. A lot of different groups that people take different interests. So, for example, if I'm part of a group that are interested in that kind of stuff, that'd be good for me. But if I'm someone like, just, done that to sort of go into a different group, people will possibly look at me and think, you know, "What's this freak doing?" Like I've seen and stuff.

Katina Michael: If I was to ask you a hypothetical, just say this was to become everyday life. Like, I've got my Blackberry here, my smart phone, which allows me to track and monitor my movements 24 x 7. What if everyone had one? Do you think … would your group accept it?

Interviewee: For technology-wise …

Katina Michael: The group that you're with.

Interviewee: … I'm not too sure. I don't think … well, I haven't seen many groups that sort of base their friendship or their belonging together on technology. Like, whether you have the fastest computer or slowest computer, best phone, worst phone, you know, stuff like that. It's more … it's just more what you have to give to people is how much they take, I guess.

Katina Michael: Common interests, maybe.

Interviewee: Yeah. Like, … yeah, if, like, because if you're hanging out for a group that's been together for a while, they sort of act a certain way, and then an outsider comes in and you either adapt to them, or you don't, and you move somewhere else, pretty much.

Katina Michael: Okay. Do you think there are any … we've mentioned social issues, cultural issues. Do you think there are any religious issues that people might have concerns with the technology-

Interviewee: Definitely. Well, I mean, me coming from a Greek Orthodox background, like, for example, earrings and tattoos and stuff like that are considered not allowed, against our religion to do it. But then again, people still do it, a lot of Orthodox people do tattoos... I've got earrings. It's just … it's just up to you. I mean, it depends how religious you are, and I mean, it's not like, you know, the religion's going to stop you from living your life, because you only live once, so you might as well try to experience as much as you can.

Katina Michael: With respect to your faith, has anyone ever told you why you should or should not be getting body piercing done or tattoos or the like?

Interviewee: No, it's just the norm … it's just the standard "You just shouldn't do it."

Katina Michael: An etiquette?

Interviewee: That's all.

Katina Michael: With no explanation?

Interviewee: … that's all you hear. And then you're getting, "It's just against the religion." It's like my mum says, "Don't do it, because it's against your religion. Don't do it because it's against your religion." But it's only happened, like, once or twice. I mentioned getting a tattoo, and she's like, "You know, you're not allowed to get it. You're not allowed, because it's against the Bible," and stuff like that. So…

Katina Michael: Very interesting.

Interviewee: … whether I do or not, it's something different.

Katina Michael: Thank you for that. Would you get implanted if it made your life easier at home, so that you could switch on your radio, or your computer, or your TV?

Interviewee: No. That's just lazy.

Katina Michael: Okay.

Interviewee: I mean, that's what they created the remotes for, pretty much.

Katina Michael: A remote for a remote possibility… [laughing]

Interviewee: Pretty much, yeah.

Katina Michael: Yeah. Would you accept an implant for prosthesis if you got sick? For example, if you had diabetes and you had to have drug delivery, automated insulin delivery in your body, would you accept an implant for that… to help you with diabetes, to regulate the diabetes?

Interviewee: Well, it depends how serious the disease I had was… like, if there was a potential cause to harm, say, my family, my mum, my dad, possibly a future wife, my kids… I would try to obviously avoid any damage as possible.

Katina Michael: If the government tomorrow said, "We're introducing a national ID scheme, and it's based on a chip implant," would you accept the ID chip?

Interviewee: Depends on what I'd do. Like, it depends on my hobbies, it depends on my job, sort of thing. If I'm working in a company that requires that, then I guess so, but if it's just, like, just for the sake of it, then probably not. I wouldn't change or get something added to me for no …

Katina Michael: For no purpose.

Interviewee: … specific reason, yeah, or an important reason.

Katina Michael: What if school said they wanted to track your attendance at school? What would you say then? Even between periods?

Interviewee: Definitely not. They're dreaming.

Katina Michael: I like it.

Interviewee: Typical answer from any teenager.

Katina Michael: Do you think in the future everyone might be carrying an implant for identification purposes?

Interviewee: No, I don't think they will for … like I said, unless they need to, pretty much. I mean, it's not that hard to carry a little card and swipe it, or to scan it, pretty much.

Katina Michael: Yeah. Have you seen any sort of sci-fi movies that you've been influenced by in the area that we've been talking about?

Interviewee: I've actually never heard of implants ever …

Katina Michael: Okay.

Interviewee: … apart from today …

Katina Michael: Okay. So I just want to say thank you for this very interesting conversation, and for your time and to reaffirm that this interview will not be published with identifying information given you are under the age of 18.

KEY TERMS AND DEFINITIONS

Amplification: The act or result of amplifying, enlarging, or extending the capabilities of the human body. May be to correct a dysfunctional organ or grant more ability to a working function.

Cochlear Implant: Is a surgically implanted electronic device that provides a sense of sound to a person who is profoundly deaf or severely hard of hearing.

Cool Dude: Is a person who is always calm, sociable, and usually has a great sense in music and can talk to anyone who approaches him/her. Ultra-confident person who does not mind being bothered but is always willing to help. Usually, but not always, up with the latest trends.

Faith: Is the substance of things hoped for, the evidence of things not seen.

Freak: Is commonly used to refer to a person with something strikingly unusual about their appearance or behaviour.

Piercing: Is a form of body modification. It is the practice of puncturing or cutting a part of the human body, creating an opening in which jewelery may be worn.

Prosthetics: Is the provision of cosmetic and/or functional artificial limbs (prostheses) for people who have had an amputation or have a congenital deficiency.

Tattoos: A form of body modification, made by inserting indelible ink into the dermis layer of the skin to change the pigment.

Chapter 7
Surveilling the Elderly:
Emerging Demographic Needs and Social Implications of RFID Chip Technology Use

Randy Basham
University of Texas – Arlington, USA

ABSTRACT

This chapter describes the usefulness of RFID (Radio Frequency Identification Device) implant technology to monitor the elderly, who are aging in place in various retirement arrangements, and who need to maintain optimal functioning in the absence of available, and on location, service or care providers. The need to maintain functioning or sustainable aging is imperative for countries experiencing rapid growth as a demographic trend for the elderly. The chapter also raises some concerns including the social acceptance or rejection of RFID implant technology, despite the utility of the device. These concerns include a variety of political, social, and religious issues. Further, the chapter also attempts to show how RFID implant technology could be used in combination with other emerging technologies to maintain physical, emotional, and social functioning among the growing population of elderly. What follows is the introduction and a partial literature review on emergent elderly needs, and on the utilization of RFID and other technologies.

INTRODUCTION

Services to the aged population have been steadily improving in developed or developing nations over a number of decades, in part due to policy and funding allocations to promote improvements in health and quality of life for the elderly, and in part due to the development of a number of tools involving some measure of advances in technology. Emerging technologies, such as the RFID tag and responder implants, may make possible the capacity to remain a functioning, productive

DOI: 10.4018/978-1-4666-4582-0.ch007

and engaged member of a contributing portion of the society of which they are members, for many more years than previously expected for those who are willing and those who have some level of access to such technology. As the demand for these services and emerging technologies are expected to increase over the next several decades in both developed and developing countries, sustainable aging may become a more commonly understood construct and social and economic reality.

LITERATURE ON THE ELDERLY AND DEMOGRAPHY

Globally, there are some alarming growth trends in the numbers of elderly and projected elderly relative to anticipated service needs, due to the rapid and belatedly anticipated swelling of the aging portions of the population. For example, for the first time in recorded history, the number of surviving elderly members of the planet will be greater than the number of living children aged five years and younger. A recent United Nations estimate suggests that the global elderly population will more than double over the next forty years. Europe is also expected to lead in the trend having the largest proportion of elderly per population during this period. However, China and India, having larger populations, also have greater total numbers of elderly than most other countries, with these numbers expected to triple during the same time frame (The Demographics of Aging, 2011).

Demographic trends in these developed and developing countries suggest, however, that fewer young people will be available to meet the labor and economic needs of their societies while servicing larger and larger segments of the population who are expected to retire over the next several decades. This is also true of Asian populations in developed areas, as well as developed Middle Eastern, African and Western cultures. As a result, immigrations trends in these areas are expected

to increase, in part due to service needs for the elderly. The term "aging in place" has emerged to describe the less mobile, more service dependent and possibly less productive and less functional, elderly. Service needs for these elderly, and others, may be generally conceptualized as falling into the categories of physical, emotional and social needs, as differentiated in the aging literature.

A global investments organization projects that the world population of above age 65 years is expected to increase from 6.9% in the year 2000 to 19.3% by the year 2050 (International Wealth Solutions, Ageing Demographics, 2008). Population growth overall is expected to slow, however, with decreases in fertility rates. The United States, a highly developed country, may be used as one example. The already large aging population, estimated at 12.3% in the year 2000, will be increasing to 21.1% by the year 2035. As a result of these demographic shifts in the proportion of the aging population, there may be insufficient numbers of available laborers to service either the needs of the elderly, or in some cases, segments of society as a whole. Employment is expected to be available in surplus, but not enough workers will be available in some developed areas to suffice these needs, resulting in expected labor shortages (Foreign-Born Workers and Baby Boomers, 2010).

There are a number of causes for the changing demographics relative to aging in industrialized first world countries. Fluctuations in birth rate are part of the issue, as with the well known baby boomers of the United States. A boom, or expansion of the number of births, began at the close of World War II in the Unites States, which has dramatically contributed to the current number of elderly. Another reason for a growing elderly proportion of the population is declining fertility rates in several large developed countries. This may be due to policy, as in the case of China where one child to a family has been the policy and expectation for some time, or due to increases in utilization of family planning services, abortion, and increased utilization of contraceptives.

Yet, in other instances, increases in longevity, due to prosperity, improvements in nutrition, health care technology and decreases in various disease morbidity and mortality rates due to public health efforts, may be contributing factors to societal aging in the more technological and industrially advanced nations. These identified causal factors may be addressed in future by social planners over the longer term, but do little to address the impending demographic shift toward large proportions of elderly in these nations. Addressing the demands placed on society over the shorter term will require the needs of the elderly to be moderated to some degree. Successful aging, according to Gilmer and Aldwin (2003) includes the following: no physical disability over the age of seventy five as rated by a physician; a good subjective health assessment (i.e. good self-ratings of one's health); length of "undisabled" life; good mental health; objective social support; self-rated life satisfaction in eight domains (marriage, income-related work, children, friendship and social contacts, hobbies, community service activities, religion and recreation or sports). Further, according to the American Association of Retired Persons in the United States, it is expected that nine out of ten of the elderly will prefer to age in the homes in which they reside, at retirement (The Demographics of Aging, 2011).

In response to the need to maintain a contributing and healthier elderly population these nations have sought various solutions, other than technology, to meet the increasing demand for more service providers and professionals to address these needs (Atul, 2006), as well as to address the more immediate need to increase the overall supply of workers to service their respective economies. These attempted workforce solutions, have included a relaxation of immigration policy (Weil, 2002), various workforce retraining programs (Carmel & Lowenstein, 2007), and changes in labor policy directed at increasing overall labor outputs (Brown & Braun, 2008). Yet the alteration of immigration policies, labor laws, and voca-

tional training initiatives has not kept pace with the service needs of the aging populations of the developed and developing countries.

Technology has been proposed to bridge the service need versus the available care provider gap for the elderly by a number of proponents (Hargreaves, 2010). Numerous tracking, monitoring and identification technologies have been proposed to promote the ongoing health and welfare of elderly members of society and to facilitate their continued functioning and contributions to a productive society. This interest is due in part to recognition of the need to have increased numbers of successful aging, among the elderly populations, or due to societal aging of the industrialized nations (Harper, 2006). Recent and emerging technological devices, may bridge the gap by simply serving as extensions of earlier interventions such as pacemakers and joint implants capable of enhancing or prolonging human functioning and independence. Some devices are capable of providing diagnostic analytics, and information that may regulate healing and functioning, These passive and implantable communication devices are capable of informing the healers and care providers and professionals thereby maintaining elderly functioning (Michael & Michael, 2013).

The costs of traditional forms of care for the elderly to the respective governments and health programs is expected to be unaffordable for most of the affected countries, especially following a period of economic downturn (Spillman, 2004). However, the introduction of various forms of monitoring and computing applications could facilitate an increased number of the elderly achieving a level of successful aging and possible continued productivity (Rajasekaran, Radhakrishnan, & Subbaraj, 2009). Self care, or collaborative self care, to avoid disability may be achieved with health education programs and applications, personal medical monitoring devices and digital communication. Mental health and personal outlook, as well as retirement financial status, may be improved by using technologies to transfer skills and expertise of the elderly to less experienced or younger associates. Social and community activities may be streamlined to correspond to the aging person's interests or hobbies and pushed technologically, or made available for the elderly, daily or as needed. Such technologies could be personalized to improve outcomes for needy seniors and the outcomes accomplished much more economically if matched in some way to their personal identity, preferences, limitations and capacities (Lazaros & Ahmadi, 2008). Interfacing these technologies into customized systems can also arguably improve their efficacy.

RFID TECHNOLOGY AND THE ELDERLY

RFID technology relies upon a communication interface of two components. The first of these components is an RFID chip, or integrated miniaturized circuit, which may be attached or imbedded (including surgically) into items, animals, or humans. It can transmit for short distances a unique identifier code, that may be registered to the item, animal, or human to which it is attached. The second component consists of an antenna or receiver that can identify the chip, though it may, or may not be, within line of sight of the receiver (Shah, 2011). The receiver may be connected to any number of data interfaces, or databases, such as consumer, commercial, communication, medical records, medical telemetry, financial, banking, tracking, or GIS (Geographic Information Systems, or locator systems), inventory, transportation, airport security, government, military, health, health monitoring, disaster response, or quarantine effort (for disease tracking and prevention), or to access demographic, service utilization, or purchasing background on the user.

There are a number of possible applications of RFID technology relative to the needs of the elderly (Radio Frequency Identification [RFID] Systems, 2011). These include the utilization of RFID devices to ensure the basic security and safety of the elderly. For example, an elderly person admitted for an emergency condition could immediately have their medical information accessed, including provider information, insurance coverage, allergies, past medical care and prescriptions (Mohammadian and Jentzsch, 2008). If the elderly person is either living alone and is confused or disoriented, or living in a skilled care facility under the same conditions, a wander alert or alarm could be triggered if the person managed to leave a safe area (Schneider, 2006). For those who opt to age in place and have few social connections, the RFID could be placed on food parcels to measure consumption as an indicator of eating and maintaining nutritional health (Isomursu, Häikiö, Wallin, & Ailisto, 2008). For those who have care providers and family, the RFID may be combined with other smart home sensor technologies to provide continuous monitoring to distant locations and assure caring families of their relatives' overall well being (Rose, 2011). Additionally, purchases made by the elderly could be supervised if needed to assure that their purchases do not exceed their income and are consistent with their needs. Payment systems for retirement benefits and the purchases could also be included to confirm that benefits are being received and utilized by the beneficiary (Allan, 2006). Furthermore, elderly persons who become lost or separated from relatives or authorities while in transit could more easily be identified and located (Landau, Werner, Auslander, Shoval, & Heinik, 2009). If the RFID device is required to authenticate the ignition of a car engine, or that the ignition will not work, then the elderly person may not be able to drive alone without the presence, or authorization, of a second agreed, or appointed relative, or care provider (Frenzel, 2001). If the chips are implanted in the elderly there are also some counter arguments for health and safety. RFID devices may cause adverse reactions such as tissue damage and migration of the device, especially if the elderly person has to have a diagnostic MRI (Magnetic Resonance Imaging) medical scan performed (Aarti, 2011). There are also a number of instances where the use of RFID technology has been shown not to be a panacea for identity verification. For example, passports containing RFID devices have been reportedly "hacked" and the code stolen, with the possibility of identity theft occurring (Naone, 2009).

RFID TECHNOLOGY AND VALUES CONFLICTS

Values conflicts are likely to occur with the deployment of technologies unfamiliar to the aging population. Other values conflicts may occur between service providers and other portions of the population, such as younger and differentially trained service providers and the elderly. There are cost allocations as well to sustain the elderly on a larger scale in anticipated times of scarcity of resources, due to historically higher national debt and declining economic markets.

Privacy, self determinism, informed consent, and moral and religious factors affecting utilization of RFID implant technology may be expected to impede, or obstruct, the acceptance of this potentially beneficial surveillance technology application. Some solutions may, as a result, be effective in one population of elderly, yet completely unacceptable in another due to cultural differences. Autonomous decision making may be more or less valued in some demographic segments, so as to make some forms of service delivery and RFID implant and other emergent monitoring technology a less preferred option.

Nearly all nations and cultures, whether developed, developing, or undeveloped would acknowledge having unique social contexts relative to the indigenous, or immigrant populations of their own portion of the globe, which must be advocated for and politically and geopolitically validated and valued. This is no less true for the

elderly sub populations, who are considered the most invested members of their local, national, or global populations. What follows are some of the more commonly held values issues, or social contexts for the elderly across a number of cultures that may be expected to emerge relative to the adoption and utilization of RFID technology.

Resistance to Change

Commonly, anything that is new or unfamiliar may be resisted by the elderly. Senior citizens are reluctant and often slow to adapt to, or accept readily, new technologies. The elderly have often seen numerous changes over a lifetime and at some point simply seem to fatigue in interest and adopt newer technologies less frequently (Gilly & Zeithaml, 1985). Interestingly, this is anticipated as the largest growing segment of the population for industrialized nations. Therefore, senior citizens will exercise greater purchasing power than younger groups that might more readily accept the technology. Seniors will also comprise a greater aggregate voice in affairs as voters in democratic societies, or constituents relative to adopting new technologies in general, and for their age group in particular. Numerous elderly may prefer to avoid novel or new technologies and could delay policy decisions for their implementation through their group influence.

Technophobia

Fear of emerging technologies is also of frequent concern for the elderly, though this phenomenon is closely related to resistance to change. The technology may simply lack familiarity (Sponselee, Schouten, Bouwhuis & Willems, 2008). Senior citizens have been witness over a lifetime to a myriad of devices, hailed as new, beneficial and benign technologies that were later found to have had some negative or harmful attribute. The elderly then, are understandably cautious about rushing in to adopt the next touted technical advance. Separately from this concern though, is often simply a lack of information on how to best use, or exploit, the new technology for their own advantage in a way that is meaningful (Vastenburg, Visser, Vermaas, & Keyson, 2008). Numerous elderly may simply not invest adequate resources, or interest, to fully use emerging and available technology that could promote optimal functioning for themselves, or their elderly family members.

Informed Consent

Independence and competence are related to understanding fully the consequences of one's actions and commitments, as well as their associated risks or benefits. The established elderly are often concerned that they will become debilitated and unable to decide their life course. Imposing an RFID chip without consent would be invasive and likely experienced as a physical assault as well as an assault on dignity (Good, 2008). Self determination, and the right to enter into or withhold permission to agreements or contracts, is highly valued in Western democracies and many of the developed and developing nations share, or closely identify with this. Surreptitious implanting of RFID chips into hand held devices, consumer products, or within one's body without thorough evaluation, foreknowledge and thoughtful consent is likely to be met with strong opposition by the elderly (Stanton, 2005). Attempts to press the implementation of RFID technology without inclusion of consent by the affected seniors may well result in substantial resistance from advocates for the elderly and the elderly themselves.

In one recent and popular media, publicized example, RFID chip testing on Alzheimer's affected elderly was begun, as a trial by the chip manufacturing company, in the United States. A number of RFID chips were provided to facilities with instruction for use. Some affected elderly were reported to have been implanted and monitored (Swedberg, 2007; Rotter, P., Daskala, B., & Compano, R., 2008). However, due to their neurological condition and diminished mental capacity, could not have sufficiently benefitted from informed consent and in the case of guardianship, their rights were likely judicially delegated to non-familial authorities, which may have not held their personal values, or interests, as primary.

Caretaker Need

Alzheimer 's disease, as it relates to the elderly and the advantages of RFID technology presents something of a contrast in acceptability when viewed from the demands placed upon familial or institutional care providers. In the absence of maintained or sustained functioning among the elderly, especially as functioning relates to self-care and decision making capacity, RFID technology may serve to provide respite for family members, or institutional caretakers. A recent published interview illustrated the acceptance, if not desirability of RFID technology, for impaired Alzheimer's sufferers, to restore some independence and quality of life for challenged and fatigued care providers (Michael, 2009). Care providers may be willing to have the convenience and sense of safety, which RFID implant technology provides for their charges with Alzheimer's disease.

Risks of Physical Harm

RFID implants are made of some combination of micro-circuitry, non organic elements, and likely trace metallic substances. They are of course foreign objects. There is some preliminary research to suggest that magnetic devices or MRI (Magnetic Resonance Imaging) may cause them to migrate within the human body. As a result of the migration, the RFID chip may become dysfunctional over time, or stop working, or have some as yet undiscovered adverse physical consequence (Steffen, Luechinger, Wildermuth, Kern, Fretz, Lange, et al., 2010). These would be reasons enough for most cognizant individuals to prefer that they not be implanted with an RFID device. The elderly, having some additional physical frailties over time, would likely be even more averse to having these embedded in their person. As the RFIDs also emit a passive radio response to an available, nearby, antenna there may be some additional concerns about harm from radio waves, or microwave damage, from being in proximity to either the microchip or the antenna device (Härmä, 2009). Those intimidated by the risk of physical harm will likely then avoid RFID chip technology.

Western Religious Opposition

Perhaps the best publicized value conflict that will serve to promote opposition to the RFID chip is its correlation with end times Christian bible prophesy as a result of its defined function, specific historical development, description and early market studies (forehead and right hand placement) on best deployment for conducting individual commerce (Noack & Kubicek, 2010). More than one billion inhabitants of earth identify themselves as Christian by belief, or faith; a large proportion is entirely opposed to the existence, much less the deployment of, the RFID chip in humans. Implantation of the chip into one's body is equated to eternal damnation from God (see Additional readings section). This global value conflict as to the utilization of the chip is the most extensive and unyielding. Certainly, senior members of the Christian faith, and senior citizens within the faith, are likely to be intractable and steadfast in their absolute opposition to the device in any fashion. Further, the problem may become exacerbated by the inclusion of decision makers for the elderly person's care that are immigrant, foreign, or from a culture not sharing their same faith, values, or beliefs (Campbell, Clark, Loy, Keenan, Matthews, Winograd, & Zoloth, 2007).

Privacy Concerns

Utilizing the RFID chip as a tracking device for various consumer items raises a number of privacy concerns. An embedded microchip can connect the purchaser and the item at the point of sale, but can also continue to identify both the item and the purchaser, or person connected with the item over time (USA Today Magazine, 2010). That is, it can establish one's whereabouts, the locations and duration of time in which the item was maintained, or in the possession of the owner, other associated purchases, acts, or actions, or even various events at which the item and the person connected to it may have been present, or in which they were involved in some way. Consider then, that the device is sufficiently small enough to be unobtrusively attached to almost anything, or several things held in combination, or separated one from another at some point, by the owner, purchaser, or person connected, or otherwise known to be connected to the RFID chip in some way. Presence of the chip could be sufficient for a person under suspicion to activate any number of electronic devices when present to include audio and video and recording devices, remotely. An elderly person, socialized to a lifetime of personal privacy and independence and raised in an historical time frame when one's word and a handshake constituted a compelling moral contract, would likely be horrified by the invasion of personal privacy (Spiekermann, 2009). Of course the level of dehumanization experienced by the unwilling, but aware, senior member of society might contribute negatively to health issues and emotional concerns, to such a degree that using the device to monitor consumption, or other behaviors might not be beneficial for the elderly in need of monitored care, or self care. Members of western societies socialized to the rights of the individual over and above those of the state, will avoid using the chip.

Fears of GPS Tracking

RFID devices may readily be included in tele-communication devices, portable and wearable medical equipment, clothing, vehicles and other equipment such that knowledge of the ongoing physical location and probable moment to moment circumstance of the person being monitored is virtually assured. Within an equipped "smart home", these can be combined with systems of acoustic monitors, video cameras and other telemetry or medical monitoring devices, to confirm constantly the location of the senior family member, or medical patient. Furthermore, the person could be tracked continuously (depending on the number and proximity of receiving antennae), while commuting, or in transit to locations outside of the senior citizen's residence (Shoval, Auslander, Cohen-Shalom, Isaacson, Landau, & Heinik, 2010). Products can also be tracked in this way to confirm the presence, or absence, of medications, or other tagged items (Rawal, 2009). Logistical information for medication and services may be improved by using integrated systems of RFID device technology. An available and model service Project Lifesaver is currently using the technology to locate and return autistic children and demented elderly to their home or place of residence (Project Lifesaver, 2012).

Fears of Asset Tracking

Many of the elderly are on fixed incomes and often these are inadequate to provide for their increasing health and personal needs. Further, assets may be seized for any number of reasons including unpaid medical debt, or care. RFID chips may be placed, or embedded within banking cards, or money, stock certificates, deeds, automobile titles, or anything of value, held, or owned by the senior person or relatives (Raths, 2009). Transactions involving assets, including placing them in a safe deposit bank vault, or residential safe, may become known to third parties, or creditors.

Assets may be cancelled, or seized without notice for such debts, providing no security or safety net for the affected senior or family members. In some instances where the debt is substantial and the assets limited, asset cancellation could extend to home foreclosure, or loss of affordable, but essential care (Premier Inc., 2009). Numerous elderly may fear that they will have less access to resources to recover from adversity.

Fears of Identity Theft

There is ample concern, even among engineers, that the RFID chip's unique numerical code, though encrypted, may be discovered or copied, placing the owner at risk of identity theft, which could result in the fabricated identify being almost undetectable. The code itself is presumed to be unique to an individual, or an item in which the RFID chip is embedded, such as a credit card. Should the encrypted code be discovered by someone other than the owner, the unique identifying number could provide access to other associated numbers for other proximal people, or items, such that the restoration of one's digital identity becomes impossible (Britt, 2007). Some elderly may be at risk of losing everything unless the RFID technology becomes virtually identity theft proof.

Transhumanist Concerns

Transhumanism refers to the enhancement of human capacities through the direct infusion of technologies into the human body and brain (Bostrom, N., 2012). Transhumanists, or those who would advocate for transhumanism, would optimistically view the benefits of technology, to prolong life, or add to the quality of life, above any potential risks, for the recipient elderly. RFID technology would be viewed as an extension or advance over earlier technologies that have provided mobility, strength, information, socialization, life extension and a range of other modifications and enhancements (HUMANITY +, 2012) including the possibilities of human machine hybrids and transfer or simulation of human brains into non-human hardware (Stephan, Michael, Michael, Jacob, & Anesta, 2012). The RFID "chip" technology provides a direct information technology interface for the wearer. That is, the device has the capacity to serve as an identifier for various databases of information about the elderly person associated with it, as well as others who wear, or are embedded with, the device. The use of RFID technology may also serve as a cueing device, which when recognized activates other devices, communicates care or medical requests on behalf of the wearer, and tracks their whereabouts and habits.

The elderly population, or many of the elderly population's current members, when competent, may not generally share the values of transhumanists. Nor, are the prevailing views and values of the elderly likely to be quickly revised, or altered to become rapidly tolerant of embedded technology. Autonomy and independence and traditional ways of doing things are well documented as mainstays of elderly shared values. Technologies are often tolerated, rather than enthusiastically accepted as functioning and health are lost by affected elderly individuals. Yet, some aspects of transhumanism and RFID technology have the capacity to prolong human independence and functioning, in terms of health infirmities (Posts tagged "Transhumanism," 2012). In terms of the shifting demographic in developed countries toward the elderly, the application of these technologies may serve to provide greater economic security and prosperity for a population, by sustaining the elderly as viable contributors to the larger society.

Solutions and Recommendations

Acceptance or rejection of RFID chip technology among the elderly in developed and developing countries relative to augmenting the many anticipated service needs for the elderly, depends greatly on developing active strategies to address some of these identified value conflicts and fears. In some

instances, working within the social context and long held values and beliefs of the elderly may mean creating some social and technological compromise or accommodations such that the technology can be deployed usefully where warranted. Senior adults who are expected to continue to remain productive, healthy and contributing to their respective societies for longer should retain their independence and legal rights as citizens, for as long as possible. This includes the right, of most adults, to decline unwanted or intrusive interventions. RFID technology is, and may continue to be, an evolving and adaptable approach to identify verification and database access for various items, animals and humans.

Many of the concerns of the elderly relative to RFID chip technology are centered on risks to privacy (surveillance, or uberveillance) including economic and health information, and threats of possible identify theft. Some concerns are related to religious beliefs and strong social values derived from the cultural contest of the elderly person's life experience. Reservations to adopting the technology also relate to concerns of being exploited in some way, in part due to the common experience of many elderly in losing capacity to remain self-reliant in later years of life.

Gaining acceptance relative to privacy concerns will likely require a combination of socialization and education. It will also require verifiable security measures that are tested and evaluated and confirmable by authoritative organizations, such as those for consumer safety and endorsed though media that are most accessible or most utilized by the representative group of potential elderly RFID technology users. Additional safeguards may be implemented within elderly residences such that RFID information is only transmitted though an encrypted communication link, in much the same was as residential internet access is currently accomplished. Consumer institutions such as banking and credit agencies may also provide linkage technology that masks identification and assures financial privacy. Though policy may follow engineering advances in this area, other less exacting forms of identification may continue to be used for lower levels of non-essential or non-critical transactions for elderly users. The RFID device could be automatically masked by any antenna device either by geographic location, or function if the RFID device is not intended to be read, allowing for some privacy from GPS location devices. Legal protections and assurances that preserve assets as the private property of the elderly, unless a crime is committed, may serve to reduce anxiety about the tracking of privately held financial assets. Furthermore, once an item (physical or financial) that may be tracked is sold to a consumer, the selling entity should no longer have an option to track the movement of that item, nor should law enforcement officials without some alternate evidence of some illegality. Perhaps accessing an RFID would require probable cause of the commission of a crime before access by law enforcement could occur.

Assurances of the lack of risks of RFIDs to render physical harm will require extensive medical testing under a variety of conditions so that the risks are clearly understood. Possible use of varying materials used to construct some specialized RFID devices to lower the risk of harm, for example, a specialty chip version might be made so that it is relatively non-reactive to MRI or other magnetic based medical scanners would also reassure the elderly. The primary physician for the elderly person can also be called on to evaluate the risks of physical harm and to educate the elderly patient so that anxiety may be further reduced over time.

Resistance to change and some technophobia from the elderly is a relatively common reaction for any number of technologies that may be adapted to, or needed by, the elderly. Social service providers and allied health professionals can contribute when possible to providing ongoing education to groups of the elderly and family members to facilitate adoption and adaptation. In some instances, training programs may be

transmitted by popular media for the elderly to provide further information and appropriate reassurances. This would also serve to improve rates of informed consent for those elderly in need of RFID technology.

Western religious opposition or Christian faith opposition to the deployment or personal utilization of the RFID chip technology should be recognized and accommodated. Those opposed to an implanted device should not be required under any circumstance to obtain one. It is likely that a number of the same oppositional religious elderly raise few objections to the magnetic strip of identifying information in their credit or banking card. Some may even carry access devices to open their employer's garage for entry that includes RFID chip technology. Most carry personal cell phones that are locatable by an embedded RFID such as a GPS traceable chip installed within the cellular phone. Additionally, for those who would benefit strongly from keeping an identity device or RFID chip on their person, there are already a number of non-removable devices used for house arrest, or tracking, that are wearable and washable without undue discomfort and that may be less offensive.

FUTURE RESEARCH DIRECTIONS

For the most part, RFID tags or devices are considered either passive or active, that is, they derive their energy to transmit a numerical or identifying code to an antenna, or reader, whenever the two are in close proximity and the reader transmits a signal which provides the energy for the device to be read. There are advantages, especially for the elderly, in developing active RFID chip technology with its own power source, or battery, to strengthen the signal, in the event that a reader is not close by and a health crisis or other adverse event befalls the elderly person. There are a number of engineering efforts proposed and ongoing, which are attempting to improve the active version of the RFID tag.

Another important research consideration is to improve the capacity of the antenna or RFID reader devices to manage a large number of signals, many of which may be coded or encrypted, when several RFID devices are in close proximity to a reader. Though the device may have greater consumer application, the improvements are needed to ensure that there is no substantial risk of confusion of identity between any numbers of closely grouped devices relative to the unique identity of the user, and that the many potential signals are linked to the appropriate and responsive data infrastructure, as quickly as feasible. This advance would render it highly unlikely that there is any signal capture, or method to replicate a unique identity, or duplicate the RFID device signal in any way.

However, the development of levels of systems of care and information access, with a variety of social and cultural, or religious values considerations and applicable legal protections for both the elderly RFID user, or potential user, and for those opting not to use the devices (whether implanted, or not) is needed. Such a development would make the technology acceptable to the broadest degree possible by the elderly, to facilitate elderly care and wellness and continuance of productivity in the absence of sufficient care providers from the younger population within those developed, or developing nations, affected by rapid sociological aging.

CONCLUSION

The phenomenon of the graying of civilization is underway in most of the developed and developing world. In these countries, most of the population will be affected socially and economically by rapid shifts in the demography of their respective nations, over the next few years, or decades. As these same countries are well advanced in industry and technology, they will need to rely on these advances to supplement or augment allocation of resources to, and the maintenance of functioning and continuing contributions of, the elderly. Suc-

cessful aging and productive and healthy aging in place will be an imperative for the longer term leadership, if not survival, of these cultures. This problem is compounded by inadequate available numbers of youth to provide care for the population, in some instances to fully provide the labor needs of these nations.

Longer term policies or strategies may be adopted which would increase fertility and childbirth rates, or perhaps some natural events such as epidemics may change the composition of the relative age of a population, or even changes in economics may mean that fewer can afford high tech medical procedures that are extending life. Over the shorter term of the next generation or two, these changes will not be timely enough to make any foreseeable substantial change.

Of necessity, surveilling the elderly is one of a number of effective strategies that may be adopted to produce a best case scenario of continuing health and productivity during senior years. Such approaches may mean that the entirety of society is gradually strengthened and may also gradually recover to optimal population norms. In some instances, retired elderly will seek residence in resort areas, and form mutual support networks within NORCs (Naturally Occurring Retirement Communities) which may or may not be technologically supported, but which are likely to support healthy aging.

RFID chip technology provides a potential method of addressing the demographic needs of the elderly in developed and developing countries. This is despite a number of limitations at present and a series of known apprehensions and values issues of the elderly concerning the deployment and wide distribution of RFID technology. Uberveillance (or the omnipresent surveillance) of the elderly can be circumvented with limited additional research and development, and with the development of specified social, legal and engineering advances, to address the anticipated values conflicts and natural apprehension of the elderly in use of this and other emergent related technology.

REFERENCES

Aarti, R. (2011). *Pros and cons of RFID technology*. Retrieved April 12, 2011, from http://www.buzzle.com/articles/pros-and-cons-of-rfid-technology.html

Allan, R. (2006). Wireless sensing spawns the connected world. *Electronic Design, 54*(7), 49–56.

Atul, G. (2006). Critical care workforce: A policy perspective. *Critical Care Medicine, 34*(3), S7–S11. PMID:16477206

Bostrom, N. (2012). *Transhumanist values*. Retrieved October 29, 2012 from http://www.nickbostrom.com/ethics/values.html

Britt, P. (2007). New push to protect your identity. *Information Today, 24*(1), 1–46.

Brown, C., & Braun, K. (2008). Globalization, women's migration, and the long-term-care workforce. *The Gerontologist, 48*(1), 16–24. doi:10.1093/geront/48.1.16 PMID:18381828

Campbell, Clark, Loy, Keenan, & Matthews, Winograd, & Zoloth. (2007). The bodily incorporation of mechanical devices: Ethical and religious issues (part 2). *Cambridge Quarterly of Healthcare Ethics, 16*(3), 268–280. doi:10.1017/S0963180107070302 PMID:17695618

Carmel, S., & Lowenstein, A. (2007). Addressing a nation's challenge: Graduate programs in gerontology in Israel. *Gerontology & Geriatrics Education, 27*(3), 49–63. doi:10.1300/J021v27n03_04 PMID:17347110

Demographics of Aging. (2011). Retrieved April 4, 2011, from http://www.transgenerational.org/aging/demographics.htm#Characteristics

Foreign-Born Workers and Baby Boomers. (2010). *Babyboomercaretaker.com*. Retrieved March 19, 2011, from http://www.babyboomercaretaker.com/baby-boomer/foriegn-born-workers-and-babyboomers.html

Frenzel, L. (2001). An evolving ITS paves the way for intelligent highways. *Electronic Design, 49*(1), 102.

Gilly, M., & Zeithaml, V. A. (1985). The elderly consumer and adoption of technologies. *The Journal of Consumer Research, 12*(3), 353–357. doi:10.1086/208521

Gilmer, D. F., & Aldwin, C. M. (2003). *Health, illness, and optimal ageing: Biological and psychosocial perspectives*. Thousand Oaks, CA: Sage Publications.

Good, N. S. (2008). *Designing for informed consent: A multi-domain, interdisciplinary analysis of the technological means to provide informed consent*. Retrieved April 5, 2011, from http://search.proquest.com/docview/304697082?accountid=7117

Hargreaves, J. S. (2010). Will electronic personal health records benefit providers and patients in rural America?. *Telemedicine and e-Health, 16*(2), 167.

Härmä, A. (2009). Ambient human-to-human communication. In H. Nakashima, H. Aghajan, & J. Augusto (Eds.), *Handbook of ambient intelligence and smart environments* (pp. 795–823). New York: Springer.

Harper, S. (2006). *Ageing societies: Myths, challenges and opportunities*. London: Arnold Publishers.

Humanity +. (2012). *Humanity + mission*. Retrieved October, 29, 2012, from http://humanity-plus.org/about/mission/

International Wealth Solutions. (2008). *Ageing Demographics*. Retrieved March, 6, 2011, from http://www.iwslimited.com/ageingdemographics.html

Isomursu, M., Häikiö, J., Wallin, A., & Ailisto, H. (2008). Experiences from a touch-based interaction and digitally enhanced meal-delivery service for the elderly. *Advances in Human-Computer Interaction*. doi:10.1155/2008/931701

Landau, R., Werner, S., Auslander, G. K., Shoval, N., & Heinik, J. (2009). Attitudes of family and professional caregivers towards the use of GPS for tracking patients with dementia: An exploratory study. *British Journal of Social Work, 39*(4), 670–692. doi:10.1093/bjsw/bcp037

Lazaros, E. J., & Ahmadi, R. (2008). Integration of supportive design features and technology. *Technology Teacher, 67*(7), 20–25.

Michael, K. (2009). Interview 14.3 the Alzheimer's carer, Mr. Kenneth Lea, Wollongong, Australia. In K. Michael, & M. G. Michael (Eds.), *Innovative automatic identification and location-based services: From bar codes to chip implants*. Hershey, PA: IGI Global. doi:10.4018/978-1-59904-795-9

Michael, K., & Michael, M. G. (2013). The future prospects of embedded microchips in humans as unique identifiers: The risks versus the rewards. *Media Culture & Society, 34*(3).

Mohammadian, M., & Jentzsch, R. (2008). Intelligent agent framework for secure patient-doctor profiling and profile matching. *International Journal of Healthcare Information Systems and Informatics, 3*(3), 8–57. doi:10.4018/jhisi.2008070103

Naone, E. (2009). RFID's security problem. *Technology Review, 112*(1), 72–74.

Noack, T., & Kubicek, H. (2010). The introduction of online authentication as part of the new electronic national identity card in Germany. *Identity in the Information Society, 3*(1), 87–110. doi:10.1007/s12394-010-0051-1

Posts Tagged Transhumanism. (2012). Retrieved October 29, 2012, from http://discombobulaated.wordpress.com/tag/transhumanism/

Premier Inc. (2009, September 8). Awards dynamic computer corporation with a group purchasing agreement to provide RFID asset tracking and management solutions for its member network. *Science Letter*.

Project Lifesaver International. (2012). *Bringing loved ones home*. Retrieved October, 28, 2012, from http://www.projectlifesaver.org/

Radio Frequency Identification (RFID) Systems. (2011). Retrieved April 12, 2011, from http://epic.org/privacy/rfid/

Rajasekaran, M. P., Radhakrishnan, S., & Subbaraj, P. (2009). Elderly patient monitoring system using a wireless sensor network. *Telemedicine and e-Health, 15*(1), 73-79.

Raths, D. (2009). RFID is finding a home in the data center. *Computerworld, 43*(18), 21.

Rawal, A. (2009). RFID: The next generation auto ID technology. *Microwave Journal, 52*(3), 58–76.

Rose, D. (2011). *Chip checkups and embedded beds: RFID and home monitoring*.

Rotter, P., Daskala, B., & Compano, R. (2008). RFID implants: Opportunities and challenges for identifying people. *IEEE Technology and Society Magazine, 27*(2), 24–32. doi:10.1109/MTS.2008.924862

Schneider, M. E. (2006). Technology helps prevent wandering, falls by elderly residents. *Internal Medicine News, 39*(10), 37. doi:10.1016/S1097-8690(06)73524-7

Shah, N. A. (2011). *RFID chip in humans*. Retrieved March 30, 2011, from http://www.buzzle.com/articles/rfid-chip-in-humans.html

Shoval, N., Auslander, G., Cohen-Shalom, K., Isaacson, M., Landau, R., & Heinik, J. (2010). What can we learn about the mobility of the elderly in the GPS era? *Journal of Transport Geography*, *18*(5), 603–612. doi:10.1016/j.jtrangeo.2010.03.012

Spiekermann, S. (2009). RFID and privacy: What consumers really want and fear. *Personal and Ubiquitous Computing*, *13*(6), 423–434. doi:10.1007/s00779-008-0215-2

Spillman, B. C. (2004). Changes in elderly disability rates and the implications for health care utilization and cost. *The Milbank Quarterly, 82*(1), 157–194. doi:10.1111/j.0887-378X.2004.00305.x PMID:15016247

Sponselee, A., Schouten, B., Bouwhuis, D., & Willems, C. (2008). Smart home technology for the elderly: Perceptions of multidisciplinary stakeholders. In M. Mühlhäuser, A. Ferscha, & E. Aitenbichler (Eds.), *Communications in computer and information science constructing ambient intelligence* (Vol. 11). New York: Springer. doi:10.1007/978-3-540-85379-4_37

Stanton, R. (2005). RFID: Ripe for informed debate. *Computer Fraud and Security, 12*, 12-14. doi:10.1016/S1361- 3723(05)70285-2

Steffen, T., Luechinger, R., Wildermuth, S., Kern, C., Fretz, C., & Lange, J. et al. (2010). Safety and reliability of radio frequency identification devices in magnetic resonance imaging and computed tomography. *Patient Safety in Surgery, 4*(2). doi: doi:10.1186/1754-9493-4-2 PMID:20205829

Stephan, K. D., Michael, K., Michael, M. G., Jacob, L., & Anesta, E. (2012). Social implications of technology: Past, present, and future. *Proceedings of the IEEE, 100*(13), 1752–1781. doi:10.1109/JPROC.2012.2189919

Swedberg, C. (2007). Alzheimer care center to carry out verichip pilot. *RFID Journal*. Retrieved November, 6, 2012 from http://www.rfidjournal.com/article/view/3340

Today Magazine, U. S. A. (2010, September 20). Are consumers being stalked by RFID tags? *USA Today Magazine*, p. 8.

Vastenburg, M., Visser, T., Vermaas, M., & Keyson, D. (2008). Designing acceptable assisted living services for elderly users. In *Proceedings of the European Conference on Ambient Intelligence* (pp. 1-12). Berlin: Springer-Vertag.

Weil, P. (2002). Towards a coherent policy of co–development. *International Migration (Geneva, Switzerland), 40*(3), 41–56. doi:10.1111/1468-2435.00196

ADDITIONAL READING

Arcelus, A., Howell, J., Goubran, R., & Knoefel, F. (2007). Integration of smart home technologies in a health monitoring system for the elderly. *21st International Conference on Advanced Information Networking and Applications Workshops* (pp. 820-825). Niagara Falls, Canada: Advanced Information Networking and Applications

Bang, S., Kim, S., Song, M., & Park, S. (2008). Toward real time detection of the basic living activity in home using a wearable sensor and smart home sensors. In *Engineering in Medicine and Biology Society 30th Annual International Conference of the IEEE*, 5200-5203. doi: 10.1109/IEMBS.2008.4650386

Basham, R. E., & Kang, S. Y. (2010). Sustainable aging: Innovations in emerging technology and service delivery. In J. Baydo. (Ed.), *National Technology and Social Science Conference Proceedings 2010* (pp. 30-40). El Cajon, California: The National Social Science Association, (NSSA), National Social Science Press.

Chan, M., Campo, E., Laval, E., & Esteve, D. (2002). Validation of a remote monitoring system for the elderly: Application of mobility measurements. *Technology and Health Care, 10*, 391–399. PMID:12368559

Chatfield, A., Wamba, S., & Tatano, H. (2010). E-government challenge in disaster evacuation response: The role of RFID technology in building safe and secure local communities. Proceedings of the *43rd Hawaii International Conference on System Science* (pp. 1-10). Honolulu, Hawaii: Hawaii International Conference on System Sciences.

Coleman, R. (1993). A demographic overview of the ageing of First World populations [Special issue]. *Applied Ergonomics, 24*(1), 5-8. Special Issue Designing for our future selves, doi: 10.1016/0003-6870(93)90152-Y. http://www.sciencedirect.com/science/article/B6V1W-47XSS7D- H3/2/c1bc3aa0de71d0e2670bae6534b41a20

Droes, R., Mulvenna, M., Nugent, C., Finlay, D., Donnelly, M., & Mikalsen, M. et al. (2007). Healthcare systems and other applications. *IEEE Pervasive Computing / IEEE Computer Society [and] IEEE Communications Society, 6*(1), 1, 59–63. doi:10.1109/MPRV.2007.12

Fisher, J. A. (2006). Indoor positioning and digital management: Emerging surveillance regimes in hospitals. In T. Monahan (Ed.), *Surveillance and security: Technological politics and power in everyday life* (pp. 77–88). New York, NY: Routledge.

Fogel, R. W. (2000). The extension of life in developed countries and its implications for social policy in the twenty-first century, population and development review (Vol. 26) Supplement: Population and Economic Change in East Asia (2000) (pp. 291-317). Published by: Population Council, URL: http://www.jstor.org/stable/3115220

Helal, A., & King, J. Bose. R., Elzabadani, H., & Kouddourah, Y. (2009). Assistive environments for successful aging. In A. Kameas, V. Callagan, H. Hagras, M. Weber, & W. Minker (Eds.), Advanced Intelligent Environments (pp. 1-26), Springer: US.

Heydon, D. (2011). RFID pharmaceutical services. Retrieved from http://www.buzzle.com/articles/rfid-pharmaceutical-services.html

http://www.sciencedirect.com/science/article/B986R-503SKKR- 3/2/eeba95b-518c5a17f6d1203b7dac17839)

Hussain, S., Schaffner, S., & Moseychuck, D. (2009). Applications of wireless sensor networks and RFID in a smart home environment. *2009 Seventh Annual Communication Networks and Services Research Conference* (pp.153-157). New Brunswick, Canada

Jones, E., Henry, M., Cochran, D., & Frailey, T. (2010). RFID Pharmaceutical Tracking: From Manufacturer Through In Vivo Drug Delivery. *Journal of Medical Devices, 4*(1), 015001. doi:10.1115/1.4000495

Kenner, A. M. (2008). Securing the elderly body: Dementia, surveillance, and the politics of Aging in Place. *Surveillance & Society, 5*(3), 252–269. http://www.surveillance-and-society.org

Ledger, W. L. (2009). Demographics of infertility. *Reproductive Biomedicine Online, 18*(Supplement 2), S11–S14. doi:10.1016/S1472-6483(10)60442-7 PMID:19406025

Lin, Y., Su, M., Chen, S., Wang, S., Lin, C., & Chen, H. (2007). A study of ubiquitous monitor with RFID in an elderly nursing home. *2007 International Conference on Multimedia and Ubiquitous Engineering* (pp. 336-340). Seoul, Korea.

Lodge, J. (2010). Kevin Warwick's experiment 1: Future identity. *Studies in Ethics, Law, and Technology, 4*(3). *Article, 9.* doi: doi:10.2202/1941-6008.1150

Lunenfeld, B. (2008). An aging world – demographics and challenges. *Gynecological Endocrinology, 2*(1), 1-3. doi: 10.1080%2F09513590701718364

McCall, C., Maynes, B., Zou, C. C., & Zhang, N. J. (2010). RMAIS: RFID- based medication adherence intelligence system. *Engineering in Medicine and Biology Society (EMBS) 32nd Annual International Conference of the IEEE (pp.* 3768-3771). Buenos Aires, Argentina doi: 10.1109/IEMBS.2010.5627529

Michael, K., Roussos, G., Huang, G. Q., Chattopadhyay, A., Gadh, R., & Prabhu, B. S. at al. (2010). Planetary-scale RFID services in an age of uberveillance. *Proceedings of the IEEE, 98*(9) (pp.1663-1671). doi: 10.1109/JPROC.2010.2050850

Michael, M. G., & Michael, K. (2010). Toward a state of uberveillance [Special section]. *Technology and Society Magazine, IEEE, 29*(2), 9–16. doi:10.1109/MTS.2010.937024

Revelation, The Holy Bible, King James Version New York: American Bible Society: 1999, Bartleby.com, 2000

Steele, R., Lo, A., Secombe, C., & Wong, C. (2009). Elderly persons' perception and acceptance of using wireless sensor networks to assist healthcare. [Special issue] Mining of Clinical and Biomedical Text and Data. *International Journal of Medical Informatics, 78*(12), 788–801. doi:10.1016/j.ijmedinf.2009.08.001 PMID:19717335

Wang, Q., Shin, W., Liu, X., Zheng, Z., Oh, C., AlShebli, B. K., et al. (2006). I-Living: An open system architecture for assisted living. *IEEE International Conference on Systems, Man and Cybernetics,* 5 (pp. 4268-4275). Taipei, Taiwan. doi: 10.1109/ICSMC.2006.384805

Weil, D. N. (1997). The economics of population aging. In M. Rosenzweig & O. Stark, (Eds.), *Handbook of population and family economics,* Vol. 1(2) (pp. 967-1014). Retrieved from http://www.sciencedirect.com/science/article/B7P62-4FF827BV/2/ c395b421e767c27a51f3d-12da5211c

Xiao, Y., Shen, X., Sun, B., & Cai, K. (2006). Security and privacy in RFID and applications in telemedicine. *Communications Magazine, IEEE, 44*(4), 64–72. doi:10.1109/MCOM.2006.1632651

Yamamoto, Y., Huang, R., & Ma, J. (2010). Medicine management and medicine taking assistance system for supporting elderly care at home. *2010 2nd International Symposium on Aware Computing (ISAC), 1*(4) (pp. 31-37). Tainan, Taiwan. doi: 10.1109/ISAC.2010.5670451

KEY TERMS AND DEFINITIONS

Aging in Place: Remaining one's present residence, while elderly with the capacity to function as independently as possible and access necessary support services in response to changing service, health or financial needs.

Developed Countries: Countries with an economic base and large gross domestic product, built largely on manufacturing and technology rather than agriculture with the potential to access adequate health care and other essential services.

Developing Countries: A country that is poorer than developed countries and whose citizens are mostly agricultural workers, but that desires and has some capacity, or support, to become more advanced socially and economically.

NORC (Naturally Occurring Retirement Community): A retirement community, or active older adult community, especially designed or geared for people who no longer work, or restricted to those over a certain age.

RFID (Radio Frequency Identification Device): A wireless microchip mechanism that may be attached to, or imbedded, to identify and manage inventory, animals, or people. Unique identifying data stored on an environmentally resistant tag several centimeters in size is passively or actively transmitted to a microchip reader via electrical or electromagnetic waves which may then access a comprehensive data file, for the item, animal, or person to which it serves as an identification tag.

Surveilling the Elderly: Surveillance, (or) the monitoring of the behavior, activities, or other changing information, relative to the needs, or status, of the elderly population, with, or without, consent, in either, an overt, or surreptitious manner.

Values Conflicts: A state of opposition between persons (including, groups, or populations), or ideas or interests; derived from the person's principles, morals, or values, or other standards of behavior relative to one's judgment or beliefs of what is important in life.

Section 4
Tracking and Tracing Laws, Directives, Regulations, and Standards

Chapter 8
Towards the Blanket Coverage DNA Profiling and Sampling of Citizens in England, Wales, and Northern Ireland

Katina Michael
University of Wollongong, Australia

ABSTRACT

The European Court of Human Rights (ECtHR) ruling of S and Marper v United Kingdom will have major implications on the retention of Deoxyribonucleic Acid (DNA) samples, profiles, and fingerprints of innocents stored in England, Wales, and Northern Ireland. In its attempt to develop a comprehensive National DNA Database (NDNAD) for the fight against crime, the UK Government has come under fire for its blanket-style coverage of the DNA sampling of its populace. Figures indicate that the UK Government retains a highly disproportionate number of samples when compared to other nation states in the Council of Europe (CoE), and indeed anywhere else in the world. In addition, the UK Government also retains a disproportionate number of DNA profiles and samples of specific ethnic minority groups such as the Black Ethnic Minority group (BEM). Finally, the S and Marper case demonstrates that innocent children, and in general innocent citizens, are still on the national DNA database, sometimes even without their knowledge. Despite the fact that the S and Marper case concluded with the removal of the biometric data of Mr S and Mr Marper, all other innocent subjects must still apply to their local Metropolitan Police Service to have their fingerprints or DNA removed from the register. This is not only a time-consuming process, but not feasible.

DOI: 10.4018/978-1-4666-4582-0.ch008

INTRODUCTION

The Police and Criminal Evidence Act of 1984 (UK) (PACE) has undergone major changes since its inception. The PACE and the PACE Codes of Practice provide the core framework of police powers and safeguards around stop and search, arrest, detention, investigation, identification and interviewing detainees (Police Home Office 2009). In the month of December 2008, post the S and Marper European Court of Human Rights ECtHR judgment, PACE underwent a review and changes were effective on the 31 December 2008, however, more changes especially on the issue of the retention of fingerprints and DNA are forthcoming. According to the Home Office the changes expected in the PACE will be to ensure that the "right balance between the powers of the police and the rights and freedoms of the public" are maintained (Police Home Office 2009). On reviewing the legal changes that have taken place since 1984 via a multitude of Acts, it can be said the United Kingdom (with the exception of Scotland) has, contrary to the claims of the Home Office, experienced a significant imbalance between the powers of the police and the rights and freedoms of the public. In the last 15 years, the rights and freedoms of the public have been severely encroached upon, and police powers significantly increased. A brief review of the major legislative impacts between 1984 and 2008 will be reviewed below. They are summarized in a timeline in Figure 1.

LEGISLATIVE CHANGES BETWEEN 1984 AND 2009

PACE was introduced in 1984, one year prior to Dr Jeffrey's discovery of DNA. Interestingly, PACE allowed for the police to ask a doctor to take a blood sample from a suspect during the investigation of a serious crime but only with their express consent. Thus a suspect had to volunteer or "agree" to a blood sample being taken, it could not be taken by force. Even after Jeffrey's discovery, there was limited use of blood samples for forensic analysis as tools and techniques were still in their infancy. The Single Locus Probe (SLP) technique which was in use in early DNA examinations had numerous limitations. While new SLP technology overcame some of these limitations, "the statistical evaluation of SLP DNA evidence brought a new set of problems, perhaps even more difficult to overcome than the preceding technical limitations" (Sullivan 1998). In sections 61-65 the original PACE classified blood samples and scrapings of cells from the inner cheek as intimate in nature. Hair samples (save for pubic hair) was the only type of non-intimate DNA sample that could be retained for forensic analysis without the permission of the suspect, and this on account of an investigation into a serious arrestable offence. Although this kind of DNA cut with scissors rarely provided enough of a good sample to conduct single locus probe (SLP) profiling, it was in the late 1980s that PCR (polymerase chain reaction) profiling could amplify and type a single strand of hair (Home Office, 2004). This is when mass screenings of DNA samples were possible. To begin with there was great contention over the admissibility of DNA evidence in a court of law but this changed as commonplace errors and procedural issues were rectified, new more modern profiling techniques were introduced, and larger databases for statistical purposes became available.

A significant moment in the fight against crime in the United Kingdom came in 1993 after a Royal Commission on Criminal Justice (Hansard 2003). The Commission was set up because there was a feeling among the community that the criminal justice system was just not working well enough to convict the guilty and exonerate the innocent. Leading up to 1993, there were a number of high profile miscarriages of justice which weakened the public's confidence in the criminal justice system,

Figure 1. Changes to U.K. Legislation 1984-2008 that have Given the Police Greater Powers and have had an Impact on Fingerprint and DNA Retention (The content in was taken from Genewatch UK (2009a) but adapted and turned into a timeline for the sake of readability)

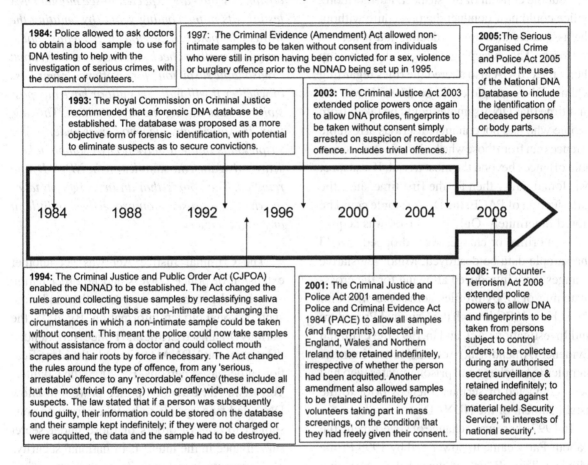

for example, the Birmingham Six, who had been jailed in 1974 for allegedly planting an IRA (Irish Republican Army) bomb that killed 21 people (BBC, 1991). One of the key recommendations coming from the Commission was the setting up of a national forensic DNA database. In the following year in 1994, the Criminal Justice and Public Order Act (CJPOA) introduced amendments to PACE and in 1995 the National DNA Database (NDNAD) was launched. At first, the Association of Chief Police Officers in England, Wales and Northern Ireland, believed that the system should have processed around 135, 000 samples in the first year, but by the end of that year only

one quarter of the original target had been loaded into the system due to significant procedural and technical teething problems related to the database. The expected annual rate was not reached until 1998 as police did not know how to fully exploit the new legislation (Lynch, 2008).

One of the fundamental changes heralded by the CJPOA was the reclassification of particular types of DNA samples from intimate to non-intimate. Authorities knew too well from their limited experience with DNA since the mid-1980s, that "richer" cellular samples were needed if a useable database of the size being projected was going to be possible. Saliva samples and mouth swabs became non-intimate samples, and it followed that non-intimate samples could be taken without the consent of the suspect. Furthermore, police could now conduct the procedure without the assistance of a trained doctor, and if needed by force. The sweeping changes did not stop there; the CJPOA also altered the rules regarding when a DNA sample could be taken. It was the first time that DNA samples could be taken from people who had not conducted serious arrestable offences but from those who had conducted recordable offences beyond the most trivial. If a suspect was found guilty then for the first time since the introduction of PACE, the DNA sample could be stored indefinitely. Only if a person was acquitted of a crime, or charges were dropped, would the sample data be destroyed. Minor legislative changes were introduced allowing for the cross-matching of DNA profiles across the whole of the U.K. in 1996 through the Criminal Procedure and Investigations Act, and in 1997 the Criminal Evidence (Amendment) Act enabled non-intimate samples to be taken from prison inmates who had been convicted of serious offences prior to the establishment of the NDNAD.

In 1997, there was a change of government, the Labour Party came to power and by 1999 Prime Minister Tony Blair announced the aggressive expansion of the NDNAD to contain some 3 million profiles by 2004. It was in 2001, post the Sept 11 attacks via the Prevention of Terrorism Act that DNA profiles which entered the database remained there indefinitely, even if the suspect was acquitted or charges were dropped. The PACE was impacted by these changes and even volunteers who had partaken in mass screenings or dragnets who had willingly provided their DNA samples remained on the database indefinitely (Beattie, 2009). In 2003, under the Criminal Justice Act of s. 10 (amending s. 63 of PACE), those who were simply arrested or detained at a police station on suspicion of a recordable offence had their DNA sample taken. According to McCartney (2006):

This enables police to take DNA samples from almost all arrestees and preempts technological advances which are expected to see mobile DNA testing kits in the coming years (by omitting the words "in police detention"). It means that a sample (usually a cheek swab) can be taken upon "reasonable suspicion" for an offence, regardless of whether it will indicate guilt or have any possibility of use during the investigation. The law, then, is explicit: anyone who comes under police suspicion is liable to have a DNA sample taken, searched against the samples on the NDNAD, and retained. The course that an investigation takes or whether a prosecution proceeds is of little, if any, significance.

The Criminal Justice Act was yet another extension of police powers and no other nation state had the same freedom to gather and store such personal citizen information. By 2005, the Serious Organised Crime and Police Act extended the uses of the NDNAD to include the identification of deceased persons. By 2008, the Counter-Terrorism Act extended police powers to allow DNA and fingerprints to be taken from persons subject to control orders or those under secret surveillance in the interests of national security.

Numerous legal analysts have been critical of the changes that PACE has undergone since 1984 - ironically the increase in police powers and the establishment of the NDNAD was originally introduced to increase public confidence in the criminal justice system and has instead eroded citizen trust in the state and impinged on the rights of every day Britons by going too far. Beattie (2009) is rather candid in her assessment of the changes, stating:

[there is] no statutory guidance for decisions about the retention of samples, no readily accessible mechanism whereby individuals can challenge the decision to retain their records (other than judicial review) and no independent oversight by a designated regulatory body.

This assessment seems to strike at the very heart of the problem. With only a judicial route at one's disposal to question current practices, an innocent citizen is left almost entirely powerless to battle against its own government. We can see no greater example of this than in the DNA sample storage of juveniles between the ages of ten and eighteen, "230,000 of whom were alleged to have been added following legislative changes in 2004, and of whom 24,000 were taken from 'innocent children' against whom no charges had been brought ..." (Lynch, 2008). An utterly disturbing statistic, and one which rightly led to the accusation of the Labour government compiling a database by stealth.

It now seems that PACE "1984" really did lay the seeds to an Orwellian state. According to the most recent Government statistics, 7.39 per cent of the UK population has their DNA profiles retained on the NDNAD (Beattie, 2009). This is an alarming figure when one considers that most other European states have less than 1 per cent of

their population on their respective DNA database, and do not keep cellular samples but rather DNA profiles alone and for a defined period of time (Table 1). The U.K. Government would possibly have us believe by these figures that they are dealing with an unusually high crime rate, but the reality is that the figures do not reveal the percentage of persons who have committed violent crimes as opposed to those who have committed petty crimes. Another problem with the NDNAD is that it is highly disproportionate in terms of its recording of citizens by ethnic background. The *Guardian* newspaper calculated that 37 per cent of black men and 13 per cent of Asian men in the nation are contained in the NDNAD, as compared to only 9 per cent of white men (Lynch, 2008). Liberty has stated that 77 per cent of young black men had records on the NDNAD in 2006 and that black people in general were almost 4 times as likely to appear on the database as white people (Rodgers, 2009).

Table 1. Characteristics of some National DNA Databases

Country (Year established)	Reference profile size	Crime-scene sample size	Suspect to scene hits	Scene to scene hits	Entry criteria for suspects	Entry criteria for convicted offenders	Removal criteria
UK (1995)	2.5 million	200,000	550,000	30,000	Any recordable offence*	Entered as suspect	Never removed, including suspects
USA (1994)	1.52 million	67,000	Figure unavailable	Figure unavailable	No suspects entered, but under revision	Depends on state law	Depends on state law
Germany (1998)	286,840	54,570	13,700	5,500	Offence leading to >1 yr in prison	After court decision	After acquittal or 5–10 years after conviction, if prognosis is good
Austria (1997)	64,740	11,460	3,200	1,350	Any recordable offence*	Entered as suspect	Only after acquittal
New Zealand (1996)	44,000	8,000	4,000	2,500	No suspects entered	A relevant offence (including ≥7 yr in prison)	Never removed, unless conviction quashed
Switzerland (2000)	42,530	7,240	4,840	5,540	Any recordable offence*	Entered as suspect	After acquittal or 5–30 years after conviction
France (2001)	14,490	1,080	50	70	No suspects entered	Sexual assault and serious crime	40 years after conviction
Finland (1999)	8,170	5,450	2,080	780	Offence leading to >1 yr in prison	Entered as suspect	Only after acquittal
Slovenia (1998)	4,820	2,360	370	80	Any recordable offence*	Entered as suspect	Depends on severity of crime
Netherlands (1997)	4,260	13,700	2,520	4,260	No suspects entered‡	Offence leading to >4 yr in prison	20–30 years after conviction
Sweden (2000)	3,980	9,860	2,500	4,750	No suspects entered	Offence leading to >2 yr in prison	10 years after release from prison

*That leads to a term of imprisonment. ‡Except when the suspect's DNA is tested for the case. Adapted from REF. 140, with additional information from Peter Schneider and Jill Vintiner (personal communications). See also BOX 3.

THE NATIONAL DNA DATABASE

The U.K. National DNA Database (NDNAD) of England and Wales was launched in April of 1995 at the Forensic Science Service (FSS) laboratory. It took several years for Northern Ireland to be included in the NDNAD. Before launching the official database the FSS trialed a small-scale forensic database to ensure the validity of such a system. The FSS began developing DNA testing in 1987 and in 1995 achieved a scientific breakthrough, inventing a chemical that enabled DNA profiling which led to the establishment of the NDNAD (FSS, 2009a). The NDNAD is the oldest and largest DNA database in the world with national legislation to foster and support its growth. The U.K. has also adopted a privatized model for forensic science services as related to the criminal justice system (Lynch, 2008). This was not always the case however, as the FSS was once an agency of the Home Office. When it became FSS Ltd. it became a profit maximizing, government-owned company under almost exclusive contract to the Home Office in forensic services to the police.

Although the legislation that enabled the police to collect DNA samples, request the FSS to process them and to store DNA profiles on the NDNAD, the annual expected growth rate was not reached until the late 1990s. As one of the main strategic objectives of the NDNAD was to demonstrate a return on investment, the Home Office set out to detect more crimes and thus reduce overall crime rates in the hope of closing the justice gap (McCartney, 2006, p. 175). In April 2000, five years after the establishment of the NDNAD, the UK government announced the DNA Expansion Programme, aimed at getting all known active offending persons onto the database which at the time was estimated to be about 3 million people. The total government investment in the program to March 2005 stood at £240.8 million which enabled police forces to increase the sampling of suspects and to recruit additional crime scene investigators, purchase the appropriate equipment,

train more police etc. (Home Office, 2005). Old samples from 1995 to 1999 were also able to be reanalyzed (McCartney, 2006, p. 176). A portion of the profiles were updated to match upgrades in the system software of the NDNAD from the standard profiling software known as SGM (Second Generation Multiplex) which had an average discrimination power of 1 in 50 million, to SGM Plus profiles which was said to reduce the chance of an adventitious match as the size of the NDNAD inevitably increased fuelled by the funding from the Expansion Programme.

An adventitious match is the possibility that two different people would have a profile that was almost identical owing to a "false positive" also know in statistics as an α (alpha) error. Thus an adventitious match shows a positive result for the matching of two persons (e.g. that of a crime scene sample, and that of a record on the NDNAD) when in actual fact there is no match at all. In the original NDNAD the risk of an adventitious match using the original SGM profiles was calculated to be 26 per cent but it has been claimed that since the introduction of the SGM Plus software, no adventitious matches have occurred (Nuffield Council, 2007). Sir Alec Jeffreys, however, has warned publicly that the genetic profiles held by police for criminal investigations are not sophisticated enough to prevent false identifications. "Dissatisfied with the discriminatory power of SGM Plus, Jeffreys recommends that following the identification of a suspect, the authority of the match should be tested by reanalyzing the sample at six additional loci" (Lynch 2008, pp. 144-145). Reanalysis of samples (whether volunteers, suspects, or those convicted) without consent, raises additional ethical questions however, even if it might indeed be able to exonerate a small number of individuals, if anyone at all.

The FSS are aware of the small possibility for an error but believe that the 10 markers currently stored on the database are sufficient (Jha 2004). In their defense FSS claim that the NDNAD is simply a type of intelligence database, and ultimately one is not convicted on mere "intelligence" but on multiple sources of evidence (Koblinsky, Liotti & Oeser-Sweat 2005, p. 273). Peter Gill of the FSS responded to Jeffreys concerns to the need to increase the number of markers for each profile by emphasizing that adventitious matches occur quite often when degraded samples are used and that the jury had to make up their mind based on numerous sources of evidence not just DNA evidence in isolation (Jha, 2004). For Jeffreys, storing "unnecessary" personal information on the NDNAD, for instance of persons who have previously been wrongly suspected of a crime, will only act to over-represent certain ethnic minorities which could lead to resentment by some citizen sub groups. The other issue that Jeffreys raises is the potential to use DNA sample information at some time in the future, and the risks associated with the potential to reveal health information from those samples; he is strongly opposed to the police gaining access to that kind of information (FSS, 2009).

Looking at some cross-sectional data of the NDNAD can provide us with a better feel for the size of this databank, which per capita, stores the largest number of DNA profiles for any given nation. By the end of March 2005, the Nuffield Bioethics Council reported that there were 3 million profiles stored on the NDNAD, an estimated 5.2 per cent of the U.K. population with 40,000 to 50,000 profiles being added monthly. Specifically, the police had retained 3,072,041 criminal justice (CJ) profiles, 12,095 volunteer profiles, and 230,538 scene-of-crime (SOC) profiles (Lynch, 2008, p. 149). The increase in loading samples of crimes was not just due to the Expansion Programme but also the legislative changes noted above via the Criminal Justice Act of 2003 and also the Serious Organised Crime and Police Act of 2005, and because of innovations in processing capabilities by the FSS. These legislative changes broadened the net of people who would now be added to the databank, in effect lowering the threshold for making it onto the NDNAD.

From the perspective of the Association of Chief Police Officers, this was a positive because it meant getting offenders onto the database earlier in their criminal careers. By the end of December 2005, the NDNAD held around 3.45 million CJ and elimination profiles and 263,923 crime scene sample profiles. At that rate it was predicted that an estimated 25 per cent of the adult male population and 7 per cent of the adult female population would eventually enter the database (Williams and Johnson 2005). More sober estimates indicate that the overall number of persons to be admitted to the NDNAD would be a little over 10 per cent of the UK population (Table 2) (Jobling & Gill, 2004, p. 745).

Current NDNAD Statistics

The most recent NDNAD statistics were made public during a parliamentary debate in October of 2009 (Hansard 2009). Here new figures from between 2007 and 2009 were tabled. Figure 2 is based on the data that was presented and shows that at the end of March in 2007, there were about 151,882 DNA profiles of persons between the ages of 10 and 15 on the NDNA which constituted about 3 per cent of all DNA profiles. There were 206,449 DNA profiles of persons between the age of 16 and 17 equating to about 5 per cent of all DNA profiles. Not counting children under the age of 10 whose DNA profiles are stored on the NDNAD, we can estimate that about 9 per cent of the profiles on the NDNAD are of persons under the age of 18. These are numbers that have the wider community, especially civil liberties groups, other self-interest groups and key non-government organizations (NGOs) expressing deep concern over the widening retention of persons for inclusion on the NDNAD. The matter has now gone through judicial review and while the UK courts refused to acknowledge the rights of innocents or those of young children or those who have been acquitted of a crime from entering the NDNAD, the European Court of Human Rights (ECtHR) ruled

Table 2. A NDNAD snapshot using year-end 2007 data

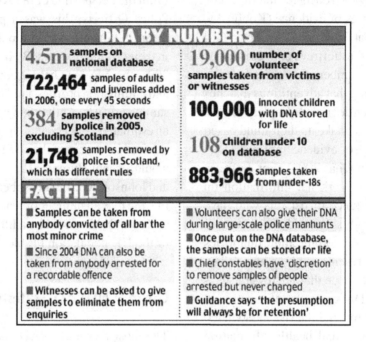

Figure 2. DNA profiles on the NDNAD by age as of end March 2007

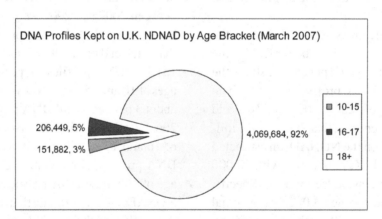

otherwise. The *S and Marper v. United Kingdom* will be the focus of the next section of this paper.

Beyond the problem of children on the NDNAD is the disproportionate number of persons of other ethnic appearance outside white Europeans who have had their DNA sample taken and analyzed and stored indefinitely. The NDNAD does not record detailed data about one's ethnicity but it does categorise an individual into one of six ethnic origins based on appearance. These categories include: White-South European, White-North European, Asian, Black, Chinese Japanese or South East Asian, Middle Eastern and one more category referred to as Unknown. At first glance the numbers in Figure 3 show that about 77 per cent of the DNA profiles on the NDNAD have come from "White-Europeans" (summing both the South and North White European categories)

Figure 3. DNA profiles on the NDNAD by ethnic appearance as of end March 2007

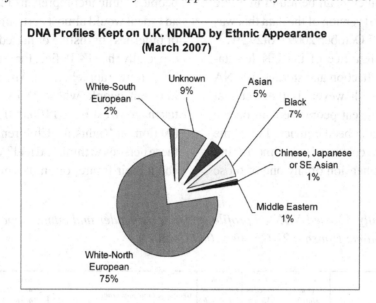

and only 7 per cent from "Blacks" and about 5 per cent from "Asians". But one should not look at these percentages on face value. Relatively speaking, when one analyses these numbers alongside census data, the truer picture emerges. Blacks and Asians do not make up the largest ethnic portion of the UK and thus a figure of 7 per cent of Blacks on the NDNAD means that more than 37 per cent of the Black male population in the UK have their DNA profile recorded on the NDNAD, and 5 per cent of "Asians" means that about 13 per cent of the Asian population have their DNA profile recorded on the NDNAD. This is compared with only 9 per cent of the total White population that is on the NDNAD.

Some groups refer to this kind of disproportionate ethnic presence on the NDNAD as institutionalized racism. Institutionalized racism can be defined as "that which, covertly or overtly, resides in the policies, procedures, operations and culture of public or private institutions - reinforcing individual prejudices and being reinforced by them in turn" (Lawrence, 1999). It is a structured and systematic form of racism built into institutions. While this researcher would not label the

disproportionate ethnic representation in the NDNAD as racism, she does acknowledge that minority ethnic populations, particularly black men, do not stand to benefit from the current UK legislation, but rather the legislation has been to the detriment of minority groups. According to National Black Police Association of the UK black men are four times more likely to be stopped and searched than white men. They are also more likely to be arrested and released without charge, let alone convicted, and without being compensated for their ordeal. The NDNAD statistics seem to suggest that black males are more likely to offend than white males, which is a fallacy. And this kind of feeling among the community of the Black Ethnic Minority (BEM) may not only provoke great mistrust in the UK police and the Government but also strong resentment toward future life opportunities and freedoms, a feeling echoed by Sir Jeffreys. It also means that less competent officers may be inclined, whether mindfully or not, to draw in ethnic minorities in general because they are the "usual" suspects in crimes (Jarrett, 2006). The most up-to-date figures on the profiles that constitute the NDNAD by gender, age and ethnicity can be found in Table 3, which is an adapted version of the data that was tabled in Hansard 27 October 2009 Col292W.

Of the greatest injustice of the UK legislation related to the collection and storage of DNA samples and profiles however, is the fact that at least 857,000 innocent people remain on the NDNAD who have not been convicted of crime and who may never be convicted of a crime. Living in this state of apprehension of any one of those people is quite incomprehensible. For some, such an ordeal would almost certainly lead to a feeling of bitterness or dislike or hatred for the State and especially the UK Police, for that individual who was wrongly apprehended. Among the one million innocent people whose DNA sample has been taken are an estimated 100,000 innocent children (Action on Rights for Children 2007). What are these persons to think and feel? What does it mean about their future, or employment opportunities

Table 3. Most recently released NDNAD profile statistics by gender and ethnic appearance (compare 2008 and 2009). Source: Hansard 27 October 2009 Col292W.

Gender	Ethnic appearance	Current age as at 31 March 2009	Profiles retained at 31 March 2009	Profiles retained at 31 March 2008
Female	Unknown	10-15	1,855	2,929
		16-17	4,116	5,099
		18+	89,878	83,266
	Asian	10-15	914	888
		16-17	1,598	1,384
		18+	33,839	27,807
	Black	10-15	2,859	2,902
		16-17	4,274	4,109
		18+	76,008	63,843
	Chinese, Japanese or SE Asian	10-15	83	93
		16-17	203	199
		18+	9,340	7,472
	Middle Eastern	10-15	70	50
		16-17	89	84
		18+	3,290	2,689
	White—North European	10-15	32,912	36,756
		16-17	52,699	53,528
		18+	813,093	701,320
	White—South European	10-15	636	617
		16-17	868	721
		18+	17,621	14,330
Male	Unknown	10-15	4,537	7,313
		16-17	10,139	13,538
		18+	382,983	360,727
	Asian	10-15	4,314	4,281
		16-17	7,625	7,275
		18+	244,229	208,640
	Black	10-15	7,277	7,038
		16-17	11,242	11,062
		18+	310,471	274,282
	Chinese, Japanese or SE Asian	10-15	216	197
		16-17	443	378
		18+	25,019	20,168
	Middle Eastern	10-15	500	323
		16-17	999	739
		18+	36,300	30,467
	White—North European	10-15	72,265	81,248
		16-17	109,096	114,249
		18+	3,115,325	2,790,097
	White—South European	10-15	1,335	1,187
		16-17	1,921	1,690
		18+	82,818	70,090

Gender	Ethnic appearance	Current age as at 31 March 2009	Profiles retained at 31 March 2009	Profiles retained at 31 March 2008
Gender not recorded	Unknown	10-15	45	46
		16-17	73	258
		18+	34,449	34,460
	Asian	10-15	18	17
		16-17	34	48
		18+	591	476
	Black	10-15	44	26
		16-17	33	34
		18+	487	376
	Chinese, Japanese or SE Asian	10-15	1	1
		16-17	2	2
		18+	83	72
	Middle Eastern	10-15	1	1
		16-17	4	2
		18+	106	94
	White—North European	10-15	336	419
		16-17	458	475
		18+	4,868	4,289
	White—South European	10-15	11	9
		16-17	5	11
		18+	164	122
TOTAL			5,617,112	5,056,313

requiring security checks? And how might their experience with Police impact them later in life? Psychologists will always point out that someone treated like a criminal may retaliate as if they were one: "[b]ecause it feels like someone is punishing us by making us feel guilty, we often have an urge to retaliate against those who do" (Stosny 2008).

But beyond the psychological repercussions on the individual stemming from what some refer to as "emotional pollution" is the effort that a person must go through to get their details removed from the NDNAD (Geoghegan, 2009), a process that was almost impossible until the S and Marper EC-tHR judgment. Since 2004, in England, Wales and Northern Ireland records are removed and DNA destroyed only under "exceptional circumstances" (Genewatch UK, 2009). And given the profiles on the NDNAD belong to individual police forces, innocents whose profiles remain on the NDNAD and who wish to have them removed need to appeal to their Constabulary, although most recently ACPO have asked officers to ignore the ECtHR ruling (Travis, 2009).

At the end of March 2009, Lord West of Spithead noted that the NDNAD contained DNA profiles and linked NDA samples from approximately 4,859,934 individuals included by all police forces, of which an estimated 4,561,201 were from English and Welsh forces (more than 7 per cent of the UK population) (Hansard, 2009). This figure should be compared with those cited on 27 October 2009 in Parliament which indicated that at the end of March in 2008 there was a total of 5,056 313, profiles on the NDNAD and as of 2009 for the same period there were 5,617,112 (See Table 3). According to the latest population statistics obtained from the Office for National Statistics (2009), there are about 61.4 million people residing in the UK, which means that the NDNAD contains profiles of more than 8.36 per cent of the total population in the United Kingdom. This figure is rather conservative an estimate when one considers that Scotland has a different legislative requirement regarding the retention of DNA profiles.

Why these specifics are important is because they indicate a number of things. First, the size of the UK databank is growing at over 560,000 profiles per annum which is in keeping with the rate of 40,000 to 50,000 samples per month. Secondly one in nine persons in England, Wales and Northern Ireland is registered on the databank. Thirdly, and more to the point, there are 507,636 DNA profiles which are of unknown persons. This either means that these samples have been collected at crime scenes and have not been individually identified alongside "known" persons or that potentially errors exist in the NDNAD itself. Here an important complementary factor must be underscored in support of the latter claim. If we are to allege that 507,636 profiles came from scenes of crime (SOC) where the individual has not been identified since April 1995 then we also need to understand that (McCartney, 2006, p. 182):

only 5 per cent of examined crime scenes result in a successful DNA sample being loaded onto the NDNAD, and only 17 per cent of crime scenes are examined, meaning that just 0.85 per cent of all recorded crime produces a DNA sample that can be tested (NDNAD, 2003/04: 23)...

Thus it is very rare for a perpetrator of a serious crime to leave body samples behind unless it is saliva on a cigarette butt or a can of drink or in more violent crimes such as sexual assaults, semen or some other bodily stain sample. In the case of some violent crimes like sexual assault,

most victims do not, and are unlikely to begin, reporting to police. Many of these who do report do so too late for DNA profiling to be an option. Of those who do report in time, the occurrence of sexual intercourse is often not an issue in dispute. The existence or non-existence of consent will be the critical matter. DNA profiling can offer nothing to resolve this problem. However, in the case of serial rapes or where there is no real doubt about identity of the assailant, DNA profiling potentially has a great deal to offer (Freckelton, 1989, p. 29).

Of Dragnets and Mass Screenings

In cases where heinous violent crimes have occurred, often of a serial nature, local police have conducted mass DNA screenings of the population in and of surrounding neighborhoods of the scene of the crime (Butler, 2005, p. 449). It becomes apparent to local police that a mass DNA screening is required when it seems that the crimes have been conducted by a single person nearby, given the trail of evidence left behind and other intelligence information. A DNA mass screening was used in the very first case where DNA was used to convict an individual. Mass screenings are now termed intelligence-led screens and the subtle change in nuance as of 1999 was of great importance to how the UK perceived its use of DNA evidence in criminal cases. In a talk on DNA technology, Lynn Fereday of the FSS said in 1999 that:

[t]he screens now are a routine method of policing. This is a major way of saving police resources. What happens is that once a crime is being investigated, and DNA evidence has been found, police immediately do a scoping of who or what area they have to screen. They decide on a select area, and they then look for volunteers in that area. One of the first cases involved a murder of the young girl using STRs ...The interesting thing about the mass screens is that although there seem to be some unease about continuing with them here, people are volunteering constantly. They volunteer for a reason, because they know they are innocent. They have nothing to fear, and we will end up with crime detection.

Of course, such comments come from an employee of the FSS. Examples of very early mass screenings in the UK can be found in DNA user conferences (Burton, 1999).

There is no denying that mass screenings have led to convictions of perpetrators who would have otherwise gone unnoticed but the statement that people volunteer because they are "innocent" or they "have nothing to fear" is not entirely true.

In her landmark paper in 2006, Carole McCartney described Operation Minstead where the police profiled 1,000 black men in South London in the hunt for a serial rapist, and then requested each of them to volunteer a DNA sample. McCartney (2006, p. 180) writes:

Of those, 125 initially refused, leading to "intimidatory" letters from the police, urging reconsideration, and five were arrested, their DNA taken post-arrest and added to the NDNAD. Such actions have raised questions of legality, with arrests only lawful with 'reasonable suspicion' of an individual having committed a criminal act. If the police are to arrest on non-compliance with a DNA request, then that casts non-compliance as a crime--a step that worries civil libertarians and may lose the spirit of cooperation essential in these circumstances.

Table 4 shows an example of a prioritisation grid to deal with DNA intelligence led screen actions. While it is an early example of a grid, and today's practices are much more sophisticated in manner, it does indicate why an individual approached to volunteer a DNA sample by the police might refuse to do so. Being targeted to donate a sample by the police in a mass screen such as Operation Minstead means you are under some suspicion and fall into one of the priority areas of concern. If you are indeed innocent of a crime, you may refuse to donate a DNA sample for any number of reasons, among which could be a basic right not to be insulted particularly by the State. An individual resident who lives in a mass screen prioritization area and meets the criteria of any number of priorities might feel like they are being presumed guilty, and may not trust technology to prove them innocent, or may even fear being accidentally matched to a crime they did not commit.

Table 4. A prioritisation grid to deal with DNA intelligence LED screen actions

	PRIORITY 1	PRIORITY 2	PRIORITY 3	PRIORITY 4
AREA 1 Resident or has strong links to within ½ mile of scene	Age:16 to 35 Recorded Assault &/or Sex Pre Cons Living without partner	Age:16 to 35 Recorded Other Pre Cons Living without partner	Age:16 to 35 Known for violent &/or criminally sexual behaviour Living without partner	Age:16 to 35 No Pre Cons Living without partner
AREA 2 Resident or has strong links to within 1 mile of scene	Age:16 to 35 Recorded Assault &/or Sex Pre Cons Living without partner	Age:16 to 35 Recorded Other Pre Cons Living without partner	Age:16 to 35 Known for violent &/or criminally sexual behaviour Living without partner	Age:16 to 35 No Pre Cons Living without partner

Source: (Burton, 1999)

Now while the police can ask any person in the UK to volunteer a DNA sample, there is some controversy related to what happens with a sample once it is analyzed and an individual is proven to be innocent. If an individual has been eliminated from enquiries then the question remains whether or not their DNA profile should be retained on the NDNAD. According to Genewatch (2009c):

[i]n these cases, people can consent to having their DNA used only for the inquiry, or give an additional signature if they agree to having their DNA profile added to the database. In Scotland volunteers can change their minds and ask to be removed from the Database, but this is not possible in England and Wales. However, the NDNAD Ethics Group recommended in April 2008 that volunteers should not have their DNA added to the Database at all, and their DNA should be destroyed when the case has ended. This recommendation is likely to be implemented because there is no evidence that adding volunteers' DNA to the database is helping to solve crimes.'

Still this practice has yet to be implemented categorically and the claim remains that innocent people should be kept off the NDNAD.

Statistics presented by the Home Office will always tout suspect to scene matches and scene to scene matches and provide the numbers of murders, rapes and car crimes where suspects are identified but it is very important to note that not all successful matches result in a conviction or even in an arrest (McCartney, 2006). So while statistics might seem to indicate that the NDNAD is returning value for money, overall crimes rates in the UK have not been reduced (Ministry of Justice, 2009), and the number of persons convicted using DNA evidence remains relatively moderate based on previous years reports. The FSS and the Government will always seek to show that the NDNAD has been an excellent evidential tool that has supported many successful prosecutions and provided important leads in unsolved "cold" cases but no matter how one looks at it, the storage of innocent persons' DNA profiles should not be permitted.

WHERE WAS THE NDNAD HEADED?

The Possibility of Blanket Coverage DNA Sampling of All Citizens

Putting the brakes on the NDNAD was not going to be easy. Several cases had been heard through various local courts but were unsuccessful in their attempts to have their clients' fingerprints and DNA samples and profiles destroyed. Of course, some scientists working in the area of forensic analysis continued to dream of databases and databanks that ideally would contain the profiles of every person in the country. This was a view maintained by scientists not only within the UK but as far as the United States and even New Zealand. Although the overwhelming feeling among this community of experts was that such a database would "obviously never be compiled" (Michaelis et al., 2008, p. 106). Still this goodwill does not halt the potential for DNA databases to become commonplace into the future. In 2005, Koblinsky et al. (p. 290) rightly predicted that more people would find themselves onto national DNA databases. They believed that it was likely:

... that legislation will be passed that will require juveniles who commit serious crimes to be included in the database. It is possible that eventually every citizen will be required to have his or her profile in a national database despite concerns about privacy issues and constitutional protections.

Such attitudes must be understood within their context. It makes sense to forensic analysts and scientific-literate commentators that a larger database would help to capture repeat offenders and thus reduce overall crime rates. Many would not debate the importance of DNA profiling for serious crimes, but there are issues with relating DNA profiling techniques in a mandatory fashion to the whole populace. Even the Nuffield Bioethics Council was allegedly supportive of the benefits of a universal database. According to Lynch et al. (2008, p. 154) the Council:

...[found] that while the balance of argument and evidence presented in the consultation was against the establishment of a population-wide database, it recommend[ed] that the possibility should be subject to review, given its potential contribution to public safety and the detection of crime, and its potential for reducing discriminatory practices.

In 2005, Koblinsky et al. (p. 163) wrote: "[a]s DNA analysis becomes more and more common in criminal investigations, there will come a day when millions upon millions of people will have been profiled." Well, we no longer have to look into the future for the fulfillment of such prophecies - they are here now. There are millions upon millions of DNA samples and profiles stored in the UK alone and the US too is now driving new initiatives on the road of mass DNA profiling (Moore, 2009). The FBI's CODIS database has 6.7 million profiles and it is expected that it will accelerate its DNA database from 80,000 new entries a year to 1.2 million by 2012 (Michaelis et al., p. 105). But it may not be criminal legislation that impacts on such outlandish figures. One day it is indeed possible that the medical research field will have such an impact on society that "… every citizen's genetic profile may be stored in a national database. There are many who are concerned about the ramifications of a government agency maintaining such records. It is essential that all DNA data can be encrypted and protected from abuse or unauthorized access" (Koblinsky et al., 2005).

Expanding databanks will clearly have an impact on civil liberties and individual privacy. And while there are those who believe such statements do a "disservice to a society suffering from a constant rise in violent crime," (Melson, 1990) the recent ECtHR ruling is proof enough that we need to reconsider the road ahead. But it is not scientists alone who are providing the impetus for even larger databanks, politicians or political commentators also are entering the debate. Former mayor of New York, Mr Rudy Giuliani had advocated taking DNA samples of all babies born in American hospitals. This idea would not take much to institute in practice, given cellular samples (blood) are already taken from babies with the permission of the parent to test for common disorders. The same practice also exists in Australia and is known as the Guthrie Test or more commonly the Heel Prick Test (Guthrie Test, 2009). Michaelis et al. (2008, pp. 100-101) comment on such a potential status of mass DNA sampling at birth but are mindful of the implications on civil liberties and privacy:

Having a databank of all American-born persons would obviously be of great benefit, not only in violent crime investigations but also in cases of missing persons, inheritance disputes, immigration cases and mass casualties such as airline crashes and terrorist acts. The obvious concerns over privacy and civil liberties, however, have caused commentators to urge caution when deciding which samples to include in the databanks.

DNA Developments and Innovations Challenging Ethical Practice

The 13 year Human Genome Project (HGP) conducted by the US Department of Energy and the National Institutes of Health has gone a long way into identifying all the approximately 20,000-25,000 genes in human DNA, and determining the sequences of the 3 billion chemical base pairs that make up human DNA. The project was and still is surrounded by a number of very challenging ethical, legal and social issues (Table 5). Points 3 and 7 in the table are of particular interest when we consider what it means for someone's DNA sample to be taken, analyzed, and stored indefinitely in a criminal databank. What kind of psychological impact will it have on the individual and forthcoming stigmatization by the individual themselves, and then by the community around them. This is particularly the case of minority groups. And what of the potential to "read"

someone's DNA and be able to make judgments on their mode of behavior based on their genetic makeup? Are persons for instance, more prone to violence because they carry particular genes? Or would some generalities based on genetics affect someone's free will and determine their future because of some preconceived statistical result?

Already under research are "DNA identikits" which can describe a suspect's physical appearance from their DNA sample in the absence of an eyewitness account. At present the FSS provide an ethnic inference service (McCartney, 2006, p. 178). The FSS used this technology in 2008 to investigate the stabbing of Sally Anne Bowman in 2005, although it was not this forensic result that ultimately led the police to her perpetrator (FSS, 2009). Used to supplement ethnic inference is the red hair test which can detect 84 per cent of red heads (McCartney, 2006, p. 181). The continued research into the HGP will inevitably determine very detailed information about a person in the future. The other problem closely related to innovations in identikits are those of advances in familial searching techniques. Given that families share a similar DNA profile, obtaining the DNA of one individual in a family, let us say "the son", can help to determine close matches with other persons in the immediate family such as the sister, mother, father or first cousin. While only identical twins share exactly the same DNA, a sibling or parent share a very close match. The technique of familial searching was also used in the Sally Anne Bowman case without success. A suspect's DNA was taken and matched against the UK NDNA but no exact matches were returned. The FSS then attempted the familial searching technique and that too did not aid their investigation. Familial searching was first used in 2002 in a rape and murder case when a list of 100 close matches was returned from the NDNAD to identify a perpetrator who had since died. DNA samples were first taken from the living relatives and then from the dead body of the offender Joe Kappen.

Table 5. Societal concerns arising from the new genetics (adapted from the Human Genome Project, 2009)

1. **Fairness in the use of genetic information by insurers, employers, courts, schools, adoption agencies, and the military, among others.** • Who should have access to personal genetic information, and how will it be used?
2. **Privacy and confidentiality of genetic information.** • Who owns and controls genetic information?
3. **Psychological impact and stigmatization due to an individual's genetic differences.** • How does personal genetic information affect an individual and society's perceptions of that individual? • How does genomic information affect members of minority communities?
4. **Reproductive issues including adequate informed consent for complex and potentially controversial procedures, use of genetic information in reproductive decision making, and reproductive rights.** • Do healthcare personnel properly counsel parents about the risks and limitations of genetic technology? • How reliable and useful is fetal genetic testing? • What are the larger societal issues raised by new reproductive technologies?
5. **Clinical issues including the education of doctors and other health service providers, patients, and the general public in genetic capabilities, scientific limitations, and social risks; and implementation of standards and quality-control measures in testing procedures.** • How will genetic tests be evaluated and regulated for accuracy, reliability, and utility? (Currently, there is little regulation at the federal level.) • How do we prepare healthcare professionals for the new genetics? • How do we prepare the public to make informed choices? • How do we as a society balance current scientific limitations and social risk with long-term benefits?
6. **Uncertainties associated with gene tests for susceptibilities and complex conditions (e.g., heart disease) linked to multiple genes and gene-environment interactions.** • Should testing be performed when no treatment is available? • Should parents have the right to have their minor children tested for adult-onset diseases? • Are genetic tests reliable and interpretable by the medical community?
7. **Conceptual and philosophical implications regarding human responsibility, free will vs genetic determinism, and concepts of health and disease.** • Do people's genes make them behave in a particular way? • Can people always control their behavior? • What is considered acceptable diversity? • Where is the line between medical treatment and enhancement?
8. **Health and environmental issues concerning genetically modified foods (GM) and microbes.** • Are GM foods and other products safe to humans and the environment? • How will these technologies affect developing nations' dependence on the West?
9. **Commercialization of products including property rights (patents, copyrights, and trade secrets) and accessibility of data and materials.**

The Risks Associated with Familial Searching and Medical Research

Familial searching has very broad ethical implications. It is conducted on the premise that a rotten apple comes from a rotten tree. Put another way, the old adage goes, "tell me who your friends are and I'll tell you who you are." Instead today, we may be making the false connection of - "tell me who your friends are and I'll tell what gene you are"! Interestingly this latter idea has formed the titled of a biology paper written by P. Morandini (2009). The point is that we return to models of reputation by association and these cannot be relied upon to make judgments in a court of law. We learnt all too well in Australia through the

Dr Haneef case, that guilt by association, even guilt by blood-line, is dangerous to civil liberties. Considered another way, some have termed this kind of association based on DNA profiles, "genetic redlining." Genetic redlining can be defined as "the differentiated treatment of individuals based upon apparent or perceived human variation" (Melson, 1990, p. 189). David L. Gollaher discusses the risks of what essentially is genetic discrimination in a 1998 paper.

Perhaps the most disturbing practice that may enter this field and make things impossible to police both in the "criminal law" arena and the "medical research" field is the deregulation and privatization of the DNA industry internationally. Future technological innovations will surely spawn the growth of this emerging industry. We have already noted the home-based DNA sampling kits available for less than 100 US dollars which come with free DNA sample databanking. It will not be long before some citizens volunteer somebody else's DNA, instead of their own, forging consent documentations and the like. The bungle with the first ever UK DNA case shows that even the police could not imagine that Pitchfork (the offender), would have conceived of asking a friend to donate a sample on his behalf. Such cases will inevitably occur in volunteer home sampling methods, as fraudsters attempt to access the DNA samples of friends, strangers or even enemies via commonplace saliva-based sampling techniques. All you need is a pre-packed buccal swab from the DNA company providing the kits and away you go. If this seems an extreme possibility to the reader, consider the "spit kits" that have been issued to public transport drivers who have been harassed by passengers by being spat at or otherwise, who can now collect the DNA samples of an alleged offender and turn them into the appropriate authorities. No consent of the donor is required here (Lynch, 2008, p. 153).

When we consider how we as a society have traversed to this point of "accepting" the construction and development of such unusually large national databanks as the NDNAD in the UK,

we can identify a number of driving forces. Some nations are at this point of almost indiscriminate storage of DNA profiles primarily due to changes in policing practices and the law, government policy, innovation in forensic science (the idea that *because we can, we should*), co-existing with venture capitalists who are backing commercial opportunities and the parallel developments in the genetic medical research field. In the case of the UK the PACE changed so much, and there was such a redefinition of what constituted a "recordable offence" that non-intimate samples could be obtained from individuals for investigation into the following offences without their consent (Roberts & Taylor, 2005, pp. 389-390):

unlawfully going onto the playing area at a designated football match; failing to leave licensed premises when asked to do so; taking or destroying rabbits by night; riding a pedal cycle without the owner's consent; allowing alcohol to be carried in vehicles on journeys to or from a designated sporting event.

Consider the Home Office's August 2008 proposal to expand police powers which included plans to set up new "short term holding facilities" (STHFs) in shopping centers to take people's DNA and fingerprints but was later quashed with the S and Marper ECtHR judgment (Genewatch UK, 2009b).

This is short of being farcical. It makes little sense to take such personal data from an individual when the profile itself cannot be used for investigative purposes. There must be some other motivation toward the sampling of persons who on occasion might find themselves charged with a petty crime and are punished by fine, penalty, forfeiture or imprisonment other than in a penitentiary. Why store such petty crime offenders' DNA profiles indefinitely on the NDNAD? Surely the action of someone who might find themselves, for instance, under the influence of alcohol and refuse to leave a licensed premise when asked to do so, is not indicative of their capacity to commit a serious felony in the future. There is a grave issue of proportionality here commensurate to the crime committed by the individual, and on the side of the crime itself, a major issue with what constitutes a recordable offence. The original PACE wording stated a "serious arrestable offence" (Ireland, 1989, p. 80) not just any old offence. As a result policing powers were increased significantly, and the individual's right not to incriminate himself or herself was withdrawn in conflict with the underpinnings of Common Law (Freckelton, 1989, p. 31).

Our legal system has traditionally eschewed forcing people to incriminate themselves by becoming the instruments of their own downfall. That principle has suffered a number of encroachments in recent years.

It is here that we need to take a step back, reassess the balance needed in a robust criminal justice system and make the necessary changes to legislation, save we get too far ahead that we find recourse a near impossibility.

CONCLUSION

When one analyses the case of Mr S and Mr Marper, one realises how short of the mark the UK Government has fallen. Instead of upholding the rights of innocent people, the retention of their fingerprint and DNA data is kept for safe keeping. Some have claimed that this initial boost in the number of samples was purposefully conducted to make the NDNAD meaningful statistically, while others believe it was in line with more sinister overtones of a surveillance state. One thing is certain, that where the courts in England did not provide any recourse for either Mr S or Mr Marper, the European Court of Human Rights ruling indicated a landslide majority in the case for both Mr S and Mr Marper to have their DNA samples destroyed, and profiles permanently deleted. One of the major

issues that has triggered this change in the collection of such personal and sensitive data have been the alleged 3,000 individual changes to the PACE Act. The watering down of laws that are meant to uphold justice, but instead are being alternatively used to abuse citizen rights, is an extremely worrying trend, and adequate solutions, despite the ECtHR ruling, are still lacking.

REFERENCES

Action on Rights for Children. (2007). *How many innocent children are being added to the national DNA database?* Retrieved from http://www.arch-rights.org.uk/issues/dna/dnabrief.htm

BBC. (1991). *Birmingham six freed after 16 years*. Retrieved from http://news.bbc.co.uk/onthisday/hi/dates/stories/march/14/news-id_2543000/2543613.stm

Beattie, K. (2009). S and Marper v UK: Privacy, DNA and crime prevention. *European Human Rights Law Review*, 2, 231.

Burton, C. (1999). The United Kingdom national DNA database. *Interpol*. Retrieved from http://www.interpol.int/Public/Forensic/dna/conference/DNADbBurton.ppt

Butler, J. M. (2005). *Forensic DNA typing: Biology, technology, and genetic of STR markers*. Oxford, UK: Elsevier.

Fereday, L. (1999). *Technology development: DNA from fingerprints*. Retrieved from http://www.ojp.usdoj.gov/nij/topics/forensics/events/dnamtgtrans6/trans-i.html

Forensic Science Service. (2009a). *Analytical solutions: DNA solutions*. Retrieved from http://www.forensic.gov.uk/html/services/analytical-solutions/dna/

Forensic Science Service. (2009b). *Sally Anne Bowman*. Retrieved from http://www.forensic.gov.uk/html/media/case-studies/f-47.html

Freckelton, I. (1989). DNA profiling: Forensic science under the microscope. In J. Vernon, & B. Selinger (Eds.), *DNA and criminal justice* (Vol. 2). Academic Press.

Genewatch, U. K. (2009a). *A brief legal history of the NDNAD*. Retrieved from http://www.gene-watch.org/sub-537968

Genewatch, U. K. (2009b). *Police and criminal evidence act (PACE) consultations*. Retrieved from http://www.genewatch.org/sub-551990

Genewatch, U. K. (2009c). *Whose DNA profiles are on the database?* Retrieved from http://www.genewatch.org/sub-539482

Geoghegan, J. (2009, October 12). Criticism for police over silence on DNA database. *Echo*. Retrieved from http://www.echo-news.co.uk/news/4673015.Criticism_for_police_over_silence_on_DNA_database/

Gollaher, D. L. (1998). Genetic discrimination: Who is really at risk? *Genetic Testing*, *2*(1), 13. doi:10.1089/gte.1998.2.13 PMID:10464593

Guthrie Test (Heel Prick Test). (2009). *Discovery*. Retrieved from http://www.discoverychannel.co.uk/homeandhealth/article.jsp?section_id=7&theme_id=23&subtheme_id=80&article_id=81&site=uk

Hansard. (1993). *Royal commission on criminal justice*. Retrieved from http://hansard.millbanksystems.com/commons/1993/jun/24/royal-commission-on-criminal-justice

Hansard. (2009). *DNA databases*. Retrieved from http://www.publications.parliament.uk/pa/cm200809/cmhansrd/cm091027/text/91027w0019.htm

Hansard. (2009). *Police: Databases*. Retrieved from http://www.publications.parliament.uk/pa/ld200809/ldhansrd/text/90505w0003.htm

Home Office. (2004). *Coldcases to be cracked in DNA clampdown*. Retrieved from http://press.homeoffice.gov.uk/press-releases/'Coldcases'_To_Be_Cracked_In_Dna?version=1

Home Office. (2005). *DNA expansion programme 2000–2005: Reporting achievement*. Retrieved from http://police.homeoffice.gov.uk/publications/operational-policing/DNAExpansion.pdf

Human Genome Project. (2009). *Human genome project information: Ethical, legal and social issues*. Retrieved from http://www.ornl.gov/sci/techresources/Human_Genome/elsi/elsi.shtml

Ireland, S. (1989). What authority should police have to detain suspects to take samples? In J. Vernon & B. Selinger (Eds.), *DNA and criminal justice*. Retrieved from http://www.aic.gov.au/media_library/publications/proceedings/02/ireland.pdf

Jarrett, K. (2006). DNA breakthrough. *National Black Police Association*. Retrieved from http://www.nbpa.co.uk/index.php?option=com_content&task=view&id=40&Itemid=58

Jha, A. (2004, September 9). DNA fingerprinting no longer foolproof. *The Guardian*. Retrieved from http://www.guardian.co.uk/science/2004/sep/09/sciencenews.crime

Jobling, M. A., & Gill, P. (2004). Encoded evidence: DNA in forensic analysis. *Nature Reviews. Genetics*, *5*(10), 745. doi:10.1038/nrg1455 PMID:15510165

Koblinsky, L., Liotti, T. F., & Oeser-Sweat, J. (Eds.). (2005). *DNA: Forensic and legal applications*. Hoboken, NJ: Wiley.

Lawrence, S. (1999, February 24). What is institutional racism? *The Guardian*. Retrieved from http://www.guardian.co.uk/uk/1999/feb/24/lawrence.ukcrime7

Lynch, M. et al. (2008). *Truth machine: The contentious history of DNA fingerprinting*. Chicago: Chicago University Press. doi:10.7208/chicago/9780226498089.001.0001

McCartney, C. (2006). *Forensic identification and criminal justice: Forensic science justice and risk*. Cullompton: Willan Publishing.

McCartney, C. (2006). The DNA expansion programme and criminal investigation. *The British Journal of Criminology, 46*(2), 189. doi:10.1093/bjc/azi094

Melson, K. E. (1990). Legal and ethical considerations. In L. T. Kirby (Ed.), *DNA fingerprinting: An introduction*. Oxford, UK: Oxford University Press.

Michaelis, R. C., Flanders, R. G., & Wulff, P. H. (2008). *A litigator's guide to DNA: From the laboratory to the courtroom*. Burlington, UK: Elsvier.

Ministry of Justice. (2009). *Population in custody*. Retrieved from http://www.justice.gov.uk/publications/populationincustody.htm

Moore, S. (2009). F.B.I. & states vastly expand DNA databases. *The New York Times*. Retrieved from http://www.nytimes.com/2009/04/19/us/19DNA.html?_r=1

Morandini, P. (2009). *Tell me who your friends are and I'll tell what gene you are*. Retrieved from http://www.siga.unina.it/SIGA2009/SIGA_2009/6_01.pdf

Nuffield Council on Bioethics. (2009). *Forensic use of bioinformation: Ethical issues*. Retrieved from http://www.nuffieldbioethics.org/bioinformation

Office for National Statistics. (2007). *Mid-2006 UK, England and Wales, Scotland and Northern Ireland: 22/08/07*. Retrieved from http://www.statistics.gov.uk/statbase/Product.asp?vlnk=15106

Office for National Statistics. (2009). *UK population grows to 61.4 million*. Retrieved from http://www.statistics.gov.uk/pdfdir/popnr0809.pdf

Parliamentary Office of Science and Technology. (2006). *Postnote: The national DNA database*. Retrieved from http://www.parliament.uk/documents/upload/POSTpn258.pdf

Police Home Office. (2009). *Police and criminal evidence act 1984 (PACE) and accompanying codes of practice*. Retrieved from http://police.homeoffice.gov.uk/operational-policing/powers-pace-codes/pace-code-intro/

Roberts, A., & Taylor, N. (2005). Privacy and the DNA database. *European Human Rights Law Review, 4*, 373.

Rodgers, M. C. (2009). Diane Abbott MP and liberty hold DNA clinic in Hackney. *Liberty*. Retrieved from http://www.liberty-human-rights.org.uk/news-and-events/1-press-releases/2009/24-09-2009-diane-abbott-mp-and-liberty-hold-dna-clinic-in-hackney.shtml

Stosny, S. (2008). Guilt vs. responsibility is powerlessness vs. power: Understanding emotional pollution and power. *Anger in the Age of Entitlement*. Retrieved from http://www.psychologytoday.com/blog/anger-in-the-age-entitlement/200805/guilt-vs-responsibility-is-powerlessness-vs-power

Travis, A. (2009, August 8). Police told to ignore human rights ruling over DNA: Details of innocent people will continue to be held: Senior officers will not get new guidance for a year. *The Guardian*. Retrieved from http://www.theguardian.com/politics/2009/aug/07/dna-database-police-advice

Williams, R., & Johnson, P. (2005). Inclusiveness, effectiveness and intrusiveness: Issues in the developing uses of DNA profiling in support of criminal investigations. *Medical Malpractice: U.S., & International Perspectives*, 545.

KEY TERMS AND DEFINITIONS

BEM: Black Ethnic Minority group. BEM has specific national or cultural traditions from the majority of the population.

DNA: Deoxyribonucleic acid (DNA) is a molecule that encodes the genetic instructions used in the development and functioning of all known living organisms and many viruses.

DRAGNETS: In policing a dragnet is any system of coordinated measures for apprehending criminals or suspects, such as widespread DNA testing, pressuring potential criminals who have committed a given act to come forward.

ECtHR: European Court of Human Rights is a supra-national or international court established by the European Convention on Human Rights.

Familial Searching: Familial searching is a second phase step conducted by law enforcement after a search on a DNA database has returned no profile matches. Familial searching attempts to find a match of first-order relatives (e.g. sibling, parent/child) based on a partial match, granting some leads to law enforcement, as opposed to no leads.

HGP: The Human Genome Project is an international scientific research project with a primary goal of determining the sequence of chemical base pairs which make up human DNA, and of identifying and mapping the total genes of the human genome from both a physical and functional standpoint.

Mass Screenings: Occur when the police encourage people residing in a given area, or encourage people who are members of a certain group to volunteer their DNA sample. Mass screenings are supposed to save police resources in apprehending the offender(s) of a criminal activity.

NDNAD: Is a National DNA Database that was set up in 1995. As of the end of 2005, it carried the profiles of around 3.1 million people. In March 2012 the database contained an estimated 5,950,612 individuals. The database, which grows by 30,000 samples each month, is populated by samples recovered from crime scenes and taken from police suspects and, in England and Wales, anyone arrested and detained at a police station.

PACE: The Police and Criminal Evidence Act 1984 (PACE) (1984 c. 60) is an Act of Parliament which instituted a legislative framework for the powers of police officers in England and Wales to combat crime, as well as providing codes of practice for the exercise of those powers.

Profiling: With respect to DNA is the banding patterns of genetic profiles produced by electrophoresis of treated samples of DNA.

Scene of a Crime: Is a location where a crime took place or another location where evidence of the crime may be found. This is the area which comprises most of the physical evidence retrieved by law enforcement personnel, crime scene investigators (CSIs) or in some circumstances forensic scientists.

SLP: The Single Locus Probe (SLP) is a technique which was in use in early DNA examinations and has numerous limitations with respect to newer more advanced techniques.

Chapter 9
ID Scanners and Überveillance in the Night–Time Economy:
Crime Prevention or Invasion of Privacy?

Darren Palmer
Deakin University, Australia

Ian Warren
Deakin University, Australia

Peter Miller
Deakin University, Australia

ABSTRACT

ID scanners are promoted as an effective solution to the problems of anti-social behavior and violence in many urban nighttime economies. However, the acceptance of this and other forms of computerized surveillance to prevent crime and anti-social behavior is based on several unproven assumptions. After outlining what ID scanners are and how they are becoming a normalized precondition of entry into one Australian nighttime economy, this chapter demonstrates how technology is commonly viewed as the key to preventing crime despite recognition of various problems associated with its adoption. The implications of technological determinism amongst policy makers, police, and crime prevention theories are then critically assessed in light of several issues that key informants talking about the value of ID scanners fail to mention when applauding their success. Notably, the broad, ill-defined, and confused notion of "privacy" is analyzed as a questionable legal remedy for the growing problems of überveillance.

INTRODUCTION

Many metropolitan and regional areas are trying to enrich night-time economies that have been traditionally centered on alcohol consumption. It is therefore not surprising that a rise in anti-social behavior, violence, serious interpersonal crime, and associated concerns over personal health, safety and environmental amenity, generate many contentious policy interventions (Hadfield et al., 2009). Governments appear keen to be seen as responsive to community concerns over the lack

DOI: 10.4018/978-1-4666-4582-0.ch009

of security in the night-time economy. However, there are considerable doubts over whether the complex range of spatial, patron-based or regulatory interventions actually changes the behavior of nightclub patrons.

The ID scanner has emerged as a key method of increasing surveillance in many night-time economies throughout Australia, the United Kingdom, Canada and the United States. An extensive report into surveillance in public places by the Victorian Law Reform Commission (VLRC) highlights the reach of contemporary digital surveillance, by illustrating that:

[i]dentification scanners record the image and written details on an individual's driving license or other identity card, including their name and address. Facial recognition software scans patrons' faces as they enter the nightclub and matches those images against a database of photos. In this way the software can be used to identify patrons who have been previously banned from a venue. The software can be shared among venues (VLRC, 2010, p. 40).

These systems use inexpensive and accessible 'technologies for a new, security-driven purpose' (Goold et al., 2010, p. 21). They are particularly attractive to large venues where the scale of patronage complicates security provision. They are appealing to both governments and private businesses for simultaneously promising improved public safety and increased revenue.

The Law, Justice and Safety Committee for the Legislative Assembly of Queensland (QLALJSC, 2010) provides a rare examination of the role of ID scanners in licensed venues. This analysis is useful, as it provides insight into how the potential benefits of this technology appear to have greater political credence than the various problems associated with information security. The perceived benefits of ID scanning identified by the Committee are:

- Aiding in detection of offenders, with the scanned information able to be retrieved from the data base and provided to police;
- Acting as a deterrent, as potential offenders know that their personal details have been recorded and can be provided to police; and
- Providing information to support a ban of the offender from that venue, and in some cases other venues as well (QLALJSC, 2010, p. 25).

The first two benefits suggest scanned data is a valuable method of enhancing the detection of offenders, or deterring potential anti-social behavior. However, there is no evidence these objectives are realized in practice (Palmer et al., 2010). Further, networked data sharing has proactive value in warning other venues of troublesome individuals identified in these systems. This enhances their deterrence capabilities amongst licensed venues with network access, but overlooks the potential displacement of anti-social behavior to surrounding areas.

Despite considerable support amongst the liquor industry (QLALJSC, 2010, p. 25), the Queensland Office of the Information Commissioner documented several concerns over the need for using personal data collection, storage and dissemination to curb problematic alcohol-fuelled behavior. Notably, current state and federal privacy laws may not apply to venues that have introduced ID scanners or companies that install and manage this technology. Table 1 summarizes various unresolved privacy issues identified in the Queensland Information Commissioner's submission that could be addressed through alternative harm reduction methods.

Despite these concerns, the report recommended licensees trading after midnight should be encouraged to install ID scanning systems with 'due regard to privacy issues and matters of natural justice' (Queensland Parliament, 2010, p.

Table 1. Summary of privacy concerns by Queensland Information Commissioner (QLALJSC, 2010, pp. 25-26)

Potential breaches of Federal privacy legislation, including arbitrary interference with individual privacy principles where venues have an annual turnover of less than $3 million per annum	Other less intrusive technologies might achieve reductions in violence and antisocial behavior, such as the rolling out of blood alcohol testing machines and colorless ultra-violet sprays
The tenuous causal link between alcohol and violence and misconceptions about this link amongst the broader public	Various other harm minimization strategies to combat violence in and around licensed premises warrant further policy consideration
The status of some venues which 'are known to be more violent than others and at particular times'	Conflicts between the financial gains from selling alcohol and principles of responsible service
The potential disproportionate policy response of ID scanners and CCTV to the general problem of alcohol-related violence, which occurs in a range of public environments other than licensed venues	The need to focus on planning, licensing and price regulation or liquor taxation laws in moderating drinking culture and assisting with minimizing alcohol-related violence
The impact of ID scanners on the privacy of most people who attend licensed venues at the expense of a small minority of people who engage in violent behavior	The lack of evidence to support the deterrence effect of ID scanners and the use of scanners as an 'all seeing eye for law enforcement' by police and liquor establishments

27), and proposed a discounted license fee structure for venues adopting this technology. However, the report is silent on what 'due regard' or 'natural justice' might entail. This demonstrates that governments consider privacy can be readily overridden when it comes to combating alcohol-related disorder. Interestingly, the Committee's interim report cautioned against recommending networked ID scanning systems until concerns over information privacy, data storage and maintenance were resolved.

An important issue raised with a system of networked ID scanners relates to privacy, in particular the collection and storage of this sensitive information. The committee recognizes that the safety of patrons and the protection of their identity documents are paramount. These issues need to be closely considered before any recommendation can be made on this matter (QLALJSC, 2009, p. 8).

The willingness to sidestep privacy implications of crime prevention technologies has considerable impacts on those administering these measures at individual sites where they are deployed, the public police, system developers and the broader community. Of particular concern

is how 'back end data' (Greenleaf, 2007, p. 7) is used once a person's individual details are entered into a scanning system. Table 2 summarizes the unanswered questions contained in the Information Commissioner's detailed submission, but remained ignored in the final report.

The widespread public and political concern over alcohol-related violence has rapidly contributed to an expansion of ID scanners in Australia's licensed venues at the expense of other citizen rights or protections. However, there is little clarity surrounding the desired regulatory and accountability relationships between the state, a scanning company or database administrator and the patron. These problems are common to the introduction of new technologies and emerging forms of e-governance. By viewing these development outside the lens of crime prevention, and as extensions of many forms of contemporary e-governance, it is possible to gain a greater appreciation of the limits of how new technology is adopted, its implications in light of contemporary developments in surveillance and überveillance and to in turn glean new and meaningful insights into possible methods of better regulating their use.

Table 2. Unresolved privacy issues identified by the Queensland Information Commissioner (QLALJSC, 2010, pp. 26-27)

Will the transfer of personal information be limited to those found guilty of a crime or misdemeanor in a licensed venue, or will it extend to anyone that has committed a crime or misdemeanor or to anyone a licensee would rather not have in their premises?	Will police be able to access the information when investigating the whereabouts of interested persons, establishing alibis in unrelated crimes, including using licensed premises databases of fingerprints as an extension of police records?
To whom can a patron complain if they find themselves unjustly placed on a blacklist, perhaps because someone used a fraudulent government ID the scanner could not detect?	Will it be shared only between licensed premises owned by the same legal entity, or between all other licensed premises, regardless of who owns them, including restaurants?
What training will be provided to licensed premise employees to ensure all personal information is handled appropriately?	What mechanisms will be put in place to ensure the information is accurate and up to date? What safeguards will surround the sharing?
Will the length of the ban be proportionate to the seriousness of the anti-social behavior?	Who will decide whether a misdemeanor is serious enough to warrant blacklisting?
What mechanisms will there be for a person to challenge their placement on a 'ban list'?	Will it be shared between interstate licensed premises? Internationally?
Will patrons be banned for non-criminal behavior?	How will the identity of the person be confirmed?

TECHNOLOGY, SECURITY AND DIGITAL GOVERNANCE

It is easy to fall into a theoretical and methodological trap that equates the introduction of a new technology with reductions in crime or anti-social behavior. However, Matthews (2009) demonstrates a seemingly obvious link between installing a CCTV camera in a car park and reduced car theft can overlook potential causal explanations that, if ignored, help reinforce the perceived success of new technological innovations.

...[I]t could be because potential offenders are deterred, more are caught and prosecuted, more people might use a car park thus making it safer or the increased publicity associated with the introduction of cameras may serve to deter or deflect potential criminals ... all these hypotheses need to be examined and assessed while it is recognized that the context – the size, location, design and the like – or the deployment of CCTV will influence the outcome (Matthews, 2009, p. 356).

Governments and private entities commonly embrace untested technologies to streamline manual bureaucratic processes, business performance, security and identity verification (Senate Standing Committee on Finance and Public Administration, 2007; Greenleaf, 2007a; Greenleaf, 2007b; Hart, 2007). Lips et al. (2009) indicate that organizational streamlining and improved service delivery are standard justifications to support new technologies and forms of surveillance, which promise 'perfect enforcement' (Mulligan, 2008) to enhance financial or interpersonal security. The benefits of heightened efficiency, accuracy, security and service delivery (Taylor et al., 2008) appear self-validating and beyond question, even if there is limited research supporting their realization.

The implications of auto-generated data on notions of responsible and participatory citizenship are only beginning to be understood. Dataveillance can generate 'particularized' as opposed to rights-based or universal citizenship (Lips et al., 2009, p. 731), with the capacity to exacerbate community 'segmentation'. When linked to quasi-criminal laws or 'multiple hybrid, civil, contractual, and administrative' legal requirements, a growing body of 'irregular citizens', including 'antisocial youth, persistent offenders, sexual and violent offenders, and suspected terrorists' (Zender, 2010, p. 389; 394-397), commonly bear the brunt of tighter risk classifications that undermine their full citizenship status. For example, in Britain the introduction of up to 1000 new offences over the last decade (Matthews, 2009, p. 315), largely targeting minor or trivial forms of incivility with the goal of preventing more serious crime (von Hirsch & Simister, 2006), combined with the erosion of conventional due process requirements that protect individuals from state intrusion, is contingent on using new forms of surveillance technology to enforce more complex forms of administrative, legal and social sorting. Increasingly, knowledge generated about 'good' and 'bad' conduct is constructed, shared and monopolized by public and private agencies that convert 'common information sources [about 'irregular citizens'] into exclusive knowledge resources' (O'Connor and De Lint, 2009, p. 59). Technology therefore consolidates sophisticated enforcement networks in promoting greater security or 'assessing and constructing citizens' digital footprints, which then constitute the basis for setting an individual's trust profile' (Lips et al., 2009, p. 730). However, these processes are given practical meaning when informed by laws and bureaucratic strategies that promote new forms of citizenship, rights and accountability while restricting an individual's capacity to formally contest these processes.

In identifying a package of reforms aimed at combating anti-social behavior through the increased surveillance of 'at risk' young people in New South Wales, Osmond (2010) documents a sophisticated automated and networked data matching system operating across several government departments, which is justified in the overall public interest of reducing crime, improving community safety and preventing identified individuals from graduating into adult criminal careers. These processes aim to produce a more effective, intensive and personalized case management system that simultaneously nullifies any countervailing privacy concerns. Those 'at risk' youth siphoned into the dataveillance network become guinea pigs for a new form of citizen oversight that bears a striking resemblance to überveillance (Michael &

Michael, 2008). Überveillance refers to a society where everyone is always watched or watchable, with individual liberty, responsibility, independence or retreat from the viewer's gaze impossible (Clarke, 2010). Increasingly, 'irregular citizens' face the allied risk their lives become defined by an intrusive web of digital surveillance and abstracted interpretations of the data, rather than their actual conduct.

Of course, the surveillance of at risk populations is by no means a new development. Thompson and Genosko (2009) offer a fascinating account of the punch card system that recorded all takeaway liquor sales by the Liquor Control Board of Ontario between 1927 and 1975. This system relied on rigid bureaucratic categories to give meaning to the abstracted data, which pigeonholed 'at risk' individuals into arbitrary categories, then justified tight paternalistic surveillance by local liquor distributors that was in turn overseen by provincial authorities. This system validated several pre-existing restrictions on the ability of individuals to purchase alcohol, such as prohibitions targeting First Nations people who did not renounce their Indian identity (R. v. Webb, 1943). A formal legal challenge in 1974 (Ontario (Liquor Control Board) v. Keupfer, 1974) helped trigger the collapse of this highly cumbersome dataveillance system. However, the eventual demise of this model under its own bureaucratic weight is secondary to the fact it took almost fifty-years to successfully challenge its autocratic power structure through the Canadian courts.

Moving beyond this historical example, automated digital sorting is now embedded within many e-citizenship and private consumer transactions, including international travel, passport and temporary visa authentication, access to public or private sector buildings or entry into major sporting events. The digitized tracking of citizens by governments occurs within a tightening of bureaucratic sorting criteria under the law. While it is debatable whether the processes of monitoring citizen activity have changed to accommodate these uses of technology (Manning, 2008), the legal power structures that underpin them have. Therefore, 'new administrative sorting practices... [that] lead to new forms of citizen segmentation: those who can be trusted according to their digital footprint, and those who cannot' (Lips et al., 2009, p. 731) are invariably informed by new legal regimes that reconfigure ideals of state accountability and citizen capacity to challenge the uses of data trails.

The logic of increased security, crime prevention or greater community protection enables digital surveillance to compile more detailed and potentially accurate information on people and their activities. However, the 'back end' assemblages of this data that sanction arbitrary classifications, interpretations, reinterpretations and reconstructions to manufacture new forms of 'truth' is increasingly unchallengeable under 'law against law' (Ericson, 2007, p. 24).

Privacy is often cited as a central legal countermeasure to challenge intrusive dataveillance and social sorting practices. However, it is virtually impossible to come to a 'satisfactory definition of 'privacy'' (New South Wales Law Reform Commission, 2009, p. 3) that can meaningfully challenge the 'truth' of a problematic surveillant assemblage. When viewed in its legal rather than popular sense, the right to privacy and its enforcement is extremely limited in protecting individuals from public and private sector agencies that gather and use personal data.

AUSTRALIAN PRIVACY LAWS

Privacy law imposes certain constraints on the capacity of state or federal public sector agencies, and the private sector, on the gathering, use, storage and dissemination of personal information. Leman-Langlois (2008, p. 113) offers a five-point definition of 'privacy' that refines these issues in a social context where personal data is commonly provided to access public or private services.

- **Control Over Information:** Including the assurance that personal information will be used according to contractual arrangements;
- **Secrecy of Information:** Including the ability to escape surveillance or protect against unwanted prying, or access to anonymity;
- **Desire to Protect Personal Space:** Involving the psychological need to retreat to non-social space (even if this might be in the public arena) to engage in individual activities;
- **Right to Keep Secrets:** Involving rules defining institutional, social, political or administrative limits to collecting and sharing information;
- **Data Security:** Including the development of appropriate technical safeguards against unauthorized access to protected information.

How these issues are operationalized as a legal right to privacy is more complex. Divisions of responsibility between Australian state and federal laws magnify these problems, to ensure that even if privacy is enshrined as a basic human right, its enforcement in practice is extremely difficult.

In Victoria the *Charter of Human Rights and Responsibilities Act* (2006) stipulates a right to privacy is enforceable against Parliament, courts, tribunals or relevant statutory authorities (*Charter of Human Rights and Responsibilities Act*, 2006, sections 38 and 6) unless a suspected breach can be validated under a competing national law. Under Section 13 a citizen has the right:

1. Not to have his or her privacy, family, home or correspondence unlawfully or arbitrarily interfered with; and
2. Not to have his or her reputation unlawfully attacked.

However, the term 'privacy' is undefined in the Charter. Schedule 1 of the *Information Privacy Act* (Victoria) (2000), equivalent national legislation (*Privacy Act* (Commonwealth), 1988) and laws governing the use of listening, optical, tracking and data surveillance technologies by public agencies to monitor private activity (*Surveillance Devices Act* (Victoria), 1999) provide some guidance on the meaning of 'privacy'. Both the state and federal laws establish base standards for the collection and use of personal information, with state law focusing on the activities of public or private agencies undertaking contracted government functions. The most compelling definition under Victorian law relates to 'private activity', which involves:

... an activity carried on in circumstances that may reasonably be taken to indicate that the parties to it desire it to be observed only by themselves, but does not include:

1. *An activity carried on outside a building; or*
2. *An activity carried on in any circumstances in which the parties to it ought reasonably to expect that it may be observed by someone else (Surveillance Devices Act (Victoria), 1999, section 3).*

Both state and federal privacy laws articulate a series of key privacy principles that are recognized under International law (*Privacy Act* (Commonwealth), 1988, Section 6C). Federal law only applies to private businesses that have an annual financial turnover of $ 3 million or more. Table 3 identifies the main 'information privacy principles' enshrined in Victorian and Commonwealth law. All government and private organizations covered by these laws must adhere to this combination of guidelines relating to the gathering, dissemination or use of personal information.

Federal law enables an industry to work with the national Privacy Commission to develop a Code of Conduct that incorporates the privacy principles. However, these restrictions can be waived under both Victorian and federal law where the 'prevention, detection, investigation, prosecution or punishment of criminal offences' is concerned (see *Privacy Act* (Commonwealth), 1988, Schedule 3 (6)(j)(i); *Information Privacy Act* (Victoria), Schedule 1, 2.1(d)-(h)).

In 2006 the biometrics industry had developed a formally approved Code of Conduct (Office of the Information Privacy Commissioner, 2006), which could be translated to the use of ID scanning technologies. However, this voluntary Code has only four signatories and has been widely criticized for offering citizens minimal protection 'beyond the default requirements of the Privacy Act' (Australian Privacy Foundation, 2010). More problematically, many common uses of surveillance technology in public places 'are likely to be beyond the reach of privacy laws' (VLRC, 2010, p. 24). Nevertheless, clear standards for regulating public and privately managed closed circuit television systems (CCTV) are common. However, as voluntary industry-based codes of practice, it is debatable how detailed guidelines developed by organizations such as the Australian Privacy Foundation (2010; Clarke, 2010), requiring community consultation to ensure CCTV use remains proportional to the social benefits it aims to produce, appropriate controls on the use, disclosure, publication and cyclical destruction of CCTV data, and periodic review of the viability of individual systems to ensure compliance with these principles, are incorporated to protect citizens from the expansion of uberveillance. It also remains debatable how these principles are enshrined or enforced in practice, or whether they impact on newly emerging forms of computerized or biometric surveillance permeating the security landscape of the public or private sector.

Table 3. Victorian and Federal information privacy principles

General Principles	Privacy Principles, Victoria Information Privacy Act 2000, Sch 1	Privacy Principles, National Privacy Act (Commonwealth) 1988, Sch 3
Data collection	The organization must clearly communicate its identity and legal purpose for gathering the information	Personal information can only be collected if it is necessary for one or more organizational functions and is collected by 'lawful', 'fair' and unobtrusive methods
Use and disclosure	Ensures that information can only be used or disclosed by the organization for specified purposes, such as 'the prevention, detection, investigation, prosecution or punishment of criminal offences or breaches of a law' or 'the protection of public revenue'	Reasonable steps to inform the individual about the identity and contact details of the organization; the right for individuals to access any information and be informed of the purposes for its collection or disclosure; the main consequences to the individual if information is not disclosed; and the primary purpose or uses of the information
Data quality and accuracy	Aimed at preserving the accuracy of personal information that has been gathered and stored	The records relating to personal information are updated accurately
Data security	Aimed at preventing unauthorized access and use	Reasonable steps to protect personal data from misuse, loss or unauthorized access
Open policies	Involve the development and communication of clear policies regarding data uses and management	Policies regarding the use and management of personal information are clear and accessible to individuals making a request
Right to access and correct data	This right applies unless access or correction compromise the rights of others or the conduct of criminal investigations	Information can be accessed and corrected by individuals providing their data unless there are justified reasons to the contrary
Trans-border data transfer	Prohibitions apply except in certain restricted circumstances	Restrictions relating to the foreign transfer of information
Bans on unique identifiers	Applicable unless these are necessary to undertake core organizational business	
Anonymity	To be preserved where practical	
Sensitive information	Citizen must provide consent on information about political preference, ethnicity, religion, trade union affiliation or criminal history	

Greenleaf et al. (2007) indicate that compliance with these complex privacy requirements is more difficult to achieve where public sector data management services are subcontracted to private service providers. However, a more problematic series of issues arises in a political context where privacy is juxtaposed against other individual or collective rights associated with human security. The recent growth of ID scanners in the Australian night-time economy offers a pertinent case study into the limits of contemporary privacy law in its own right, and as a counterpoint to more punitive conceptions of anti-social behavior that are governed by tighter laws to promote public safety that are impervious to legal challenge or based on popular misconceptions about the effectiveness of new surveillance technologies.

LIQUOR LICENSEES' VIEWS OF PRIVACY

After a spate of violent incidents in and around the nightclub precinct of the Victorian regional city of Geelong between December 2006 and February 2007, the community's police, venue licensees and local council sought concrete measures to regain control over disorderly night-time economy. In the ensuing months, several venues piloted ID scanners, and by November 2007, ten inner city venues participating in a voluntary Liquor Accord designed to improve practices associated with alcohol service had adopted this technology. Throughout 2008 and 2009, we conducted a series of in-depth interviews with local police, councilors, venue licensees and door staff examining perceptions of the effectiveness of ID scanners in reducing violence and the processes surrounding their normalization as a precondition of entry into all 'at risk' venues trading after 1 am (Palmer et al., 2010). While there was general acceptance ID scanners encouraged more behavioral accountability amongst nightclub patrons, the countervailing issue of privacy is of most interest to the present discussion.

Most respondents recognized the importance of privacy in relation to this form of surveillance. However, the interviews revealed few practical insights into how privacy is assured given the dominant purpose of this technology is to enhance personal safety in licensed venues. The concept of privacy was universally considered secondary or marginal to the overriding concern of combating serious violence and anti-social behavior in the Geelong nightclub precinct. As such, the popular conception of privacy as a right that is justifiably conceded or overridden by the more prominent concern of promoting safety is a consistent thread emerging in our data.

Those who question the privacy implications of ID scanners and the associated issues of data storage, maintenance and use, are most likely to be viewed as potential sources of violent or anti-social behavior. Under this view, privacy is not only considered an inconvenience by many respondents, but a sign that the person seeking to challenge the integrity of the ID scanning system is a troublemaker that should be denied entry into a participating venue.

People that walk in the place, if they've got any concerns about privacy – stay home, don't go out, because I don't care who gets my details, I haven't done nothing wrong. I don't think that should worry anyone. There's more than scanners if you want to get details of someone, let's be honest.

This view sidesteps the more pertinent issues of how privacy is protected according to current legislative requirements or the idea of responsible on-site administration of this technology. Those

questioning the privacy integrity of ID scanning are showing a distinct mistrust of a system that is considered to have discernible benefits in promoting greater security in the night-time economy. This degree of mistrust is equated with anti-social behavior. Therefore, it is better for people who have concerns over privacy to stay home, as to do otherwise demonstrates an intention to test a clearly beneficial system designed to enhance safety for all nightclub patrons.

Similarly, patrons who do not possess identification and attempt entry or express dismay at being rejected, it's a further sign a person is likely to 'cause trouble'. This view shows how the technology provides a paternalistic and protective benefit, which can be justified as 'law', while any attempt to use privacy to contest this new 'law' or failing to produce ID is considered to undermine the entire viability of this innovative safety measure.

…[T]he ones that come out without ID are the ones I think come out for not a good time … to cause trouble. They've got all these fanciful reasons in their heads with big imaginations as to why we'd want it, but it's a safety tool. We just say provide law … you have to do it … all the clubs are doing it. If you go to the club down the road you are going to have the same problems there. You have to have your ID on you. If you get hit by a car we need to contact somebody.

Leaving aside the stated concerns over road safety, the integrity of ID scanning is reinforced by its in-built technological protections, making it impervious to criticism. Any concerns over data integrity, such as possible misuses of data for marketing purposes, are again indicators of mistrust. By stressing how this element of privacy is preserved, the value of using scanned data to identify and ban troublesome patrons becomes self-validating.

[T]he first week we had a couple of people ring up complaining about where it's going to be used,

but there are only 2 people at both ends that have got access to the scanner information, so we don't use it for any marketing purposes ... [I]n fact I don't even know how to use it for stuff like that, so no one [here] has access to go into scanned details, so on a Monday if any of the managers has got a problem with that guy and we've got footage, we ban them ...

Complaints about privacy issues were considered rare. When they did emerge, licensees indicated they were countered by reminding patrons this technology is necessary to increase their safety. Eighteen-months after being first introduced, the process, and the safety justifications underlying it, became normalized: 'everybody does it ... [the] first thing everyone does at the door is pull it [their ID] out. This reinforces the attitude that privacy is considered an inconvenience when compared with improved safety in nightclubs. Trust is pivotal to interpreting how licensees view privacy. Trust applies to both the technological protections where database records are cleared every '28 days or 30 days' and the licensees' direct assurances to patrons of the integrity of this surveillance system.

... [E]specially the ones from out of town who don't know about it and don't want to use it. 'How do I know your not going to use it [the information]?' 'You don't know, but I'm telling you', and I can't remember if it's 28 days or 30 days where it overrides itself ... if they don't do anything wrong, you've got no problems.

Although different venues have deployed different scanning technologies in Geelong, all systems can be configured to restrict a licensee's ability to access or alter patron data. Technical requirements allowing for 'password protected' systems to restrict unauthorized access are considered to equate with privacy. This ensures only trusted or senior employees can access and modify individual patron records. Many of these issues relate to good business practice, which further

validates their integrity, even if some venue managers have consciously 'avoided having to think about' these issues.

I guess as licensees we were committed at the start that they wouldn't be used for any untoward reasons. Other than the last person to scan as they go through you need a password to go back and look. So you can't have dodgy security guys and door people going back looking up your address.

Privacy guidelines are communicated to all patrons upon entering each venue. However, the extent to which they are enforced or monitored for compliance remains unclear. All venues involved in this study referred to policies that inform patrons of the privacy implications of ID scanning, but this does not necessarily preclude sharing scanned information with police to investigate criminal behavior. There is some sensitivity about data sharing even for the purposes of detecting a crime within a licensed venue. This is arguably due to the informality of the protocols for sharing scanned data amongst police and other venue operators. This lack of clarity extends to the motives to justify sharing scanned data with police.

...[W]e all have our own privacy policy at the front door, or should have. I think you'd find that all the venues should have a privacy policy somewhere as you walk in. We all agreed ... [that] if police wanted to come back and have a look at something they'd be made available. As far as I know everyone's abided by that.

Most ID scanning systems are administered under a contractual arrangement between a licensed venue and the hardware and software manufacturers. As such, there is little incentive for individual licensees to consider the privacy implications of this technology, or interrogate how external providers administer the technology to protect individual privacy rights. For many venue operators adherence to lawful privacy requirements were considered beyond their direct responsibility. Their time is consumed by other licensing requirements that are more closely linked to business performance and improved safety. This reinforces the inconvenience privacy creates in the on site administration of ID scanning systems and how trust underpins the licensing role more generally to promote safety even through contentious technological methods.

But if it's tightly controlled and if we're responsible enough to hold a liquor license we should be responsible enough. There are probably some out there that like setting them up on databases, but they could get fined for that – there's tough laws about that.

To supplement this data, several observations at late night venues where ID scanners have been installed were conducted during the research period. Figure 1 indicates that around 2/3rds of the 324 patrons interviewed considered ID scanners an effective method of improving venue safety. This confirms similar findings from a study in Denver, Colorado, which suggests that patrons attending late-night venues express few concerns over information (Holloman & Ponder, 2007). The vague legal environment that enables this process to occur in a largely unregulated context has several further implications, given the apparent lack of concordance between our respondents' conceptions of privacy and other non-technological methods of improving safety within licensed venues.

DISCUSSION AND FUTURE RESEARCH DIRECTIONS

Trust, privacy, improved human security and the regulation of surveillance technologies are linked by a conflicting series of citizenship issues. When applied to technology through an e-governance perspective, the focus on trust examines how gathered data is used to promote safety and validated by those deploying increased surveillance at the expense of other countervailing rights or accountability mechanisms. Although technology can promote greater accuracy in monitoring human behavior and incorporate various measures to promote secure access and storage, privacy as a key conceptual vehicle to protect citizens has varied legal and cultural meanings. The 'back end' uses of data cause numerous problems, yet reinforce the overriding security benefits of this technology amongst those licensees deploying ID scanners as a condition of entry into 'at risk' venues. This reinforces the findings of Holloman

Figure 1. Patron views on the effectiveness of ID scanners

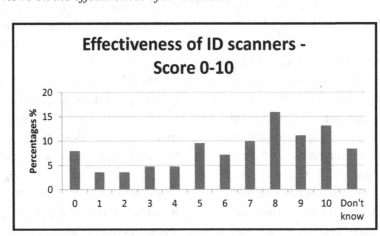

and Ponder's examination of ID scanners in the US city of Denver, which encapsulates the problem of juxtaposing privacy with safety, while reinforcing a profound ambivalence towards privacy issues amongst citizens who attend licensed premises.

... [W]hile there is a general apathy about having one's driver's license scanned, most people trust that the places scanning their licenses are not saving the information past the door. Of the people worried about their information being stored, very few believe that it is being used outside of the scope of ensuring a safe and legal atmosphere within the establishments ... This general unawareness is a sign that businesses are not divulging what is actually happening when an ID is scanned, and that is a gross abuse of trust ... when [these businesses] don't consider their customers' safety and privacy concerns, they are sending a very strong message that they don't think about morality (Holloman & Ponder, 2007, p. 45).

For Goldsmith (2005), a lack of trust can undermine notions of consensus-based policing, which in turn undermines citizen security. However, under a crime prevention framework, issues of trust are focused solely on the individual, with technology offering accurate methods of compiling digital trust profiles of citizen behavior (Lips et al., 2009). Our research indicates the institutional trust profiles, which encompass the public police who endorse the deployment of this technology, venue licensees, their security personnel or door staff, and the system administrators, involve several implied dimensions that are barely questioned, and although given adequate protection under current Australian state and federal privacy laws, appear to have limited direct applicability to ID scanners. This implied trust is indicative of a broader problem relating to the enforcement of privacy law in Australia (Greenleaf et al., 2007), but is compounded by equating privacy concerns with potential disorder. This schism between the

legal right to privacy and the cultural conception of privacy as a nuisance that compromises good order or the well-intentioned use of technology to promote safety, generates two key problems.

First, privacy is secondary or peripheral to the overarching goal of improved physical safety. This is facilitated by current privacy legislation which determines a 'legal purpose' for data collection by using extremely unclear criteria. This logic reinforces the 'success' of ID scanners in regaining order within the contemporary night-time economy, while undermining the rights of all nightclub patrons, including those who are well-behaved. The popular belief that only those who might cause trouble are affected by the privacy implications of ID scanning is misconceived, yet is viewed as somehow contradicting the law or undermining the more pressing concern for improved safety. This turns the privacy question on its head to reinforce the trust in ID scanning as 'the solution' to the problem of alcohol-related violence and disorder. As such, any privacy objections are irrelevant as the system is foolproof and contains several in-built technical and administrative protections. However, the foolproof nature of these systems is undermined by human methods of administering the technology, with observational data indicating a substantial proportion of nightclub patrons are rarely scanned before entering 'at risk' venues.

Second, these issues feed the broader political endorsement of ID scanners, which further relegates privacy to something that can be conceded in the quest to improve human security. This technological determinism prevails regardless of any substantive proof ID scanners actually reduce violence and anti-social behavior in the night-time economy. The endorsement of ID scanners by the QLALJSC (2010) in the face of an extensive list of unanswered privacy issues, exemplifies how problematic elements of e-governance are filtering into a more punitive regulatory landscape. Osmond (2010), Zedner (2010) and Matthews (2009) highlight how the political quest to 'solve problems' associated with crime and disorder through more problematic legal criteria is reshaping the very meaning of the criminal law. These issues are compounded where accountability for multiple public or private agencies is concerned. When new technology is married to legal standards aimed at ensuring organizational compliance, the incremental erosion of individual rights emerges in several ways. As Thompson and Genosko (2009) indicate, paternalistic motives justifying complex bureaucratic social sorting procedures are not necessarily confined to new or innovative forms of technology. Rather, the lack of capacity to formally challenge the 'back end' surveillant assemblages through countervailing criminal, administrative or human rights protections reinforces the impermeability of data about people as the 'truth' associated with their behavior. With privacy subordinated by safety, this legal and political impermeability overplays the value of technology in promoting human welfare, while raising a more problematic series of legal issues that are difficult to contest when data equates with truth.

Future research into the connections between safety, technology, überveillance and the construction of surveillant assemblages should focus away from simple cause and effect relationships between new technologies and crime prevention. Rather, the more complex issues associated with e-governance, citizenship and competing rights justify closer analysis. This requires developing research methods that unravel how policies relying on the value of new technology to protect the community can equally respect individual and collective rights to public and private organizational accountability, transparency in the uses of personal information and citizen capacity to challenge government interference through meaningful legal and conceptual measures.

CONCLUSION

There is no doubt that technology contributes to überveillance to streamline contemporary governance. However, the mentalities driving the acceptance of new technologies that are built on notions of cause and effect 'success' lead to the development of new and highly selective citizenship criteria. Governments have always done this without necessarily relying on the use of technology to enhance human surveillance. The foregoing discussion illustrates governments, private entities and citizens have a skewed and unclear perception of the value and meaning of privacy as a legal right worthy of protection when contrasted with the seemingly greater demands for human security. Future research and policy should investigate how competing rights to security and privacy might unwittingly favor the expansion of technology to promote greater citizen compliance, while simultaneously reducing the availability of legal and cultural mechanisms that challenge complex social sorting procedures driving contemporary e-citizenship.

ACKNOWLEDGEMENT

Criminology Research Council Grant 42/08-09 funded the research documented in this chapter. The authors sincerely thank Emma McFarlane for her diligence in conducting and transcribing the key stakeholder interviews in this project, and Bo Hedwards for her comments on earlier drafts of this chapter.

REFERENCES

Australian Privacy Foundation. (2010). *APF policy statement re- visual surveillance, including CCTV.* Retrieved May 9, 2011, from http://www.privacy.org.au/Papers/CCTV-1001.html

Clarke, R. (2010, Summer). What is überveillance? (And what should be done about it?). *IEEE Technology and Society Magazine*, 17–25. doi:10.1109/MTS.2010.937030

Ericson, R. (2007). *Crime in an insecure world*. Cambridge, UK: Polity Press.

Goldsmith, A. (2005). Police reform and the problem of trust. *Theoretical Criminology*, *9*(4), 443–470. doi:10.1177/1362480605057727

Goold, B., Loader, I., & Thumala, A. (2010). Consuming security? Tools for a sociology of security consumption. *Theoretical Criminology*, *14*(1), 3–30. doi:10.1177/1362480609354533

Greenleaf, G. (2007a). Australia's proposed ID card: Still quacking like a duck. *University of New South Wales Law Research Series*. Retrieved November 2, 2010, from http://papers.ssrn.com/sol3/papers.cfm?abstract_id=951358&download=yes

Greenleaf, G. (2007b). Access all areas: Function creep guaranteed in Australia's ID card bill (no. 1). *Computer Law & Security Report*, *23*(4), 332–341. doi:10.1016/j.clsr.2007.05.006

Greenleaf, G., Waters, N., & Bygrave, L. A. (2007). Implementing privacy principles: After 20 years its time to enforce the privacy act. *University of New South Wales Law Research Series*. Retrieved November 2, 2010, from http://papers.ssrn.com/sol3/papers.cfm?abstract_id=987763

Hadfield, P., Lister, S., & Traynor, P. (2009). This town's a different town today: Policing and regulating the night-time economy. *Criminology & Criminal Justice*, *9*(4), 465–485. doi:10.1177/1748895809343409 PMID:20401315

Hart, C. (2007). Micro-chipping away at privacy: Privacy implications created by the new Queensland driver license proposal. *Queensland University of Technology Law and Justice Journal*, *7*(2), 305-324.

Holloman, K. R., & Ponder, D. E. (2007). Clubs, bars, and the driver's license scanning system. In K. R. Larsen, & Z. A. Voronovich (Eds.), *Privacy in a transparent world* (pp. 40–47). Boulder, CO: Ethica Publishing.

Leman-Langlois, S. (2008). Privacy as currency: Crime, information and control in cyberspace. In S. Leman-Langlois (Ed.), *Technocrime: Technology, crime and social control* (pp. 112–138). Collumpton, UK: Willan Publishing.

Lips, A., Miriam, B., Taylor, J. A., & Organ, J. (2009). Identity management, administrative sorting and citizenship in new modes of government. *Information Communication and Society*, *12*(5), 715–734. doi:10.1080/13691180802549508

Manning, P. K. (2008). A view of surveillance. In S. Leman-Langlois (Ed.), *Technocrime: Technology, crime and social control* (pp. 209–242). Cullompton, UK: Willan Publishing.

Matthews, R. (2009). Beyond 'so what?' criminology: Rediscovering realism. *Theoretical Criminology*, *13*(3), 341–362. doi:10.1177/1362480609336497

Michael, M. G., & Michael, K. (2006). National security: The social implications of the politics of transparency. *Prometheus*, *24*(4), 359–363. doi:10.1080/08109020601029912

Mulligan, C. M. (2008). Perfect enforcement of law: When to limit and when to use technology. *Richmond Journal of Law and Technology*, *14*(4), 1–49.

New South Wales Law Reform Commission. (2009). *Invasion of privacy, 120*. Retrieved November 1, 2010, from http://www.lawlink.nsw.gov.au/lawlink/lrc/ll_lrc.nsf/pages/LRC_r120toc

O'Connor, D., & De Lint, W. (2009). Frontier government: The folding of the Canada-US border. *Studies in Social Justice*, *3*(1), 39–66.

Office of the Information Privacy Commissioner. (2006). *Explanatory statement: Approval of biometrics institute privacy code*. Retrieved June 3, 2010, from www.privacy.gov.au/materials/types/download/9197/6797

Ontario (Liquor Control Board) v. Keupfer. (1974). *Dominion Law Reports* (3d). *47*, p. 326.

Osmond, C. (2010). Anti-social behavior and its surveillant inter-assemblage in New South Wales, Australia. *Surveillance & Society*, *7*(3/4), 325–343.

Palmer, D., Warren, I., & Miller, P. (2010). ID scanners in the night-time economy. In K. Michael (Ed.), *Proceedings of the 2010 IEEE International Symposium on Technology and Society (ISTAS '10)* (pp. 234-241). Wollongong, Australia: University of Wollongong.

Queensland Legislative Assembly Law and Justice Safety Committee (QLALJSC). (2009). *Inquiry into alcohol-related violence, interim report, 72*. Retrieved June 3, 2010, from http://www.parliament.qld.gov.au/view/committees/documents/lcarc/reports/Report%2073.pdf

Queensland Legislative Assembly Law and Justice Safety Committee (QLALJSC). (2010). *Inquiry into alcohol-related violence, 74*. Retrieved June 3, 2010, from http://www.aic.gov.au/crime_types/violence/alcohol%20and%20drug%20related%20violence.aspx

R. v. Webb. (1943). *Canadian Criminal Cases* (C.C.C), *80*, pp. 151.

Senate Standing Committee on Finance and Public Administration. (2007). *Human services (enhanced service delivery) bill 2007*. Retrieved November 2, 2010, from http://www.aph.gov.au/senate/committee/fapa_ctte/completed_inquiries/2004-07/access_card/report/index.htm

Surveillance Devices Act (Victoria), 1999.

Taylor, J. A., Lips, M., & Organ, J. (2008). Identification practices in government: Citizen surveillance and the quest for public service improvement. *Identity and Information Society*, *1*(1), 135–154. doi:10.1007/s12394-009-0007-5

Thompson, S., & Genosko, G. (2009). *Punched drunk: Alcohol, surveillance and the LCBO, 1927-1975*. Blackpoint, Canada: Fernwood Publishing.

Victorian Law Reform Commission (VLRC). (2010). *Surveillance in public places final report, 18*. Retrieved November 1, 2010, from http://www.lawreform.vic.gov.au/wps/wcm/connect/justlib/law+reform/home/completed+projects/surveillance+in+public+places/lawreform+-+surveillance+in+public+places+-+final+report

von Hirsch, A., & Simister, A. P. (Eds.). (2006). *Incivilities: Regulating offensive behavior*. Oxford, UK: Hart Publishing.

Zedner, L. (2010). Security, the state, and the citizen: The changing architecture of crime control. *New Criminal Law Journal*, *13*(2), 379–403.

ADDITIONAL READING

Armitage, R., Smyth, G., & Pease, K. (1999). Burnley CCTV evaluation. *Crime Prevention Studies*, *10*, 225–249.

Ashworth, A., & Zedner, L. (2008). Defending the criminal law: Reflections on the changing character of crime, procedure, and sanctions. *Criminal Law and Philosophy*, *2*(1), 21–51. doi:10.1007/s11572-007-9033-2

Cradduck, L., & Mccullagh, A. (2007). Identifying the identity thief: Is it time for a (smart) Australia card? *International Journal of Law and Information Technology*, *16*(2), 125–158. doi:10.1093/ijlit/eam008

Cross, J. T. (2005). Age verification in the 21st century. Swiping away your privacy. *The John Marshall Journal of Computer & Information Law*, *232*, 363–410.

Fussey, P. (2008). Beyond liberty, beyond security: The politics of public surveillance. *British Politics*, *3*(1), 120–135. doi:10.1057/palgrave.bp.4200082

Hadfield, P. (2008). From threat to promise: Nightclub security, governance and consumer elites. *The British Journal of Criminology*, *48*(4), 429–447. doi:10.1093/bjc/azn015

Haggerty, K. D., & Ericson, R. (2000). The surveillant assemblage. *The British Journal of Sociology*, *51*(4), 605–622. doi:10.1080/00071310020015280 PMID:11140886

Krevor, B., Capitman, J. A., Oblak, L., Cannon, J. B., & Ruwe, M. (2003). Preventing illegal tobacco and alcohol sales to minors through electronic age-verification devices: A field effectiveness study. *Journal of Public Health Policy*, *24*(3/4), 251–268. doi:10.2307/3343372 PMID:15015859

Kulinski, T., & Monk, B. (2008). The use of ID reader-authenticators in secure access control and credentialling. In *Proceedings of the 2008 IEEE International Conference on Technologies for Homeland Security* (pp. 246-251), Quincy FL: The Printing House. Retrieved, June 3, 2010 from http://ieeexplore.ieee.org/xpls/abs_all.jsp?arnumber=4534458

Kypri, K., Jones, C., McElduff, P., & Barker, D. (2011). Effects of restricting pub closing times on night-time assaults in an Australian city. *Addiction*, *106*(2), 301–310. doi:10.1111/j.1360-0443.2010.03125.x PMID:20840191

Lea, J., & Stenson, K. (2007). Security, sovereignty and non-state governance from below. *Canadian Journal of Law and Society*, *22*(2), 9–27.

Leman-Langlois, S. (Ed.). (2008). *Technocrime: Technology, crime and social control*. Cullompton, UK: Willan Publishing.

Levi, R. (2009). Making counter-law: On having no apparent purpose in Chicago. *The British Journal of Criminology*, *49*(2), 131–149. doi:10.1093/bjc/azn080

Marx, G. (2007). Rocky bottoms: Techno-fallacies of an age of information. *International Political Sociology*, *1*(1), 83–110. doi:10.1111/j.1749-5687.2007.00006.x

Michael, M. G., Fusco, S.-J., & Michael, K. (2008). A research note on ethics in the emerging age of überveillance. *Computer Communications*, *31*(6), 1192–1199. doi:10.1016/j.comcom.2008.01.023

Michael, M. G., & Michael, K. (2010). Toward a state of überveillance. *IEEE Technology and Society Magazine*, (Summer): 9–16. doi:10.1109/MTS.2010.937024

Murakami Wood, D. (2009). The surveillance society: Questions of history, place and culture. *European Journal of Criminology*, *6*(2), 179–194. doi:10.1177/1477370808100545

Newton, A., & Hirschfield, A. (2009). Measuring violence in and around licensed premises: The need for a better evidence base. *Crime Prevention and Community Safety*, *11*(3), 171–188. doi:10.1057/cpcs.2009.12

Squires, P. (Ed.). (2008). *ASBO nation: The criminalization of nuisance*. Bristol, UK: The Policy Press.

Svedberg Helgesson, K. (2011). Public-private partners against crime: Governance, surveillance and the limits of corporate accountability. *Surveillance & Society*, *8*(4), 455–470.

Taylor, N. (2011). A conceptual legal framework for privacy, accountability and transparency in visual surveillance systems. *Surveillance & Society*, *8*(4), 471–484.

Tootle, H. (2006). The application of critical social theory to national security research. *Prometheus*, *24*(4), 405–411. doi:10.1080/08109020601029995

Waiton, S. (2010). The politics of surveillance: Big brother on prozac. *Surveillance & Society*, *8*(1), 61–84.

Welsh, B. C., & Farrington, D. (2009). Public area CCTV and crime prevention: An updated systematic review and meta-analysis. *Justice Quarterly*, *26*(4), 716–745. doi:10.1080/07418820802506206

Westin, A. F. (2003). Social and political dimensions of privacy. *The Journal of Social Issues*, *59*(2), 431–453. doi:10.1111/1540-4560.00072

Wilson, D., & Sutton, A. (2004). Watched over or over-watched? Open street CCTV in Australia. *Australian and New Zealand Journal of Criminology*, *37*(2), 211–230. doi:10.1375/acri.37.2.211

KEY TERMS AND DEFINITIONS

Dataveillance: The tendency for governments and private businesses to monitor populations and consumer behaviour, then use the resulting computerised data to develop formal policies. Frequently, decisions about public or private service provision are based on the data about a person, rather than the person themselves (see surveillant assemblage).

Due Process: The series of rules, procedures and rights designed to protect the individual against the power of the state, which constrains the activities of state agencies in dealing with citizens under the criminal law. Examples of rules that involve due process include the right to silence, the right to legal representation during police questioning and the obligation on police to have clear evidence of criminal behaviour before entering and searching private property.

e-Governance: The use of computerised methods of data collection and sorting to streamline government and public service delivery.

ID Scanner: An electronic device comprising various technologies, including a portable camera, image scanner, biometric fingerprint reader and computer, designed to create a replica of a person's identity documents to ensure authorised entry into public or private premises. In the night-time economy, these devices enable a person's identity to be collected and stored in a licensed venue and/or a computer network to enable security personnel to prevent undesirable or banned patrons from gaining entry. These technologies are considered to minimise the prospect of disorder and violence occurring within hotels and nightclubs due to their potential deterrence effects, and their ability to enhance the rapid identification and detection of offenders.

Night-Time Economy: The development of urban precincts to enable increased commercial trade, largely through entertainment, restaurants and licensed venues, that are specifically promoted to operate outside of daytime business hours.

Privacy: The series of legal protections governing the use of personal information for public and private service delivery.

Public Policing: Government agencies and agents with specific legal powers to help promote order and investigate crime for the public benefit.

Security: The range of human and technological measures designed to prevent losses to governmental agencies, private businesses and the community. Security is commonly linked to preventing crime, rather than prosecuting suspected offenders after a crime has been committed.

Surveillant Assemblage: This involves the activities of multiple agencies (state, non-state and hybrid) engaged in surveillance and the various processes, forms and purposes of conducting surveillance, involving the consolidation of data through information sharing networks or other means (such as the sale of access to database information or secret court orders demanding technology companies transfer megadata to state agencies) to create more intensive and comprehensive surveillance and data sorting capacities. This data can then be used to inform policing practices to target or distinguish between 'desirable' and 'undesirable' people in a range of settings, and is further intensified by the use of computerised surveillance technologies to enforce such divisions.

Technological Determinism: The belief that automated technologies can solve complex social problems, including crime.

Chapter 10
Global Tracking Systems in the Australian Interstate Trucking Industry

Jann Karp
C.C.C. Australia, Australia

ABSTRACT

Technology, trucking, and the surveillance of workers in the workplace: helpful or a hindrance? Technological advances are produced by the creative ideas individuals: these ideas then become selling items in their own right. Do tracking devices effectively regulate traffic breaches and criminality within the trucking industry? The data collection was conducted in the field while the authors rode as a passenger with truck drivers on long-haul trips. The complexities of tracking systems became more apparent as the authors listened to the men and placed their narratives in a broader context for a broader audience. The results of the work indicated that the Global Positioning System (GPS) has a role in the management of the industry as a logistics tool, but that there are limitations to the technology. The drivers use the devices and also feel the oppressive oversight when managers use the data as a disciplinary tool.

INTRODUCTION

Rounded out, it resembled a tiny earth, because its hinged wires traced the same pattern of intersecting circles that I had seen on the globe in my schoolroom–the thin black lines of latitude and longitude (Sobel, 1996, p.1).

GPS is technology that provides accurate satellite navigation and thus, accurate time sheets. This "accuracy" may be affected by weather, distance, buildings and satellite failure. The aim is to reduce operational and compliance costs and improve business profitability. The emphasis is on "real time" recording of distance, time and

DOI: 10.4018/978-1-4666-4582-0.ch010

costs. Driving behavior is recorded – stop, start, braking, speed – engine speed versus tachometer records versus road conditions. In other words road and traffic conditions are not recorded. The fieldwork was conducted in a male environment and the company owners and boards were, in the majority, male. Men are doing the driving and men are doing the watching – different types of work. One reason for men doing the work is explained:

That's why we use men exclusively for long distance telegraphy ... because they naturally press down hard. They have a strong touch. Women wouldn't naturally press down hard and are therefore not adaptable to long distance work (Wile, cited in Winston, 1998, p. 56).

The companies selling the tracking devices are men, so the quote above seemed to support a theme that men are controlling this industry and perhaps believe that "heavy handed" control is necessary. Drivers in Australia seem helpless in the face of increasing use of technology as the interface between employer and employee.

Managers and employees are faced with the reality of electronic monitoring of communications and collection and use of information about employees (Mello, cited in Tabak and Smith 2005, p.173).

There has been little or no research by public institutions into drivers' use of this technology, the design of which emphasizes "real time" surveillance. Business organisations have their own "othered" populations to control, namely their employees. Snider wrote:

Since the earliest days of capitalism businesses have been obsessed with finding ever more sophisticated (and intrusive) mechanisms to manage, discipline and ultimately eliminate human employees from the production process. Business has thus been a major player in the development of surveillance technologies, constantly commissioning studies to tell those at the top better ways to control their workforce (Snider, 2011, p. 5).

The trucking industry (also referred to as the transportation or logistics industry) involves the transport and distribution of commercial and industrial goods. Trucks fitted with GPS are commercial vehicles – trucks, semi-trailers, dump trucks – and are used in industries such as mining. The Australian trucking industry provides an essential service; drivers transport large quantities of raw materials and import and export products. The usual destinations include docks and distribution centres and, in the case of building materials, construction areas. Trucks are important to the construction industry where large amounts of rocks, dirt, concrete, and other building materials are used. There are extensive economic infrastructures, investments and costs from capital, wages and government taxes.

The use of satellite communication is now being introduced as emission standards are being taxed. The Federal government will be able to access which companies are contributing to pollution, based on the use of diesel calculated by distance travelled.

The technology we are discussing is run through the satellite tracking of a device that is located on the truck. This tracking device is placed on the prime mover, the trailer and if there is more than one trailer then each trailer will have a device. One driver could have three or more separate tracking devices on the truck he is driving. The media is now talking about technology providing driverless trucks. The trucks use GPS technology to navigate autonomously around a pre-defined course from loading units to dump locations, including waste dumps, stockpiles and crushers (Asia in Focus in Asia Pulse, Sydney, accessed 2 Nov 2011).

Snider (2011) again notes that the:

... elite claims are most likely to come fully clothed in the latest legitimising concepts, such as modernity, efficiency, or prosperity. This, too, enhances their chances of being heard in public arenas. Key elite groups, particularly those that own and/or control central instruments of economic and political power, play important roles in structuring which knowledges are produced ...They become not political ideas, contentious issues that are by definition open to everyone and debatable but "common-sense" assumptions, things "everyone knows" (p. 107).

Another example of the driverless truck follows. The media reports the claim that the technology is available and the reader is left to understand that no longer are toys "self-propelling objects" but trucks which now no longer need a driver:

Rio Tinto lifts order for Komatsu driverless trucks to 150 (ASX:RIO) Rio Tinto has significantly boosted its plans for driverless trucks at its Western Australian mines, increasing its order from 10 trucks to 150. Rio Tinto said on Wednesday it would receive at least 150 of the driverless trucks over four years from 2012 under a new agreement signed with Japanese mining and construction equipment manufacturer Komatsu (TSE:6301). The trucks will be used in Rio Tinto's Pilbara iron ore mines, and controlled from its operations centre in Perth, over 1,500 kilometres away (Asia in Focus in Asia Pulse, Sydney, accessed 2 Nov 2011).

As Snider (2001) suggests, the media claims and the technological reports legitimize the process of tracking and surveillance innovation. There is no discussion about introducing the technology and whether or not workers are entitled to have a choice about the systems. Instead the GPS allows the logistics manager, the people in the office, to watch in real time, the movement of their trucks;

a fait accompli. This record of the truck being driven can be seen on a screen and is recorded so that a printout can be retrieved which will allow owners, regulators, police, union officials and the driver to examine, investigate, observe and discuss what the driver has done through his working shift/shifts over a period of months. This would also encourage the purchase of more "driverless trucks".

I conducted research in the field between 2008 and 2011. Narratives of twenty-five professional career drivers were recorded from the passenger seat. The men had worked in the industry for an average of twenty-five years. Twenty-one drivers agreed to a recorded interview; the others were recorded as hand notes. The drivers were guaranteed anonymity. They also understood that they did not have to answer questions that they felt were too intrusive. I also clearly explained that, as a civilian, I had a mandatory reporting obligation concerning serious or indictable offences.

Sitting in the truck for long periods and listening to the drivers allowed me to observe, write notes and gather my thoughts about their workplace. But my main objective was to record their conversations about their workplace.

This is what I observed that a tracking system cannot record. Drivers maneuver trucks, they don't just drive forward. The truck may be stopped but the driver is working. The driver will assist with loading and unloading operations. They will reverse a truck, tip out a load of product such as wheat, and then sweep out the rear of a B-double. The GPS records the truck as stationary – resting? The driver may be on a weighbridge being told by the unloading logistics employee to reverse or go forward, tonnes of truck at a time as the truck lines up to have the produce it is delivering or loading weighed.

The estimations are completed by the logistics teams, the driver, the person controlling the arrival of other trucks. Many people are involved, all communicating via radio with the driver. The estimate of weight, before unloading and on loading is recorded. The time of arrival and unloading recorded. The driver is, however, responsible for complying with the weight restrictions and will receive a fine if overloaded. The driver talks to the inspectors, police, road traffic authority, logistic managers and owners as he tries to comply with multi-regulatory systems that cover how the load is covered, and safe distribution of weight. The driver is also responsible for preventing loss and damage to the goods carried, and the truck. The driver will be required to know the appropriate route to drive for the truck he is driving (e.g. roads on which a B-double can travel) and also the best approach to the delivery point. If the driver is carrying explosive or dangerous goods then another level of responsibility is required.

The reliability of positioning systems undergoes ongoing technical development. According to research carried out by Tampere University of Technology in Finland it is possible to improve the positioning accuracy and reliability of navigational systems by using digital road maps in addition to the main navigation system (Asia Pulse, accessed 17/11/2011). The media, companies and other advertising mediums like to present GPS and tracking devices as authorities on positioning identification (Business Wire, VANTAA, Finland, accessed 16 November 2011). Perhaps these systems are more vulnerable to errors than we realise. Interstate truck drivers are knowledgeable; they have to be to complete the tasks already mentioned. However, they are not going to be able to gain sufficient education around the errors involved in tracking devices, particularly when they are discussed in the following manner.

The aim of our study and development work was to utilize the valuable information contained in digital road maps in order to accurately determine the position of a vehicle. To achieve this we developed an algorithm for accurate sensor-based car navigation. Our approach requires a standard

speed sensor as available in most cars through the OBD interface, one MEMS high performance gyro, and digital street maps …

explains Mr Pavel Davidson, inertial navigation specialist, from the Department of Computer Systems of Tampere University of Technology Finland (www.tut.fi accessed Nov 2011).

Map-aided dead reckoning navigation system has smaller position errors… several different map-matching algorithms already exist, but they are not well adapted to situations in which Global Positioning Systems navigation is not available or is not reliable, as in many urban environments. The study shows that a map-aided dead reckoning navigation system has significantly smaller position errors than a traditional dead reckoning navigation system. With the aid of a street map [author's emphasis], cross-track errors can be eliminated any time when the road segment is correctly identified, while along-track errors can be reduced after the turn. Therefore, the position accuracy depends very much on the vehicle route.

"Our solution is based on a recursive implementation of Monte-Carlo-based statistical signal processing, also known as particle filtering", Mr Davidson describes (www.vtitechnologies.com):

The basic principle is to use random samples – also referred to as particles – to represent the posterior density of the car position in a dynamic state estimation framework where road map information is used. Since particle filters have no restrictions on the type of models and noise distribution, the velocity and heading measurement errors can be modelled accurately (Davidson, 2011).

The above quotes are examples of the confusing language that a driver may or may not understand. But what we do know is that the driver is being tracked for long distances and long periods of time. A driver is an interstate driver if he drives five hundred kilometres or more or is away from home overnight. For example, a driver can travel from Sydney to Melbourne (a distance of 860 kilometres, estimated driving time 11 hours) or Sydney to Darwin (nearly 4,000 kilometres) which could take 47 hours to drive. Sydney to Brisbane is a 12-hour drive, and all these trips involve "real time" surveillance. The time taken to complete a trip is dependent on road conditions, weight carried and/or truck maintenance. Of course the driver then turns around to go home.

Snider writes:

In the workplace employers have pretty much carte blanche with regard to surveillance of employees outside Victoria and New South Wales,[1] which have both enacted workplace privacy laws (Watts, 2009). Elsewhere employers can "record keystrokes, log emails sent and received, screen emails for offensive or inappropriate content, take snapshots of the desktop at set times and track programs run by users" (Watt 2009:7). It is too early to gauge the effectiveness of the laws enacted in Victoria and New South Wales since only four cases challenging surveillance of workers under the New South Wales Workplace Surveillance Act 2005[2] have been pursued, with no convictions registered even though all four parties involved had union membership (2011 p 8.)

As mentioned previously, men's business, men's worksites involves counting, statistical analysis and recording keeping of data that is meant to either prove or improve business efficiency. The Australian Government takes counting in the trucking industry to include truck movements, container movements and travel distance movements, and describes these movements in units. The units are used to calculate a level of truck work contributing to, and costing the national economy. Do these calculations become part

of the reasons behind the surveillance of truck workers? The companies that are measuring the drivers' behaviors are contributing to the Australian data collection. The value of costs involved in the infrastructure of the industry are calculated in units providing, for example, a mathematical model for road taxes, developments, fuel excise and tax rebates. Included in the following quote is a description of part of the counting process:

Count of TEUs divided by the number of trucks. TEUs per truck are a measure of truck efficiency; it encapsulates the 40ft/20ft dimension difference and is consistent with other wharf related TEU measures. For example, suppose on a given day: 10 trucks each make a trip to the port terminal empty but leave the terminal with 2 TEUs; and 10 trucks each make a trip to the port terminal with 2 TEUs but leave the terminal empty. Total TEUs moved = 40; total number of trucks = 20. So average TEUs per truck (for a two way in and out) trip is 2 (Bureau of Infrastructure, Transport and Regional Economics 2011, p.3).

I have included the above information, as there is one vital statistic that has not been determined and that is the number of truck drivers involved in interstate trucking. The men involved in the work are not counted even though they have health issues, they battle economic frustration, and they are often members of unions and associations. They are subjected to a high level of surveillance and regulatory practices but it appears no agencies have calculated their numbers. The latest information seems to be a record in 2007 of possible numbers excluding two states of Australia (Personal Communication, National Transport Commission 2012).

The rationale used for GPS is that drivers can be made to comply with fatigue regulations. The law will dictate how long a driver can drive and the tracking system will determine, in real time, if the driver is complying. This is presented as a feasible safety measure. However, the most simplistic rationale often fails to capture the reality of the workplace, for example, a traffic hold-up.

GPS allows the company to track the drivers, but it also provides information to smaller companies and owner/drivers concerning fuel usage, speed, gear optimisation, engine idle time, location, direction of travel, and amount of time spent driving. In New South Wales the Heavy Vehicle Driver Fatigue Schedule includes the following offences and penalties for drivers breaking fatigue laws Road Transport (Safety and Traffic Management) Act 1999 (NSW):

- Duties of employers, prime contractors and operators: $253
- Duties for schedulers: $253
- Duties of loading managers: $253
- Standard hours of solo drivers (minor): $197
- Standard hours of solo drivers (major): $422
- How driver must record information in work diary: $253
- Changing work and rest hours option: $253
- Information that driver must record in work diary: $253

The previous offences are recorded by police and road traffic authority regulators.

However, drivers do have positive opinions concerning tracking devices. The introduction of technology will record the times that products are dropped and picked up. They can prove and relay information concerning their compliance with the logistical demands of an employer. The companies and managers who employ the drivers can also be subjected to severe penalties as follow:

- Employer/prime contractor/operator not take reasonable steps to assure scheduler complies with duties: $2,868.

- Prime contractor/operator business practices do not comply with requirement: $2,868.
- Consignor/consignee not take reasonable steps to ensure other parties not cause offence: $2,868(NSW Traffic Act Effective 1 July 2009).

Such offences, though, are more difficult to detect/prove and the companies have more resources to plead "not guilty" and avoid a conviction. There are no dedicated enforcement officers to combat company crimes in this area. The driver can argue with the company and provide records if obtained to prove their role in non-compliance events.

Drivers do report negative aspects of GPS. A driver spoke to me recently saying that he had been on the road for nearly 14 hours but that he was approximately two hours away from home. He wanted to get home for the Friday afternoon preparations for a weekend at home. He had complied with all required legal breaks but although he had been on the road he had not travelled the distance that he had calculated that would have him at home with his family. He can't stop and rest because that delays his arrival home and he is not at all tired. He has a tracking system that will record his illegal attempt to go over time and make it home if he keeps driving. So he is supposed to stop and have a half hour rest, meaning rest and not worry about his family. Will he break the regulations and know his company can prove it and report his infringement? (pers. comm., October 2011).

Another example of relying on a GPS to prove the owner/driver was disadvantaged follows, and rather than relying on personal relationships and trusting people the owner/driver had proof of illegality. The owner/driver had put his truck in for a major service to repair a particular item in the engine. The truck had to be moved from the service centre to another manufacturing site for the work. The GPS records proved the work had not been carried out, as the truck had not moved from the first location. The owner/driver however, had been given an account for work supposedly not carried out.

Truck drivers get little physical, emotional or spiritual comfort from face-to-face contact with people. They do get contact with others via technology: phones, CB radios, listening to the radio, and computers when stationary. Technology has improved drivers' ability to contact each other through digital media services. They are also now accessing a wider range of communication services as digital communication is also being introduced. However, these men do work in a mobile and isolated workplace and their work surveillance technology places them in minute-by-minute contact with the sense of being watched. I was in the field and watched a black dot on a screen move as a manager tracked where one of his trucks was geographically located. I have seen printouts attached to pay advice slips and job sheets. These men know that they have contact with their bosses, but it is a different type of contact to that experienced by other workers.

CONCLUSION

When time is broke and no proportion kept!

So it is in the music of men's lives (Richard 11 Act V Scene 5)

Employee monitoring and surveillance, both covert and overt, has become the most common means of disciplining the interstate trucking workforce. Herbert and Tuminaro record that "[t]echnological dehumanisation, whether intentional or unintentional has already led to employee anger and protests" (Herbert & Tuminaro 2009, p. 390).

Anthropologists, sociologists and criminologists have a duty to record the impact of technology on the working culture and individual workers' rights. The development of legislation and social policy directions must include the interdependence of technology in the workplace. Employers may voluntarily develop public accountability around the introduction of surveillance technology and data keeping but Australian companies are not declaring this in the public arena (Herbert & Tuminaro, 2009). Drivers will continue to develop strategies to use the data recorded to their advantage, but the perception of this writer is that there is a conversation about the negative effects of this "real time data" collection which is not appearing in the public discourse. Perhaps individual concerns about privacy, control, and civil rights might only become part of the greater public discourse when more influential workforces are affected by the introduction of surveillance in their workplace.

REFERENCES

Bureau of Infrastructure, Transport and Regional Economics. (2001). *Australia*. Canberra, Australia: Australian Government Statistical Report.

Business Wire. (n.d.). *Vantaa Finland*. Retrieved from www.vtitechnologies

Davidson, P. (2011). *Website*. Retrieved from www.vtitechnologies

Herbert, W., & Tuminaro, A. K. (2009). Symposium: The impact of emerging technologies in the workplace: Who's watching the man (who's watching me)? *Hofstra Labor & Employment Law Journal, 25*, 355–393.

Pulse, A. Sydney. (2011). *VTI technologies: SCC1300 combined sensor improves positioning accuracy in map-matching applications.* Retrieved from www.vtitechnologies

Shakespeare. (n.d.). *Richard* (11 Act V Scene 5). Retrieved from www.opensourceshakespeare.org

Snider, L. (2001). Crimes against capital: Discovering theft of time social justice. *Law, Order, and Neoliberalism, 28*(3), 105–120.

Snider, L. (2011). *Criminalizing the algorithm? Stock market crime in the 21st century.* Paper presented at the ANZ Conference of Criminology. Geelong, Australia.

Sobel, D. (1995). *Longitude.* London: Fourth Estate.

Tabak, F., & Smith, W. (2005). Privacy and electronic monitoring in the workplace: A model of managerial cognition and relational trust development. *Employee Responsibilities and Rights Journal, 17*(3), 173–189. doi:10.1007/s10672-005-6940-z

University of Finland. (2011). Retrieved from www.tut.fi

Winston, B. (1998). *Media and technology: A history from the telegraph to the internet.* New York: Routledge.

KEY TERMS AND DEFINITIONS

Control: Is considered as an element of management. Social control carries different connotations however, and sometimes management controls may become a form of social control. Social controls are societal and political mechanisms or processes that regulate individual and group behaviour in an attempt to gain conformity and compliance.

Employee Monitoring: The ability to know what an employee is doing, where they are, and whether or not they are achieving their goals. For example, time stamps of various kinds record an employee's work hours. Increasingly, however, employees' location details are also being stored.

Inspectors: An administrative position that has the ability to scrutinise schedules, logbooks and other records related to employees, trip data and trucking deliveries.

Long-Haul Trucking: Typically trips that require a driver to be on the road farther than 200-300 kilometres from the driver's home terminal. Drivers operate a truck with a sleeper unit and in many instances are gone for days at a time. Some long-haul trips might require a driver to be away from home for up to three weeks.

Oppression: Is the exercise of authority or power in a burdensome, cruel, or unjust manner. Drivers who feel oppressed usually have feelings of being heavily burdened, mentally or physically.

Oversight: Usually, but not always, conducted by a stakeholder who exercises authority over another. Oversight by government or agencies happens when there are regulations that need to be adhered to.

Radio: Is the wireless transmission of signals through electromagnetic radiation of a frequency significantly below that of visible light, in the range from about 30 kHz to 300 GHz.

Safety: The condition of being protected against physical, social, financial, emotional, occupational, psychological or other types or consequences of damage, accidents, harm or any other event which could be considered non-desirable.

Trucking Companies: Accept cargo for road transport. Truck drivers operate either independently working directly for the client or through freight carriers or shipping agents. Some big companies (e.g. grocery store chains) operate their own internal trucking operations to get their products from the distribution centre to the retail outlet.

ENDNOTES

[1] Snider (2011) includes in her footnotes the following useful information. According to Watt (2009), "New South Wales first introduced legislation to regulate covert video surveillance of employees in 1998…The act sought to balance an employer's right to use video surveillance to investigate unlawful activities and an employee's right to privacy… In 2005, the Government enacted the Workplace Surveillance Act… the Act regulates both overt and covert surveillance by video, computer and tracking devices such as GPS… [there is a] requirement to provide prior notice of monitoring… In 2007 Victoria enacted legislation governing the use of surveillance devices in designated areas of the workplace".

[2] Originally created in 1998, the Workplace Video Surveillance Act was designed to monitor employees through video surveillance. In 2005 that was extended by the State Government to regulate camera surveillance as well as computer systems and tracking devices under the Workplace Surveillance Act (see Bartier Perry, 2005). Though employers are supposed to give 14 days' notice, there are loopholes and debates this year to determine whether or not notice to employees is even necessary anymore.

Chapter 11
Tracking Legislative Developments in Relation to "Do Not Track" Initiatives

Brigette Garbin
University of Queensland, Australia

Kelly Staunton
University of Queensland, Australia

Mark Burdon
University of Queensland, Australia

ABSTRACT

Online behavioural profiling has now become an industry that is worth billions of dollars throughout the globe. The actual practice of online tracking was once limited to individual Websites and individual cookies. However, the development of new technologies has enabled marketing corporations to track the Web browsing activities of individual users across the Internet. Consequently, it should be no surprise that legislative initiatives are afoot throughout the world including the United States (US), the European Union (EU), and Australia. These different jurisdictions have put forward different methods of regulating online behavioural profiling and Do Not Track initiatives. Accordingly, this chapter overviews legislative developments and puts forward a typology of different legislative initiatives regarding the regulation of online behavioral profiling and Do Not Track issues. Particular focus is given to the Australian situation and whether existing Australian privacy law is sufficient to protect the privacy interests of individuals against the widespread use of online behaviour profiling tools.

INTRODUCTION

"Do Not Track" initiatives have emerged as a popular legislative response to the difficult problem of privacy concerns in relation to online behavioural profiling. For example, there is a significant amount of legislation before the US Congress dealing with online behavioural profiling. Currently the FTC is entitled to take action in order to protect consumer rights when a business engages in unfair or deceptive practices, or more specifically, where they do not adhere to their

DOI: 10.4018/978-1-4666-4582-0.ch011

own privacy policies. The proposed legislation offers varying degrees of state involvement in the behavioural advertising industry, from simply the introduction of a mandatory mechanism to elect whether or not to be tracked, to the more complex and encompassing privacy rights and obligations enumerated by the Obama administration in the White Paper. Both the White Paper and the Commercial Privacy Bill of Rights Act propose a "safe harbor" program by which companies could keep or design their own privacy policies, which would be approved and subsequently enforced by the FTC as an alternative to their adherence to legislation. Initiatives have also been undertaken in the EU and Canada. At present, little action has taken place in Australia but given the worldwide interest in "Do Not Track" it would seem unlikely that inaction will suffice.

Consequently, this Chapter examines current legal initiatives to identify the complex issues that arise out of online behavioural profiling and subsequent Do Not Track proposals. The second section provides an overview of how online behavioural profiling operates, the privacy concerns that arise and highlights recent contemporary controversies. The next two sections detail Do Not Track legislative initiatives that have recently taken place in the United States (US) and outline developments in the EU, Canada and New Zealand. These are followed by an overview of recent Australian developments while the final section provides a typology of Do Not Track regulatory approaches and concludes with suggested recommendations for legislative improvements based on the analysis of jurisdictional approaches and recent Australian developments.

HOW ONLINE BEHAVIOURAL PROFILING OPERATES

Online behavioural profiling is "the practice of tracking an individual's online activities in order to deliver advertising tailored to the individual's

interests" (Federal Trade Commission [FTC], 2009). The actual practice of tracking was once limited to the installation of traditional cookies that record the websites a user visits (Wall Street Journal, n.d.). However, marketing and advertising companies are now employing a range of new tools such as flash cookies, third-party cookies and beacons in order to track the online behaviour of individuals (Electronic Privacy Information Centre [EPIC], n.d.). *Third party cookies* are the primary mechanism used for online tracking. These cookies are operated by a "third party", the advertising or marketing company, as opposed to the actual domain a web user is visiting, and place its cookies on the domain that a user is browsing. Generally speaking, third-party cookies will be placed by advertising network domains, allowing them to construct a "profile" of an online user based on their browsing activities that is subsequently used for the purpose of delivering targeted advertisements (Duhig, 2012). Online behavioural tracking has become a burgeoning industry precisely because of the potency of advertising that it provides for (Phillips, 2010). A user who chooses to remove cookies can still have their data accessed as a result of *flash cookies*, devices that re-install deleted cookies. *Beacons* are used by online tracking companies to track a user's every movement on a website, including what is typed and where the user is moving the mouse. The data that people are accessing or browsing on a webpage can be collected in real-time, and then be aggregated with other data about a particular user, including their location, income, hobbies and so on.

The aggregation can be primarily conducted in two ways depending on what information is being collected by the relevant cookie. First, by aggregating data around the Internet Protocol (IP) address of the device that is being used to access the web page. In this situation, it may or may not be possible to identify and aggregate information to an individual as data is being aggregated to a device (e.g. a computer or smart phone) rather than a person. However, it is potentially a relatively simple task to ascertain an identity from an IP address.

Second, aggregation is also completed by aggregating data around a specific individual identifier. For example, Target (US) is able to aggregate data about an individual because it assigns a unique code – a Guest ID number – to all customers who transact or visit the Target website, which forms the basis for data aggregation (Duhig, 2012). Google operates in a similar fashion in relation to any web user that has a Google account and is logged in to that account (Google, 2012).

Privacy Concerns

There are several privacy concerns linked to tracking and the collection of data for the purposes of online behavioral profiling. Individual users are often unaware that their internet usage history is being collected and tracked. The issue of consent is therefore important and is often raised by groups opposed to online tracking. They argue that the information collected online is not information that consumers voluntarily consent to being shared with tracking companies and online advertising businesses (EPIC, n.d.). Public sentiment would seem to support this argument, with a US Gallup poll revealing that 67% of internet users do not believe advertisers should be allowed to match ads to your specific interests based on websites you have visited (Morales, 2010). Similarly, a recent Australian survey undertaken by the Centre for Critical and Cultural Studies, the University of Queensland found that only 36% of 965 respondents were comfortable with tailored advertising as a concept. Of these, 39% were uncomfortable with the idea that their information would be shared across websites (Andrejevic & Arnott, 2011).

Furthermore, users are potentially at a distinct disadvantage as they may be unable to easily access internet browsing information collected about them, or correct any inaccuracies, leading to a concern that online tracking companies have

little transparency and are consequently unaccountable (EPIC, n.d.).

A further danger arises from the development of "digital dossiers" of aggregated data which are used by corporations and governments to make decisions that direct affect individual livelihoods. These dossiers are used as if the information collected *is* the person when in reality the aggregated data merely provides a potentially inaccurate snapshot of an individual's online life (Solove, 2001). Yet these dossiers can be used in real life for inclusion in marketing and advertising streams based on the perceived socio-economic status of certain communities of individuals. However, with inclusion also comes exclusion, which can lead to the development of segregated communities in which those individuals and families with less economically attractive digital dossiers are effectively excluded from access to certain marketing information (Turow, 2011).

The type of information collected is also another troubling aspect of online tracking. Because the industry is largely self-regulated, there appears to be almost no practical legal limits on what data can be collected and how it can be used (Phillips, 2010). Perhaps a more extreme example of this is the ability of advertisers to track people with health problems such as bipolar disorder, through the tracking company Healthline. This then allows advertisers to target these people with ads related to bipolar disorder or other sensitive medical ailments on the assumption that an individual is content for such knowledge to be disclosed or attributed (EPIC, n.d.).

Finally, there exists growing concern that through the process of data collection and the creation of a "profile" of an online user, their identity will be revealed. It is argued by marketers that online browsing data is anonymous because it identifies web browser related statistics rather than individuals (Phillips, 2010; Ohm, 2010). However a Wall Street Journal researcher has explained that the aggregation or collection of 33 "bits" of information about a particular user will be enough to expose their identity and de-anonymise the data collected (Narayanan, 2010). When it is taken into account that certain websites transmit roughly 26 "bits" of information about a user, it becomes clear that these privacy concerns are not without merit (Narayanan, 2010). This point is further encapsulated by mass data aggregation processes such as those figured around the use of third party cookies or beacons.

Current Mechanisms for Preventing Tracking

The Consumer Federation of America and Consumers Union argues that "there is a fundamental mismatch between the technologies of tracking and targeting and consumers' ability to exercise informed judgment and control over their personal data" (Comments to the FTC, 2008).A study by Carnegie Mellon University also criticises the current internet privacy tools designed to protect consumers from online behavioural profiling, labelling them hard for the average user to understand and configure (Cranor et al., 2011). The study tested several tools, including tools that block access to advertising websites, tools that set cookies indicating a user's preference to opt out of online behavioural profiling, and privacy tools that are in-built into web browsers (Cranor et al., 2011). Among the problems reported by study participants and researchers were:

- Communication issues in terms of the user being notified of the purpose of a tool and the way in which to configure it;
- Lack of feedback which would allow a user to be aware of whether or not the opt-out was working; and
- A tendency of some tools to cause websites to stop working or operate with limited functionality (Cranor et al., 2011).

As a result of these findings, the report concluded that the self-regulated status quo of online behavioural profiling is fundamentally flawed and insufficient for empowering users to protect their privacy online. Similarly, in 2010, the Federal Trade Commission (FTC) released a report which stated, among other things, that "industry efforts to address privacy through self-regulation have been too slow ... and have failed to provide adequate and meaningful protection" (FTC, 2010).

In response to these perceived failings, Do Not Track legal initiatives have been proposed to further regulate online behavioural profiling. One suggested model involves enabling users to configure their web browsers to send a Do Not Track header with HTTP requests, signalling that they do not want to be tracked (Cranor et al., 2011). Most recently, the US legislature has announced a framework for new privacy regulations, which would include a Consumer Privacy Bill of Rights whereby a set of enforceable practices would be negotiated with industry, consumer protection and privacy advocates. A "Do Not Track" agreement, recently signed by a group of advertising networks and leading internet companies such as Google and Yahoo, is a first step towards this new model, and will lead to the adoption of DNT features integrated into web browsers. The FTC will be responsible for oversight and enforcement of the agreement's terms (Gallagher, 2012). These developments are covered in greater depth in the next section.

Recent Controversies

During 2012, Google has been implicated in several Do Not Track controversies. On February 17, the Wall Street Journal revealed that Google had intentionally circumvented the privacy settings of Apple's web browser, Safari, for the purpose of setting its third-party advertising cookies (Angwin, 2012). Safari automatically prevents installation of cookies from ad networks and other third parties. However Google worked around this by exploiting a loophole in Safari that allows cookies to be placed when a user interacts with a website in some way, for example by filling out a form. Google added Javascript coding to some of its ads that caused Safari to think that invisible forms were being submitted to Google, thus paving the way for Google to install a temporary cookie on a phone or computer (EPIC, n.d.). Upon testing the 100 most-visited US websites for the cookies placed by Google's display ad network on a Safari web browser, the Journal discovered that 22 of the websites installed the Google tracking code (Angwin, 2012). A day later, it was also revealed that Google had bypassed the cookie settings of Internet Explorer users by piggybacking on a "nuance" with P3P specifications (Musil, 2012).

Google's actions highlight how easily privacy settings in browsers can be bypassed, and this perhaps lends weight to the argument that legal reform, as well as technological reform, is necessary to ensure that online privacy standards are both difficult to circumvent and legally enforceable (Brodkin, February 2012).

Prior to the Safari and Internet Explorer revelations, Google had already been facing intense scrutiny and questions over its privacy practices. In January 2012, Google announced that from March 1 2012, it would consolidate its data from across its services (which include Gmail, Google+ and YouTube to name a few) to create a single merged profile for each user (EPIC, n.d.). The policy change was marketed as a way for Google to provide a more complete, transparent and integrated service experience for users. However, it was met with considerable international opposition, particularly from the EU, whose privacy officials asked Google to "pause" its changes until it could ensure the privacy of EU citizens (Angwin, 2012). Several EU data protection agencies have reached the conclusion that the policy violates the European Data Protection Directive in several ways. It was contended that Google's new privacy policy:

- Was not in accordance with the EU law regarding data transparency;
- Utilised the data of individuals in order to hand it over to third parties; and

- Provided inadequate notification and consultation prior to the implementation of the policy (Brodkin, March 2012).

The Electronic Privacy Information Centre (EPIC) has also been vocal in its opposition, and has filed a lawsuit against the FTC on the grounds that Google's new policy violates a consent order the company signed with the Commission in March 2011 after the Google Buzz controversy (which allegedly bars Google from opting users into services). Google, however, believes that it will withstand the legal challenge because the FTC consent order relates to the company's sharing of information with third parties, which the new privacy policy will not affect (Johnston, 2012). In response to EU objections, Google is similarly confident that the policy "respects all European data protection laws and principles" (Brodkin, March 2012).

US LEGISLATIVE DEVELOPMENTS

A number of Do Not Track initiatives have commenced in parallel across the US in recent years. These initiatives include the self-regulatory regimes of the Digital Advertising Alliance (DAA); the Obama Administration's Consumer Privacy Bill of Rights; federal legislation introduced in both the House and the Senate; and California state legislation. While self-regulation has tended to dominate as the preferred approach to online privacy, new legislation would introduce stricter rules and harsher penalties for companies failing to comply with industry codes of conduct. These codes would be enforceable by the FTC, state attorney generals, and in some cases by citizens as a private right of action.

The Digital Advertising Alliance

The DAA consists of a number of different advertising and marketing companies and groups, including the America Association of Advertising Agencies, the Association of National Advertisers, the Council of Better Business Bureaus the Direct Marketing Association and the Interactive Advertising Bureau. The regulations prescribed by the DAA are largely based on self-regulation, but the DAA uses monitoring programs and public complaints to oversee breaches of the regulations, which may be reported to government agencies if not remedied (American Association of Advertising Agencies [AAAA], 2011).

The DAA has released two sets of self-regulatory principles governing online tracking. The first set of principles, released in 2009, was the *Self-Regulatory Principles for Online Behavioural Advertising*. This included measures such as educating consumers, transparency in privacy notices, website-based consumer control over whether third parties or internet service providers (ISPs) monitor their activity, data security and accountability safeguards and prohibitions against collecting sensitive data – that is, data related to children under 13 or information related to health or finances (Digital Advertising Alliance [DAA], 2009). While this document originally endorsed website-based consumer controls, facilitating site-by-site opt-outs of online tracking, in February 2012 the DAA announced that they were beginning work to bring into force a browser-based mechanism with the same purpose (DAA, 2012).

Supplementing the *Self-Regulatory Principles* was the 2011 release of the *Self-Regulatory Principles for Multi-Site Data*. This document applies principally to ISPs and third-party data monitors. It adopts self-regulatory principles which attempts to prohibit the use of multi-site data without permission from the consumer, except for the purpose of operating the business, for market research or when the data will be de-identified within a "reasonable" time (DAA, 2011). It absolutely prohibits the use of multi-site data to determine eligibility for employment, credit, insurance or health care treatment. Furthermore, it bans third parties or ISPs from collecting sensitive information, such

as data related to the activity of children online, social security or financial information, or medical records (DAA, 2011).

The Consumer Privacy Bill of Rights

The Consumer Privacy Bill of Rights was proposed by the Obama Administration in February 2012, as part of the larger policy paper *Consumer Data Privacy in a Networked World: A Framework for Protecting Privacy and Promoting Innovation in a Networked World* (The White Paper) (White House, 2012). The paper expounds a Consumer Privacy Bill of Rights, calls for consultation with commercial stakeholders to develop an enforceable code of conduct and establishes a Safe Harbor program for businesses. Notably the report does not call for a browser-based Do Not Track mechanism, although it endorses the efforts of private groups to develop such a mechanism (White House, 2012).

The Consumer Privacy Bill of Rights forms the main part of the report, and includes as its centre seven consumer rights:

1. **Individual Control:** Companies (including search engines and third-party data brokers) should request permission from users to collect their information at the time that data collection begins. Consumers have the right to refuse tracking and recording of their data.

2. **Transparency:** Privacy policies should be prominently displayed on a website and easy to read and understand. Emphasis should be given to terms which allow a website to collect information in excess of what is necessary for the given transaction.

3. **Respect for Context:** Companies must disclose the purpose of data collection at the time of collection, and only use the data for the disclosed purpose. If companies wish to reuse the data for a different purpose, they may only do so if they seek permission from the consumer first ("individual choice") and

are clear about what they will be reusing the data for ("appropriate transparency").

4. **Security:** Companies must assess the level of security that is appropriate to protect the kind of data they collect. This is a matter of discretion for the company, but some protection will always be necessary.

5. **Access and Accuracy:** Commercial companies and websites should allow consumers to correct their own personal information online. The process of correction should not raise additional security concerns.

6. **Focused Collection:** Companies should only collect the kind and amount of personal information that they need to deliver their services or accomplish their purposes. Once personal data is no longer required, it should be disposed of or de-identified securely.

7. **Accountability:** Companies that collect personal data should ensure that their employees and subsidiaries handle this data securely and in accordance with the Privacy Bill of Rights. These companies also have an obligation to ensure that any third party given access to the data also uses it securely and appropriately.

The report also calls for multi-stakeholder processes to develop their own codes of conduct tailored to specific industries. These codes of conduct would be enforceable and reviewable by the FTC. If approved by the FTC, companies with their own codes of conduct would be given "safe harbor" from the provisions of any future statutory Consumer Bill of Rights, and would only be held liable to their own codes of conduct (White House, 2012).

The Commercial Privacy Bill of Rights Act of 2011 Bill

Senators Kerry and McCain introduced the Commercial Privacy Bill of Rights bill to congress in 2011. Like the Consumer Privacy Bill of Rights,

this bill aims to protect consumer interests by ensuring that collected data and personally identifiable information is protected and disposed of appropriately; that consumers are given clear information and a choice as to whether they are tracked online; and that only information necessary for carrying out business is collected online (ss 201-203).

Like the Consumer Privacy Bill of Rights, this bill does not require a browser-based Do Not Track mechanism. It instead requires each individual website to alert consumers to their privacy policy before asking permission to track their activity. However, the general opt-out provision is supplemented by a provision whereby consumers must actively opt in to the collection of "sensitive personally identifiable information" (s 202 (a) (3) (A)). This includes information such as medical records, religious information, or data likely to cause economic or physical harm if released.

The Commercial Privacy Bill of Rights Act does not create a private cause of action enforceable by citizens. The regulations prescribed by the bill must be enforced only by the FTC or the Attorney General of a state.

The Do-Not-Track Online Act of 2011 Bill

In 2011, Senator Rockefeller introduced the Do-Not-Track Online Act of 2011 bill. The main purpose of this bill was to implement a "mechanism by which an individual can simply and easily indicate whether the individual prefers to have personal information collected by providers of online services" (s 2(a) (1)).

The FTC would be given power to implement the mechanism and enforce observance of the choices made by consumers. If a consumer were to use the mechanism to opt out of having their personal information collected, only necessary, anonymous or de-identified data could be collected (s 2(b) (1)). Both the FTC and state attorney generals would be given the power to bring a civil action against companies that did not adhere to the obligations. However, this bill would not give rise to a private right of action.

Unlike many of the other bills concerning online tracking, this bill does not provide any guidelines regarding the safe storage, collection or use of data that the consumer has given companies permission to record.

The Do Not Track Me Online Act Bill, 2011

In February 2011, Representative Speier introduced the Do Not Track Me Online Act bill to the House. This bill calls for the FTC to "promulgate regulations ... that establish standards for the required use of an online opt-out mechanism to allow a consumer to effectively and easily prohibit the collection or use of any covered information and to require a covered entity to respect the choice of such consumer to opt-out of such collection or use" (s 3(a)).

The Bill incorporates many of the principles of the Privacy Bill of Rights, requiring companies to notify consumers when their data is being collected, and to respect the decision by consumers to forego tracking and targeted advertising. It also requires that privacy policies and data collection policies are transparent and easily accessible; that consumers have access to the personal information collected about them (although there is no mechanism legislated to allow them to correct it); and that only the kind of data a consumer would reasonably expect to be collected in the course of their relationship with a website should be gathered (s 3).

The bill enables the FTC to prescribe regulations governing the specific uses of personal data, and also empowers it to audit companies and enforce the provisions of the bill. It creates a civil cause of action for which the Attorney General or agents of a state can prosecute (ss 4, 5).

California Legislation – Senate Bill No. 761

On February 18, 2011 Senator Lowenthal introduced Senate Bill 761, to add a section to the California Business and Professions Code. This addition is closely based on Speier's Do Not Track Me Online Bill, containing many of the same principles related to data protection, use and collection (California Senate, 2011, s 22947.45(b) (2)).

However, this bill does not call for any sort of broad Do Not Track mechanism, recommending instead that individual websites provide a method for consumers to opt out of data collection and use. This bill also gives rise to a private civil cause of action, allowing citizens to press charges against companies that breach prescribed regulations for damages up to $1,000 (California Senate, 2011, s 22947.45(d)). Finally, it absolutely prohibits the sale, sharing or transfer of personal data, unless that is the nature of a commercial transaction undertaken (California Senate, 2011, s 22947.45(c)).

OTHER JURISDICTIONAL DEVELOPMENTS

The following section overviews Do Not Track developments in the European Union (EU), New Zealand and Canada.

European Union Law

EU law has generally favoured an opt-in approach to online tracking and behavioural profiling. This represents a markedly different approach than that taken in proposed US legislation, and has led EU authorities to reject the self-regulatory regimes of the online behavioural profiling industry. The foundations of European online privacy law are currently found in the Data Protection Directive of 1995 and the E-Privacy Directive of 2002. However, in early 2012 the European Union released a new proposal for a privacy framework known as the General Data Protection Regulation, encompassing the principles of previous directives, as well as some new rights and obligations.

Data Protection Directive 1995

The Data Protection Directive was introduced in 1995 and deals broadly with the processing of personal data and consumer privacy rights. It requires that any processing (art 2(b)) of personal data be specifically consented to, unless that processing is necessary to perform a contract between the person and the company collecting data, the processing is in the public interest, or it is a legal requirement (art 7). It calls for personal data to be collected only for "specified, explicit and legitimate purposes" (art 6(1)(b)) and absolutely prohibits the processing or collection of data regarding race, politics, religion, trade unions, health or sexuality (art 8(1).

The Directive further outlines several rights for consumers in relation to information collected about them. These include the right to:

- Be told who has access to their data and for what purpose;
- Access and edit incorrect information;
- The erasure of data that has already been processed; and
- Object to the collection or sharing of information about them (art 10(a) (b), 14).

It also requires data processing companies to notify government authorities and data subjects of when and how their information will be processed, and to carry out this processing securely and confidentially (art 16-17). Finally, it prohibits data processors from sharing personal data with third-party countries not subject to the directive, unless that country is found to have "adequate" protections in place, or the consumer has given informed consent to have their information shared (art 29).

Article 29 Working Party

Article 29 of the Data Protection Directive creates a Working Party, whose purpose is to deliver rulings on the adequacy of privacy protections developed by private advertising groups. In December of 2011, the Working Party declared that under EU law there is a presumption that people do not wish to have their data collected or processed (Article 29 Working Party, 2011, p. 6). It therefore requires that users actively opt in to any collection or processing of their information, including the placement or use of cookies on a consumer's computer.

E-Privacy Directive 2002

The E-Privacy Directive was introduced in 2002 to supplement the provisions of the Data Protection Directive in relation to providers of publicly available electronic communications services.

The E-Privacy Directive therefore differs from the Data Protection Directive in that it does not apply universally but only applies to the telecommunications sector. The E-Privacy directive nonetheless gives more attention to technologies developed or propagated since 1995, such as cookies and spam. Cookies are considered in article 5(3) of the Directive. As it was originally passed, this section simply required companies placing cookies to inform the user of the purpose of any data processing and give them the "right to refuse". This section was amended in 2009 to only allow the placement of cookies "on condition that the subscriber or user concerned has given his or her consent" (art 2(5)(3)). The standard was therefore raised in 2009 to require explicit and specific opt-in consent to the placement of cookies and the collection of consumer information.

Article 6 of the Directive requires that data which has been processed by a publicly available electronic communications service provider and is therefore no longer needed must be de-identified and erased, unless it needs to be retained for billing. It may also be retained for direct marketing purposes or to provide value-added services; however, this use must be consented to by the user and this consent must be capable of being withdrawn at any time (art 6(3)).

General Data Protection Regulation 2012

The General Data Protection Legislation proposed in January 2012 is a clarification of the general rules provided under the Data Protection and E-Privacy Directives. However, the proposal is for a set of Regulations, rather than a Directive, making its provisions directly binding on all the countries of the European Union, without the need to transpose them into national law. The rights conveyed on consumers and the obligations of data processors are largely unchanged, with a few exceptions. Under the Regulations, consent must still be explicitly given for any data processing and consumers have a right to know who has access to their data and what kind of processing it will undergo (art 6(1)(a), 15). Similarly, the Article 29 Working Party of the Data Protection Directive is replaced under Article 64 of the Regulations by a newly founded, European Data Protection Board.

However, the General Data Protection Regulation also provides for a number of new situations and definitions. These include:

- A "right to be forgotten" when there are no legitimate reasons for a company to retain personal data and a person no longer wants their data to be processed (art 17);
- An obligation on data controllers to provide "transparent and easily accessible" information to consumers about their data collection policies (art 11); and
- An obligation for data processors to maintain documentation of all the processing operations they are responsible for (art 28).

This regulation in the currently under discussion by the EU and is expected to take effect by 2015.

EASA Best Practice Recommendations

In April 2011 the European Advertising Standards Alliance (EASA) released its *Best Practice Recommendations* (BPR), a set of non-binding, self-regulatory principles intended to guide companies engaged in online behavioural profiling. This guide contains many of the same principles as the EU directives and regulations, including notice, informed consent, and special regulations for sensitive information.

It recommends specifically the implementation of a mechanism that allows a user to give informed consent to third-party tracking by linking them to a User Choice Site. This site would enumerate the privacy policy and data collection practices of third party advertising companies (EASA, 2011, pp. 12-13). However, this method of informing the consumer and seeking consent has been rejected by the Article 29 Working Party as inconsistent with European law.

In a 2011 decision, the Article 29 Working Party held that the opt-out scheme proposed by EASA did not satisfy current EU legislation (Article 29 Working Party, 2011, p. 6). They declared that European Directives require opt-in consent, prohibiting any website or company from collecting data before informed consent is given. The method proposed by EASA would most likely result in the processing of some information before the user was able to opt out of the collection, and as such did not provide sufficient protection of consumer's online privacy rights.

New Zealand

The New Zealand privacy framework is regulated by the Privacy Act 1993. It operates in a similar fashion to the Australian legislation, containing privacy "principles" rather than prescriptive rules. From 2008-2011, the New Zealand Law Commission (NZLC) conducted a four stage review of New Zealand Privacy Law. Stage Four, released on 2 August 2011, represented the culmination of the process and reviewed the Privacy Act 1993 with a view to updating and amending it. The key changes recommended by the NZLC included expanding the powers of the Privacy Commissioner, introducing mandatory data breach notification laws and clarifying the privacy requirements for cross-border outsourcing (NZLC, 1993). It was also put forward that the Privacy Commissioner ought to have the power to issue compliance notices to organisations (rather than merely responding to complaints) and conduct privacy audits of organisations when necessary (NZLC, 1993, p. 1)

Finally, it was recommended that organisations that outsource personal information to another agency or organisation for processing or storage remain fully accountable for the storage, use and disclosure of that personal information. Furthermore, where a New Zealand agency or organisation discloses personal information to an overseas entity, the disclosing agency or organisation will be required to take "reasonable steps" to ensure that the information disclosed will be "subject to acceptable privacy standards" (NZLC, 1993, p. 3).

Interestingly, the Commission did not recommend any changes to the Privacy Act to accommodate direct marketing and online behavioural profiling. The Commission's report and recommendations are currently awaiting government response.

Canada

Canada has two federal privacy laws - the Privacy Act, which took effect in 1983, and the Personal Information Protection and Electronic Documents Act ("PIPEDA") of 2000. The Privacy Act applies to the personal information handling practices of federal government departments and agencies. PIPEDA sets out the ground rules for how private

sector organisations may collect, use or disclose personal information in the course of commercial activities. Under the law, individuals are granted rights to access and request correction of the personal information collected by companies about them.

PIPEDA provides that the knowledge and consent of the individual are required for the collection, use or disclosure of personal information, except where inappropriate. It also stipulates that personal information should only be kept "as long as it is needed" (Office of the Privacy Commissioner of Canada, 2010). With regards to what is considered "personal information" for the purposes of PIPEDA, a contextual approach is generally adopted and it is worth noting a 2003 finding in which it was concluded that the information stored by temporary and permanent cookies was deemed to be personal information. Where an IP address can be associated to an identifiable individual, this is also considered personal information (Office of the Privacy Commissioner of Canada, 2012, p. 23). In a 2011 report, the Office of the Privacy Commissioner considered whether PIPEDA needed to be updated to respond to challenges faced by online tracking, profiling and targeting. (Office of the Privacy Commissioner of Canada, 2011). However other than suggesting changes to what is considered valid consent under the Act, the Report made only general observations and proposed to consider amendments at the upcoming second mandatory 5-year review of PIPEDA.

AUSTRALIAN DEVELOPMENTS

Australia's privacy framework is primarily governed at the federal level by the Privacy Act 1988 (Cth) (hereafter *Privacy Act*). The Act contains a set of Information Privacy Principles (IPPs) for public sector agencies and a set of National Privacy Principles (NPPs) for private sector organisations. The collection, use, disclosure, storage and destruction of personal information are dealt with under these privacy principles. The Office of the Privacy Commissioner (OPC) (now the Office of the Australian Information Commissioner [OAIC]) was also established by the *Privacy Act*.

In 2008, the Australian Law Reform Commission (ALRC) released a significant review of privacy law and practice (ALRC, 2008), to which the Australian government announced a two-stage response. The first stage was released in October 2009, in the form of an exposure draft of amendments to the *Privacy Act* which was considered by the Senate Finance and Public Administration Legislation Committee (Senate Finance and Public Administration Legislation Committee, 2010). The key purpose of the exposure draft was to replace the NPPs and IPPs with uniform principles, termed Australian Privacy Principles (APPs), applicable to both the public and private sector. The Senate Environment and Communications Reference Committee also released a report recommending Privacy Act amendments in 2011, entitled *The Adequacy of Protections for the Privacy of Australians Online*. As of April 2012, the changes proposed in both the government's response and the Committee's report has not been implemented into the *Privacy Act*.

In response to the privacy concerns posed by online behavioural profiling, the Committee recommended that the OAIC in consultation with web browser developers, internet service providers and the advertising industry should develop a code which includes a "Do Not Track" model following consultation with stakeholders (Senate Finance and Public Administration Legislation Committee, 2010). In this respect, it expressed a preference to a model similar to that which the Federal Trade Commission proposed to the US advertising industry. However, no action has thus far been taken by the OAIC on this point.

As regards the current application of the *Privacy Act* to online behavioral profiling, the Act may apply but even if it does, the coverage of application may not be universal to all websites.

The first issue to resolve is whether the information collected for the purpose of online behavioral profiling is personal information. Under s 6(1) of the Act, personal information is information in which an individual's identity is apparent or reasonably ascertainable. As highlighted previously, browsing history information may not automatically make an individual's identity apparent, especially if the user is not signed in to an online account. In these situations, any browsing aggregation is likely to be conducted around an IP address. Accordingly, whilst it is possible to identify a specific device used for browsing, it may not necessarily mean that the identity of an individual is possible or reasonably ascertainable, as required by the definition of personal information under the *Privacy Act 1988* (s 6(1). This has certainly been the argument put forward by organisations which employ direct marketing or behavioral advertising techniques as they have argued that information collected for behavioral targeting cannot be classified as "personal" for the purposes of the *Privacy Act*. It has also been argued that the use of web proxies and wireless piggybacking prevent IP addresses from being identified with a user or device with total certainty.

However, as has previously been discussed in this research paper, the aggregation of data over time may enable identification of particular individuals and thus render the information personal information. Whilst it is the case that the Privacy Commissioner has not specifically determined whether an IP address is personal information or not, it should be noted that both US (Klimas v Comcast Cable Communications Inc., 2006) and Canadian (Canadian Federal Privacy Commissioner, 2009) authorities have deemed it so, as has the Queensland Privacy Commissioner in guidance notes (Office of the Information Commissioner, 2012). An IP address on its own is unlikely to be considered personal information. However, if the IP address is used as a means to aggregate data, then it is more likely to be considered personal information as the collation of data around a specific data point will make it more likely that an individual's identity is reasonably ascertainable. Furthermore, the ability to conduct organisational aggregation is context specific, that is to say, the ability to aggregate will be judged on a case by case basis that examines the aggregation ability of the organisation in question. This point again re-emphasises the likelihood that the Act would apply to major online marketing or advertising corporations as these organisations will have significant capabilities to undertake sophisticated data aggregation.

If the data collected is deemed to be personal information, then NPP2, which partially relates to direct marketing, is likely to guide where the use of personal information for targeted advertising will be permitted, under certain conditions. Under NPP2 personal information can be used for direct marketing where:

- It is impracticable to obtain consent from individuals;
- The individual must not have made a request not to receive direct marketing; and
- The individual must be informed in each communication of their ability to request the ceasing of the marketing (Privacy Act 1988, Schedule 3).

The use of "Do Not Track" mechanisms may be of relevance at his juncture and it raises several questions in relation to NPP2. For example, if an individual has their browser setting to not allow tracking, does that mean they have made a request not to receive direct marketing? Furthermore, if tracking is taking place, does the tracking organisation also have to inform the individual of their ability to request the ceasing of tracking and targeted advertising? These issues have yet to be addressed and it is therefore currently unclear the extent to which the Act applies to online behavioral profiling.

NPP 2 broadly has the effect that any information used or disclosed by an organisation must be within the parameters for which it was collected. It could be argued that the use of web browsing history is collected for the directly related purpose of profiling and it could reasonably be expected that the individual would expect the collecting organisation would use browsing information for that purpose. However, as studies have demonstrated, the understandings and expectations of individuals in relation to the use of their browsing information for online behavioral profiling are by no means clear (Andrejevic & Arnott, 2011). It is therefore equally unclear to what extent NPP2 actually applies to online behavioral profiling and how it applies.

Furthermore, under NPP4.2, an organisation must take reasonable steps to destroy or permanently de-identify personal information if it is no longer needed for any purpose for which the information may be used or disclosed. The application of NPP4.2 is limited in the context of online behavioural advertising as the creation of profiles requires the continued collection and iterative review of previously collected browsing information. Accordingly, online marketers may always find a business use for collected online browsing data which thus negates some of the individual protections afforded by NPP4.2.

One final point should also be noted regarding exemptions to the Act. Since 2000, the *Privacy Act* has made small businesses (defined as those businesses having an annual turnover of $3 million or less) exempt from compliance with its requirements (Privacy Amendment (Private Sector) Act, 2000). It is estimated this exemption covers at least 94% of actively trading Australian businesses (Senate Environment Committee, 2011). Given that a growing number of these businesses are conducting online transactions with customers, holding and using significant quantities of personal information in the process, small businesses operating in the online context pose substantially greater risks to personal privacy in comparison to the old offline model. It is possible therefore that the Act may only apply to large-scale commercial online marketing companies which have an annual turnover of over three million dollars and may not apply to a large number of individual websites that nonetheless collect and track user browsing information.

ANALYSIS AND RECOMMENDATIONS

In the last substantive section of this chapter, a typology of different Do Not Track approaches is put forward to differentiate regulatory frameworks applied in different jurisdictions. The section concludes with an overview of recommendations for improvement.

Typology of Regulatory Approaches

First, it is necessary to cover some background regarding the development of intra-jurisdictional information privacy law frameworks. The implementations of information privacy laws have taken essentially different tracks despite their similar origins. That in itself is not surprising as a right to privacy is not perceived as an absolute right and thus the interpretation of the emphasis given to an individual's right to control his or her personal information is in competition with other social rights and interests. The application of information privacy legal regimes is likely to be a matter of contestable discussion amongst different legislative jurisdictions. As such, information privacy laws are manifestations of political processes which have implications for the implementable scope of such laws (Bennett & Raab, 2006). Jurisdictional information privacy laws therefore reflect the wider social, legal and policy values of individual jurisdictions (Swire & Litan, 1998).

The US attitude towards information privacy law is based on a sectoral regime and as such, is focussed towards certain types of industries or various types of sensitive information (Reidenberg, 1999). In conjunction with this are a handful of laws that have been implemented that have arisen from specific circumstances, ranging from the use of driver licence information for stalking purposes to the protection of videos borrowed from video stores (Drivers Privacy Protection Act, 1994). Furthermore, these sectoral divisions are emphasised by the fact that some federal privacy laws have been replicated at state level while others have not (Reidenberg, 1992). Not surprisingly therefore, the US approach to information privacy has been much criticised for its inconsistency of approach and application, particularly in relation to the manner in which information privacy is dealt with in other regimes as part of a comprehensive legal framework (Gellman, 1997, 195).

Comprehensive frameworks, such as those found in the EU, Canada, New Zealand and Australia, adopt an entirely different approach to the regulation of information privacy – essentially by establishing broad information privacy rights for individuals and stipulating requisite obligations for all organizations regarding the collection, storage and use of personal information. The type of information covered also has wide application and is purposefully construed in a broad sense - see for example, the definition of "personal data" found in the Data Protection Directive or the definition of "personal information" detailed in the *Privacy Act*. In conjunction with these definitions, supervisory authorities are given a wide discretion to regulate and monitor the actions of organisations and potential infringements against individuals.

It should therefore be no surprise that different jurisdictions have put forward different methods of regulating online behavioral profiling and Do Not Track initiatives.

Based on the summary of approaches to online consumer privacy summarised previously, three systems of regulatory application can be identified. The first, adopted by New Zealand, Australia and currently the US, is a predominant approach of self-regulation, in which coalitions of advertising companies or companies themselves are responsible for developing and adhering to their own privacy policies and codes of conduct. The second approach is co-regulation which refers to industry self-regulation initiatives that are overseen or ratified by government agencies. Canada is a current example and the US appears to be moving towards this approach as recent legislative proposals assume a much greater oversight role by the FTC. The third approach, adopted by the European Union, is a prescriptive system of mandatory regulation which is enforced by an independent body, typically a data protection commissioner.

These three systems are not absolute, and no jurisdiction entirely uses one approach to the total exclusion of the other. They can rather be considered a spectrum with self-regulation dominating at one end, and state regulation dominating at the other, as represented by Figure 1.

Australia has a largely self-regulated industry of targeted advertising. The best example of self-regulation in the Australian context is the *Australian Best Practice Recommendation for Online Behavioural Advertising*, developed by the Australian Association of National Advertisers (AANA) in March 2011. These recommendations have been designed by stakeholders in the online advertising industry, and are therefore tailored to their desire to engender consumer trust and to provide them the flexibility to carry on their business relatively freely rather than providing meaningful legal protections or redress.

The most frequently raised argument in favour of self-regulation in Australia is that it is the system that can best adapt to and keep up with technological advances (Senate Environment Committee, 2011, p. 13). The AANA, in its submission to the Senate Environment and Communications Reference Committee, cited it's Code of Ethics (applicable to all Australian advertisers) to argue that self-regulation "provides a flexible mechanism to meet the challenges of ever evolving advertising and marketing practices, media environment as well as consumer expectations" (Senate Environment Committee, 2011, p. 13). The Communications Council also submitted that its online privacy guidelines and its proposed standards on online behavioral profiling are examples of the effectiveness of self-regulation as a tool for enhancing online privacy.

Another perceived advantage of a self-regulatory approach is that it allows those parties with significant interests at stake to have their initiatives accepted or potentially incorporated into legislative amendments dealing with online privacy. Given that it has long been recognised that the challenges posed by online behavioral profiling will most likely require more than just legislative changes, it could be argued that a self-regulatory approach operating in tandem with a strong legislative framework will ultimately be more effective than pure state regulation in providing meaningful privacy protection for online consumers. Such an argument would have the approach adopted by Canada over the EU.

However, a predominantly self-regulatory approach for privacy protection tends to go hand-in-hand with a relatively weak legislative framework,

Figure 1. Regulatory approaches to Do Not Track

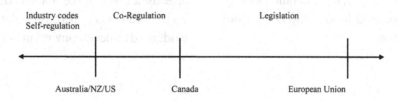

as has been frequently argued about the US legal framework. This statement could also be applied to Australia, where the legislative framework historically lacks meaningful teeth and is practically inapplicable in many instances (Greenleaf, 2011). Furthermore, the Privacy Commissioner's powers to enforce privacy rights are limited, and there is little capacity to undertake forward looking initiatives, such as the development of industry-wide privacy codes as developed in Canada.

An approach that is too far towards the self-regulated end of the spectrum arguably leaves consumers more vulnerable and dependent on the individual policies and practices of particular organisations (Hoofnagle, 2006, 379). It is frequently claimed by companies such as Google that the risk of organisations acting self-interestedly at the expense of privacy is mitigated by their need to gain the trust of consumers (Senate Environment Committee, 2011, p. 14). However, the high number of instances of improper use of personal data, and the apparent slow speed with which the advertising industry has developed privacy initiatives, must cast at least some doubt on the merits of such claims.

With that in mind, the substantive part of this note will conclude by looking at suggested recommendations to improve the scope of the *Privacy Act* in relation to online behavioural profiling.

Recommendations for Australian Do Not Track Initiatives

Both the Senate Environment and Communications Reference Committee Report and the Australian Government, in its response to the ALRC's report, have made a number of proposals to improve the efficacy of Australia's privacy law framework. Those proposals that are relevant to the issue of Do Not Track are detailed briefly next and are supplanted by developments from different jurisdictions.

Strengthening the Self-Regulatory Framework

The Senate Environment and Communications Reference Committee found that, at present, many organisations that manage browsers, social networking sites and other web 2.0 sites are exempt from the operation of the *Privacy Act* due to the "small businesses" exemption. Therefore, in these instances the privacy of Australians online is largely dependent on the individual policies and practices of particular organisations. In response to the ALRC's report, the Australian Government proposed to extend the powers of the Privacy Commissioner to request the development of industry-wide privacy codes where it is considered in the public interest to do so. If such a request was not complied with, the Government also proposed that the Commissioner should be vested with the power to develop and impose an adequate code following consultation with stakeholders. Under these proposals, self-regulation would still largely be the regulatory mode of choice, but it would at least be underpinned by the Privacy Commissioner who could take enforcement actions in circumstances where industry has failed to effectively self-regulate. Such an approach is more in line with the co-regulatory methods adopted in Canada and reflects the changes currently taking place in the US.

Those companies currently subject to the Small Business exception under the *Privacy Act* and not party to any self-regulatory regime will face the cost of introducing new privacy infrastructure on their networks, and of training staff in the proper use and protection of consumer information. Thus the economic concerns of Australian business should be considered in the implementation and content of mandatory rules. The self-regulations already adopted by the AANA could provide a basis for future work on a co-regulatory approach as evidenced by developments in the US and Canada.

In many ways, the regulatory approach adopted by the EU provides the highest level of protection for individuals. However, the adoption of such an approach in Australia would require significant amendment to the underlying philosophy and application of information privacy law. For example, the shift from an opt-out approach to an opt-in would in itself have a number of consequences as highlighted by the reluctance of some EU member states to fully implement the E-privacy Directive.

Enhancing the Powers of the Privacy Commissioner

The change in regulatory focus would also require enhanced powers for the Privacy Commissioner. The Canadian Privacy Commissioner acts as an ombudsman with authority to investigate complaints made by Canadian citizens and report on whether there has been a violation of the Privacy Act or PIPEDA. The Commissioner has also proved willing to become involved in enforcing and auditing the privacy policies of the industry. This is a statutory power conferred by PIPEDA. The Commissioner has also worked with IAB Canada and a variety of online advertising interests to develop a self-regulatory framework for the industry, which was released in August 2011 (IAB Canada, 2011). This framework represents another element of cooperation between state and private sectors, having been developed after frequent and extensive consultation with the office of the Privacy Commissioner. Canada therefore offers an example of a jurisdiction in which the protections offered by self-regulation and by state regulation are closely balanced against each other.

It would also appear that successive Canadian Commissioners have been more willing to take a wider view of their role than their Australian counterparts and have demonstrated a greater willingness to become involved in contemporary privacy controversies. The Senate Committee Report, the ALRC and the OAIC itself have recommended that the statutory powers of the Privacy Commissioner be strengthened. It would appear that any substantive change in the law is effectively predicated on the enhancement of statutory powers if Australian privacy law in this area is to have any "teeth".

However, it should also be noted that the findings of the Canadian Privacy Commissioner are not legally binding, and therefore in this respect the US enforcement approach, which centres around the FTC protecting consumer rights when organisations breach their own privacy policies, may be the strongest approach to adopt. It should nonetheless be noted that whilst the FTC has developed a meaningful jurisprudence in the area of corporate responsibilities for privacy protection, it cannot not be considered on the same lines as specific information privacy commissions, such as those in operation in comprehensive information privacy law frameworks.

Providing Meaningful Consent

Under the *Privacy Act*, the restrictions on the collection, use and disclosure of personal information can potentially be circumvented in circumstances where the consent of the individual is acquired (Schedule 3). The Australian Privacy Foundation submitted to the Senate Environment and Communications Reference Committee that the "cure-all" effect of consent on individual privacy is not proportionate to the ease with which consent can be obtained. The Foundation gave the example of an individual being forced to "consent" to "unspecific privacy invasive practices, bundled with pages of other terms and conditions, when signing up for a social networking site" (Australian Privacy Foundation, 2010, p. 2) to illustrate this point. The Committee considered that while the *Privacy Act* has allowed for consent to justify the waiver of privacy rights in the offline sphere, perhaps this approach is inappropriate in the online context (Senate Environment Committee, 2011, p. 31). Liberty Victoria made submissions to this effect, arguing that the fundamental differences between offline and online transactions requiring consent rendered the consent justification somewhat untenable for the latter. They pointed to several distinguishing features, including that:

- Australian law often does not cover online transactions, and that consequently the collected data may be used for purposes, or disclosed to other organisations, not envisaged by the consumer;
- Third parties may be collecting the transactional data; and
- Electronic data is rarely deleted, and is more accessible to a greater number of people and organisations than offline data.

In order to meet the challenges presented by online transactions and more effectively deal with complaints about the misuse of privacy consent forms, the Committee recommended an expansion of the Privacy Commissioner's complaint-handling role under s 21(1) (ab) of the *Privacy Act*. Additionally, it recommended that the OAIC examine the issue of consent in the online context and subsequently develop guidelines on the appropriate use of privacy consent forms for online services. At present, the OAIC has not developed such a guideline.

As regards other jurisdictions, in terms of offering consumers greater control over the collection, use and disclosure of their personal information, the EU model is exceptional, as it comprises a comprehensive framework that applies across all industry sectors which is enforced by active regulators (Burdon, 2011, p. 85). Furthermore, it employs an opt-in consent mechanism, which sets a much higher standard to satisfy than the opt-out mechanisms used in Australia, where the default setting is to allow collection and disclosure of personal information until the individual elects to opt-out of such practices. A suggested middle ground could be to require opt-in consent for the collection and dissemination of sensitive information, such as biometric data, race, sexu-

ality, religion, financial and health records. This is arguably a more appropriate approach, as it still allows opt-out consent to operate in many circumstances, but also takes into account the potentially more damaging ramifications of the misuse of sensitive information and thus sets a higher bar for organisations to satisfy if they wish to collect and use such information. A similar approach has also been suggested for US Do Not Track proposals.

Reducing the Scope of the Small Business Exemption

The Senate Committee recommended amendment of the small business exemptions to ensure that those businesses which hold substantial quantities of personal information, or which transfer personal information offshore, are subject to the requirements of the *Privacy Act*. A related recommendation suggested that the *Privacy Act* be amended to provide that all Australian organisations which transfer personal information overseas must ensure that the information will be given at least equivalent protections to those afforded under Australia's privacy framework. These seem practical suggestions to improve the efficacy of the *Privacy Act* in relation to online behavioral profiling.

Regulating Transborder Information Flows

One of the inherent limitations when attempting to regulate online behavioural profiling is that the Australian Parliament can only enact privacy laws relating to companies incorporated in Australian or with an Australian link .Regarding the latter, the *Privacy Act* currently applies where the act or practice of an organisation relates to the personal information of an Australian citizen or permanent resident, and the organisation carries on business in Australia and collects or holds the information in Australia (s 5B). In its submission to the 2010

Senate inquiry into the exposure drafts, the OAIC submitted that the requirement for information to have been collected in Australia is ambiguous, because in a situation where an individual in Australia submits information over the internet to an organisation based overseas, it is uncertain whether the overseas organisation has collected the information at the point of upload (Australia), thereby making it subject to Privacy Act provisions, or whether it has been collected wherever the recipient organisation is based (OPC, 2010).

Despite the exposure draft amendments attempting to clarify this issue, the Committee recommended that item 19(3)(g)(ii) of the amendments be altered to provide that an organisation has an Australian link if it collects information from Australia, thereby enhancing the scope of the Act's extra-territorial operation to ensure that information collected from Australia in the online context is protected (Senate Environment Committee, 2011, p. 46).

The Committee also recommended that the *Privacy Act* be amended so that all Australian organisations that transfer personal information offshore would be fully accountable for protecting the privacy of that information. The Committee was of the view that this would help to avoid situations where small companies could engage in cross-border data transfers with no responsibility to ensure that the privacy of those to whom the information relates would be protected (Senate Environment Committee, 2011, pp 48-49). Again, this could add significant practical protections to data collected from Australian citizens for the purpose of online behavioral profiling.

Developing A New Online Privacy Statute?

Most of the proposed amendments in the Australian context envisage amendments to the *Privacy Act* rather than the implementation of a new piece of legislation to deal specifically with the unique challenges raised by online privacy protection.

While any changes to current online privacy law will necessarily require amendments to the *Privacy Act*, the introduction of an entirely new statute would allow for clarification of the separate rules regarding online privacy as distinct from general privacy. The enormous growth of online behavioral profiling in the last decade and the potential for the further economic expansion of the internet may also necessitate a distinct Online Privacy Act. A more specific act will make the rules related to online privacy easier to find and follow, both for consumers and for businesses. However, such an initiative would require a radical rethink of Australia's privacy law framework given the core principles of Australian privacy law that is enshrined through the *Privacy Act*.

CONCLUSION

It is currently unclear exactly what application the Australian *Privacy Act* will have regarding the collection of user browsing activity for the purpose of online behavioral profiling. The first issue to resolve is whether the information collected for the purpose of online behavioral profiling is personal information. Second, the applicability and coverage of National Privacy Principle 2 needs to be clarified in relation to Do Not Track mechanisms. Third, clarification is required about whether the secondary use of data profiling purposes meets the requirements of National Privacy Principle 2. Finally, the application of the small business exemption may have a debilitating effect on coverage under the Act.

The typology developed in this chapter highlights that different jurisdictions have put forward different methods of regulating online behavioral profiling and Do Not Track initiatives. Three broad approaches are apparent:

- The predominantly self-regulatory approach adopted in Australia and currently in the US;

- The co-regulatory approach of Canada; and
- The prescriptive, legislative approach of the EU.

These three systems are not absolute, and no jurisdiction entirely uses one approach to the total exclusion of the other. They can rather be considered a spectrum with self-regulation dominating at one end, and state regulation dominating at the other. Australia has a largely self-regulated industry of targeted advertising. However, a predominantly self-regulatory approach for privacy protection tends to go hand-in-hand with a relatively weak legislative framework, as has been frequently argued about the US legal framework. This statement could also be applied to Australia. An approach that is too far towards the self-regulated end of the spectrum arguably leaves consumers more vulnerable and dependent on the individual policies and practices of particular organisations.

The issue of online behavioral profiling and Do Not Track legal responses are garnering worldwide interest. Developments are happening apace and it is likely that some form of US legislation or regulation will be implemented within the next two years. At the same time, ongoing EU developments involving the continuing implementation of the E-Privacy Directive and discussions relating to the new Data Protection Regulation will ensure that the issue of online behavioral profiling will never be far from the policy table. How Australia will respond to these developments is as yet unclear. However, it would appear from global initiatives that there is a distinct move away from predominant self-regulatory approaches to more nuanced, legislative options. Given the global nature of online behavioral profiling, it is likely that Australia will have act in some form or another as maintaining the status quo for the sake of it may not be a viable option.

REFERENCES

Act, P. 1988 (Cth) (Australia)

ALRC. (2008). *For your information: Australian privacy law and practice: Report no. 108*. Retrieved from www.alrc.gov.au/publications/report-108

Amendment, P. (Private Sector) Act 2000 (Cth) (Australia)

American Association of Advertising Agencies. (2011). *Digital advertising alliance begins enforcing next phase of self-regulatory program for online behavioural advertising*. Retrieved from http://www.aaaa.org/news/press/Pages/052311_digital_next.aspx

An act to add Section 22947.45 to the Business and Professions Code, relating to business. State of California Senate. 761. (2011)

Andrejevic, M., & Arnott, C. (2011). *Internet privacy research*. Retrieved from http://cccs.uq.edu.au/documents/privacy-report.pdf

Angwin, J. (2012, February 17). Google's iPhone tracking. *Wall Street Journal*. Retrieved from online.wsj.com/article_email/SB10001424 052970 2048804045772253 80456599176- %0A%0AlMyQjAxMTAy MDEwNjExNDYyWj .html?mod=wsj_share_email#articleTabs%3Darticle %26project%3DS AFARITRACKINGCODE0212

Article 29 Working Party. (2011). Opinion 16/2011 on EASA/IAB Best Practice Recommendation on Online Behavioural Advertising Adopted on 8 December 2011, 02005/11/EN WP188

Australian Association of National Advertisers. (2011, March). *Australian best practice recommendation for online behavioural advertising*. Retrieved from http://www.easa-alliance.org/page.aspx/386

Australian Government. (2009). *First stage response to ALRC privacy report*. Retrieved from http://www.dpmc.gov.au/privacy/reforms.cfm

Australian Privacy Foundation. (2010). *Submission to senate committee on environment, communications and the arts' inquiry into the adequacy of protections for the privacy of Australians online*. Retrieved from http://www.privacy.org.au/Papers/Sen-OLP-100813.pdf

Bennett, C. J., & Raab, C. D. (2006). The governance of privacy: Policy instruments in global perspective (2nd ed.). Cambridge, MA: MIT Press.

Brodkin, J. (2012a). Web privacy standards: Easy to break, hard to enforce. *ARS Technica*. Retrieved from http://arstechnica.com/tech-policy/news/2012/02/web-privacy-standards-easy-to-break-hard-toenforce.ars

Brodkin, J. (2012b). Google privacy change taking effect today is illegal, EU officials say. *ARS Technica*. Retrieved from http://arstechnica.com/tech-policy/news/2012/03/google-privacy-change-taking-effect-today-is-illegal-eu-officials-sayers

Burdon, M. (2011). Contextualising the tensions and weaknesses of information privacy and data breach notification laws. Santa Clara Computer and High-Technology Law Journal, 27(1), 63–129.

Burdon, M., & Telford, P. (2010). The conceptual basis of personal information in Australian privacy law. Murdoch Elaw Journal, 17(1), 1–27.

Canada, I. A. B. (2011) *Canada's advertising industry releases self-regulation framework for online behavioural advertising that ensures transparency, education, choice and accountability for consumers*. Retrieved from http://www.iabcanada.com/pr-news/oba-self-regulatory-framework

Commercial Privacy Bill of Rights Act of 2011, U.S. Senate. 799, 1. (2011)

Consumer Federation of America and Consumers' Union. (2008, April 11). *Comments to the FTC concerning the proposed online behavioural advertising self-regulatory principles*. Author.

Cranor, L., Leon, P., Ur, B., Balebako, R., Shay, R., & Wang, Y. (2011). Why Johnny can't opt out: A usability evaluation of tools to limit online behavioural advertising. *Carnegie Mellon University*. Retrieved from www.cylab.cmu.edu/research/techreports/2011/tr_cylab11017.html

Digital Advertising Alliance. (2009). *Self-regulatory principles for online behavioural advertising*. Retrieved from http://www.iab.net/public_policy/behavioral-advertisingprinciples

Digital Advertising Alliance. (2011). *Self-regulatory principles for multi-site data*. Retrieved from http://www.aboutads.info/resource/download/Multi-Site-Data-Principles.pdf

Digital Advertising Alliance. (2012). *White House, DOC and FTC commend DAA's self-regulatory program to protect consumer online privacy: DAA announces plans to expand program consumer choice mechanisms*. Retrieved from http://www.aaf.org/default.asp?id=1322

Do Not Track Me Online Act, U.S. House of Representatives. 654, 1. (2011)

(2011). Do-Not-Track Online Act of 2011. U.S. Senate., 913, 1.

Drivers Privacy Protection Act of 1994 18 USC § 2725

Duhig, C. (2012, February 19). How companies learn your secrets. *New York Times*. Retrieved from http://www.nytimes.com/2012/02/19/magazine/shopping-habits.html?_r=1

EASA. (2011). *Best practice recommendation on online behavioural advertising*. Retrieved from http://www.easa-alliance.org/page.aspx/386

Electronic Privacy Information Centre. (n.d.). *Online tracking and behavioural profiling*. Retrieved from http://epic.org/privacy/consumer/online_tracking_and_behavioral.html

EPIC. (2012). *Google's circumvention of browser privacy settings*. Retrieved from http://epic.org/privacy/google/tracking/googles_circumvention_of_brows.html

European Union. (1995). *Directive 95/46/EC of the European parliament and of the council of 24 October 1995 on the protection of individuals with regards to the processing of personal data, and on the free movement of such data OJ L 281/31*. Author.

European Union. (2002). *Directive 2002/58/EC of the European parliament and of the council of 12 July 2002 concerning the processing of personal data and the protection of privacy in the electronic communications sector (directive on privacy and electronic communications) OJ L 201/37*. Author.

European Union. (2009). *Directive 2009/136/EC of the European parliament and of the council of 25 November 2009 amending directive 2002/22/EC on universal service and users' rights relating to electronic communications networks and services, directive 2002/58/EC concerning the processing of personal data and the protection of privacy in the electronic communications sector and regulation (EC) no 2006/2004 on cooperation between national authorities responsible for the enforcement of consumer protection laws. OJ L 337/11*. Author.

European Union. (2012). *Proposal for a commission regulation (EC) no 0011/2012 of 25 January 2012 on the protection of individuals with regard to the processing of personal data and on the free movement of such data (general data protection regulation) COM(2012)*. Author.

Federal Trade Commission. (2009). *Self-regulatory principles for online behavioural advertising*. Retrieved from http://www.ftc.gov/os/2009/02/P085400behavadreport.pdf

Federal Trade Commission. (2010). *Protecting consumer privacy in an era of rapid change*. Retrieved from http://online.wsj.com/public/resources/documents/PrivacyReport_FINAL.pdf

Gallagher, S. (2012, February 23). White House announces new privacy 'bill of rights', do not track agreement. *ARS Technica*. Retrieved from http://arstechnica.com/tech-policy/news/2012/02/white-house-announces-new-privacy-bill-of-rightsdo-not-track-agreement.ars

Gellman, R. (1997). Does privacy law work? In P. Agre & M. Rotenberg (Eds.), Technology and privacy: The new landscape (pp. 193–218). Cambridge, MA: MIT Press.

Google. (n.d.). *Privacy policy*. Retrieved from http://www.google.com/policies/privacy/

Greenleaf, G. (2001). Tabula rasa: Ten reasons why Australian privacy law does not exist. The University of New South Wales Law Journal, 24(1), 262–269.

Healthline. (n.d.). *Privacy policy*. Retrieved from http://www.healthline.com/health/privacy-policy

Hoofnagle, C. J. (2006). Privacy self-regulation: A decade of disappointment. In J. K. Winn (Ed.), Consumer protection in the age of the 'information economy' (pp. 381–404). Burlington, UK: Ashgate.

IPGP. (n.d.). *IP address lookup*. Retrieved from http://www.ipgp.net/

Johnston, C. (2012, January 26). *Google's new privacy policy could anger FTC*. Retrieved from http://arstechnica.com/gadgets/news/2012/01/pascals-wager-googles-new-privacy-policy-could-anger-ft.cars

Klimas v Comcast Cable Communications Inc. 2006 C.A.6 (Mich.)

Lee, M. (2011). Privacy commissioner pushes for new powers. *ZDNet*. Retrieved from http://www.zdnet.com.au/privacy-commissioner-pushes-for-powers-339319337.htm

Lovells, H. (n.d.). *EU modifies cookie rules*. Retrieved from http://ehoganlovells.com/rv/ff0001f-bca7d4cfee09193f6c241b40f4ea39f24/p=32

Morales, L. (2010, December 21). US internet users ready to limit online tracking for ads. *Gallup*. Retrieved from http://www.gallup.com/poll/145337/internet-users-ready-limit-online-tracking-ads.aspx

Musil, S. (2012). Microsoft: Google bypassed ie privacy settings too. *CNET*. Retrieved from http://news.cnet.com/8301-1009_3-57381371-83/microsoft-google-bypassed-ie-privacy-settings-too/

Narayanan, A. (2010). *Do not track explained - 33 bits of entropy*. Retrieved from http://33bits.org/2010/09/20/do-not-track-explained/

New Zealand Law Commission. (2011). *Key recommendations*. Retrieved from http://www.lawcom.govt.nz/sites/default/files/publications/2011/08/key_recommendations_-_for_report_release.pdf

New Zealand Law Commission. (2011). *Review of privacy*. Retrieved from http://www.lawcom.govt.nz/project/review-privacy?quicktabs_23=report

Office of the Information Commissioner. (n.d.). *IP addresses, Google analytics and the privacy principles*. Retrieved from http://www.oic.qld.gov.au/files/InformationSheets/Information%20Sheet%20-%20privacy,%20IP%20addresses%20and%20Google%20Analytics.pdf

Office of the Privacy Commissioner. (2010). *Submission to senate finance and public administration legislation committee inquiry into exposure drafts of Australian privacy amendment legislation*. Retrieved from http://www.privacy.gov.au/materials/types/download/9565/7125

Office of the Privacy Commissioner of Canada. (2005). *Canadian federal privacy commissioner's findings in PIPEDA case summaries #2005-319*. Retrieved from http://www.priv.gc.ca/cf-dc/2005/319_20051103_e.cfm

Office of the Privacy Commissioner of Canada. (2009). *Canadian federal privacy commissioner's findings in PIPEDA case summaries #2009-010*. Retrieved from http://www.priv.gc.ca/cf-dc/2009/2009_010_rep_0813_e.cfm

Office of the Privacy Commissioner of Canada. (2009). *Privacy legislation in Canada*. Retrieved from http://www.priv.gc.ca/fs-fi/02_05_d_15_e.cfm

Office of the Privacy Commissioner of Canada. (2010a). *Report on the 2010 office of the privacy commissioner of Canada's consultations on online tracking, profiling and targeting, and cloud computing*. Retrieved from http://www.priv.gc.ca/resource/consultations/report_201105_e.pdf

Office of the Privacy Commissioner of Canada. (2010b). *Report on the 2010 office of the privacy commissioner of Canada's consultations on online tracking, profiling, targeting and cloud computing*. Retrieved from http://www.priv.gc.ca/resource/consultations/report_201101_e.pdf

Ohm, P. (2010). Broken promises of privacy: Responding to the surprising failure of anonymization. UCLA Law Review. University of California, Los Angeles. School of Law, 57(6), 1701–1778.

Phillips, N. (2010, October 5). Inside the cookie monster - Trading your online data for profits. *The Australian*. Retrieved from http://www.smh.com.au/technology/technology-news/inside-the-cookie-monster--trading-your-online-data-for-profits-20101004-164ee.html

Raab, C. (1999). From balancing to steering: New directions for data protection. In R. Grant & C. J. Bennett (Eds.), Visions of privacy: Policy choices for the digital age (pp. 68–96). Toronto, Canada: University of Toronto Press.

Reidenberg, J. (1999). The globalisation of privacy solutions: Movement towards obligatory standards for fair information practices. In R. Grant & C. J. Bennett (Eds.), Visions of privacy: Policy choices for the digital age (pp. 217–230). Toronto, Canada: University of Toronto Press.

Reidenberg, J. R. (1992). Privacy in the information economy: A fortress or frontier for individual rights? Federal Communications Law Journal, 44(2), 195–243.

Senate Environment and Communications Reference Committee. (2011, April 7). *The adequacy of protections for the privacy of Australians online*. Retrieved from http://www.aph.gov.au/Parliamentary_Business/Committees/Senate_Committees?url=ec_ctte/online_privacy/report/index.htm

Senate Finance and Public Administration Legislation Committee. (2010). *Inquiry into exposure drafts of Australian privacy amendment*. Retrieved from http://www.privacy.gov.au/materials/types/submissions/view/7125

Solove, D. (2001). Privacy and power: Computer databases and metaphors for information privacy. Stanford Law Review, 53(6), 1393. doi:10.2307/1229546 doi:10.2307/1229546

Swire, P., & Litan, R. (1998). None of your business: World data flow, electronic commerce, and the European privacy directive. Washington, DC: Brookings Institution Press.

The Video Privacy Protection Act of 1998, 18 USC § 2710.1994.

Turow, J. (2011). The daily you. New Haven, CT: Yale University Press.

Wall Street Journal. (2011, May 30). A short guide to cookies. *Wall Street Journal*. Retrieved from http://online.wsj.com/public/page/0_0_WP_3001.html?currentPlayingLocation=158&currentlyPlayingCollection=Tech&currentlyPlayingVideoId={92E525EB-9E4A-4399-817D-8C4E6EF68F93}

White House. (2012). *Consumer data privacy in a networked world: A framework for protecting privacy and promoting innovation in a networked world*. Retrieved from http://www.whitehouse.gov/sites/default/files/privacy-final.pdf

KEY TERMS AND DEFINITIONS

Accountability: In ethics and governance, accountability is answerability, blameworthiness, liability, and the expectation of account-giving. As an aspect of governance in an organisation it encompasses the obligation to report, explain and be answerable for resulting consequences.

Beacons: Is an intentionally conspicuous device designed to attract attention to a specific location. Beacons today in common smart phone apps might serve a different function, to surveil individuals without their knowledge.

Behavioural Profiling: Is an intelligence capability that is intended to help corporations to accurately predict and profile the characteristics of consumers who would otherwise be unknown by name, or identifier.

Civil Law: The law that apply to the citizens of a city or state as opposed to international law. Civil law is synonymous with common law.

Cookies: Is a small piece of data sent from a website and stored in a user's web browser while a user is browsing a website. When the user visits the same website again using the same browser browses, the data stored in the cookie is sent back to the website to notify the website of the user's previous activity.

Consent: Refers to the provision of approval or agreement, particularly and especially after thoughtful consideration.

Data Aggregation: In statistics, aggregate data describes data combined from several measurements. When data are aggregated, groups of observations are replaced with summary statistics based on those observations.

Data Transparency: The ability to easily access and work with data no matter where they are located or what application created them. It is the assurance that data being reported are accurate and are coming from an official source.

Do Not Track Header: Is the proposed HTTP header field DNT that requests that a web application disable either its tracking of an individual user. The Do Not Track header was originally proposed in 2009 by researchers Christopher Soghoian, Sid Stamm, and Dan Kaminsky.

E-Privacy: The protection of email from unauthorized access and inspection is known as electronic privacy. In countries with a constitutional guarantee of the secrecy of correspondence, email is equated with letters and thus legally protected from all forms of eavesdropping.

Information Privacy: Also known as data privacy, is the relationship between collection and dissemination of data, technology, the public expectation of privacy, and the legal and political issues surrounding them.

Jurisdiction: Is the practical authority granted to a formally constituted legal body to make pronouncements on legal matters and, by implication, to administer justice within a defined area of responsibility. The term is also used to denote the geographical area or subject-matter to which such authority applies.

P3P: The Platform for Privacy Preferences Project is a protocol allowing websites to declare their intended use of information they collect about web browser users.

PIPEDA: The Personal Information Protection and Electronic Documents Act is a Canadian law relating to data privacy. It governs how private sector organizations collect, use and disclose personal information in the course of commercial business.

Privacy Policies: Is a statement or a legal document that discloses some or all of the ways a party gathers, uses, discloses and manages a customer or client's data.

Self-Regulation: Regulating oneself without the need for regulatory controls at the industry or government level.

Third Party: One other than the principals involved in a transaction.

Transborder Flows of Personal Data: Means movements of personal data across national borders.

Chapter 12

Uberveillance, Standards, and Anticipation:
A Case Study on Nanobiosensors in U.S. Cattle

Kyle Powys Whyte
Michigan State University, USA

Paul B. Thompson
Michigan State University, USA

Monica List
Michigan State University, USA

Lawrence Busch
Michigan State University, USA

John V. Stone
Michigan State University, USA

Daniel Buskirk
Michigan State University, USA

Daniel Grooms
Michigan State University, USA

Erica Giorda
Michigan State University, USA

Stephen Gasteyer
Michigan State University, USA

Hilda Bouri
Michigan State University, USA

ABSTRACT

Uberveillance of humans will emerge through embedding chips within nonhumans in order to monitor humans. The case explored in this chapter involves the development of nanotechnology and biosensors for the real-time tracking of the identity, location, and properties of livestock in the U.S. agrifood system. The primary method for research on this case was an expert forum. Developers of biosensors see the tracking capabilities as empowering users to control some aspects of a situation that they face. Such control promises to improve public health, animal welfare, and/or economic gains. However, the ways in which social and ethical frameworks are built into standards for the privacy/access, organization, adaptability, and transferability of data are crucial in determining whether the diverse actors in the supply chain will embrace nanobiosensors and advance the ideals of the developers. Further research should be done that explores the possibilities of tripartite standards regimes and sousveillance in relation to nanobiosensors in agrifood.

DOI: 10.4018/978-1-4666-4582-0.ch012

INTRODUCTION

Uberveillance is the ability to track an item, handling, or life form through a nexus of its identity, location, and properties in real time through embedded radio frequency identification (RFID) chips—technologies that use radio waves to exchange data between a reader and an electronic tag attached to an object (Michael & Michael, 2010). Historically, uberveillance follows after dataveillance, which is "the systematic use of personal data systems in the investigation or monitoring of the actions of one or more persons" (Clarke, 1988, p. 498). One of the marked distinctions between the two types of tracking is that uberveillance "takes that which is static and discrete in the dataveillance world, and makes it constant and embedded" (Michael & Michael, 2010, p. 9). Uberveillance is made possible by emerging technologies like RFID in combination with data management software that promises to create an effect of almost omnipresent monitoring of subjects in which the chips are embedded.

Yet omnipresence does not entail omniscience; facts and information do not become actionable by themselves without applying additional premises, judgments, and assumptions that are based on varying combinations of values and interests. The large amount of data and the integration of values and interests create more numerous possibilities for misinformation, misinterpretation, and information manipulation (Michael & Michael, 2006). This suggests that the development of standards including standards of information analysis, provenance, access, and granularity, are areas where values and interests are integrated with data, which can generate relations of control over the variables being tracked. Systems of analysis can serve to guide and discipline the monitored subjects and to create desirable or preferred interpretations of their behavior.

The emerging literature on uberveillance focuses on tracking humans through RFID chips embedded in humans. However, uberveillance of humans will also emerge through embedding chips within nonhumans in order to monitor humans (e.g. chips in packaging can be used to monitor the activities of those who transport the packages). Consequently, uberveillance should also be explored within technologies that allow human actors to be evaluated and controlled, for example, through constant tracking of animals, products, transactions, and handlings in supply chains.

The case explored in this paper involves the development of nanotechnology and biosensors (nanobiosensors) for the real-time tracking of the identity, location, and properties of livestock in the U.S. agrifood system. Biosensors promise many dramatic real-time applications, from monitoring of blood parameters to detect the presence of metabolic or infectious diseases, to cortisol levels in cattle as one potential (and controversial) measure of animal welfare. In the U.S. agrifood system, nanobiosensors could be integrated into the broader initiatives to improve national food traceability.

The primary method for research on this case was an expert forum. The method is modeled on scientific committee processes in which individuals with complementary domains of specialization convene to develop an integrated statement of what is known about a given problem, and to identify key areas for further research. The approach has been generalized as a method for sustainability science (Carpenter et al., 2009). Over the last thirty years, this method has been extended to an array of ethics and values problems in the medical arena by the Hastings Center of Garrison, NY (Callahan, 1999).

Our workshop, funded by a National Science Foundation (NSF) grant titled "Anticipatory Workshop on Agrifood Biosensors", was held at Michigan State University (MSU) in December 2010. It included specialists in biosensor design and development, zoonosis in livestock, animal production, regulation, tracing and tracking technology, ethics of emerging technology, public engagement and environmental justice and specialists in both technical and social dimensions of standards development.

Based on the workshop, attendance at conferences and public meetings by the authors, and academic literature research, we advance the following claims. Developers of biosensors see the tracking capabilities idealistically as empowering users to control some aspects of a situation that they face. Such control promises to improve public health, animal welfare, and/or economic gains. However, the ways in which social and ethical frameworks are built into standards for the privacy/access, organization, adaptability, and transferability of data are crucial in determining whether the diverse actors in the supply chain will embrace nanobiosensors and advance the ideals of the developers.

One key part of governing uberveillance is to unlock the capacity of anticipatory participation mechanisms to provide early social and ethical guidance to technology developers and regulators. Indeed, a recent report of the National Nanotechnology Initiative Workshop on "Nanotechnology-Enabled Sensing" notes that "These systems will be ubiquitous" and that "deploying and understanding their potential benefits and risks will require developers and educators to engage citizens in proactive and ongoing conversations" (NNCO 2010, p. 2).

We begin in the Background section by describing some of the research that led us to organize the expert forum. We then explore the information gathered during the expert forum and through research of the relevant literatures on the health and economic goals of nanobiosensors. A discussion of the importance of considering how biosensors will be received by actors in the supply chain is followed by evaluation of those actors. We recommend that the public benefits of nanobiosensors should be explored through early

participation processes that address key social and ethical issues. We conclude by featuring the possibilities for "tripartite standards regimes" in the governance of uberveillance and for further research on "sousveillance."

BACKGROUND

The definition of nanotechnology is itself socially contested. Nanomaterials are defined as materials from 1-100 nanometers in physical size. Very broadly, the term nanotechnology has been applied to any functional device using materials at this scale. The difference of working at this scale is that properties such as physical strength, magnetic and optical qualities, chemical reactivity, and electric conductivity are size dependent. For example, many chemical reactions occur through bonds that are formed between the surfaces of materials. When particle size is reduced and total mass remains constant, total surface area is increased, which implies that reactivity, as a function of mass, increases.

For example, while silver in all forms has weak antibiotic properties, nano-sized particles of silver can be used as a practical and effective antibiotic agent in circumstances where the sheer mass of larger sized particles of silver would make it impossible for it to be used for the same purpose. Though this understanding of nanotechnology is sufficient for the case study explored in this paper, it should be noted that there are guidance documents and science initiatives that have placed additional qualifications on what should be considered nanotechnology (Luoma, 2008; NNI, 2008).

The impetus to explore nanotechnology in agrifood systems began with research on agrifood nanotechnologies conducted through an NSF Nanoscale Interdisciplinary Research Team (NIRT) grant that included three of the authors of this article (Thompson, Busch, and Stone) who are part of the Center of the Study of Standards in Society

(CS3) at Michigan State University. The study is the Building Capacity for Social and Ethical Research and Education in Agrifood Nanotechnology," SES-0403847, 8/2004—7/2009). The NIRT project was guided by the following questions:

1. What lessons can be derived from the controversy over genetically modified (GMOs) for scientists and engineers working at the nanoscale?
2. How is nanoscale work affecting agricultural and food science?
3. How do standard setting issues involve and shape values for nanotechnology, especially in the agrifood sector?

We will review briefly some of the relevant background information gained through researching these questions because this helps to frame some of the reasons for the expert forum on nanobiosensors in the agrifood sector.

The principal lesson learned from the debate over GMOs is that social controversies capable of affecting the development and implementation of an emerging technology do not self-organize around a single axis of debate. Rather, a number of contested themes, promoted by a number of distinct social actors, interact (Thompson & Hannah, 2008). The controversy itself provides a strategic opportunity for social movements organized around very diverse themes to become aligned with the economic and political motives of other groups. This creates an opportunity for the emergence of "hyper-controversy," a situation in which parties have an interest in maintaining the public appearance of controversy and debate (David, 2008).

In the GMO debate, scientists and industry chose to address only a few of the key substantive issues, ignoring others. By the mid-1990s most leading U.S. environmental groups had decided not to make an issue out of GMOs (Burkhardt, 2008), but globally, environmental groups allied with other civil society organizations focused on a broad array of social justice and democratization issues. This led to a public discussion that created a climate of mistrust about the industry's willingness to address even food safety and environmental quality (Gaskell, 2008). Although it is far from clear that nanoscience will face the same array of issues and interests that congealed in the reaction to GMOs, the nanoscience community places its future at risk by relying on a simplistic understanding of the sources and dynamics of opposition to emerging technology (Busch & Lloyd, 2008).

Nanotechnology does not appear to be having an impact on the organization and conduct of agricultural and food science comparable to that of genetic engineering. Perhaps because agricultural science has traditionally been far more multidisciplinary and applied than other domains, these nanoscale research projects have not affected the structure, practice and organization of research that have been predicted for nanoscience (Busch, 2008; Thompson, 2010). Nanoscale materials are being developed in food and agricultural chemistry, but this development is fully consistent with longstanding approaches to the control of chemical potency and reactivity through manipulation of chemical structure and catalysis (Hannah & Thompson, 2008).

However, ambiguity in the definition of nanotechnology and about how to distinguish novel from normal research and capabilities create both confusion and an opening for resistance movements focused on food issues. Because chemical reactions occur at the nanoscale, it is possible to re-label many food industry projects in food chemistry under the heading of nanotechnology. Food industry firms were inconsistent in this respect between 2004 and 2008, though at this juncture major retail-oriented firms appear to have decided against such re-labeling. However, pressure groups focused on food industry issues have responded by claiming that the food industry is concealing its use of nanotechnology (Thompson, 2010).

BENEFITS OF NANOBIOSENSORS IN LIVESTOCK TRACEABILITY SYSTEMS

Nanotechnology is a key element in the development of active biosensors. A biosensor is an analytical device that is composed of a biological sensing element (bioreceptor) in close proximity to a transducer, and the interaction between the bioreceptor and the target analyte is converted into an electronic signal for direct reporting (Zhang et al., 2009). Biosensors are attachable to human skin as a way of monitoring metabolic properties. Applications include the reporting of health problems. They are also attachable to non-living objects for monitoring their physical and chemical properties while they transition through complex supply chains. Applications include reporting on food spoilage, measuring glucose in diabetic patients and detecting health concerns in soldiers. One company, Shimmer, offers a wearable wireless sensor platform that can engage in complex motion sensing, vital signs and biophysical sensing and environmental and ambient sensing (Shimmer, 2013).

In terms of nanotechnology, a biosensor may use nano-transducing materials such as nanoparticles, nanotubes, nanostructures, and nano-wires. Nanobiosensors are capable of rapid, highly sensitive and highly specific detection that can be designed to be easy to use, portable, field-operable, and inexpensive (Pal & Alocilja, 2009). It is very likely that nanobiosensors will be adapted for traceability systems in ways that seek to enhance food safety and animal and plant health monitoring. This would follow trends toward using other tracking technologies in livestock production such as RFID (Harrop et al., 2006; Look, 1998; Trevarthen et al., 2006; Trevarthen et al., 2008; Michael et al., 2009) Technology developers see in nanobiosensors the idealistic possibility of increasing our control over items, handlings, and life-forms in ways that could greatly benefit us. In

the agrifood sector, the traceability of contaminated food and potentially diseased animals has been projected as a high priority (Holman, 2006; Kuzma, et al., 2006).

Traceability is the ability to follow or study in detail the history of a certain activity or process. In the U.S., the term has been applied to the ability to trace the transfer of livestock in food production processes. Food traceability systems usually involve product identification, data to trace, product routing, and traceability tools (Reggatieri et al., 2007). Traceability systems can be implemented by government agencies, non-profit organizations, businesses, or combinations of these three. In the wake of bovine spongiform encephalopathy (BSE, also known as mad cow disease), foot and mouth disease and bovine tuberculosis (BTB) outbreaks, more accurate and timely traceability has become both a public health imperative and an economic priority for the livestock sector (Bickel, 2010; Popper, 2007; White, 2007).

The public health gains of a traceability system are primarily tied to improvements in the speed and capacity to respond to disease outbreaks or food safety recalls. In the food animal industry, disease monitoring has become increasingly important due to the industrialization of animal production, which tends to concentrate large numbers of animals in ways that raise the potential for a disease outbreak with a broad rather than localized impact on both animal and human populations.

A traceability system equipped with nanobiosensors in livestock could potentially monitor for important pathogens, detect metabolic disease, enhance reproductive efficiency, improve product quality, maintain animal welfare, screen inputs (feed, air, water), and be linked to animal identity and location. Examples are many. Biosensors could be placed in milk collection systems for detecting chemical residues or zoonotic pathogens (that is animal diseases that can be transmitted to humans). A biosensor could monitor in real time as milk is being collected, checking for chemical, biological, and pharmaceutical residues of interest that would then be gated into different collection vessels. One tank could contain the suspected residues, another would have safe products.

Nanobiosensors could measure nutrient output in feces or urine to monitor animal nutrient status. Biosensors could also be used to measure feed intake, water intake and core body temperature as an early indicator of adverse health events. Measurements of blood parameters would give early warning of impending metabolic diseases. Nanobiosensor signals could be filtered and translated by computer and summarized as management advice or warning messages. In terms of animal welfare, biosensors could be used to monitor for markers of "bad welfare" such as cortisol or other physiological markers.

The potential economic motives include avoidance of the costs of managing a disease outbreak, lost production, suppressed demand for livestock products, lost export markets, indirect losses in related industries, and the costs specific to responding to the outbreak and preventing its spread (Elbakidze, 2007). Early-adopting producers or early adopting countries should expect economic gains from being part of a comprehensive traceability system. An example of this is producers in Argentina and Uruguay who are voluntarily adopting traceability requirements in order to access E.U. and U.S. markets.

The credence attributes of an agricultural product cannot be directly identified by consumers by seeing or using the product. Credence attributes of beef may include grass or forage fed, no supplemental hormones given, no antibiotics used, GMO grain fed, source verified, locally produced, animal welfare certified, and conforming to fair trade standards. Though the last two may be undetectable using nano-biosensors, it may be plausible for others. Product labeling and packaging can be used to demonstrate the credence attributes that belong to a particular product, and this can impact consumer confidence (Cho & Hooker, 2002). The uberveillance of nanobiosensors could be used

to create more robust labeling and packaging. Frank Yiannas - Walmart's VP for Food Safety - recently called upon meat industry representatives at the Global Food Safety Conference to "break with tradition" and support, among other things "increased surveillance and the use of more new technologies in surveillance of foodborne disease" (Strak, 2011).

The key component of animal traceability systems is animal identification. Tracking animal products back to the animal of origin is well within the technological capability of industrial food processors, hence traceability can be implemented to the extent that individual animals can be uniquely identified immediately before slaughter. Unique identification was accomplished in Europe through the use of electronically readable numbered ear tags over a decade ago (Pettitt, 2001). In the United States, animal identification is not mandatory by Federal regulation. However, at the state level, Michigan has enacted legislation requiring mandatory RFID identification of cattle in response to public health and economic concerns. In contrast to these simple numerical identification systems, nanobiosensors may allow for nearly instantaneous detection and tracking of animal disease, enabling animal health authorities to take timely preventive measures such as quarantine. Nanobiosensors may have a role as a complementary or integral part of such identification in traceability systems.

Disease destroys basic resources of livestock production processes, for example by mortality of breeding or productive animals, decreasing the efficiency of productive processes, as well as the productivity of the resources employed (e.g. reduced feed efficiency). Disease may either reduce the quantity and/or quality of output (e.g. lower beef quality, reduced milk yield, hide damage). Losses are also related to additional costs incurred to avoid or reduce the incidence of disease (vaccination, treatment). Zoonotic diseases are especially important, as they are direct causes of the detriment to human well-being (McInerney, 1996). The Michigan Department of Agriculture initiated a pilot RFID-based tracking system in 2001 in order to support the eradication of bovine tuberculosis. This program was initiated in the high-risk area/ infected zone located in Northeast Michigan. On January 9, 2006, the Michigan Commission on Agriculture adopted a policy mandating RFID for all cattle in Michigan. The commission charged the Michigan Department of Agriculture (MDA) with developing an implementation plan and set the mandate to begin March 1, 2007 (MDA, 2011). Bovine tuberculosis is a public health concern due to its zoonotic potential. Milk-borne infection has historically been the principal route of transmission to humans. Transmission from cattle to humans can also occur though aerosols and through exposure of abattoir employees to infected cattle (Grange & Yates, 1994).

Due to long term eradication efforts, BTB has been nearly eradicated from the US cattle population. Other zoonotic diseases, however, have recently caused significant economic losses. A recent example is the case of BSE reported in a dairy cow in the state of Washington. Within days of the 2003 BSE announcement, 53 countries, including major markets (Japan, Mexico, S. Korea and Canada) banned imports of U.S. cattle and beef. U.S. beef exports were valued at $3.95 billion and accounted for 9.6% of U.S. commercial beef production in 2003. In 2004, even though Mexico and Canada partially reopened, exports declined by 82%. U.S. beef industry losses arising from the loss of beef and offal exports during 2004 were estimated at $3.2 – $4.7 billion (Coffey et al., 2005).

Economic losses are not uniquely caused by zoonotic diseases, as evidenced by the negative economic impact of Bovine Respiratory Disease (BRD) in feedlot cattle in the U.S. In 1999, the National Animal Health Monitoring System (NAHMS, 2000) conducted a study of feedlots with a 1000- head or larger capacity, in the 12 top cattle feeding states. In this sample, most feedlots (97.4%) reported an overall BRD incidence of 14.4%. While the disease generally presents a low

mortality rate (1%), economic losses also include decreased weight gain and significant costs related to necessary treatment, such as antimicrobials, and non-steroidal anti-inflammatory drugs (Snowder et al., 2006). The economic cost associated with BRD to the U.S. beef industry has been estimated at US$750 million annually (Griffin, 1997).

These examples clearly demonstrate the need for efficient and effective animal identification and traceability systems in the U.S. However, even where mandatory identification and tracking are implemented, economic losses may still be significant, as the spread of the disease may be faster than the capacity for detection and control allowed by the current tracking technology. For example, the foot and mouth disease outbreak in the United Kingdom in 2001 caused the slaughter (for disease control) of 594,000 cattle, 3,334,000 sheep, and 145,000 pigs. At the time of the outbreak, some productive animals in the UK were individually identified, however, in spite of this, losses added up to approximately $10 billion (Thompson et al., 2002).

SOCIAL AND ETHICAL DIMENSIONS OF NANOBIOSENSORS IN LIVESTOCK

Nanobiosensors can be cast as an emerging form of uberveillance that will improve traceability to achieve public health and economic gains. Yet the capability for uberveillance in U.S. agrifood does not automatically produce such outcomes. To make the real time data collection actionable, standards must be created for what data will be emphasized and how they will be organized and analyzed, and who will have access to which aspects. In short, standards are a means by which we construct social realities; they are partial orderings of products, processes, practices, and people (Busch, 2011). Moreover, though the ideals for developing nanobiosensors focus on tracking non-humans, the ways in which the data are standardized have implications for humans. More specifically, real time tracking monitors both the animals and actors in the supply chain, and it is the latter who will be subject to evaluation.

Because humans will be the focus of assessment via disease, animal movement, and welfare monitoring, among the other possibilities mentioned, there is the possibility for concern about "whose" standards will be promulgated. We recognize that standards are shaped by cultural, ethical, political and strategic, as well as technical considerations, and this perspective guides our work in this area. It is not possible to talk about the potential gains without facing the difficulties involved in developing standards for how to organize, distribute, and access the information without privileging powerful interests and harming or exploiting certain actors and groups of actors in the supply chain.

In the U.S. context, we can take previous experience with RFID systems as a cue for how data management systems and access issues make it hard to take advantage of the public benefits of the technology. While mandatory tracking systems utilizing tags and either barcode readers or RFID devices have become widely utilized in Europe, Australia and Japan (Mousavi et al., 2002; Attaran, 2006; Trevarthen, 2007; Hall. 2010), U.S. cattle producers have resisted attempts to impose similar practices in the United States (Stecklow, 2006; Popper, 2007).

Producers have argued that systems for tracking animal movements proposed by the United States Department of Agriculture (the USDA) had unacceptably high error rates, did not operate at the speed of commerce, and did not lend themselves to easy adaptation into current production settings. Some have objected to the sheer cost of the electronic ear tags. Because beef producers perceive themselves to be in competition with other livestock commodity producers, they are resistant to procedures that entail extra cost, even when costs are imposed on an industry wide basis. Producers also express the opinion that they derive no economic benefit from tracking efforts, their

cultural lifeways are not recognized fairly, and they mistrust the intention of individual animal identification.

Cattle producers tend to view data on the movement of cattle as both proprietary and personal. Producers express concern that beef processors could gain unfair price advantages if they were able to access data on movement of individual cattle. Furthermore, some evidence suggests that social stigma associated with animal disease as well as complications associated with social goods derived from hunting or other outdoor recreational activities also affect participation in tracking programs. Additionally, such concerns can be expected to vary widely across cultural and historical contexts. For example, some Native American interests in the beef industry have suggested that animal tracking in Indian Country extends a historical association with federal monitoring of, and intervention in, tribal affairs (Stecklow, 2006).

There may also be practical concerns about nanobiosensors. What is the safety of nanobiosensors? Is there a possibility some nano-materials will remain in the animal beyond the slaughterhouse? Will livestock production associated with nanobiosensors require labels? Is the data produced too much for livestock producers to handle? Can harms to the animal occur during implantation and attachment of devices (Trevarthen, 2006)?

Current plans to promote greater acceptance of tracking emphasize the coupling of record-keeping and data ownership and management functions to the collection of monitoring data in the hopes that improved decision making can be translated into an economic advantage for producers (Elbakidze, 2007). Standards are generally considered to be a convenient, neutral, and benign means for handling issues of technical compatibility. However, if one thinks of social power as the ability to set the rules that others (must) follow, then standards represent a form of codified power reflecting the interests of those groups with the greatest access to, and influence within, standards-setting processes. Whether nanobiosensors can begin to fulfill the ideals of its designers depends largely on how the social and ethical dimensions of standards setting will be addressed.

RECOMMENDATIONS

Uberveillance through nanobiosensors certainly promises the ideals held by developers. However, realizing the benefits of these ideals is not likely unless the concerns of other actors in the supply chain are taken seriously. We propose that future development of nanobiosensors should also include processes that engage these actors in ways that elicit productive dialogue on the social and ethical contexts in which they are embedded. We refer to this process as "anticipatory governance".

Anticipatory governance is often associated with science and technology studies (STS). Broadly, its goals are to increase democratic decision-making in technology design processes and to facilitate improved institutionalization of the emerging technology through activities that engage members of the public with upstream scientists, engineers, and policymakers (Barben et al., 2008; Karinen & Guston, 2010). Anticipatory governance is not about forecasting the future outcomes of an emerging technology. Rather, it is about creating opportunities for qualitative feedback from diverse perspectives in order to better contextualize the possible pathways of the emerging technology. Anticipatory governance is preparatory, not predictive. One approach to enable feedback is to have a spectrum of participants engage in a deliberative discussion in which they comment critically on different future scenarios of the emerging technology. Scenarios have been used in activities such as consensus conferences (Guston, 1999; Fischer, 2001; Dubreuil et al., 2002; Kleinman et al., 2007; Kleinman et al., 2009).

We see anticipatory governance as referring to a set of diverse engagement activities imple-

mented over time and in relation to one another that facilitate the exchange of social and ethical information and concerns between technology developers and the actors who will likely be impacted by the emerging technology. In this way, the activities should provide an interface among experts, policymakers and members of the public. Informal community building is a means of linking the different anticipatory governance activities and by increasing the involvement of actors in the agrifood supply chain who will ultimately be the first to have to adopt and adapt to the introduction of nanobiosensors. Our approach has three distinct yet mutually complementary ideas.

First, anticipatory governance engages the full range of stakeholders who are developing biosensors and work in the agrifood supply chain - in addition to members of the general public. Our conception of anticipatory governance, then, involves several groups of actors: upstream experts, supply chain actors, members of the public and the policymakers responsive to the latter's attitudes and views. We are not only interested in reactions of members of the public who may be affected by the technology, but also in understanding the ethical and social frameworks of the developers of the technology and those in the supply chain who are likely to be the first to use the technology if it reaches implementation.

Second, anticipatory governance can be especially effective when each of these groups of stakeholders takes an active interest in discussion about the social and ethical dimensions of biosensors. One way of fostering interest is if members of the different sets of stakeholders form communities that are engaged in dialogue over these issues in ways that build over time. A community of practice is an informal group of professionals that interacts regularly to share wisdom and best practices of the discipline (Lave & Wenger, 1991; Cox, 2005). Members of communities of practice are in regular contact with each other in ways that facilitate sharing.

We believe that developing something like a community of practice among professions that includes upstream experts and supply chain stakeholders who are either developing or will be the first to use or adapt to biosensors can serve the goals of anticipatory governance. The activities conducted in the proposed research activity should engage each of these groups and encourage future interactions among them. In the case of nanobiosensors, such little knowledge exists of their possibilities and social implications that more information and interest was needed to begin. We convened experts with knowledge of distinct phases in the supply chain and product development process to begin modeling such a community of practice in the December workshop. In the context of anticipatory governance, "anticipatory communities of practice" do not solely share "wisdom of a trade," but discuss their shared conceptions of the social and ethical significance of the work that they do. The future workshops proposed by this project will further extend and cultivate this community of practice.

Third, the resources created within such a community of practice can be used to interface with members of the public and policy-makers sensitive to public opinions and concerns. In this activity, a shared understanding of the technical potential for ethical betterment (e.g. managing disease for human and animal health) and of the potential for ethically sensitive unintended outcomes (e.g. breach of privacy or concentration of economic power) can be deployed within a context that has the potential to shape the policy making process. For example, lessons learned and social capital formed during the early phases of anticipatory governance can be transferred to settings where regulatory decisions for animal disease management can be made.

In this way, our approach to anticipatory governance involves activities that will build communities of practice and dialogue among stakeholders that include subject matter and process experts, individuals and firms at each juncture in the supply chain, the trade organizations that represent such firms, the interested public such as consumer groups, animal protectionists or environmental advocates, and officials from key public agencies. We began with the community of upstream experts because they are in the position to make decisions right now that impact the trajectory of technical development. Beginning to form a community of practice with upstream experts is important for making it known that there will be fora for the consideration of social and ethical dimensions of their projects.

A key theoretical issue for any anticipatory governance activity, from stakeholder workshops to consensus conferences, concerns how participants with different backgrounds, stakes and experiences can communicate using familiar language and concepts (Norton, 2005). The participants must have some way of identifying and having partial ownership over the language and concepts, even if each participant does not do so in the same way. Technical standards for biosensors circumscribe a key site for scenario development in anticipatory governance activities. Decisions on standards may reflect considerations that range from the intellectual property holdings of key participants in the standard setting process to purely arbitrary choices made solely for short-term expediency. Yet far beyond simply getting the technology to work, these standards may have crucial long-term impacts on power, social relationships, and the personal and commercial practices of both users and others who are indirectly affected by the implementation of the technology.

We hypothesize that the ethical dimension of nanobiosensor standards can be articulated by close attention to the way that a final design affects the ability of a user to retain control over access to, and transfer of, information. This control has to do with whether information can be easily alienated from (or alternatively remains embedded within) a situation or practice, in our case, the practice of livestock production. Only someone at the actual site of production would have access to a host of potentially useful data. But one could imagine that access to the data would go beyond the farm gate.

Biosensors could report to commercial databases (the cloud) as opposed to having data stored or analyzed locally. This may even be part of initial designs. As such, this information is currently controlled by producers. But while biosensors can make new data available to producers, they can also do so in a way that makes that and other information about production practice accessible to food chain actors (sale barns, beef processors, retailers, consumers) at remote locations. As such, there are factors inherent in nanobiosensor design that could affect the broader institutional context of livestock production.

FUTURE RESEARCH DIRECTIONS

Serious Games, Uberveillance, and Conformity Assessment in the Tripartite Standards Regime

One potentially fruitful area for future uberveillance research concerns the emerging application of serious gaming within the tripartite standards regimes (TSR). Researchers at MSU's Center for the Study of Standards in Society (CS3) have coined the term tripartite standards regime in reference to an emerging regime of governance that consists of standards-setting, accreditation and certification by third parties (Loconto, Stone & Busch, 2011). Standards are means by which we construct social realities; they are partial orderings of products, processes, practices, and people, and they serve as exemplary measures against which these are judged (Busch, 2011). To this extent, standards may be viewed not only as devices of social control but reflections of social power arrangements as well, where social power may be seen as the ability to set the rules that others must follow. Standards may be seen as a form of codified social power reflecting the interests of groups with greatest social access to standards-setting and certification processes.

In order for formal standards to create and keep the ordering that is intended by their use, a number of elements are employed: (1) processes for certifying compliance to the standards, (2) processes for accrediting the certifiers who audit the standards, and (3) relatively clear sanctions for violation of these standards. Generally referred to as conformity assessment, these processes traverse and integrate the public and private sectors domestically and internationally. As Loconto and Busch (2010) note, the TSR comprises a novel but increasingly powerful form of governance that aims to produce security through the market. What may appear as an institutionalization of mistrust in the global market economy - represented by the need for constant conformity assessment and auditing - is also a strategy of self-governance that pre-empts state-led regulation of markets. The TSR is fundamental to this movement towards "government at a distance" that is part and parcel of the neoliberal shift from government to governance, and it is through this entanglement of standards, intermediaries and enabling technologies into the supply chains themselves that an alternative form of self-governance emerges. One might reasonably ask how the tools of uberveillance might be integrated for deployment within the TSR to assess the conformity of people and objects and utilize this information to reward and/or sanction (and thus control) them in specific contexts (e.g., supply chains).

The term serious game refers to "an activity among two or more independent decision-makers seeking to achieve their objectives in some limiting context ... these 'serious games' have an explicit and carefully thought-out purpose" (Abt, 1970). More recently, the enabling technological components of this concept have been further developed, for example, through the Wilson International Center for Scholars' "Serious Games Initiative" to spur creation of serious game technologies to address pressing policy and management issues, and through coordinated graduate research and training programs, such as that offered at MSU, established to advance state of the art knowledge about societal effects of digital game applications and integrated technical systems. In the private sector, the FX Palo Alto Laboratory (FXPAL) in Palo Alto, California, is presently integrating "serious game" technologies - high-end game engines with robust sensor media and data importation - to enable remote monitoring and control of a "virtual factory world". In such a world one can visit a machine to read its sensors and conduct real-time virtual inspections of employees, processes and inventories. Such systems are presently being deployed in the confectionery and biscuit processing industries (Glaberson, 2011).

Clearly, these systems could be deployed within the TSR to verify and perhaps even certify conformity to given standards, including technical specifications for, and tolerances within, objects and mechanical (and biological?) systems as well as process standards governing the behaviors of the human agents who interact with and otherwise "perform" those systems. Indeed, to what extent might firms, supply chains, and third-party certifiers integrate serious game technologies into their conformity assessment processes? Would doing so alter the dynamics of the TSR? And would doing so constitute a new or alternative application of uberveillance as a tool of TSR governance at a distance?

Sousveillance

Uberveillance could be conceptualized as implemented through top down monitoring systems by run regulators. Nanotechnologies allow for tracking elements of a given food, livestock, and environmental supply chain that provide feedback of information to employees of regulatory agencies. These agencies are then able to better ensure enforcement of regulations or standards - and to take mitigation steps to protect public health, maintain public order, and to penalize those who violate standards. This scenario fits into what Ganascia (2010) refers to as "surveillance society."

The societal embodiment of the "Panopticon" prison described by Foucault (1975), surveillance society is characterized as a "centralized, hierarchical social structure, and localized in a physical building." It allows for a more complete level of oversight and regulation of social life and the processes of production in society.

Sousveillance is presented as representing a new social form that has emerged from the widely available communications technologies. It has been most popularly characterized as the ability of citizens to broadcast messages and images, often in direct defiance of government agencies and authorities of the state (Mann et al., 2003). Examples include the "Youtube," "Facebook," and "Twitter" broadcasts of violations of civil rights by police in Europe and the United States, and even more dramatically the popular protests and attempted violent crackdowns in the Middle East in 2010 and 2011. Ganascia (2010, p. 496) characterizes an emerging sousveillance society as "equally distributed, strictly egalitarian and delocalized over the entire planet." The increasing omnipresence of recording devices and the ability to make their information available to the public constitutes a society where the "Inforsphere is structured as a huge Catopticon," where, in contrast to the Panopticon, there is "transparency of society; … equality, which gives everybody the ability to watch, and consequently control, everyone else; total communication, which enables everyone to exchange with everyone else" (Ganascia 2010, p. 497).

The greater traceability that is possible through nanobiosensors in animal agriculture and other food tends to be viewed from the perspective of *surveillance*, or improved regulatory oversight of food and agricultural standards. The development of these sensoring technologies, however, may be more usefully seen in the context of sousveillance, or the emergence of Ganascia's (2010) Catopticon society. As the widely watched, secretly taped "Youtube" video of clearly sick dairy cattle being abused before slaughter demonstrates (Humane

Society, 2008), the livestock industry, and indeed the food industry in general, may already be subject to the pressures of Catopticon society. While the emergence of nanobiosensors allowing for traceability in the food system could empower regulators with new data for *surveillance*, in a climate of government budgetary limits, it is not inconceivable that nanobiosensor generated information could be made available to the public for the purpose of sousveillance of the food system – where popular pressure is exerted on food producers based on how food is produced and processed.

CONCLUSION

The claims presented in this paper are preparatory in nature. We explored how the development of standards including standards of information analysis, provenance, access, and granularity are areas where values and interests are integrated with data, which can generate relations of control over the variables being tracked. Systems of analysis can serve to guide and discipline the monitored subjects and to create desirable or preferred interpretations of their behavior. The emerging literature on uberveillance focuses on tracking humans through RFID chips embedded in humans. However, uberveillance of humans will also emerge through embedding chips within nonhumans in order to monitor humans (e.g. chips in packaging can be used to monitor the activities of those who transport the packages).

The case explored in this paper involves the development of nanotechnology and biosensors (nanobiosensors) for the real-time tracking of the identity, location, and properties of livestock in the U.S. agrifood system. Biosensors promise many dramatic real-time applications, from monitoring of blood parameters to detect the presence of metabolic or infectious diseases, to cortisol levels in cattle as one potential (and controversial) measure of animal welfare. In the U.S. agrifood system,

nanobiosensors could be integrated into broader initiatives to improve national food traceability.

The primary method for research on this case was an expert forum. The method is modeled on scientific committee processes in which individuals with complementary domains of specialization convene to develop an integrated statement of what is known about a given problem, and to identify key areas for further research.

Developers of biosensors see the tracking capabilities idealistically as empowering users to control some aspects of a situation that they face. Such control promises to improve public health, animal welfare, and/or economic gains. However, the ways in which social and ethical frameworks are built into standards for the privacy/access, organization, adaptability, and transferability of data are crucial in determining whether the diverse actors in the supply chain will embrace nanobiosensors and advance the ideals of the developers. One key part of governing uberveillance is to unlock the capacity of anticipatory participation mechanisms to provide early social and ethical guidance to technology developers and regulators. We believe there is also ample opportunity for future research on tripartite standards regimes and sousveillance.

REFERENCES

Abt, C. (1970). *Serious games*. New York, NY: Viking Press.

Attaran, M. (2006). The coming age of RFID revolution. *Journal of International Technology and Information Management*, *15*(4), 77–87.

Barben, D., Fisher, E., Selin, C., & Guston, D. H. (2008). Anticipatory governance of nanotechnologies: Foresight, engagement and integration. In E. J. Hackett, O. Amsterdamska, M. Lynch, & J. Wajcman (Eds.), *The new handbook of science and technology studies*. Cambridge, MA: MIT Press.

Bickel, A. (2010, February 12). Ag department drops livestock tracing program. *Tribune Business News*.

Burkhardt, J. (2008). The ethics of agri-food biotechnology: How can an agricultural technology be so important? In K. David, & P. B. Thompson (Eds.), *What can nanotechnology learn from biotechnology? Social and ethical lessons for nanoscience from the debate over agricultural biotechnology and GMOs*. Amsterdam, The Netherlands: Academic Press. doi:10.1016/B978-012373990-2.00003-0

Busch, L. (2008). Nanotechnologies, food, and agriculture: Next big thing or flash in the pan? *Agriculture and Human Values*, 25(2), 215–218. doi:10.1007/s10460-008-9119-z

Busch, L. (2011). *Standards: Recipes for reality*. Cambridge, MA: MIT Press.

Busch, L., & Lloyd, J. R. (2008). What can nanotechnology learn from biotechnology? In K. David, & P. B. Thompson (Eds.), *What can nanotechnology learn from biotechnology? Social and ethical lessons for nanoscience from the debate over agricultural biotechnology and GMOs*. Amsterdam, The Netherlands: Academic Press. doi:10.1016/B978-012373990-2.00013-3

Callahan, D. (1999). The hastings center and the early years of bioethics. *Kennedy Institute of Ethics Journal*, 9(1), 53–71. doi:10.1353/ken.1999.0001 PMID:11657314

Carpenter, S. R., Mooney, H. A., Agard, J., Capistrano, D., DeFries, R. S., & Díaz, S. et al. (2009). Science for managing ecosystem services: Beyond the millennium ecosystem assessment. *Proceedings of the National Academy of Sciences of the United States of America*, 106(5), 1305–1312. doi:10.1073/pnas.0808772106 PMID:19179280

Cho, B., & Hooker, N. H. (2002). *A note on three qualities: Search, experience and credence attributes* (Working Paper: AEDE-WP-0027-02). Columbus, OH: The Ohio State University. Retrieved from http://aede.osu.edu/resources/docs/pdf/CC8A1957-93C9-42AF-AE91E1288D-D1EEED.pdf

Coffey, B., Mintert, J., Fox, S., Schroeder, T., & Valentin, L. (2005). The economic impact of BSE: A research summary (Kansas State University Agricultural Experiment Station and Cooperative Extension Service). *Livestock Economics*, MF-2679.

Cox, A. (2005). What are communities of practice? A comparative review of four seminal works. *Journal of Information Science*, 31(6), 527–540. doi:10.1177/0165551505057016

David, K. (2008). Socio-technical analysis of those concerned with emerging technology, engagement and governance. In K. David, & P. B. Thompson (Eds.), *What can nanotechnology learn from biotechnology? Social and ethical lessons for nanoscience from the debate over agricultural biotechnology and GMOs*. Amsterdam, The Netherlands: Academic Press. doi:10.1016/B978-012373990-2.00001-7

Dubreuil, G. H., Bengtsson, G., Bourrelier, P. H., Foster, R., Gadbois, S., & Kelly, G. N. et al. (2002). A report of TRUSTNET on risk governance - Lessons learned. *Journal of Risk Research*, 5(1), 83–95. doi:10.1080/13669870110039916

Elbakidze, L. (2007). Economic benefits of animal tracing in the cattle production sector. *Journal of Agricultural and Resource Economics*, 32(1), 169–180.

Foucault, M. (1975). *Surveiller et punir*. Paris, France: Gallimard.

Ganascia, J. (2010). The generalized sousveillance society. *Social Sciences Information. Information Sur les Sciences Sociales, 49,* 489–506. doi:10.1177/0539018410371027

Gaskell, G. (2008). Lessons from the bio-decade: A social science perspective. In K. David, & P. B. Thompson (Eds.), *What can nanotechnology learn from biotechnology? Social and ethical lessons for nanoscience from the debate over agricultural biotechnology and GMOs.* Amsterdam, The Netherlands: Academic Press. doi:10.1016/B978-012373990-2.00012-1

Glaberson, H. (2011, February 7). US chocolatier develops virtual factory world. *Confectionery News.* Retrieved from www.confectionerynews.com/content/view/print/356965

Grange, J. M., & Yates, M. D. (1994). Zoonotic aspects of mycobacterium Bovis infection. *Veterinary Microbiology, 40,* 137–151. doi:10.1016/0378-1135(94)90052-3 PMID:8073621

Griffin, D. (1997). Economic impact associated with respiratory disease in beef cattle. *The Veterinary Clinics of North America. Food Animal Practice, 13,* 367–377. PMID:9368983

Guston, D. H. (1999). Evaluating the first U.S. consensus conference: The impact of the citizens' panel on telecommunications and the future of democracy. *Science, Technology & Human Values, 24*(4), 451–482. doi:10.1177/016224399902400402

Hall, D. (2010). Food with a visible face: Traceability and the public promotion of private governance in the Japanese food system. *Geoforum, 41*(5), 826–835. doi:10.1016/j.geoforum.2010.05.005

Hannah, W., & Thompson, P. B. (2008). Nanotechnology, risk and the environment: A review. *Journal of Environmental Monitoring, 10,* 291–300. doi:10.1039/b718127m PMID:18392270

Harrop, P., & Napier, E. (2006). *Food and livestock traceability.* London: IDTechEx.

Holman, M. (2006). *The nanotech report* (4th ed.). Lux Research, Inc.

Humane Society. (2008). *Slaughterhouse investigation: Cruel and unhealthy practices.* Retrieved from http://www.youtube.com/watch?v=zhlhSQ5z4V4

Karinen, R., & Guston, D. H. (2010). Toward anticipatory governance: The experience with nanotechnology. *Sociology of the Sciences Yearbook, 27*(4), 217–232.

Kleinman, D., Powell, M., Grice, J., Adrian, J., & Lobes, C. (2007). A toolkit for democratizing science and technology policy: The practical mechanics of organizing a consensus conference. *Bulletin of Science, Technology & Society, 27*(2), 154–169. doi:10.1177/0270467606298331

Kleinman, D. L., Delborne, J. A., & Anderson, A. A. (2011). Engaging citizens: The high cost of citizen participation in high technology. *Public Understanding of Science (Bristol, England), 20*(2), 221–240. doi:10.1177/0963662509347137

Kuzma, J., & VerHage, P. Woodrow Wilson International Center for Scholars, & The Pew Charitable Trusts. (2006). Nanotechnology in agriculture and food production: Anticipated applications. Washington, DC: Project on Emerging Nanotechnologies.

Lave, J., & Wenger, E. (1991). *Situated learning: Legitimate peripheral participation.* Cambridge, UK: Cambridge University Press. doi:10.1017/CBO9780511815355

Loconto, A., & Busch, L. (2010). Standards, techno-economic networks, and playing fields: Performing the global market economy. *Review of International Political Economy, 17*(3), 507–536. doi:10.1080/09692290903319870

Loconto, A., Stone, J. V., & Busch, L. (2011). Standards, certifications, and accreditations. In G. Ritzer (Ed.), *Encyclopedia of globalization*. Hoboken, NJ: Wiley-Blackwell.

Look, G. (1998). *Auto ID makes tracks in livestock traceability*. ID Systems - The European source for auto ID. Retrieved 13 May 2013 from http://www.idsyseuro.com/cow0498.htm

Luoma, S. N. (2008). *Silver nanotechnologies and the environment*. Philadelphia, PA: The Pew Charitable Trusts.

Mann, S., Nolan, J., & Wellman, B. (2003). Sousveillance: Inventing and using wearable computing devices for data collection in surveillance environments. *Surveillance & Society, 1*(3), 331–355.

McInerney, J. (1996). Old economics for new problems - Livestock disease: Presidential address. *Journal of Agricultural Economics, 47*(1-4), 295–314. doi:10.1111/j.1477-9552.1996.tb00695.x

Michael, K., & Michael, M. G. (2009). *Innovative automatic identification and location-based services: From bar codes to chip implants*. Hershey, PA: IGI Global. doi:10.4018/978-1-59904-795-9

Michael, K., & Michael, M.G. (2010, Summer). A note on uberveillance. *IEEE Science and Technology Magazine*, 9-16.

Michael, M. G., & Michael, K. (2006). National security: The social implications of the politics of transparency. *Prometheus, 24*(4), 359–363. doi:10.1080/08109020601029912

Michigan Department of Agriculture. (2011). Retrieved 10 February 2011 from http://www.michigan.gov/mda/0,1607,7-125--156829--,00.html

Mousavi, A., Sarhadi, M., Lenk, A., & Fawcett, S. (2002). Tracking and traceability in the meat processing industry: A solution. *British Food Journal, 104*(1), 7–19. doi:10.1108/00070700210418703

NAHMS (National Animal Health Monitoring System). (2000). *Treatment of respiratory disease in U.S. feedlots*. Retrieved 13 February 2011 from http://www.aphis.usda.gov/animal_health/nahms/feedlot/downloads/feedlot99/Feedlot99isTreatResp.pdf

National Nanotechnology Coordination Office (NNCO). (2010). *Nanotechnology-enabled sensing: Report of the national nanotechnology initiative workshop*. Arlington, VA: NNCO.

NNI (National Nanotechnology Initiative). (2008). *Strategy for nanotechnology-related environmental, health, and safety research*. Subcommittee on Nanoscale Science, Engineering, and Technology Committee on Technology National Science and Technology Council.

Norton, B. G. (2005). *Sustainability: A philosophy of adaptive ecosystem management*. Chicago: University of Chicago Press. doi:10.7208/chicago/9780226595221.001.0001

Pal, S., & Alocilja, E. C. (2009). Electrically active polyaniline coated magnetic (EAPM) nanoparticle as novel transducer in biosensor for detection of bacillus anthracis spores in food samples. *Biosensors & Bioelectronics, 24*(5), 1437–1444. doi:10.1016/j.bios.2008.08.020 PMID:18823768

Pettitt, R. E. (2001). Traceability in the food animal industry and supermarket chains. *Revue Scientifique et Technique-Office International Epizootics, 20*(2), 584–597. PMID:11548528

Popper, D. E. (2007). Traceability: Tracking and privacy in the food system. *Geographical Review, 97*(3), 365–388. doi:10.1111/j.1931-0846.2007.tb00511.x

Reggatieri, A., Gamberi, M., & Manzini, R. (2007). Traceability of food products: General framework and experimental evidence. *Journal of Food Engineering, 81*, 347–356. doi:10.1016/j.jfoodeng.2006.10.032

Snowder, G. D., Van Vleck, L. D., Cundiff, L. V., & Bennett, G. L. (2006). Bovine respiratory disease in feedlot cattle: Environmental, genetic, and economic factors. *Journal of Animal Science, 84*, 1999–2008. doi:10.2527/jas.2006-046 PMID:16864858

Stecklow, S. (2006, June 21). Twisting trail: U.S. falls behind in tracking cattle to control disease, USDA plans voluntary system after the industry divides on making one mandatory, a mad cow's unknown origins. *Wall Street Journal.*

Strak, J. (2011). Walmart asks meat industry to break with tradition. *Meatingplace.* Retrieved 13 May 2013 from http://www.traincan.com/feb22a-2011.asp

Thompson, D., Muriel, P., Russell, D., Osborne, P., Bromley, A., & Rowland, M. et al. (2002). Economic costs of the foot and mouth disease outbreak in the United Kingdom. *Rev. Sci. Tech. Off. Int. Epiz., 21*(3), 675–687. PMID:12523706

Thompson, P. (2010). Agrifood nanotechnology: Is this anything? In U. Fiedeler, C. Coenen, S. R. Davies, & A. Ferrari (Eds.), *Understanding nanotechnology: Philosophy, policy and publics.* Amsterdam: IOS Press.

Thompson, P. B., & Hannah, W. (2008). Food and agricultural biotechnology: A summary and analysis of ethical concerns. *Advances in Biochemical Engineering/Biotechnology, 111*, 229–264. doi:10.1007/10_2008_100 PMID:18496654

Trevarthen, A. (2005). *The importance of utilising electronic identification for total farm management: A Case study of dairy farms on the south coast of NSW.* University of Wollongong Thesis Collections.

Trevarthen, A. (2007). The national livestock identification system: The importance of traceability in e-business. *Journal of Theoretical and Applied Electronic Commerce Research, 2*(1), 49–62.

Trevarthen, A., & Michael, K. (2006). *Beyond mere compliance of RFID regulations by the farming community: A case study of the Cochrane dairy farm.* Paper presented at the The Sixth International Conference on Mobile Business. Toronto, Canada.

Trevarthen, A., & Michael, K. (2008). *The RFID-enabled dairy farm: Towards total farm management.* Paper presented at the 7th International Conference on Mobile Business. Barcelona, Spain.

United States Congress. (2011). 111 P.L. 353: *The FDA Food Safety Modernization Act,* 1-4-2011 (124 Stat 3885)

White, L. (2007). The chase to trace. *National Provisioner,* 1/2001.

Zhang, D., Carr, D. J., & Alocilja, E. C. (2009). Fluorescent bio-barcode DNA assay for the detection of salmonella enterica serovar enteritidis. *Biosensors & Bioelectronics, 24*(5), 1377–1381. doi:10.1016/j.bios.2008.07.081 PMID:18835708

ADDITIONAL READING

Anders, S. M., & Caswell, J. A. (2009). Standards as Barriers Versus Standards as Catalysts: Assessing the Impact of HACCP Implementation on U.S. Seafood Imports. *American Journal of Agricultural Economics, 91*, 310–321. doi:10.1111/j.1467-8276.2008.01239.x

Anderson, B. (2007). Hope for nanotechnology: anticipatory knowledge and the governance of affect. *Area, 39*(2), 156–165. doi:10.1111/j.1475-4762.2007.00743.x

Bainbridge, W., & Roco, M. (2006). Progressive Convergence. In W. Bainbridge, & M. Roco (Eds.), *Managing nano-bio-info-cogno innovations* (pp. 1–7). Springer Netherlands. doi:10.1007/1-4020-4107-1_1

Bainbridge, W. S., & Roco, M. C. (2006). Reality of Rapid Convergence. *Annals of the New York Academy of Sciences*, *1093*(1), ix–xiv. doi:10.1196/annals.1382.001 PMID:17354289

Bennett, I., & Sarewitz, D. (2006). Too Little, Too Late? Research Policies on the Societal Implications of Nanotechnology in the United States. *Science as Culture*, *15*(4), 309–325. doi:10.1080/09505430601022635

Brown, N., Rappert, B., & Webster, A. (2000). *Contested futures: a sociology of prospective techno-science*. Aldershot, England: Ashgate.

Busch, L. (2000). The Moral Economy of Grades and Standards. *Journal of Rural Studies*, *16*, 273–283. doi:10.1016/S0743-0167(99)00061-3

Busch, L. (2010). Standards, Law and Governance. *Journal of Rural Social Sciences*, *25*(3), 56–78.

Caldwell, L. (2002). Public Administration - the New Generation. In E. Vigoda (Ed.), *Public Administration: An interdisciplinary Critical Analysis* (pp. 151–176). New York: Marcel Dekker Inc.

Delborne, J. A., Anderson, A. A., Kleinman, D. L., Colin, M., & Powell, M. (2009). Virtual deliberation? Prospects and challenges for integrating the internet in consensus conferences. *Public Understanding of Science*, published online in October 2009.

Dietz, T., & Stern, P. C. (2008). *Public participation in environmental assessment and decision making*. Washington, DC: National Academies Press.

Douglas, H. (2005). Inserting the Public into Science. In P. Weingart, & S. Maasen (Eds.), *Democritization of Expertise? Exploring Novel Forms of Scientific Advice in Political Decision-Making* (Vol. 24, pp. 153–169). Netherlands: Springer.

Einsiedel, E. F., & Eastlick, D. L. (2000). Consensus Conferences and Deliberative Democracy. *Science Communication*, *21*(4), 323–343. doi:10.1177/1075547000021004001

Guston, D. H. (2007). The Center for Nanotechnology in Society at Arizona State University and the Prospects for Anticipatory Governance. In N. M. S. Cameron, & M. E. Mitchell (Eds.), *Nanoscale: Issues and Perspectives for the Nano Century* (pp. 3377–3392). Hoboken, NJ: Wiley. doi:10.1002/9780470165874.ch21

Guston, D. H., & Sarewitz, D. (2002). Real-Time Technology Assessment. *Technology in Society*, *24*(1-2), 93–109. doi:10.1016/S0160-791X(01)00047-1

Irwin, A. (2006). The Politics of Talk: Coming to Terms with the New Scientific Governance. *Social Studies of Science*, *36*(2), 299–320. doi:10.1177/0306312706053350

Irwin, A., & Wynne, B. (Eds.). (1996). *Misunderstanding Science? The Public Reconstruction of Science and Technology*. Cambridge: Cambridge University Press. doi:10.1017/CBO9780511563737

Konefal, J., & Hatanaka, M. (2010). The Michigan State University School of Agrifood Governance and Technoscience: Democracy, Justice, and Sustainability in an Age of Scientism, Marketism, and Statism. *Journal of Rural Social Sciences*, *25*(3), 1–17.

Lessig, L. (1999). *Code and Other Laws of Cyberspace*. New York: Basic Books.

Macnaghten, P., Kearnes, M. B., & Wynne, B. (2005). Nanotechnology, Governance, and Public Deliberation: What Role for the Social Sciences? *Science Communication*, *2005*(27).

Meidinger, E. (2009). Private Import Safety Regulations and Transnational New Governance. In C. Coglianese, A. M. Finkel, & D. Zaring (Eds.), *Import Safety: Regulatory Governance in the Global Economy* (pp. 233–256). Philadelphia, PA: University of Pennsylvania Press.

Rehmann-Sutter, C., & Scully, J. L. (2010). Which Ethics for (of) the Nanotechnologies. In M. Kaiser (Ed.), *Governing Future Technologies* (pp. 233–252). Dordrecht: Springer.

Roco, M. C., & Bainbridge, W. (2001). *Societal Implications of Nanoscience and Nanotechnology*. Boston: Kluwer Academic Publishers. doi:10.1007/978-94-017-3012-9

Schot, J., & Rip, A. (1997). The Past and Future of Constructive Technology Assessment. *Technological Forecasting and Social Change, 54,* 251–268. doi:10.1016/S0040-1625(96)00180-1

Shimmer. (2013). Wireless Sensor Platform for Wearable Applications. Retrieved 13 May 2013 from www.shimmer.research.com.

Spinardi, G., & Williams, R. (2005). The Governance challenges of Breakthrough Science and Technology. In C. Lyall, & J. Tait (Eds.), *New Modes of Governance: Developing an Integrated Policy Approach to Science, Technology, Risk and the Environment* (pp. 45–68). Aldershot: Ashgate.

Thompson, P. B., Harris, C. A., Holt, D., & Pajor, E. A. (2007). Livestock Welfare Product Claims: The Emerging Social Context. *Journal of Animal Science, 85,* 2354–2361. doi:10.2527/jas.2006-832 PMID:17504960

Whyte, K. P., & Thompson, P. B. (2010). A Role for Ethical Analysis in Social Research on Agrifood and Environmental Standards. *Journal of Rural Social Sciences, 25*(3), 79–98.

KEY TERMS AND DEFINITIONS

Biosensors: A device that detects, records, and transmits information regarding a physiological change or process.

Controversy: A dispute, especially a public one, between sides holding opposing views.

Genetically Modified Organisms: Is an organism whose genetic material has been altered using genetic engineering techniques. Organisms that have been genetically modified include microorganisms such as bacteria and yeast, insects, plants, fish, and mammals.

Nanomaterials: Is a field that takes a materials science-based approach on nanotechnology. It studies materials with morphological features on the nanoscale.

Nanoscale: Is usually defined as smaller than a one tenth of a micrometer in at least one dimension, though sometimes includes up to a micrometer.

Outbreak: Is a term used in epidemiology to describe an occurrence of disease greater than would otherwise be expected at a particular time and place. It may affect a small and localized group or impact upon thousands of animals across an entire region. Two linked cases of a rare infectious disease may be sufficient to constitute an outbreak.

Panopticon: Is a type of institutional building designed by English philosopher and social theorist Jeremy Bentham in the late 18th century. The concept of the design is to allow a watchman to *observe* (-opticon) *all* (pan-) inmates of an institution without their being able to tell whether they are being watched or not.

Policymakers: Someone who sets the plan pursued by a government or business.

Public Health: Is concerned with threats to health based on population health analysis. Public health incorporates the interdisciplinary approaches of epidemiology, biostatistics and health services.

Standardization: Is the process of developing and implementing technical standards. The goals of right standardization can be to help with independence of single suppliers, compatibility, interoperability, safety, repeatability, or quality.

Supply Chain: Is a system of organizations, people, activities, information, and resources involved in moving a product or service from supplier to customer. Supply chain activities transform natural resources, raw materials, and components into a finished product that is delivered to the end customer.

Traceability: Is the ability to verify the history, location, or application of an item by means of documented recorded identification.

Zoonotic Diseases: Are caused by infectious agents that can be transmitted between (or are shared by) animals and humans. This can include transmission through the bite of an insect, such as a mosquito.

Section 5
Health Implications of Microchipping Living Things

Chapter 13
Microchip–Induced Tumors in Laboratory Rodents and Dogs:
A Review of the Literature 1990–2006

Katherine Albrecht
CASPIAN Consumer Privacy, USA

ABSTRACT

This chapter reviews literature published in oncology and toxicology journals between 1990 and 2006 addressing the effects of implanted radio-frequency (RFID) microchips on laboratory rodents and dogs. Eleven articles were reviewed in all, with eight investigating mice and rats, and three investigating dogs. In all but three of the articles, researchers observed that malignant sarcomas and other cancers formed around or adjacent to the implanted microchips. The tumors developed in both experimental and control animals and in two household pets. In nearly all cases, researchers concluded that the microchips had induced the cancers. Possible explanations for the tumors are explored, and a set of recommendations for policy makers, human patients and their doctors, veterinarians, pet owners, and oncology researchers is presented in light of these findings.

PROBLEMS WITH MICROCHIP IMPLANTATION – AND WHY THEY MATTER

Since their introduction in the late 1980s, implantable microchips have become the industry standard for identifying mice and rats used in laboratory research. Animal shelters and veterinarians now routinely inject microchips into dogs and cats. More recently, there has been a push to implant microchips into people for security and building access, to manage medical records, and to identify elderly patients.

American workers at the now-defunct City-Watcher surveillance company (VeriChip Corp., 2006) and officials with the Mexican Attorney General's office (Applied Digital Solutions, 2004) have been microchipped. Concern that the practice could spread has raised the specter of Big Brother and prompted lawmakers in three states to pass

DOI: 10.4018/978-1-4666-4582-0.ch013

laws preventing the forced or coerced implantation of microchips in human beings. California, Wisconsin, and North Dakota have all passed laws banning forced or coerced microchip implantation in human beings. See: California SB 362 (2007), Wisconsin AB 290 (2005), and North Dakota SB 2415 (2007).

There is now an ongoing debate regarding the safety of the chips. As a result of lobby pressure combined with heavy advertising by Schering Plough for its HomeAgain pet recovery system, close to 5% of the United States' estimated 164 million dogs and cats have now been chipped (Banfield the Pet Hospital, 2005). Animal shelters around the United States are routinely chipping dogs and cats before releasing them for adoption, and governments, including those of Portugal, Singapore, Bangkok, Los Angeles County, and El Paso, Texas, have passed ordinances requiring that all dogs under their jurisdiction be microchipped. El Paso has extended the chipping mandate to cats and ferrets.

In addition, horses around the nation are also being chipped, and the USDA recently approved the use of equine radio-frequency identification (RFID) injectable transponders as part of the National Animal Identification System (NAIS). The National Animal Identification System (NAIS) is a national premises registration, animal identification, and animal tracing program for owners of livestock. NAIS is a national program run by the United States Department of Agriculture (USDA), but is being implemented primarily at the state level.

As for human beings, an estimated 300 Americans and 2,000 people worldwide have been implanted with microchip transponders. This chipping apparently proceeded with the full consent of the implantees until early 2007, when the VeriChip Corporation began implanting Alzheimer's patients and their caregivers with microchips as part of a research study. These patients have reduced mental capacity and are unlikely to understand what is being done to them.

It appears that few people undergoing microchip implantation have been told about the potential health risks associated with the device. In fact, up until September 2007, almost three years after FDA approval, no mention had been made by the company or the FDA in relation to the well-established, though generally under-reported, finding that the microchip caused cancer in laboratory mice and rats.

Microchip-Induced Cancer in Mice and Rats

In at least six studies published in toxicology and pathology journals between 1996 and 2006, researchers found a causal link between implanted microchip transponders and cancer in laboratory mice and rats. The tumors were typically sarcomas, including fibrosarcomas. Other cancers found included rhabdomyosarcoma, leiomyosarcoma, malignant fibrous histiocytoma, mammary gland adenocarcinoma, malignant schwannoma, anaplastic sarcoma, and histiocytic sarcoma.

In almost all cases, the tumors arose at the site of the implants and grew to surround and fully encase the devices. In several cases the tumors also metastasized or spread to other parts of the animals, including the lungs, liver, stomach, pancreas, thymus, heart, spleen, lymph nodes, and musculature of the foreleg.

The tumors generally occurred in the second year of the studies, or after half a lifetime's exposure to the implant. At the typical time of tumor onset the animals were in middle to advancing age. The exception to this was the Blanchard (1999) study, in which genetically modified mice developed fast-growing cancers well before six months.

The percentage of mice and rats developing microchip-induced tumors in the six studies reviewed ranged from 0.8% to 10.2%. Several researchers, including Elcock et al. (2001), Le Calvez et al. (2006), and Tillmann et al. (1997) suggest that the actual rate of tumor formation may have been higher than was reported in their studies, since they examined only visible lesions and thus may have missed microscopic changes that signaled the onset of additional tumors around the implants.

Elcock et al. (2001) write, "It should be noted ... that these tumor incidences only approximated the potential incidence of microchip-induced tumors for these studies. The original intent of the studies was to characterize the toxicological profile of the chemical test substance in question, therefore tissue surrounding the animal-identification microchips was not examined microscopically unless there was a gross lesion. Thus, small pre-neoplastic or neoplastic lesions may have been missed" (p. 488).

A similar observation was made by Le Calvez et al. (2006). In their study 4.1% of animals developed visibly detectable tumors. However, researchers suspected the actual incidence of cancer may have been higher, had they looked at tissue samples. Tillmann et al. (1997) also write that "only implantation areas with macroscopic findings have been examined microscopically, so that possible pre-neoplastic lesions could have been missed" (p. 200).

Microchip-Induced Cancer in Dogs

In addition to the six studies that identified cancer in rodents, two studies evaluated cancerous tumors (fibrosarcoma and liposarcoma) that developed in dogs at the site of microchip implants. In one case, the tumor was attached to the implant. In the other case, the tumor completely encased the microchip.

Microchip Studies in Which No Cancer Was Found

Included in this review are three studies, one involving dogs, one involving rats, and one involving mice, in which none of the animals developed cancer from the microchip implant. Though these studies were originally presented as evidence that

implantable microchip devices were safe, they suffer from methodological limitations that call their statistical validity into question. These limitations include the small number of animals used and the short duration of the studies. Those issues are discussed at further length in this document.

Overall Cancer Incidence

Tables 1 and 2 summarize the results of the 11 studies reviewed in this Chapter. Table 1 lists the cancer incidence from eight studies where cancer was found in connection with a microchip implant. Table 2 lists details from the three studies in which no cancer was found.

Animals Used in the Research

Toxicology and carcinogenicity researchers rely on laboratory animals to help determine which substances are safe and which are potentially harmful. Since most substances that cause cancer in humans also cause cancer in mice and rats, these animals can serve as an early indicator that a substance may not be safe for use in humans.

Several different strains of laboratory mice and rats were evaluated in the rodent studies reviewed in this report and several breeds of dog were included in the dog studies reviewed. A listing of the animals involved in each research study has been provided in Table 3.

Animals in the first group of prior studies developed microchip-induced tumors. Animals in the second group did not develop tumors. The third group of studies pertain to dogs that developed cancer around or attached to microchip implants.

Rodents used in laboratory studies are specially bred for uniformity and hardiness. They are utilized in cancer studies for their ability to respond to carcinogenic substances while remaining relatively free from spontaneous tumors that are unrelated to carcinogenic test substances.

Table 1. Studies that found microchip-induced cancer

Author(s)	Species	# of Animals	Length of Implant Exposure	Developed Cancer
Le Calvez et al., 2006	mice	1,260	2 years	4.1%
Vascellari et al., 2006	dog	N/A	7 months (at age 9)	1 dog
Vascellari et al., 2004	dog	N/A	18 months (at age 11)	1 dog
Elcock et al., 2001	rats	1,040	2 years	0.8%
Blanchard et al., 1999	mice	177	6 months	10.2%
Palmer et al., 1998	mice	800	2 years	2.0%
Tillmann et al., 1997	mice	4,279	lifespan	0.8%
Johnson, 1996	mice	2,000	2 years	~1.0%

Table 2. Studies that did not find microchip-induced cancer

Author(s)	Species	# of Animals	Length of Implant Exposure	Developed Cancer
Murasugi et al., 2003	dogs	2	3 days	none observed
		2	3 months	
		2	1 year	
		2	3 years	
		1	6 years	
Ball et al., 1991	rats	10	2 weeks	none observed
		10	3 months	
		10	6 months	
		10	1 year	
Rao & Edmondson, 1990	mice	10	3 months	none observed
		10	15 months	
		74	2 years	
		39	< 2 years	

Table 3. Animals examined in the studies, identified by breed or strain

Author(s)	# of Animals	Type of Animal Studied	Developed Cancer
Le Calvez et al., 2006	1,260	B6C3F1 mice	4.1%
Elcock et al., 2001	1,040	Fischer 344 rats	0.8%
Blanchard et al., 1999	177	*p53+/-* transgenic mice	10.2%
Palmer et al., 1998	800	B6C3F1/CrlBR VAF/Plus mice	2.0%
Tillmann et al., 1997	4,279	CBA/J mice	0.8%
Johnson, 1996	2,000	B6C3F1 mice and CD1 ("albino") mice	~1.0%
Murasugi et al., 2003	9	Beagle; mixed breed dogs	none observed
Ball et al., 1991	40	Sprague-Dawley rats	none observed
Rao & Edmondson, 1990	140	B6C3F1 mice	none observed
Vascellari, 2006	1	French bulldog	1 dog
Vascellari, 2004	1	Mixed breed dog	1 dog

The B6C3F1 mouse was the most commonly used mouse in these studies, appearing in four of the eight rodent studies. The Handbook of Carcinogen Testing (Milman & Weisburger, 1994) states that National Toxicology Program studies use the B6C3F1 mouse almost exclusively for cancer research because of its desirable characteristics. The Handbook describes the mouse as "hardy, easy to breed, disease resistant, and [having] a low spontaneous tumor incidence at most sites" (p. 353).

The p53+/- mouse contains a genetic mutation in the p53 gene which normally sends protein to help repair damaged cells. In these mice, one allele, or portion of the gene has been deleted, thus increasing their susceptibility to cancer caused by genotoxins, or substances that damage genetic material. p53+/- mice are not known to develop spontaneous cancers in the first six months of life and are expected to only develop cancer in the presence of genotoxins. The high rate of cancer development around the microchip implant in p53+/- mice at less than six months suggests that the implant may have genotoxic attributes.

The CBA/J mouse is an inbred strain that is widely used as a general purpose laboratory animal. It suffers from hereditary blindness, making it of interest to vision researchers, and it is often selected for other studies because of its low incidence of mammary tumors (The Jackson Laboratory). The CD-1 (albino) mouse is described as a "general multipurpose model [for] safety and efficacy testing, aging, surgical model, [and] pseudopregnancy" (Charles River Laboratory, 2007, p. 15).

The Sprague-Dawley rat is described as "a general model for the study of human health and disease" and an "excellent model for toxicology, reproduction, pharmacology, and behavioral research areas." They have a life span of 2.5 – 3.5 years (Ace Animals, Inc., 2007).

The Fischer 344 rat is described as the "most widely used inbred rat strain, particularly for toxicology and teratology" studies (Simonsen Laboratories, 2007).

Microchips Used in the Research

The glass used to encapsulate the microchip is known as "bioglass," a material widely used in animal studies due to its insolubility and apparent biocompatibility (Vascellari et al., 2004). Bioglass is comprised primarily of "silicon, sodium, calcium, potassium, magnesium, iron, and aluminum" and has been classified in the silicon sodium group (Vascellari et al, 2004, p. 188; citing Jansen et al., 1999).

The microchip transponder comes prepackaged in a sterile 12-gauge injection needle attached to an implantation device supplied by the manufacturer. Once the transponder is embedded in the body, it can be interrogated by a reader device that emits radio-frequency energy. This energy stimulates the embedded transponder, causing it to emit a signal that is captured by the scanner and translated into an identification code.

The microchips used in these studies were obtained from several distributors, including BioMedic Data Systems, Inc., Destron Fearing, and Merial, as indicated in Table 4.

REVIEW OF STUDIES: MICROCHIP-INDUCED CANCER IN LABORATORY RODENTS AND DOGS: EIGHT STUDIES FROM 1996 TO 2006

Le Calvez et al., 2006

Subcutaneous microchip-associated tumours in B6C3F1 mice: A retrospective study to attempt to determine their histogenesis. -Experimental and Toxicologic Pathology. 2006; 57:255–265.

Table 4. Microchip implants used in the studies, identified by brand name or supplier

Author(s)	Microchip Used	Developed Cancer
Le Calvez et al., 2006	BioMedic Data Systems Inc.	4.1%
Elcock et al., 2001	BioMedic Data Systems Inc.	1.0%
Blanchard et al., 1999	BioMedic Data Systems Inc.	10.2%
Palmer et al., 1998	Unspecified	2.0%
Tillmann et al., 1997	BioMedic Data Systems Inc.	0.8%
Johnson, 1996	BioMedic Data Systems Inc.	~1.0%
Murasugi et al., 2003	LifeChip; Destron Fearing.	none observed
Ball et al., 1991	BioMedic Data Systems Inc.	none observed
Rao & Edmondson, 1990	BioMedic Data Systems Inc.	none observed
Vascellari, 2006	Merial Indexel® (Digital Angel)	1 dog
Vascellari, 2004	Merial Indexel® (Digital Angel)	1 dog

Most of the animals with microchip-associated tumors died prematurely ... due to the size of the masses [or] the deaths were spontaneous and attributed to the masses. (p. 258)

One of the most potentially serious disadvantages of the microchip implantation is the possibility that foreign-body-induced tumours may develop ... (p. 256)

Summary

Microchips were implanted into 1,260 experimental mice for identification purposes. Two years later, 4.1% of the mice had developed malignant (cancerous) tumors at the site of the microchip implantation (Table 5). The cancers were directly attributed to the microchips. In one subgroup, the cancer rate among the chipped mice was 6.2%.

Study Design and Key Findings

1,260 mice were separated into groups for use in three oral carcinogenicity studies. The first study involved 550 mice, 110 of which received only a microchip implant. The other 440 received a microchip implant along with a low, medium, or high dose of a chemical test substance in their feed.

Two years later, 34 of the mice (6.2%) had developed malignant (cancerous) tumors around or adjacent to the microchip. These tumors occurred across groups, appearing in control mice as well as mice that had received the ingested chemical. Researchers plainly identified the microchip as the cause of the tumors.

The second study involved 600 mice. 120 received only a microchip, while the other 480 received a microchip combined with varying doses of a chemical compound in their feed. Two years later, 14 out of the 600 mice (2.3%) had developed cancerous tumors related to the microchip. For the test group of 480 mice, these tumors were deter-

Table 5. Le Calvez et al. 2006 study summary

Author(s)	# of Animals	Species	Study Length	Developed Cancer
Le Calvez et al., 2006	1,260	mice	2 years	4.1%

mined to be unrelated to the ingested compound. In the third study, 110 mice were implanted with a microchip and received no other intervention. Four of these animals (3.6%) developed a tumor around the microchip.

The researchers suggest the actual cancer rate may have been higher than reported, as they tested for cancer only when visible abnormalities were seen in the mice. Smaller tumors in the early stages of development that were not yet visible to the naked eye may have been missed. According to the authors, "as these were only sampled and examined histologically when gross abnormalities were noted, it is possible that early reaction could have been missed. These incidences may therefore slightly underestimate the true occurrence" (p. 258).

Additional Findings

- All the cancerous masses found either contained the microchip or were adjacent to it. An empty capsule where the microchip had been was frequently identified as the origin of the tumor. The researchers wrote:

All sarcomas were characterized by a poorly delineated, non-encapsulated, densely cellular mass, located in the subcutis but frequently infiltrating the panniculus muscle and various layers of the skin with occasional ulcerations. A round-to-oval empty space of 2 mm diameter corresponding to the cast of the microchip was frequently seen and associated with a vestigial fibrous capsule and/ or a focus of necrosis. (p. 261)

- Tumors were initially identified by morphology as fibrosarcoma (17 cases), rhabdomyosarcoma (12 cases), leiomyosarcoma (2 cases), malignant fibrous histiocytoma (3 cases), mammary gland adenocarcinoma (2 cases), and other sarcomas (16 cases). Researchers later redefined the tumors as "sarcomas not otherwise specified (NOS) with a large myofibroblastic component" (p. 255) after additional testing. A sarcoma is a malignant tumor of soft tissue that connects, supports or surrounds other structures and organs of the body.

- Once initiated, the tumors grew rapidly. Most of the animals that developed microchip-associated tumors died prematurely as a result of the tumors.

- Four microchip-related cancers metastasized (spread) to the lungs, liver, stomach or pancreas.

- Many of the implants migrated from the original implantation site on the back of the mice to cause cancer at other locations in the body. Nineteen percent of the cancers found involved microchips that had migrated from the back to the limbs, abdomen, or head of the mice.

- A test procedure known as desmin staining found that the tumors often infiltrated nearby muscle tissue and that there was "an extensive cavernous network of capillaries within the tumour, especially around the hole left by the microchip." (p. 261)

Study Details

- The study was conducted at MDS Pharma Services in L'Arbresle, France.

- Animals used in the study were B6C3F1 mice from Charles River Laboratory.

- Microchip implants were from BioMedic Data Systems Inc. and were described as "hermetically sealed in a cylindrical inert glass capsule measuring 12 mm in length and 2 mm in diameter and partially covered on a length of 5 mm by a porous polypropylene polymer sheath as an antimigration measure." (p. 255)

Vascellari, Melchiotti, & Mutinelli, 2006

Fibrosarcoma with typical features of postinjection sarcoma at site of microchip implant in a dog: Histologic and immunohistochemical study. -Veterinary Pathology. 2006; 43:545–548

Reports on adverse reactions to vaccination and microchips are strongly encouraged to deepen the current knowledge on their possible role in tumorigenesis . . . the cause and effect relationship between exposure (injection) and outcome (sarcoma) is still to be defined and is a matter of discussion for experts. (p. 547)

Summary

A 9-year-old bulldog developed a cancerous tumor (fibrosarcoma) adjacent to a microchip implant approximately seven months after being implanted with the device (Table 6). Researchers attributed the tumor to either the microchip or to vaccinations at the site, and called for better reporting of adverse reactions to microchip implants and vaccinations.

Overview

In September 2003, Leon, a 9-year-old male French bulldog was implanted with a microchip for identification purposes. In April 2004 (8 months later) Leon's owner detected a lump measuring 3 cm x 3 cm (1.2 x 1.2 inches) in the implant area. The mass was surgically removed and subjected to laboratory analysis whereby it was identified as a high-grade infiltrative fibrosarcoma – a ma-

lignant and fast-growing form of cancer. It was found attached to the microchip. Leon later died from complications that his owner attributes to the cancer.

The microchip is implanted into dogs through an injection procedure involving a 12-gauge needle. The researchers suggest the tumor may be a form of post-injection sarcoma, involving an inflammatory reaction around an injection site that predisposes the tissues to tumor development. The researchers note that "irritation, inflammation, and/or wounds [promote] tumor development. Virtually anything that causes a local inflammatory reaction may potentially be responsible for neoplastic initiation [i.e., abnormal proliferation of cells]" (p. 546).

The authors attributed the cancer to either the microchip or to vaccinations the dog had received at the same site. They wrote: "It is difficult to establish which was the primary cause of the neoplastic growth, because the dog had received several rabies vaccines and the microchip was detected close to but not included in the mass" (p. 547).

The investigators conclude by stating that "reports on adverse reactions to vaccination and microchips are strongly encouraged to deepen the current knowledge on their possible role in tumorigenesis [causing tumors]," calling it "a matter of discussion for experts" (p. 547).

It should be noted that a complete physical exam found nothing other than the detected lump to indicate that Leon had developed cancer. No evidence of inflammation or sepsis were found at the site of the implant. Had Leon's owner not insisted on a microscopic evaluation of the unusual growth, his cancer might never have been detected.

Table 6. Vascellari et al. 2006 study summary

Author(s)	Animal Involved	Chip Exposure Time	Cancer Developed
Vascellari, et al., 2006	9-year-old French bulldog	7 months	Fibrosarcoma

Study Details

- The evaluation was conducted by Dr. Marta Vascellari of the Instituto Zooprofilattico Sperimentale delle Venezie at Viale dell'Universita in Legnaro, Italy, with associates Erica Melchiotti and Franco Mutinelli.

- The microchip was manufactured by Digital Angel, the parent company of the VeriChip Corporation, and distributed by Merial under the Indexel® brand, through Lyon, France. Digital Angel's website states: "Digital Angel manufactures implantable RFID chips used in pets around the world ... In Europe, our product is distributed by Merial in some countries under the Indexel® brand. For more information, visit merial.com." (Source: http:// www.digitalangelcorp.com/dac_pets.asp. Accessed July 23, 2007.)

- Merial's website states: "Merial is a world-leading animal health company. We are a forward-looking company with a proven track record, producing pharmaceutical products and vaccines for livestock, pets and wildlife." (Source: http://www.merial. com/our_company/index.asp. Accessed July 23, 2007.)

Vascellari et al., 2004

Liposarcoma at the site of an implanted microchip in a dog. -The Veterinary Journal. 2004; 168:188–190

The intact microchip was found completely embedded within the mass . . . [and] a diagnosis of low-grade liposarcoma was made. (p.188)

Veterinary surgeons are . . . encouraged to check the microchips that have been implanted in pets at least annually, such as when they come in for vaccinations, and report any adverse reaction. (p. 190)

Summary

An 11-year-old dog developed a cancerous tumor (liposarcoma) around a microchip that had been implanted approximately 19 months earlier. The tumor was removed and the dog recovered (Table 7).

Overview

In April 2000, a male mixed-breed dog was implanted with a microchip for identification purposes. In November 2001 (19 months later) the dog's owner detected a firm, painless lump at the implant site measuring 10 x 6 cm (approximately 4 x 2.5 inches). The lump was examined by a veterinarian who determined that the microchip was completely embedded within the mass.

In April 2003, the tumor was surgically removed under general anesthesia. Upon microscopic examination, it was identified as a malignant liposarcoma, an aggressive and invasive type of cancer that can metastasize to the lungs, liver, and bone. The researchers note that liposarcoma is uncommon in dogs. Prior to the surgery, the dog had shown no visible signs of cancer other than the unusual lump. Blood tests run on the dog, including a complete pre-operative blood count and serum biochemistry analysis, did not detect that the mass was malignant. Thoracic radiographs (chest X-rays) were also normal. Had there not been a microscopic evaluation of the unusual growth, the cancer might not have been detected.

Table 7. Vascellari et al. 2004 study summary

Author(s)	Animal Involved	Chip Exposure Time	Cancer Developed
Vascellari et al., 2004	11-year-old mixed breed dog	19 months	liposarcoma

Study Details

- The evaluation was conducted by Dr. Marta Vascellari and Franco Mutinelli of the Instituto Zooprofilattico Sperimentale delle Venezie, Histopathology Department, in Legnaro, Italy, together with veterinary surgeons Romina Cossettini and Emanuela Altinier of Porcia, Italy.

- The microchip was manufactured by Digital Angel, the parent company of the VeriChip Corporation. It is distributed by Merial under the Indexel® brand. Researchers state that the implant "consists of a sealed glass capsule containing a chip and a coil . . . [and is] equipped with an anti-migrational capsule, located in the anterior part of the microchip."

Elcock et al., 2001

Tumors in long-term rt studies associated with microchip animal identification devices. -Experimental and Toxicologic Pathology. 2001; 52:483–491

Electronic microchip technology as a means of animal identification may affect animal moribundity and mortality [i.e., illness and death rates], due to the large size and rapid growth of microchip-induced tumors as well as the occurrence of metastases. (p. 491)

Most tumors arising from foreign bodies are malignant . . . and have a rapid growth rate, killing the animal in a matter of weeks. (p. 491)

Summary

Microchips were implanted into 1,040 rats for identification purposes. After two years, just under 1% of the rats developed malignant tumors (malignant schwannoma, fibrosarcoma, anaplastic sarcoma, and histiocytic sarcoma) surrounding the implants. The researchers attributed the tumors to the presence of the microchip, and referred to them as "microchip-induced" (Table 8).

Study Design and Key Findings

A group of 1,040 rats was implanted with microchip transponders and then divided into two random groups. Half were exposed to an ingested chemical compound at high, medium, and low doses; the other half received no compound. By the end of the second year, eight of the rats that received the compound, or 0.77%, had developed malignant tumors at the site of the microchip implant.

Though the affected rats had all been dosed with a test substance, the tumor incidence was distributed across dose groups and showed no test-substance-related trends. Stated slightly differently, higher levels of chemical compounds in the animals' feed did not correspond to higher tumor rates.

Further clarifying that the tumors had arisen in response to the microchips, not the test compound, the investigators wrote: "the process of differentiating microchip-induced tumors from suspected compound-related tumors was fairly easy in the cases described here, for all contained the embedded microchip device" (p. 491).

Table 8. Elcock et al. 2001 study summary

Author(s)	# of Animals	Species	Study Length	Developed Cancer
Elcock et al., 2001	1,040	rats	2 years	0.8%

Additional Findings

- The microchip-induced tumors were identified as malignant schwannoma, fibrosarcoma, anaplastic sarcoma, and histiocytic sarcoma. All diagnoses were confirmed with immunohistochemistry.

- All masses were confined to the area of microchip implantation and contained embedded microchips.

- Some masses were extremely fast-growing, enlarging as much as 1 cm per week. Several tumors metastasized to regions including the lungs, thymus, heart, lymph nodes, and musculature.

- Five of the eight affected animals died as a direct result of the microchips.

- All tumors occurred in the second year of the study. The average age at tumor onset was 585 days, or approximately one year and seven months. (The average life span of a rat is two to three years.)

- The researchers write that: "Although the resulting tumor rate was observed to be low, the overall health of the affected rats was compromised due to tumor size and the occurrence of metastases, leading to early sacrifice" (p. 484). In other words, the animals' health was so poor due to large, malignant tumors spreading through their bodies that researchers were forced to kill them prematurely.

Study Details

- The study was conducted by Laura E. Elcock of Bayer Corporation in Stilwell, Kansas. Other investigators were Barry Stuart, Bradley Wahle, Herbert Hiss, Kerry Crabb, Donna Millard, Robert Mueller, Thomas Hastings and Stephen Lake. The results were peer-reviewed by an independent pathologist.

- Animals used were Fischer 344 laboratory rats.

- Microchip implants were from BioMedic Data Systems Inc.

Blanchard et al., 1999

Transponder-induced sarcoma in the heterozygous p53+/- mouse. -Toxicologic Pathology. 1999;27(5):519-527

There was an unequivocal association between the [microchip implant] transponder and sarcoma that was unrelated to drug treatment. (p. 526)

The presence of the foreign body [microchip transponder] may elicit tissue reactions capable of generating genotoxic byproducts. (p. 526)

Summary

177 genetically modified mice were implanted with microchips for identification purposes as part of a chemical compound study. After six months, 18 of the mice (10.2%) had developed malignant tumors ("undifferentiated sarcomas") around the microchip (Table 9). The tumors occurred in both experimental and control animals. The researchers reported an "unequivocal association" between the implants and the cancer.

Table 9. Blanchard et al. 1999 study summary

Author(s)	# of Animals	Species	Study Length	Developed Cancer
Blanchard et al., 1999	177	mice	6 months	10.2%

Study Design and Key Findings

A group of 177 transgenic p53+/- mice were implanted with microchips as part of a six-month study to investigate the toxicity of various chemical compounds. After six months, 18 of the mice (10.2%) developed malignant tumors ("undifferentiated sarcomas" p. 520) around the microchip. The tumors occurred in both control animals and animals that had received the test compound. The authors wrote that "these masses were not related to test substance administration; they were observed in controls as well as dosed animals" (p. 520).

Of the 177 total mice studied, 56 died before researchers made a link between the microchip and the tumors. The tissue surrounding the implants in the remaining 121 mice was microscopically analyzed.

Researchers discovered that the tumors arose at the microchip's plastic anchoring barb and then expanded to eventually surround the entire microchip. They state: "It appeared that tumor(s) arose in the mesenchymal tissue surrounding the polypropylene component of the transponder, initially involving the barbed area and then in some cases extending completely around the entire transponder site" (p. 523). Further, mass development was often observed to begin at the glass-polypropylene interface (p. 521).

The mice used in this study were transgenic p53+/- mice, specially bred to lack part of the tumor suppressor gene known as p53. In normal mice, p53 regulates cell growth and causes potentially cancerous cells to destroy themselves. Missing a part of this gene makes mice more susceptible to cancer from genotoxins, or toxic substances that affect genetic material. Despite their greater tendency to develop cancer when exposed to genotoxins, p53+/- mice typically do not develop spontaneous tumors in the absence of genotoxins. When they do develop tumors, it is generally an indication that a genotoxin is present.

The researchers write that:

[D]eletion of a single allele of this tumor suppressor gene in mice appears to be without effect on the development of spontaneous tumors, at least during the first year of life, but it imparts exquisite sensitivity to the mutational and carcinogenic effects of genotoxic chemicals. (p. 524)

The glass and polypropylene components of the BioMedic transponder device used in the study are generally assumed to be free from genotoxic materials (mutagenic and/or cytoxic components), so an observation of no tumors would be predicted by this model (p 525). Because the glass capsule and polypropylene sheath around the microchip implant are generally considered not to be genotoxins, the mice should not have responded to their presence by developing cancers. Researchers did not expect this outcome, writing: "the observation of transponder implantation site sarcomas in 18/177 (10%) of the animals studied was surprising."

Additional Findings

- "Membrane endothelialization, inflammation, mesenchymal basophilia, dysplasia, and sarcoma were considered unequivocal [unmistakable] responses to the transponder" (p. 523).

- The masses increased in size rapidly. One mass measuring ½" wide in the fifteenth week of the study grew to 2" just ten weeks later (p. 520).

- The researchers "have subsequently replicated this finding in 2 separate studies with the p53+/- mouse where transponder implantation site sarcomas were also observed." Their article does not indicate whether these studies have been published.

Study Details

- The study was conducted by Kerry Blanchard, Curt Barthel, Henry Holden, Roger Moretz, Franklin Pack, and Raymond Stoll of the Department of Toxicology and Safety Assessment at Boehringer Ingelheim Pharmaceuticals in Ridgefield, Connecticut, along with John French and Raymond Tennant of the Laboratory of Environmental Carcinogenesis at the National Institute of Environmental Health Sciences in North Carolina..

- Animals used were transgenic p53+/- mice, specially bred to lack part of the tumor suppressor gene known as p53. These mice have an increased susceptibility to cancer from genotoxins (compounds which affect genetic material) but are not known to develop tumors spontaneously in the absence of a carcinogen.

- Microchips used were IMI® implants from BioMedic Data Systems. The microchip is described as encased in a glass capsule and partially encased in a polypropylene sheath.

Palmer et al., 1998

Fibrosarcomas associated with passive integrated transponder implants. -Toxicologic Pathology. 1998;26:170

All tumors were observed . . . at or near the implantation site . . . [the tumors] were attached to the implant or partially or totally encased the implant. (p. 170)

Summary

800 mice were implanted with microchips for identification purposes. After two years 2% of the mice had developed cancerous tumors (malignant fibrosarcomas) around the implants (Table 10).

Study Design and Key Findings

The article is a short, one-page writeup, around 350 words in length. The following is known based on the information provided:

800 mice were implanted with a microchip transponder for identification purposes as part of "a 104-week dietary study" lasting two years. Between weeks 79 and 105, 16 of the mice developed "subcutaneous tumors associated with the implanted transponder." The tumors occurred in both control and treated animals and were judged unrelated to the test material. The tumors were identified as malignant fibrosarcomas.

All of the tumors occurred at or near the implantation site and were "attached to the implant or partially or totally encased the implant." The larger tumors commonly had areas of necrosis and hemorrhage with inflammation, and some of the tumors invaded adjacent skeletal muscle. In addition, two of the mice developed metastases in which the cancer spread either to the lymph nodes or to the lungs.

Study Details

- The study was conducted by T. Palmer, J. Nold, M. Palazzolo, and T. Ryan at Covance Laboratories, Inc. in Madison, Wisconsin.

- Animals used were B6C3F1/CrlBR VAF/ Plus mice.

Table 10. Palmer et al. 1998 study summary

Author(s)	# of Animals	Species	Study Length	Developed Cancer
Palmer et al., 1998	800	mice	2 years	2.0%

- Microchips used are identified as "passive integrated transponder implants used for identification." No additional information is provided.

Tillmann et al., 1997

Subcutaneous soft tissue tumours at the site of implanted microchips in mice. -Experimental and Toxicologic Pathology. 1997; 49:197 – 200

The neoplasms induced in the present investigation are clearly due to the implanted microchips. (p. 200)

Further information on [tumors] induced by microchips, e.g., experiments on their chemical components (glass and polypropylene cap), or the physical presence of the implant alone are necessary. (p. 200)

Summary

4,279 mice were injected with microchip implants for identification purposes. Of these, 36 developed malignant tumors (fibrosarcoma and malignant fibrous histiocytoma) that were "clearly due to the implanted microchips" (p. 200). Control animals as well as experimental animals developed the tumors (Table 11).

Study Design and Key Findings

4,279 CBA/J mice were implanted with microchips for identification purposes as part of a study examining the influence of X-ray radiation and chemical carcinogen exposure on offspring. A sample of male mice was exposed to these carcinogens once or twice, then mated with untreated females. Their offspring were then studied to see if they had increased cancer susceptibility.

By the conclusion of the study, 36 of the mice had developed tumors around the microchip. Implant-related tumors were identified as fibrosarcomas with "extensive local invasion of the surrounding tissues" and malignant fibrous histiocytoma with "zones of necrosis and high mitotic activity" (p. 198).

Significantly, twice as many females developed cancers as male mice, though the females had not been exposed to the experimental treatment. 1.2% of the females and 0.5% of the males developed tumors in the chip implantation area. The authors wrote that "the different generation and treatment groups showed no influence on tumour incidence," meaning that the tumors were unrelated to the x-ray treatment or other experimental factors.

The authors caution that the study may have underestimated the actual rate of tumor formation, since only tumors that were visible to the naked eye were examined microscopically. Tumors at an earlier stage of development may have been missed.

Study Details

- The study was conducted by Thomas Tillmann, Kenji Kamino and Ulrich Mohr at the Institute of Experimental Pathology at the Hannover Medical School in Hannover, Germany. Other researchers included C. Dasenbrock, H. Ernst, and

Table 11. Tillmann et al. 1997 study summary

Author(s)	# of Animals	Species	Study Length	Developed Cancer
Tillmann et al., 1997	4,279	mice	lifespan	0.8%

G. Moraweitz of the Fraunhofer Institute of Toxicology and Aerosol Research in Hannover, Germany; E. Campo and A. Cardesa of the Department of Anatomic Pathology at the University of Barcelona in Barcelona, Spain; and L. Tomatis of the Instituto per L'Infanzia in Trieste, Italy.

- An acknowledgment at the end of the article states: "This study was supported by the European Union: EV5V-CT92-0222."
- Animals used in the study were CBA/J mice.
- The implants used were "glass-sealed devices with a polypropylene cap" obtained from BioMedic Data Systems, Inc. (European distributor PLEXX BV, Elst. The Netherlands).

Johnson, K., 1996

Foreign-body tumorigenesis: Sarcomas induced in mice by subcutaneously implanted transponders. -Toxicologic Pathology. 1996; 33(5):619. Abstract #198

Investigators using ... implanted devices need to be aware of foreign-body tumorigenesis [cancer development] when evaluating the results of long term studies using mice.

Summary of Study

A two-year Dow Chemical study of 2,000 mice found an approximately 1% incidence of sarcomas surrounding microchip implants used for identification purposes (Table 12). The tumors appeared in both control and experimental animals. This was consistent with a diagnosis of foreign-body-induced sarcoma.

Study Design and Key Findings

This report was based upon a series of five oncogenicity (cancer) studies involving 2,000 B6C3F1 mice and CD1 ("albino") mice. Each study consisted of 400 mice that had been implanted with a microchip for identification purposes: 300 of the mice received test chemicals in their feed at low, medium, and high dose levels, and 100 control mice received no test chemical. After two years, just under 1% of the mice developed "incidental" subcutaneous sarcomas that incorporated the implanted microchip. Both treated and control animals developed the tumors at approximately the same rate, ruling out the test substance as the cause of these tumors.

The tumors were identified as connective tissue cancers, or fibrosarcomas, and appeared typical of foreign-body-induced sarcomas. The tumors typically appeared after more than one year post-implantation. Only gross lesions were examined.

In a telephone interview, Johnson (personal communication, October 13, 2007) also reported occasional adverse events related to the microchips, which were implanted between the shoulder blades. "Occasionally some would be inserted too deep, the needle that put them in was probably held at the wrong angle. We had a few early in the studies that would migrate out if the wound wasn't healing properly, and we had a few that gave up functioning, but those were all pretty rare events," he said.

Table 12. Johnson 1996 study summary

Author(s)	# of Animals	Species	Study Length	Developed Cancer
Keith Johnson, 1996	2,000	mice	2 years	~1.0%

Study Details

- The research was conducted at the Toxicology Research Laboratory, The Dow Chemical Company, Midland, MI by Keith Johnson.
- Animals used in the study were B6C3F1 mice and CD1 ("albino") mice.
- Microchip implants were from BioMedic Data Systems Inc.

MICROCHIP STUDIES IN WHICH NO CANCER WAS FOUND

Murasugi et al., 2003

Histological reactions to microchip implants in dogs -The Veterinary Record. 2003 (Sept 13); 328

As the mean lifespan of dogs as companion animals increases, long-term evaluation of the safety and biological stability of implants is necessary. (p. 328)

Summary

Nine dogs were implanted with microchips and observed for adverse outcomes over periods of three days to three years. One dog was exposed to the implant for six years. The chips and surrounding tissue were removed and examined microscopically (Table 13). Inflammation and encapsulation had occurred, but no tumors or cancerous changes were found.

Study Design and Key Findings

Nine dogs (one female beagle, six female crossbreeds, and two male crossbreeds) were implanted with Destron Fearing LifeChip microchips. At selected time periods, the implants and a surrounding 2x2x2 cm cube of tissue were surgically removed from each dog and microscopically evaluated. The evaluations took place on the following schedule (Table 14).

After three days, a rim of inflammatory cells, blood congestion, and newly formed capillaries had developed around the implants. At three months, a capsule composed of connective tissue, elastic and collagen fibers had surrounded the implant. At twelve months, the encapsulation was complete and no inflammation was observed. The evaluations at 36 and 72 months were similar to those made at 12 months.

The researchers summarized these findings as follows:

a foreign body reaction to the subcutaneously implanted microchips was observed [initially] ... followed by ... the development of a thin capsule in close contact with the microchip. The inflam-

Table 14. Key findings in Murasugi et al. 2003

# of Dogs Evaluated	Length of Microchip Exposure
2	3 days
2	3 months
2	1 year
2	3 years
1	6 years

Table 13. Murasugi et al. 2003 study summary

Author(s)	Species	# of Animals	Length of Microchip Exposure	Developed Cancer
Murasugi et al., 2003	dogs	6	≤ 1 year	none observed
		3	3–6 years	

matory reactions disappeared three months after implantation, and enclosure of the microchip by a capsule consisting of fibroblasts, collagen fibres and elastic fibres was complete after 12 months. No marked difference was observed . . . 36 or 72 months after implantation, compared with those 12 months after implantation. (p. 329)

The researchers concluded that "[t]hese findings suggest that implanted microchips are likely to function safely throughout a dog's lifetime, without causing further histological [microscopic] changes".

Concern over the Statistical Validity of the Study Findings

Although the authors conclude that "implanted microchips are likely to function safely throughout a dog's lifetime", the absence of cancerous changes in a small sample of dogs exposed to microchips for a limited period is not sufficient evidence to conclude that microchip implants are safe for long-term use. Problems with this study include the small number of dogs examined and the short time of their exposure to the microchip.

A small sample size of just nine dogs lacks the statistical power to detect an effect that may be *in the order of a percentage point or less. Statistical validity* is the degree to which an observed result, such as a difference between two measurements, can be relied upon and not attributed to random error in sampling and measurement (National Women's Health Resource Center). *Sample size* is what gives a study statistical power, or accurate and valid predictive ability.

Dr. Elise Whitley and Dr. Jonathan Ball (2002), experts on medical statistics, explain the importance of sample size in medical studies designed to prove the safety of a device and rule out an adverse effect. They write:

The ideal study for the researcher is one in which the power is high. This means that the study has a high chance of detecting a difference between groups if one exists; consequently, if the study demonstrates no difference between groups the researcher can be reasonably confident in concluding that none exists in reality. The power of a study depends on several factors, but as a general rule higher power is achieved by increasing the sample size.

It is important to be aware of this because all too often studies are reported that are simply too small to have adequate power to detect the hypothesized effect. In other words, even when a difference exists in reality it may be that too few study subjects have been recruited . . . the erroneous conclusion may [then] be drawn that there is no difference between the groups. This phenomenon is well summed up in the phrase, 'absence of evidence is not evidence of absence'. In other words, an apparently null result that shows no difference between groups may simply be due to lack of statistical power, making it extremely unlikely that a true difference will be correctly identified." [Emphasis added]

In this case the "difference" described is the difference between the rate of cancer formation in dogs that have and have not been microchipped. The present study assumes that the difference between these populations is zero or non-existent, but the sample size lacks the statistical power to draw that conclusion.

To determine whether microchips are safe in dogs would require the statistical power of a much larger sample, in the order of hundreds or even thousands of dogs. Although such studies have not yet been conducted, researchers could draw on the existing population of microchipped dogs in the United States to reach more statistically valid conclusions about the implant's safety and long term effects.

In addition to the small sample size used, a further problem with this study is the short duration of time the dogs were in contact with the implants. Of the nine dogs studied, six had the implant removed within a year or less and only one dog retained the implant for six years. The researchers do not state the age of the dogs at the time they were implanted.

In mouse and rat studies, the onset of microchip-induced cancer typically did not occur until the second year after implantation. Very few tumors were seen in the first year of the study when the animals were in adolescence and early adulthood; most tumors arose during middle age and older for those animals. If dogs develop adverse microchip reactions at a comparable rate, we would not expect to see an onset of tumors in dogs until they, too, reached middle age and beyond. This would correspond to roughly six years of age, given that the average life span of the domestic dog is 12.8 years. When looking at dogs, it is important to take into account the wide variation in life span across breeds, with the average bulldog living just nine years, while the average chihuahua has a 15 year life expectancy (McCullough, 2007).

The two microchip-induced cancers reported in dogs (Vascellari et al., 2006, 2004) occurred in 9-year-old and 11-year-old dogs after exposure times of seven months and 19 months, respectively. Given the small number of reported cases, it is difficult to draw conclusions about the development of microchip-induced tumors in dogs, but it could be that older dogs are more susceptible to the possible cancer-inducing effects of implants than younger dogs. Future research could help determine the role of an animal's age and the duration of microchip exposure.

Study Details

- The study was conducted by E. Murasugi, H, Koie, M. Okano, T. Watanabe, and R. Asano, of the Department of Veterinary Medicine, College of Bioresource Sciences at Nihon University in Fujisawa, Kanagawa, Japan.

- An acknowledgment at the end of the article states, "We would like to thank Dainippon Pharmaceutical for providing the microchips".

- Microchip implants were described as "LifeChip injector; Destron Fearing. The microchips were approximately 2 mm in diameter and 11 mm long and contained an IC recording a unique identity number . . . [the microchips] are made of biocompatible glass and polypropylene".

Ball et al., 1991

Evaluation of a microchip implant system used for animal identification in rats. -Laboratory Animal Science. 1991;41(2):185—186

Summary

40 rats were implanted with subcutaneous microchips and evaluated for adverse reactions. The tissue surrounding the implants was evaluated after periods ranging from two weeks to one year. No palpable masses or visible tissue reactions were observed (Table 15).

Study Design and Key Findings

This was one of the original studies undertaken to evaluate what was then referred to as "a new microchip-based animal identification system" being marketed to laboratory researchers by Bio-Medic Data Systems, Inc. The goal of the study was to evaluate the safety and effectiveness of implanted microchip transponders for laboratory animal identification.

For this study, 20 male and 20 female Sprague-Dawley rats were injected with microchip implants and observed for adverse reactions. At weeks 2, 12, 26, and 52, five rats of each sex were sacrificed

Table 15. Ball et al. 1991 study summary

Author(s)	Species	# of Animals	Length of Microchip Exposure	Developed Cancer
Ball et al., 1991	rats	10	2 weeks	none observed
		10	3 months	
		10	6 months	
		10	1 year	

(killed). The microchips and surrounding tissue from each rat were examined macroscopically and through histopathologic examination.

Although the researchers reported the development of "thin rims of immature fibrous connective tissue with occasional subacute inflammatory cells present in the subcutis 2 weeks after implantation" (p. 185-186) and later found that "very thin rims of mature fibrous connective tissue were seen surrounding the implant sites" (p. 186) they did not find any cancerous changes. They concluded that the implant was a "reliable, easy-to-use, non-adverse identification system" (p. 186).

Concern over the Design and Statistical Validity of the Study

Although the authors conclude that the implanted transponders "produced no adverse clinical or histopathological side effects in the rats," the findings must be evaluated in light of the short time span for which the rats were implanted and the small sample size used.

Of the 40 rats used in this early study, none were in contact with the implants for longer than one year. Later researchers, however, found that cancerous tumors generally occur in the second year of exposure. When Elcock et al. (2001) examined a much larger sample of rats (n = 1,040), for example, they found a nearly 1% incidence of microchip-induced cancer, all of which occurred during the second year of the study. The average age of the animals at tumor onset in that study was 585 days, or approximately one year and seven months. Johnson (1996) similarly found that

tumors in mice develop during the second year of exposure. The only exception to the late onset of tumors in the studies reviewed here was the Blanchard et al. (1999) study in which 10.2% of mice developed cancer within six months of implantation. These findings were atypical, however, and may be attributable to the type of genetically altered mouse used in that study.

The absence of cancerous tumors in the present study — in which animals were examined after only 2 weeks, 3 months, 6 months, and 1 year of implant exposure — is in accord with the findings of other researchers. It is neither surprising nor anomalous, nor does it rule out the possibility that microchip-induced tumors may develop in rats after a longer exposure period.

Another problem with this study is the small number of animals that were evaluated. A sample size of 40 rats lacks the statistical power to detect a small effect. This was the case in the Murasugi et al. dog study discussed earlier, and the same discussion of sample size and statistical power is applicable here.

When Elcock et al. (2001) conducted a subsequent study using a much larger sample of Fischer 344 rats (n = 1,040), they found a nearly 1% incidence of tumor formation. Due to the larger sample size, those results have greater statistical validity than those of the present study.

Study Details

- The study was conducted by D.J. Ball from Boehringer Ingelheim Pharmaceuticals, Inc. in Ridgeford, Connecticut, and associates. Additional authors include G. Argentieri, R. Krause, M. Lipinski, and R. I. Robinson from the Sandoz Research Institute of East Hanover, New Jersey; R.E. Stoll from Cetus Corporation of Emeryville, California; and G.E. Visscher from Roche Dermatologics in Nutley, New Jersey.

- The researchers thanked BioMedic for contributing to the study: "We would like to thank BioMedic Data Systems, Inc. of Maywood, N.J. for the implants and associated electronic equipment".
- Animals used were Sprague-Dawley rats.
- Microchips used were from BioMedic Data Systems, Inc., Maywood, New Jersey. The chip was described as a miniature transponder hermetically sealed in an inert glass capsule with a polypropylene sheath that covered one end of the transponder.

Rao and Edmondson, 1990

Tissue reaction to an implantable identification device in mice. -Toxicologic Pathology. 1990; 18(3):412–416

Summary of Study

140 mice were implanted with subcutaneous microchips and evaluated for adverse reactions. The tissue surrounding the implants was examined after periods ranging from three months to two years. No neoplastic (abnormal tissue growth) reactions were observed (Table 16).

Study Design and Key Findings

The study was published in 1990, when implantable microchips were first being introduced to laboratories for animal identification purposes. The goal of the project was to "determine the tissue reaction [from the implant], especially its potential to cause subcutaneous sarcoma, and the stability and reliability of a glass-sealed permanent identification device" implanted in mice (p. 412 – 413).

Researchers implanted 140 B6C3F1 mice with a microchip at approximately six weeks of age. Ten mice of each sex were evaluated at 3 months and at 15 months. The remaining animals were evaluated either as they died or upon being sacrificed at 24 months.

Histologic examination presented a connective tissue capsule of variable thickness around most of the implants, especially in the area of the glass surface of the chips. Around the polypropylene cap of the transponder, inflammatory reactions were detected but no neoplasms observed. From a summary of the Rao & Edmondson study included in Tillmann et al. (1996), p. 200. The capsule that formed around the polypropylene cap of the device contained minimal to mild inflammatory reaction with lymphocytes, macrophages, and a

Table 16. Rao & Edmondson 1990 study summary

Author(s)	Species	# of Animals	Length of Microchip Exposure	Developed Cancer
Rao & Edmondson, 1990	Mice	20	3 months	none observed
		20	15 months	
		72	2 years	
		28	Less than 2 years*	

* Evaluated prior to study conclusion due to death of the animals

few plasma cells and neutrophils. Researchers noted that "Chronic granulomatous inflammation . . . was also observed around the polypropylene cap of 2 implants" (p. 414).

Though no cancer was found, there were other problems with the implants. According to the researchers, two of the implants were "lost" and four of the devices "failed." Three of these failures were attributed to microscopic cracks in the weld connecting the antenna leads to the microchip, and one was caused by "leakage of the glass capsule resulting in fluid accumulation around the microchip" (p. 413). One device lodged in the subcutaneous tissue over the lumbar vertebrae and was pushed out slowly through the scar tissue of the injection site during the tenth month of the study.

In addition to the lost or failed transponders, seven of the transponders were discovered in the abdominal cavity of the animals rather than in the subcutaneous tissue where they should have been located. Researchers did not know whether the devices had migrated into the abdominal cavity and eventually fixed in the perirenal tissue, or whether lab technicians had accidentally injected the devices into the abdomen.

Concern over the Design and Statistical Validity of the Study

Given the small sample of animals exposed to the microchip for a full two years, this study may suffer from similar statistical validity problems as the Murasugi et al. (2003) and Ball et al. (1991) studies previously discussed.

Tillmann et al. (1997) point out this deficiency in their writeup, stating that the lack of tumor findings by Rao and Edmondson could be explained "by the low number of 140 B6C3F1 mice used by Rao and Edmondson" (p. 200).

Study Details

- The study was conducted by Ghanta Rao and Jennifer Edmondson at the Division of Toxicology Research and Testing at the National Institute of Environmental Health Sciences National Toxicology Program in North Carolina.
- Animals used were B6C3F1 mice.
- Microchips used were obtained from BioMedic Data Systems, Inc. They were described as a glass sealed 12 x 2 mm cylindrical device with a snug-fit biocompatible polypropylene cap covering a 5 mm length of the device. There are two holes in the polypropylene cap. The purpose of the polypropylene cap is to elicit mild tissue reaction and immobilize the device at the site of the implantation (p. 413).

DISCUSSION, RECOMMENDATIONS, AND CONCLUSION

Discussion

Cancerous tumors formed around or adjacent to implanted microchips in eight of the 11 studies reviewed in this report. In six of those studies, researchers clearly identified a causal link between the implanted microchip transponder and cancer. In three studies where cancer was not found, methodological shortcomings undermined the studies' validity. Either too few animals were studied to draw a valid conclusion, or the animals were not in contact with the microchip long enough for tumors to develop, in the way predicted by other models.

The tumors generally occurred in the second year of the studies, after more than one year of exposure to the implant. At the typical time of tumor onset the animals were in middle to advancing age. The exception to this was the Blanchard

(1999) study, in which genetically modified mice developed fast-growing cancers well before six months.

In almost all cases, the tumors arose at the site of the implants and grew to surround and fully encase the devices. In several cases the tumors also metastasized to other parts of the animals, including the lungs, liver, stomach, pancreas, thymus, heart, spleen, lymph nodes, and musculature.

In addition to the tumors, researchers described other adverse reactions stemming from the use of the microchips, including migration, incorrect insertion, failure, and loss. These adverse reports appeared in studies which did and which did not find cancer.

Issues related to the studies, including several proposed explanations for the cancer findings, the breed and species of animals used, the relevance of this research to implanted microchips in human beings, other adverse reactions reported in the studies, and the possible under-reporting of cancer and other adverse events are discussed in detail next.

Explanations for the Tumors

At the present time, there is no definitive, universally accepted explanation for the formation of malignant tumors around implanted microchips in mice, rats, and dogs. Among some of the explanations that have been proposed are foreign-body tumorigenesis; post-injection sarcoma; possible genotoxic properties of the implant; and the radio-frequency energy emissions from the transponder or reader. Each hypothesis is addressed in this section.

Foreign-Body Tumorigenesis

The presence of the microchip, a subcutaneous foreign body, may cause cellular changes that can lead to cancer.

It is known that implanted foreign bodies can cause cancer both in animals and humans. Mc-

Carthy et al. (1996) reported on a liposarcoma in a dog where a glass foreign body had lodged 10 years previously. Brand and colleagues (1975) observed that rodents are particularly susceptible to developing tumors in response to foreign bodies and produced a large body of research on the topic. Compelling evidence indicates that foreign-body tumorigenesis is also operative in humans (Jennings et al., 1988), as discussed later in this paper.

Foreign-body-induced tumors can pose serious threats to animal health. Elcock et al. (2001) report from their review of the literature that most tumors arising from foreign bodies are malignant mesenchymal neoplasms with a rapid growth rate, killing the animal in a matter of weeks (p. 491).

Brand's research revealed that the size and surface of the foreign body are the key characteristics affecting tumor development. Although it may seem counter-intuitive, prior research shows that foreign bodies with smooth, continuous surfaces are actually more carcinogenic than those with rough, scratched, or porous surfaces.

The surface of the foreign body determines, in part, the length of the period of active inflammation. Rough, irregular surfaces have a longer active inflammatory phase before the foreign body is encapsulated in fibrous tissue. The extended period of inflammation is associated with lower rates of tumor development. In contrast, smooth surfaces have a shorter inflammatory period and thus are more likely to lead to tumors (Elcock et al., 2001, p. 490).

The microchip implant has both a smooth, homogeneous surface in the glass capsule and a rougher portion coated in the polypropylene sheath that is "characterized by scratches, ridges, and other irregularities" (Ball et al.).

In relation to the microchip implant, Elcock et al. write: "A chronic foreign body such as the electronic microchip, surrounded by a rim of mature fibrous connective tissue with little or no active inflammation may ... be more tumorigenic than one with ongoing active inflammation" (p. 490).

On the basis of these prior observations from the literature, it might be predicted that the cancer would form around the smooth portion of the implant first. However, Blanchard et al. (1999) reported that tumors in their study arose at the microchip's "plastic anchoring barb" and then expanded to eventually surround the rest of the device. They write: "It appeared that tumor(s) arose in the mesenchymal tissue surrounding the polypropylene component of the transponder, initially involving the barbed area and then in some cases extending completely around the entire transponder site" (p. 523). Further study is needed to better understand this issue.

Post-Injection Sarcoma

Inflammation from the chip-injection procedure may cause cellular changes that can lead to cancer.

[I]rritation, inflammation, and/or wounds [promote] tumor development. Virtually anything that causes a local inflammatory reaction may potentially be responsible for neoplastic [cancer] initiation (Vascellari et al., 2006, p. 546).

The microchip implant procedure involves the insertion of a 12-gauge needle into an animal's flesh to deliver the device. That procedure alone may be problematic, as research indicates that inflammation resulting from injections can predispose tissues to developing cancer. The resulting malignancies are known in the veterinary literature as post-injection sarcomas.

Vascellari et al. (2006) suggest that the tumor they evaluated in a French bulldog may have been this type of post-injection sarcoma, caused either by the injection of the microchip or by injection of vaccines that the dog received at the same site.

In light of the potential for post-injection sarcomas to develop in dogs, it would seem prudent to reduce inflammatory injection reactions in dogs (and cats) as much as possible. Given these

findings, veterinarians should identify the location of microchip implants in chipped animals and avoid using the same site for vaccinations or other injections.

Possible Genotoxic Properties of the Implant

The glass capsule or polypropylene sheath surrounding it may have carcinogenic or genotoxic properties, or its presence within the host may give rise to genotoxic byproducts. In the Blanchard study over 10% of p53+/- mice developed malignancies around the implants. This finding puzzled the researchers, as the mice they used were genetically modified to develop tumors specifically in response to mutagens and genotoxins (toxic substances that affect genetic material). However, the component materials of the transponders are "widely used in genotoxicity studies" and are not known to be mutagens or genotoxins.

This discrepancy suggested to the researchers that something other than a foreign-body reaction or an injection response may be involved in the microchip-induced cancers they found. The researchers suggest that "the presence of the foreign body may elicit tissue reactions capable of generating genotoxic byproducts." They provide technical descriptions of several processes through which this may occur on page 526 of their study.

It is unclear whether the suspected genotoxic byproducts were produced by the implant directly or through processes occurring in the surrounding tissues of the host animals – or a combination of the two. The mice used in the Blanchard study were genetically modified to lack a portion of the p53 gene that normally aids in the repair of damaged cells. The higher rate of malignancy seen in these animals may result from their inability to repair cellular damage resulting from the implant.

The Blanchard report does not evaluate the biocompatibility of the polypropylene polymer sheath, but it does note that the observed tumors arose in the tissue surrounding the polypropylene component of the transponder. (As previously noted, the tumors began at the microchip's plastic anchoring barb and expanded to eventually surround the rest of the device.) This suggests another possibility: that "leachates," or substances leaching from the implant into the surrounding tissue, may be involved in the tumorigenesis (Blanchard, et al., p. 525).

A literature review to assess the safety of the polymer sheath was beyond the scope of this report but would contribute to a fuller discussion of microchip-induced tumors.

Radio-Frequency Energy Emissions from the Transponder or Reader

The radio-frequency energy involved with the transponder may somehow contribute to tumor formation. Blanchard et al. also raised the possibility that "energy from the signal transmitted by the transponder [may be] carcinogenic" (p. 525). Though there is a tendency to think of the glass encapsulated transponders as biologically inert, the reality is that these implants are radio-frequency energy transponders designed to pick up and amplify electromagnetic radiation (EMF) within the body. The long-term effects of having a reactive, foreign-body capsule in the body designed to absorb and respond to electromagnetic energy are unknown.

Based on a review of published accounts, it appears the role that EMF radiation may play in the development of microchip-induced tumors has not been well studied. Blanchard et al. believe that "these variables warrant further examination" (p. 525).

Differences between Species

An important factor to consider when interpreting animal studies is whether findings in one breed or species of animal are applicable to other animals or to humans. This section examines that issue.

Possible Difference in Tumor Susceptibility between Different Strains of Mice

In studies where microchip-induced malignant tumors were found, the percent of mice affected ranged from a low of 0.8% in the CBA/J mouse to a high of 10.2% in the p53+/- mouse. This wide variation suggests that different strains of mice may have different degrees of susceptibility to cancer from the implants.

Le Calvez et al. (2006), Palmer et al. (1998), and Elcock et al. (2001) all suggest a strain difference, with Palmer and Elcock observing that no implant-induced sarcomas have been reported in the CD-1 mouse strain, for example. However, Elcock et al. suggest that it may be difficult to rule out cancer in the CD-1 mice studied, since "small pre-neoplastic or neoplastic lesions may have been missed" in the absence of microscopic evaluation (p.489).

Johnson (1996), whose study of both B6C3F1 mice and CD1 mice found a ~1% overall incidence of microchip-induced tumors, believes that CD1 mice in his study "probably did" develop foreign-body sarcomas around the implanted microchips, writing in a 2007 email correspondence: "I do not specifically recall whether or not CD-1 mice developed foreign-body sarcomas around implanted microchips. I believe they probably did, but at slightly lower incidence than B6C3F1 mice, as our experience was that CD-1 mice were somewhat shorter lived (due a disease named systemic amyloidosis) and these tumors were generally seen after a long time on study" (Personal communication, October 15, 2007).

Nonetheless, it appears that different strains of mice may develop microchip-induced cancers at differing rates.

Tumor Susceptibility across Species

It has long been observed that different species have differing levels of susceptibility to foreign-body tumors. As reported in Rao and Edmondson (1990) who cite Brand, KG (1982), evaluation of prior research shows that mice, rats, and to some extent, dogs are more susceptible to foreign body tumorigenesis than guinea pigs, chickens, and hamsters, for instance.

The fact that rodents and dogs have developed cancer in response to implants does not necessarily mean that humans will do the same. Blanchard et al. caution that "blind leaps from the detection of tumors to the prediction of human health risk should be avoided" (p. 526). In humans, fibrotic scar formation proceeds at a much slower rate than in rodents, which might indicate that humans are more resistant to foreign-body-induced tumors than rats and mice, suggest Elcock et al. (p. 491).

Humans are Susceptible to Foreign-Body Carcinogenesis

Nevertheless, according to Elcock's summary of the literature on foreign-body tumorigenesis, any inert substance inserted into the body for long periods can produce neoplasia (abnormal tissue growth), including in humans (p. 489). Vascellari et al. (2004) note that foreign-body-induced sarcomas, including osteosarcomas, rhabdomyosarcomas, hacmangiosarcomas, and liposarcomas, have been described in humans, although with a low prevalence (p. 190).

Most of the malignant, microchip-induced tumors in rodents reviewed in the present report were classified as sarcomas – soft tissue cancers that afflict the muscles, tendons, fibrous tissues, fat, blood vessels, and nerves. The following is a brief description of this type of cancer in human beings from Blake Morrison (2003) of Baylor University Medical Center:

Soft tissue sarcomas are a diverse group of neoplasms that arise in the connective tissues throughout the body. They account for approximately 1% of adult malignancies and 7% to 15% of pediatric malignancies. About 50% to 60% of sarcomas occur in the extremities [the arms and legs], and although they are rare, they are responsible for more deaths than testicular cancer, Hodgkin's disease, and thyroid cancer combined. These tumors are notorious for recurring and metastasizing—often with devastating results—despite apparently complete resection. ...The National Cancer Institute's Cancer Surveillance, Epidemiology, and End-Result (SEER) Program in 1996 reported 6400 new cases of soft tissue sarcoma, including 3500 in males and 2900 in females (2), for a male-to-female ratio of about 1.2:1.

Sarcomas can arise in human beings in scar tissue as a result of "foreign body implantation" among other causes, according to Kasper et al. (2004).

Jennings et al. (1988) reviewed published research involving six cases of angiosarcoma and 40 cases of sarcomas of other types associated with foreign-body material in humans. They found that these cases "provide compelling evidence that solid-state [foreign-body] tumorigenesis is operative in humans," and note that "implanted foreign material ... should be considered capable of inducing virtually any form of sarcoma in humans" (Jennings et al., 1988).

Jennings et al. describe each of the three cases investigated in their study as "a high-grade tumor, which metastasized and led to the death of the patient" (p. 2443). In commenting on cases from the prior literature, they observe that the malignancies developed between four months and 63 years after exposure to the foreign body, and that the foreign-body related sarcomas "appear to be highly aggressive, both morphologically and biologically" (p. 2443).

Other researchers have also found highly aggressive sarcomas and carcinomas developing in humans around or near implants, including pacemakers (Biran et al., 2006; Rothenberger-Janzen et al., 1998; Rasmussen et al., 1985), vagus nerve stimulators (Cascino et al., 2007), and orthopedic implants (Keel et al., 2001). Based on these findings, researchers recommend that all material near implants that is removed from patients should be carefully examined for cancerous changes.

In another case, surgical threads found within and near a malignant tumor were believed to have induced tumorigenesis (Martin-Negrier et al., 1996). The researchers cite Brand's animal studies showing that the physical presence and not the chemical components of the implant of foreign bodies may be responsible for tumorigenesis, and point out that the most critical factor in the induction of these sarcomas is the formation of a fibrous capsule around the foreign body. They note that, "in our case the persistence of a foreign body ... and the presence of large extensive fibrosis areas in the tumor seem to be in agreement with this possibility."

Brand et al. (1975), reporting on rodent studies, note that removing the foreign body may not be enough to prevent the development of cancer once the tumorigenesis process is already underway. They write:

As reported in the literature and infrequently observed in our laboratory, removal of the [foreign body] implant from the tissue capsule during the late preneoplastic period does not always abort development of tumors from the remaining empty capsule ... However, removal of the [foreign body] left a solid collagenous, possibly even calcifying or ossifying, scar that failed to resolve and therefore acted like [foreign body] material. The latter explanation may underline the occurrence of scar-related sarcomas in man, as reported in the literature. (p. 283)

Other Adverse Reactions to the Implants

Several studies incidentally reported other problems related to the microchips, including migration (shifting location in the body), incorrect insertion, failure to work, and loss from the body.

Migration

Despite the presence of the polypropylene sheath designed to anchor the implanted microchip, chip migration appears to be an ongoing problem. Le Calvez et al. found that microchips that had migrated from the initial implantation site accounted for 19.3% of the tumors they observed. Although the devices were originally injected into the backs of the animals, the microchip-associated tumors were later found in the limbs (4/52), the abdominal region (4/52), and the dorsal head (1/52) (p. 259).

Murasugi et al. reported no cases of migration in their study of nine dogs. However, Jansen et al. (1999) found that about half of the transponders inserted into the shoulders of beagle dogs in a four-month study had migrated to some extent. Reports from veterinarians also indicate that migration is a problem in dogs. In the United Kingdom, a voluntary registry of adverse reactions to microchip implants has been maintained by the British Small Animal Veterinary Association (BSAVA) for several years. Migration is the most common problem reported to the BSAVA, with "the elbow and shoulder being the favourite locations of wayward microchips" (BSAVA, 2004). The BSAVA reports that "[i]t is surprising how quickly some microchips migrate," noting that microchips have been found in a different location as little as one week after implantation or up to ten years later (BSAVA, 2003). Over 180 cases of migration have been reported to the BSAVA since 1996.

Injection Error

Occasionally, due to technician error, implants are injected into the wrong site on animals. Rao and Edmondson reported that 5% (7 of 140) of the microchips used in their study were later found in the perirenal area (in the abdominal cavity, surrounding the kidneys) instead of in the correct implant area just under the skin on the back. They surmise that the implants either had migrated or had been injected incorrectly directly into the abdomen. Johnson reported similar problems, stating, "occasionally some would be inserted too deep, the needle that put them in was probably held at the wrong angle" (Johnson, personal communication, 2007).

Like migration, the danger of incorrect injection also poses a risk to pets. The BSAVA cautions that technicians must be properly trained to perform the implant procedure, citing a "disastrous" incident in 2004 where an attempt to implant a struggling kitten resulted in its sudden death. A post-mortem examination later revealed that the microchip had been accidentally inserted into the kitten's brainstem (BSAVA, 2004). In another case a cat suffered severe neurological damage when a microchip was accidentally injected into its spinal column (Platt et al., 2006).

Failure and Loss of Transponder

Other problems with the microchips include failure to function, in which the microchip ceases to respond to a query from the reader device, and loss, where the microchip exits the body. Rao and Edmondson reported that four of the 140 implants used in their study failed due to microscopic cracks in the weld connecting the antenna leads to the microchip or leakage of the glass capsule resulting in fluid accumulation around the microchip (p. 413).

Rao and Edmondson also reported that an additional two of the 140 microchips in their study were lost, including one microchip lodged in the subcutaneous tissue over the lumbar vertebrae that was pushed out slowly through the scar tissue of the injection site during the tenth month after implantation.

In the Tillmann study, 1.5% of 4,279 (approximately 64) implanted microchips had to be substituted with new transponders when they either ceased functioning or were lost from the body and later found in the softwood of the cages. Most of the losses occurred in the first two days

after implantation, but some occurred as long as seven months later.

Johnson also reported that failure and loss was an issue with the implants, stating: "We had a few early in the studies that would migrate out if the wound wasn't healing properly" (Johnson, personal communication, 2007).

Adverse Reactions Likely Under-Reported

It is likely that the true rate of microchip adverse reactions in the studies was higher than reported, since the purpose of the articles was to discuss microchip-induced cancer, not other complications. One indication that this may be the case is Johnson's personal communication (2007) reporting failure, loss, and migration, as discussed prior. Though these events did occur, they were not reported in his original published report and were only solicited in response to a specific query. It is possible that other investigators may have likewise neglected to mention such reactions when they did occur.

Adverse reactions to microchips implanted in dogs and cats may also be substantially underreported. The BSAVA (2003) reported that

2003 saw a marked increase in the number of reports received through the Adverse Reaction Reporting Scheme. It is significant that several reports were received from some quite small practices while many larger practices filed no reports at all. This suggests that there is an element of under reporting which may be happening for a variety of reasons.

Anecdotal evidence supports the proposition that adverse reactions are underreported in the veterinary and oncological literature. A review of Internet discussion boards reveals the following posts by dog owners who believe their pets have suffered adverse reactions from implants (Dogster's, 2007):

My mothers dog "Buddy" actually lost his life to a "large" malignant sarcoma that was located on his back by the chip. It was removed once, but aggressively grew back and quickly took his life. I strongly believe this Chip is what took his life.

My cocker spaniel, Cooper ... has two microchips in him. The first one quit working, so he was implanted with a second one.

My dogs problem with microchip - swelling area around microchip, even to about 4 cm big, it goes away after a course of AB.

Jack was microchipped at his first vet visit when we got him - oh so many years ago ... I'm wondering - now that he is a senior citizen, I feel a small lump where the microchip was implanted - I am assuming it's only scar tissue and my vet has backed that up ...

... when Myrl was microchipped, the vet was very rough and he bled a LOT. She kind of stabbed him with the injector and he yelped and his white fur turned red. It was horrible.

None of these incidents appears to have been formally reported to any agency or decision-making body, and a review of the literature indicates that none has been investigated or written up by the academic veterinary community. Similarly, although reports of chip-related neurological damage and infection in horses have begun to appear on the Internet (see, for example, Dutch Group Nijhof.), few, if any, reports of adverse microchip reactions in horses have been written up in the literature.

Even when pet owners contact veterinarians and researchers to report their adverse experiences, they often find it difficult to get a response. Jeanne, the owner of Leon, the bulldog whose chip-related tumor is described by Vascellari et

al. (2006), reports her frustration at how difficult it was to get anyone to pay attention to what had happened. Her quest to tell Leon's story became almost a full-time endeavor as she searched the globe for a veterinary oncologist willing to look at the evidence and investigate the tumor (McIntyre, 2007; "Jeanne," personal communication, September 2007). Jeanne has maintained an updated file of articles that can be found at http://www.noble-leon.com/resourcesAdvanced/microchips.html. In the Additional References section are some of the articles she cites that have been published since this paper was originally written.

It is clear that a better mechanism for reporting adverse effects is needed and that veterinary oncologists and others need to open a better dialog with members of the public around these important issues.

What Do These Findings Mean for People?

As discussed previously, it is known that humans are susceptible to foreign-body carcinogenesis, though they appear to be less susceptible than rodents. As a foreign body, the microchip implant could potentially give rise to tumors within human beings.

The long-term effects of implanted microchips in human beings are presently unknown. Although the VeriChip implant received FDA approval for use as a medical device in October 2004, when the VeriChip Corporation became a publicly traded company in early 2007, its SEC registration statement disclosed that only 222 people in the United States had been implanted with its product. (Source: VeriChip Form S-1 Registration Statement (Amendment No. 7) see: http://www.sec.gov/Archives/edgar/data/1347022/000119312507024937/ds1a.htm p.92)

With only a few years of data available on a very small number of people, it is difficult to draw definitive conclusions about the safety of the device. If humans follow a similar pattern of microchip-induced cancer development to that observed in mice and rats, we would not expect to see implant-induced malignancies until half a lifetime's exposure, or approximately 30-40 years.

This researcher is aware of no formal follow-up procedure to evaluate the health effects or long-term safety of implanted microchips in human patients. The lack of a formal evaluation procedure and a means of publicly reporting adverse reactions that is well-understood by patients and other implantees means that such reactions could be occurring and yet be unreported to the public or to the FDA.

There is a further consideration in this day of increasing carcinogen exposure. Recent research indicates that exposure to multiple carcinogens, even within safe levels, can result in cancer development at rates that exceed what would be expected from the individual carcinogens alone. This has been called the "toxic cocktail" effect. For a discussion of research regarding this effect see Trivedi, 2007.

The microchip-induced tumors observed in the Elcock et al. study described in this paper may have been an example of the toxic cocktail effect. In that study, only rats exposed to a test chemical developed malignant tumors around the microchips. However, even rats exposed to a very low dose of the chemical compound developed the malignancies. It may be that the microchip, when combined with even small doses of a chemical compound, worked together to bring about a cancerous response.

It is estimated that every day we are exposed to 75,000 artificial chemicals (Trivedi, 2007). It would therefore seem prudent to avoid unnecessary or elective exposure to additional potential cancer-causing agents – such as implanted foreign bodies – either in ourselves or in our pets.

RECOMMENDATIONS

The following recommendations are proposed for physicians, policy-makers, veterinarians, pet owners, and veterinary researchers in light of research findings on microchip implants.

For Implanted Human Patients and Their Doctors

There are many unanswered questions about the safety of microchip implants in human beings, but what we know from animal studies is disquieting. In light of the fact that microchip implants cause serious adverse reactions in animals, the practice of chipping human beings should be immediately discontinued until the tumorigenesis process is more fully understood.

In addition, all patients, members of the public, and medical volunteers who have been implanted with microchips to date (an estimated 300 people in the United States and 2,000 people worldwide) should be immediately informed in writing of the causal link between microchips and cancer in rodents and dogs. Implanted individuals should be offered a procedure for microchip removal at the expense of the facility that provided the implant, should they choose to have the device removed. Following the advice of Jennings et al. (1988, p. 2444) that "all material removed from patients in proximity to foreign implants should be examined histologically," the tissue surrounding all removed implants should be preserved for later histological analysis.

Physicians whose patients chose to retain the microchips should routinely examine the tissue surrounding the implant for swelling, inflammation, evidence of chip migration, and pain. Any unusual sensations, lumps, or other abnormalities should be analyzed for cancerous or pre-cancerous changes. All adverse reactions, whether related to cancer or other problems, should be immediately reported to the FDA for disclosure in the public record.

For Policy-Makers

Given the clear, causal link between microchip implantation and malignant tumors in laboratory rodents and dogs, it is strongly recommended that policy makers reverse all policies that mandate the microchipping of animals under their jurisdiction or control. These include ordinances passed by state and local authorities, policies implemented at animal shelters, and formal positions adopted by animal welfare, affinity, and interest groups across the United States and around the globe.

It is the opinion of this researcher that mandatory microchipping ordinances should be repealed and replaced with a voluntary system of microchipping at the discretion of pet owners. Any pet owner who chooses to have a microchip implanted in his or her animal should be fully informed of the potential risks of the procedure. No one should be forced by law or otherwise coerced into implanting an animal against his or her conscience or medical judgment.

For Veterinarians

Veterinary offices are one of the most common places where implant procedures are performed. Since veterinarians are often the primary point of contact for pet owners on the topic of microchipping, veterinarians should familiarize themselves with the research findings and carefully consider the potential for adverse reactions before recommending implants for their patients.

Pet owners should be clearly advised of the research linking the microchip to cancer in rodents and dogs when seeking advice about the chipping procedure or choosing to have it done to their pets.

In the case of animals that have already been implanted, Vascellari et al. suggest that veterinary surgeons should routinely palpate the tissue surrounding microchip implants as part of routine medical care. Any lumps or inflammation should be investigated for cancerous or pre-cancerous

changes. To avoid the complicating risk of injection-related sarcoma, veterinarians should avoid administering vaccines or other injections at or near the site of an implanted microchip.

Finally, veterinarians should advise pet owners to routinely examine the site of the implanted microchip themselves and immediately report any abnormalities.

For Pet Owners

There have been no large-scale, statistically valid, clinically controlled, experimental studies involving microchip implants in dogs and cats, so we know very little about their long-term safety. However, the fact that we have not seen an epidemic of cancers in pets would suggest that only a small number will be impacted. As the chip-removal procedure is likely to be both costly and invasive, pet owners may wish to leave the implanted microchips intact within their animals unless a problem surfaces.

Owners of pets that have been implanted should regularly check the area for any abnormal lumps or swelling. If something unusual is found, it should be reported immediately to a veterinarian, and tests should be done to rule out cancer. The pet owner may be the key to detecting a problem in the early stages and saving the life of a pet. In the two cases where dogs developed tumors around and attached to implants, it was the owners' astute eye and probing fingers that found it, not the veterinarian. The only indication that there was a problem was the lump; all other laboratory tests came back within normal ranges.

If a pet is not currently microchipped, it may be best to keep it that way. It is the opinion of this researcher that all further implantation of pets should be halted until the existing population of chipped dogs is carefully assessed for adverse reactions, including cancer. There are other ways to ensure a pet is returned to its owner in the event it goes missing. A well-made collar and a clear, legible tag with the owner's contact information are effective tools that have worked for generations of pet owners.

For Veterinary Oncology Researchers

There is fertile ground for additional research in this area. Indeed, systematic study would add greatly to our understanding of the process of tumorigenesis as related to microchip implants. Other than preliminary research involving a very small number of animals (e.g., Ball et al.; Rao and Edmondson), there have been no studies to date that have systematically examined the development of microchip-induced sarcomas as a research goal in itself. Almost all of the cancers reported herein arose incidentally, in the course of other research.

One important direction for future research would be to explore the role of the electromagnetic energy received and transmitted by the transponder. This could help isolate whether the tumors stem from a foreign-body reaction to the external surface of the microchip alone (i.e. glass capsule and polypropylene sheath) or whether some characteristic of the device in its capacity as a radio-frequency transponder could be partially or fully responsible for the tumors. A study could be designed to investigate the role of radio-frequency energy by implanting some animals with intact transponder devices and others with empty capsules, or capsules filled with an inert substance of the same mass as the current contents of the glass capsule. In each of these groups, animals could also be exposed to different levels of energy from the reader. Although these studies would help to answer a number of the questions raised, for reasons of conscience the author does not personally endorse the use of animals for this type of experimentation.

Proposal to Create a National Registry

The research community and society at large should take advantage of the fact that there are already millions of chipped dogs in the U.S. Rather than conducting further, potentially painful and invasive studies on dogs and other animals, we can use the animals that are already chipped to learn more about how living creatures respond to these devices.

Doing so would require the creation of a central registry for reporting adverse reactions to microchips, including cancer. A registry could be created in one of the following ways:

- Dogs undergoing treatment for cancer could be voluntarily reported to an independent registry set up for this purpose. This could be done through a form similar to that used by the British Small Animal Veterinary Association. Their 2-page "Microchip Adverse Reactions Reporting Form" can be found at http://www.bsava.com/VirtualContent/85185/adverse_reaction.pdf. Because microchip-induced cancer may metastasize and lead to cancer in other parts of the body, it is important to rule out the microchip as the source of cancer in dogs. Veterinarians would report the chip status of all dogs with cancer under their care, and a statistical analysis could be made to determine whether chipped dogs have a higher overall incidence of cancer than their non-chipped counterparts.
- On a voluntary basis, veterinarians disposing of the remains of chipped animals could remove the microchip and surrounding tissue and send it to a laboratory for histological analysis.

Done on a large scale, these measures would provide important data that could be used to assess the safety of microchip implants in dogs. Establishing national registries for adverse reactions and evaluation of tissue samples would provide a more systematic way of assessing the risk than the current state of relying on case-by-case, anecdotal reports alone.

CONCLUSION

The body of research reviewed in this report indicates a clear causal link between microchip implants and cancer in mice and rats. It also appears that microchips can cause cancer in dogs–and that they have done so in at least one case, and quite likely in two. These findings raise a red flag about the continued use of microchips in both animals and human beings.

As the Associated Press reported, this concern is shared by some of the nation's most respected cancer researchers.

"There's no way in the world, having read this information, that I would have one of those chips implanted in my skin, or in one of my family members," said Dr. Robert Benezra, head of the Cancer Biology Genetics Program at the Memorial Sloan-Kettering Cancer Center in New York. He added, "[g]iven the preliminary animal data, it looks to me that there's definitely cause for concern."

Dr. George Demetri, director of the Center for Sarcoma and Bone Oncology at the Dana-Farber Cancer Institute in Boston, agreed. Even though the tumor incidences were "reasonably small," in his view, the research underscored "certainly real risks" in RFID implants, adding that the tumors can be "incredibly aggressive and can kill people in three to six months."

Dr. Chand Khanna, a veterinary oncologist at the National Cancer Institute, said that the evidence "does suggest some reason to be concerned about tumor formations." All of the cancer specialists agreed that the animal study findings should be disclosed to anyone considering a chip implant.

On the basis of these findings, physicians, patients, veterinarians, and pet owners may wish to carefully consider whether the benefits of implants are worth the potential health risks such implants appear to pose. It is the opinion of this researcher that further microchipping of pets or human beings should be immediately discontinued.

REFERENCES

Ace Animals Inc. (n.d.). *Sprague Dawley*. Retrieved September 27, 2007 from http://aceanimals.com/SpragueDawley.htm

Applied Digital Solutions Inc. (n.d.). *Press release: Applied digital solutions' VeriChip corporation confirms that the attorney general of Mexico and some of his staff have been 'chipped'*. Retrieved October 24, 2007, from http://www.adsx.com/pressreleases/2004-07-14

Ball, D. J. et al. (1991). Evaluation of a microchip implant system used for animal identification in rats. *Laboratory Animal Science, 41*(2), 185–186. PMID:1658454

Banfield: The Pet Hospital. (2005, November 3). *Banfield applauds legislation promoting open pet microchip technology where all scanners read all chips*. Author.

Biran, S. et al. (2011). Development of carcinoma of the breast at the site of an implanted pacemaker in two patients. *Journal of Surgical Oncology, 11*(1), 7–11. doi:10.1002/jso.2930110103 PMID:219300

Blanchard, K. T. et al. (1999). Transponder-induced sarcoma in the heterozygous p53+/- mouse. *Toxicologic Pathology, 27*(5), 519–527. doi:10.1177/019262339902700505 PMID:10528631

Brand, K. G. (1982). Cancer associated with asbestosis, schistosomiasis, foreign bodies or scars. In *Cancer—A comprehensive treatise: Etiology: Chemical and physical carcinogenesis* (pp. 671–676). New York: Plenum Press.

Brand, K. G. et al. (1975). Etiological factors, stages, and the role of the foreign body in foreign body tumorigenesis: A review. *Cancer Research, 35*(2), 279–286. PMID:1089044

British Small Animal Veterinary Association. (2004). *Microchip report 2004*. Retrieved October 13, 2007 from http://www.bsava.com/VirtualContent/85185/Adverse_Reaction_Report_2004PDF.pdf

British Small Animal Veterinary Association. Microchip Report 2003. (2003). Retrieved October 13, 2007 from http://www.bsava.com/VirtualContent/85185/Adverse_Reaction_Report_2003PDF.pdf

Cascino, G. D. et al. (2007). Breast cancer at site of implanted vagus nerve stimulator. *Neurology, 68*(9), 703. doi:10.1212/01.wnl.0000256035.72892.a7 PMID:17325283

Charles River Laboratory. (2007). *Research models and services catalog 2007*. Author.

Dogster's For the Love of Dogs Blog. (n.d.). *Comments posted in response to: VeriChip Corporation microchips linked to tumors, former presidential candidate Tommy Thompson linked to company*. Retrieved October 18 2007 from http://dogblog.dogster.com/2007/09/13/verichip-corporation-microchips-linked-to-tumors-former-presidential-candidate-tommy-thompson-linked-to-company/

Dutch Group Nijhof. (n.d.). *Wel of niet chippen?* Retrieved October 13 2007 at http://www.invisio.nl/antichip/

Elcock, L. E. et al. (2001). Tumors in long-term rat studies associated with microchip animal identification devices. *Experimental and Toxicologic Pathology, 52*, 483–491. doi:10.1016/S0940-2993(01)80002-6 PMID:11256750

Jackson Laboratory. (n.d.). *The jax mice data sheet: Strain name CBA/J*. Retrieved October 10, 2007 from http://jaxmice.jax.org/strain/000656.html

Jansen, J. A. et al. (1999). Biological and migrational characteristics of transponders implanted into beagle dogs. *The Veterinary Record, 145*, 329–333. doi:10.1136/vr.145.12.329 PMID:10530881

Jennings, T. et al. (1988). Angiosarcoma associated with foreign body material: A report of three cases. *Cancer, 62*(11), 2436–2444. doi:10.1002/1097-0142(19881201)62:11<2436::AID-CNCR2820621132>3.0.CO;2-J PMID:3052791

Johnson, K. (1996). Foreign-body tumorigenesis: Sarcomas induced in mice by subcutaneously implanted transponders. *Toxicologic Pathology, 33*(5), 619.

Kasper, D. L. et al. (2004). *Soft tissue sarcomas. Harrison's principles of internal medicine* (16th ed.). New York: McGraw-Hill Professional.

Keel, S. B. et al. (2001). Orthopaedic implant-related sarcoma: A study of twelve cases. *Modern Pathology, 14*(10), 969–977. doi:10.1038/modpathol.3880420 PMID:11598166

Le Calvez, S., Perron-Lepage, M.-F., & Burnett, R. (2006). Subcutaneous microchip-associated tumours in B6C3F1 mice: A retrospective study to attempt to determine their histogenesis. *Experimental and Toxicologic Pathology, 57*, 255–265. doi:10.1016/j.etp.2005.10.007 PMID:16427258

Lewan, T. (2007, September 8). Chip implants linked to animal tumors. *Associated Press*.

Martin-Negrier, M.-L. et al. (1996). Primitive malignant fibrous histiocytoma of the neck with carotid occlusion and multiple cerebral ischemic lesions. *Stroke*, *27*, 536–537. doi:10.1161/01. STR.27.3.536 PMID:8610325

McCarthy, P. E. et al. (1996). Liposarcoma associated with glass foreign body in a dog. *Journal of the American Veterinary Medical Association*, *209*, 612–614. PMID:8755980

McCullough, S. (2004). *Senior dogs for dummies*. New York: John Wiley & Sons.

McIntyre, L. (2007). *French bulldog is catalyst for investigation of microchip-cancer connection*. Retrieved September 8, 2007 from http://www. spychips.com/blog/2007/09/rfid_cancer_story_ catalyst_lov.html

Milman, H. A. (1994). *Handbook of carcinogen testing* (2nd ed.). Park Ridge, NJ: Noyes Publications.

Morrison, B. A. (2003). Soft tissue sarcomas of the extremities. [Bayl Univ Med Cent]. *Proc*, *16*(3), 285–290. PMID:16278699

Murasugi, E. et al. (2003, September 13). Histological reactions to microchip implants in dogs. *The Veterinary Record*, 328. doi:10.1136/ vr.153.11.328 PMID:14516115

National Women's Health Resource Center. (n.d.). *Clinical trials overview*. Retrieved October 10 2007 from http://findarticles.com/p/articles/ mi_m0PWR/is_2005_August_9/ai_n17214944/p

Palmer, T. E. et al. (1998). Fibrosarcomas associated with passive integrated transponder implants. *Toxicologic Pathology*, *26*, 170.

Platt, S. et al. (2006). Spinal cord injury resulting from incorrect microchip placement in a cat. *Journal of Feline Medicine and Surgery*, *9*(2), 157–160. doi:10.1016/j.jfms.2006.07.002 PMID:16982206

Rao, G. N., & Edmondson, J. (1985). Tissue reaction to an implantable identification device in mice. *Toxicologic Pathology*, *18*(3), 412–416. doi:10.1177/019262339001800308 PMID:2267501

Rasmussen, K. et al. (1985). Male breast cancer from pacemaker pocket. *Pacing and Clinical Electrophysiology*, *8*(5), 761–763. doi:10.1111/j.1540-8159.1985.tb05891.x PMID:2414760

Rothenberger-Janzen, K., Flueckiger, A., & Bigler, R. (1998). Carcinoma of the breast and pacemaker generators. *Pacing and Clinical Electrophysiology*, *21*(4), 769–771. doi:10.1111/j.1540-8159.1998. tb00137.x PMID:9584311

Simonsen Laboratories. (n.d.). *Fischer 344*. Retrieved October 5 2007 from http://www.simlab. com/products/fischer344.html

Tillmann, T. et al. (1997). Subcutaneous soft tissue tumours at the site of implanted microchips in mice. *Experimental and Toxicologic Pathology*, *49*, 197–200. doi:10.1016/S0940-2993(97)80007- 3 PMID:9314053

Trivedi, B. (2007). Toxic cocktail. *New Scientist Magazine*, *2619*, 44–47. doi:10.1016/S0262- 4079(07)62217-9

Vascellari, M. et al. (2004). Liposarcoma at the site of an implanted microchip in a dog. *Veterinary Journal (London, England)*, *168*, 188–190. doi:10.1016/S1090-0233(03)00121-7 PMID:15301769

Vascellari, M., & Mutinelli, F. (2006). Fibrosarcoma with typical features of postinjection sarcoma at site of microchip implant in a dog: Histologic and immunohistochemical study. *Veterinary Pathology*, *43*, 545–548. doi:10.1354/ vp.43-4-545 PMID:16846997

VeriChip Corp. (2006, March 1). *Press release: RFID gets under their skin*. Retrieved October 24, 2007 from http://www.verichipcorp.com/news/1141257941

ADDITIONAL READING

Cancer-related Definitions from the Centre for Cancer Education. University of Newcastle upon Tyne, UK Online at: http://cancerweb.ncl.ac.uk/omd/index.html Retrieved 1 June 2007.

Carminato, A., Vascellari, M., Marchioro, W., Melchiotti, E., & Mutinelli, F. (2011). Microchip-associated fibrosarcoma in a cat. [see www.ncbi.nlm.nih.gov.]. *Veterinary Dermatology*, (May): 2011. PMID:21535253

Daly, M. K., Saba, C. F., Crochik, S. S., Howerth, E. W., Kosarek, C. E., & Cornell, K. K. et al. (2008). Fibrosarcoma adjacent to the site of microchip implantation in a cat. [see www.ncbi.nlm.nih.gov.]. *Journal of Feline Medicine and Surgery*, *10*(2), 202–205. doi:10.1016/j.jfms.2007.10.011 PMID:18313963

Dorland's Illustrated Medical Dictionary. Online at: www.mercksource.com/pp/us/cns/cns_hl_dorlands.jsp?pg=/pp/us/common/dorlands/dorland/dmd_a-b_00.htm Retrieved 1 June 2007.

Hicks, D. G., & Bagley, R. S. (2008). Imaging diagnosis--spinal injury following aberrant microchip implantation. [see www.ncbi.nlm.nih.gov.]. *Veterinary Radiology & Ultrasound*, *49*(2), 152–153. doi:10.1111/j.1740-8261.2008.00342.x PMID:18418996

Joslyn, S. K., Witte, P. G., & Scott, H. W. (2010). Delayed spinal cord injury following microchip placement in a dog. [see www.ncbi.nlm.nih.gov]. *Veterinary and Comparative Orthopaedics and Traumatology; V.C.O.T*, *23*(3), 214–217. PMID:20422127

Linder, M., Hüther, S., & Reinacher, M. (2009). In vivo reactions in mice and in vitro reactions in feline cells to implantable microchip transponders with different surface materials. [see veterinary-record.bvapublications.com.]. *The Veterinary Record*, *165*, 45–49. doi:10.1136/vetrec.165.2.45 PMID:19596675

Pessier A.P., Stalis I.H., Sutherland-Smith M., Spelman L.H., Montali, R.J. (1999). Soft tissue sarcomas associated with identification microchip implants in two small zoo animals. *Proc Amer Assoc Zoo Vet*, 139-140.

Platt, S., Wieczorek, L., Dennis, R., & De Stefani, A. (2007). Spinal cord injury resulting from incorrect microchip placement in a cat. [see www.sciencedirect.com.]. *Journal of Feline Medicine and Surgery*, *9*(2), 157–160. doi:10.1016/j.jfms.2006.07.002 PMID:16982206

Schutt, L. K., & Turner, P. V. (2010). Microchip-associated sarcoma in a shrew (Suncus murinus). [see www.ncbi.nlm.nih.gov.]. *Journal of the American Association for Laboratory Animal Science; JAALAS*, *49*(5), 638–641. PMID:20858367

Siegal-Willott, J., Heard, D., Sliess, N., Naydan, D., & Roberts, J. (2007). Microchip-Associated Leiomyosarcoma In An Egyptian Fruit Bat (Rousettus Aegyptiacus). [see www.ncbi.nlm.nih.gov] [Abstract] [www.batconservancy.org] [Full Text]. *Journal of Zoo and Wildlife Medicine*, *38*(2), 352–356. doi:10.1638/1042-7260(2007)038[0352:MLIAEF]2.0.CO;2 PMID:17679525

Smith, T. J., & Fitzpatrick, N. (2009). Surgical removal of a microchip from a puppy's spinal canal. [see www.ncbi.nlm.nih.gov.]. *Veterinary and Comparative Orthopaedics and Traumatology; V.C.O.T*, *22*(1), 63–65. PMID:19151873

Sura, R., Schwartz, D. R., Goldman, B. D., & French, R. A. (2009). Transponder associated neoplasia in damaraland mole-rats (Crytomys damarensis). *Veterinary Pathology*, *46*(5), 1020-1083, see vet.sagepub.com (See case number 85.)

van der Burgt, G., & Dowle, M. (2007). Microchip insertion in alpacas. *The Veterinary Record, 160*(6), 204. See veterinaryrecord.bvapublications.com or www.nal.usda.gov (Abstract). (Discusses the death of a 6-month-old alpaca due to the microchip implant procedure).

Wachtman, L. M., Pistorio, A. L., Eliades, S., & Mankowski, J. L. (2006). Calcinosis circumscripta in a common marmoset (Callithrix jacchus jacchus). [see aalas.publisher.ingentaconnect.com]. *Journal of the American Association for Laboratory Animal Science; JAALAS, 45*(3), 54–57. PMID:16642972

Whitley, E., & Ball, J. (2002). Statistics review 4: Sample size calculations. *Critical Care (London, England), 6*(4), 335–341. doi:10.1186/cc1521 PMID:12225610

KEY TERMS AND DEFINITIONS

Adenocarcinoma: A form of cancer that involves cells from the lining of the walls of many different organs of the body. Breast cancer is a type of adenocarcinoma.

Anaplasia: Reversion of cells to an immature or a less differentiated form, as occurs in most malignant tumors.

Angiosarcoma: A malignant tumor originating from blood vessels.

Cancer: A general term for more than 100 diseases that are characterized by uncontrolled, abnormal growth of cells. Cancer cells can spread locally or through the bloodstream and lymphatic system to other parts of the body. (See also: *malignant*)

Carcinogen: An agent capable of initiating the development of malignant (cancerous) tumors. May be a chemical, a form of electromagnetic radiation or an inert solid body.

Carcinogenicity: The tumor-producing/cancer cell-producing potency of an agent.

Fibroblast: Resident cell of connective tissue.

Fibrosarcoma: Malignant tumor derived from connective tissue cells.

Foreign Body: Anything in the tissues or cavities of the body that has been introduced there from without, and that is not rapidly absorbable.

Genotoxin: A toxin (poisonous substance) which harms the body by damaging DNA molecules, causing mutations, tumors, or neoplasms. A substance that can mutate and damage genetic material. (Also *genotoxicant*)

Histiocyte: Long-lived resident macrophage (immune-related cells) found within tissues.

Histiocytoma: A tumor composed of histiocytes.

Histochemical: Study of the chemical composition of tissues by means of specific staining reactions.

Histology: The study of cells and tissue on the microscopic level.

Histopathology: The science concerned with the study of microscopic changes in diseased tissues.

Immunohistochemistry: Histochemical localization of immunoreactive substances using labelled antibodies as reagents.

Induce: To bring on; to effect; to cause.

Inert: Refers to a substance which will not chemically react with anything under normal circumstances.

Leiomyosarcoma: A malignant tumor of smooth muscle origin.

Liposarcoma: A malignant tumor that may be composed of fat cells.

Macrophage: Relatively long lived phagocytic cell of mammalian tissues. In response to foreign materials may become stimulated or activated. Macrophages play an important role in killing of some bacteria, protozoa and tumor cells, release substances that stimulate other cells of the immune system and are involved in antigen presentation.

Malignant: Tending to become progressively worse and to result in death. Having the properties of anaplasia, invasion, and metastasis, said of tumors.

Malignant Fibrous Histiocytoma (MFH): A deeply situated tumor, especially on the extremities of adults.

Malignant Tumor: A mass of cancer cells. These cells have uncontrolled growth and will invade surrounding tissues and spread to distant sites of the body, setting up new cancer sites, a process called metastasis.

Mesenchymal: Relating to the mesenchyme, embryonic tissue of mesodermal origin. The mesoderm is the middle of the three germ layers and gives rise to the musculoskeletal, blood, vascular, and urinogenital systems, to connective tissue (including that of dermis) and contributes to some glands.

Metastasis: The transfer of disease from one organ to another due either to the transfer of pathogenic microorganisms (for example, tubercle bacilli) or to transfer of cells, as in malignant tumors. The capacity to metastasize is a characteristic of all malignant tumors.

Metastases: A growth of abnormal cells distant from the site primarily involved by the disease process.

Metastasize: To spread to another part of the body, usually through the blood vessels, lymph channels, or spinal fluid.

Mitosis: A method of indirect division of a cell, consisting of a complex of various processes, by means of which the two daughter nuclei normally receive identical complements of the number of chromosomes characteristic of the somatic cells of the species.

Mitotic: Pertaining to mitosis.

Morphology: The configuration or structure (shape).

Moribundity: In a dying state; dying; at the point of death.

Mutagen: An agent that can cause an increase in the rate of mutation, includes X-rays, ultraviolet irradiation (260 nm), and various chemicals.

Necrosis: Morphological changes indicative of cell death.

Neoplasia/Neoplasm: New and abnormal growth of tissue, which may be benign or cancerous.

Oncology: The study of diseases that cause cancer.

P53 gene: A gene which encodes a protein that regulates cell growth and is able to cause potentially cancerous cells to destroy themselves.

Rhabdomyosarcoma: Malignant tumor (sarcoma) derived from striated muscle.

Sarcoma: Malignant tumor of soft tissue (tissue that connects, supports or surrounds other structures and organs of the body). Soft tissue includes muscles, tendons, fibrous tissues, fat, blood vessels, and nerves.

Schwannoma: A neoplasm [new and abnormal growth of tissue] originating from Schwann cells (of the myelin sheath) of neurons.

Teratology: The branch of embryology and pathology that deals with abnormal development and congenital malformations (i.e., the study of birth defects).

Toxicology: The scientific study of the chemistry, effects, and treatment of poisonous substances.

Tumor: An abnormal mass of tissue that results from excessive cell division that is uncontrolled and progressive, also called a neoplasm. Tumors perform no useful body function. They may be either benign (not cancerous) or malignant.

Tumorigenesis: The production of tumors.

Validity: The extent to which a measurement, test, or study measures what it purports to measure.

Section 6
Socio–Ethical Implications of RFID Tags and Transponders

Chapter 14
Privacy and Pervasive Surveillance:
A Philosophical Analysis

Alan Rubel
University of Wisconsin – Madison, USA

ABSTRACT

This chapter analyzes some tools of pervasive surveillance in light of the growing philosophical literature regarding the nature and value of privacy. It clarifies the conditions under which a person can be said to have privacy, explains a number of ways in which particular facets of privacy are morally weighty, and explains how such conceptual issues may be used to analyze surveillance scenarios. It argues that in many cases, surveillance may both increase and decrease aspects of privacy, and that the relevant question is whether those privacy losses (and gains) are morally salient. The ways in which privacy diminishment may be morally problematic must be based on the value of privacy, and the chapter explains several conceptions of such values. It concludes with a description of how some surveillance technologies may conflict with the value of privacy.

INTRODUCTION

The potential for continuous, contextual information gathering about individuals, referred to as *pervasive surveillance* or *uberveillance,* adds to the growing list of privacy issues with which contemporary societies must contend, including expanded legal authority for surveillance, growth of relational databases and an industry dedicated to filling them, ease of information sharing in social networks, surveillance initiatives in the service of public health, and sophisticated sensing technologies. Commentary lamenting privacy loss is common, dating back over a century, and many of us are familiar with commentary dismissing concerns about privacy, either on the grounds that we already have no privacy or that we cannot make legitimate claims to it.

DOI: 10.4018/978-1-4666-4582-0.ch014

The purpose of this paper is to analyze privacy claims in the context of pervasive surveillance, drawing on a growing philosophical literature on privacy. Specifically, I address problems related to the concept of privacy itself and problems in determining whether privacy loss is morally important. I will begin by describing a small part of the pervasive surveillance terrain, using examples that will help illustrate several important conceptual and moral problems. I then address the notion of privacy loss itself, offering an account that accommodates the broadest array of conceptual issues. Although it is obvious that pervasive surveillance technologies diminish aspects of privacy, they cannot destroy privacy altogether. In addition to diminishing privacy, pervasive surveillance can actually serve to protect certain aspects of privacy. Indeed, the more important issue is whether pervasive surveillance undermines or protects *morally salient* aspects of privacy. Put another way, whether there are privacy harms, whether privacy claims are impinged, and the extent to which objections to privacy loss are justified depends on the features of that loss. To address that issue, I outline several ways in which privacy loss may be morally weighty and apply the framework developed to some of the ways surveillance technologies may be deployed. I then describe the relationship between the value of privacy and rights to privacy, and conclude by noting the limitations of the analysis offered and directions for further work.

BACKGROUND

Pervasive surveillance, or "uberveillance"—a term developed by Michael and Michael to denote the intersection between automatic location identification, contextual information gathering, and implantable devices—is difficult to pin down precisely (Michael & Michael, 2007). Roughly, the notion is one of widespread and well-integrated information gathering that tracks persons or objects in many areas, and incorporates contextual information. The degree to which contextual information may eventually be incorporated into surveillance systems, the ability for people to create new uses for technologies, and individual willingness to be surveilled is difficult, if not impossible, to predict. This is not an attempt to offer an overarching vision for the direction and future of pervasive surveillance. Rather, in this background section I will draw on the work of others who have analyzed the technological landscape in greater detail and point out some possibilities for pervasive surveillance in different arenas, offering examples of pervasive surveillance technologies that will provide a foundation for the discussions of privacy and claims to privacy in the following sections.

There are any number of technologies that can be developed or deployed as part of pervasive surveillance. A useful starting point is Radio Frequency Identification Devices (RFID). Katina and M.G. Michael have written extensively on RFIDs that can be attached or embedded into objects (for example products in a supply chain for tracking purposes), into animals (for example, into pets or livestock for identification purposes), or into people (for example, into employees for access purposes) (Clarke, 2007; Michael & Michael, 2007). Such devices may be passive, merely providing a unique identifier when scanned by a fixed or mobile reader, or active, recording and/or transmitting information about the condition of the object, animal, or person to which the device is attached or in which it is embedded. The primary implantable devices until now have been passive RFIDs that allow for personal identification and tracking (Kosta & Bowman, 2011; Rotter, Daskala, & Compano, 2008). However, future devices may be able to monitor physiological states of the implantee (Kosta & Bowman, 2011); indeed, one important provider of RFID devices for medical purposes has announced the development of an implantable RFID that can monitor blood glucose levels (Business Wire, 2007). That in turn raises the possibility of devices that can also monitor other conditions.

Other technologies may also be understood as elements of pervasive, contextual surveillance. Currently, cell phones can be tracked via transmission tower information or Global Positioning System (GPS) satellite; the consumer data industry collects information about purchasing, credit, and finances; and closed-circuit television (CCTV) monitors many of our movements. Such technologies can be used for monitoring, or observation in real time, as well as tracking, or recording and analyzing persons over time. Although the various aspects of surveillance may be discrete, we can certainly imagine that they (and other analogous systems) could become further integrated into one or more large surveillance systems tracking much of our lives, or the lives of others. While many of us may be nervous about such a large, integrated system, imagining the whole thing fully-formed and functioning runs the risk of glossing smoothing over the particular ways in which surveillance can be deployed. That is, moving conceptually to the idea of a society in which surveillance is pervasive, contextual, and integrated makes it difficult to assess ways in which privacy will affect individuals along the way, and it can lead us to pass over analysis of individual privacy claims in favor of broad commentary on society generally. Hence, in describing the terrain here the focus is on several ways in which pervasive surveillance might be applied to groups of people, and in the following sections I will explain how we might assess the privacy of people in those groups. To put this differently, there are all sorts of ways in which a state of affairs might be undesirable or morally deficient without anyone's rights being impinged. A society might fail to adequately coordinate important aspects of its economy, for example by failing to standardize units of measure. That would be a terrible state of affairs, but not rights-violating. Likewise, the mere fact (if

it *is* a fact) that pervasive surveillance within a society is undesirable or bad does not by itself tell us whether anyone has a claim that pervasive surveillance not occur. Societal benefits may, however, *underwrite* or *provide justification* for the existence of rights, as in accounts of free expression rights based on the necessity of such rights for democratic institutions.

Note that the categories deployed here divide the conceptual terrain by arena. Often privacy rights and regulations are often a function of the sort of activity in which surveillance takes place (Nissenbaum, 2010). There are different norms and laws surrounding, for example, health information privacy, education information privacy, privacy in financial transactions, and so forth. But there are other ways to categorize surveillance. Another possibility is to categorize surveillance by the good it is intended to advance. Masters and Michael divide categories of RFID implants in such a way, analyzing uses in terms of whether surveillance serves is a mechanism of *control* (e.g., restrict who can enter or leave a space, identify who has been in a particular location), *convenience* (e.g., location-based assistance services, keyless entry systems), or *care* (e.g., linking medical records, implantable monitoring devices) (Masters and Michael, 2005). These categories may overlap, as devices that aid medical care may do so by making some aspects of care more convenience or by controlling persons' access to various things. Moreover, those categories will intersect with the arenas articulated below, such that use of pervasive surveillance in, e.g., the commercial arena may be a matter of control or convenience.

Commercial Arena

The commercial arena is ripe for pervasive surveillance. RFID is already being integrated into products for better supply chain management (Angeles, 2005; Asif & Mandviwalla, 2005; Michael & McCathie, 2005). More important, however, is the potential for consumer monitoring. Commentators have pointed out the potential advantages that implanted devices could have for consumers, who could avoid carrying cash and credit cards, and there have been some extremely limited cases of implants being used by clubs for their patrons (Michael & Michael, 2011a). It is difficult to see such a practice becoming widespread in the near future. More likely, and more intriguing, is the use of tracking/recording devices in automobiles. Such devices could be used to monitor driving habits and to serve as a recorder of events preceding accidents (Filipova-Neumann & Welzel, 2010; Iqbal & Lim, 2007). There have already been case studies involving installed microchips and GPS systems to study driving behaviors, and determination of risk factors so as to charge rates according to individual risk (Michael et al., 2010). Another application would be using records related to accidents to help determine fault and liability. This kind of monitoring is already beginning, with insurers in some cases providing rate incentives for consumers who have monitoring devices placed in their cars. Hence, there is some consumer choice in the matter, and at the moment there may be some benefit to the consumer in adopting the technology.

A further aspect of consumer monitoring uses on individuals' locations to determine major points of interest—work, home, places of worship, social setting, and so forth—to make inferences about important (and potentially commercially relevant) aspects of their lives (Stephan et al., 2013). Such inferences may extend to social status, friendly relationships, family life, and religion (among other things) (Michael & Michael, 2011b).

Another way in which such technologies can enter the commercial arena is in employee monitoring. RFID is currently widely used to restrict access, measure time spent in various places (at desk, in other departments, on breaks, etc.), and monitor employee time-in and -out (Kurkovsky, Syta, and Casano, 2011). There is the potential for much greater monitoring. Kurkovsky et. al. explain the potential for monitoring employee

location with greater granularity (e.g., down to place in a room), co-location with other employees, comparison of work habits with other employees, and so forth (Kurkovsky et al., 2011).

Penal Supervision

One of the most obviously useful, and most likely to be implemented, applications of pervasive surveillance will be in the context of penal supervision (Stephan et al., 2013). Prisoners are of course already extensively monitored, and use of tracking devices is common for persons under house arrest. RFID is used to track prisoners within institutions, and there are cases of implanting microchips into inmates (Brady, 2008; Michael & Michael, 2011a). Another possibility would be using such implants for people on parole, probation, or subject to preventative detention or monitoring. The potential reasons for using such surveillance in this context are clear enough. They could save the labor costs associated with present forms of monitoring, they could allow more complete record-keeping of supervisees' movements, and they are potentially more difficult to thwart or evade than in-person monitoring and record-keeping. Prisoners could receive benefits as inducement to receive implants, perhaps being afforded home confinement instead of incarceration. People already under supervision could be more easily tracked to meetings with parole supervisors, work release sites, home, and so forth. One commentator has suggested that RFID could be used to ensure that sex offenders remain in sanctioned places (Rosenberg, 2007). Given the high correlation between prison and drug use, one can imagine using such devices to ensure that supervisees take drug tests, monitor attendance at 12-step or treatment meetings, or detect signs of drug use. If devices able to monitor physiological states become feasible, as per Verichip's device to monitor blood glucose levels, the possibility of monitoring drug use would no doubt be explored in a penal context.

Children

Another group for whom tracking technologies could be applied is children. This could be driven by fears of losing them or for the need to monitor their well-being. Use of cell phones to monitor children is well-established, and a number of other mechanisms for tracking are being explored (Michael & Michael, 2007). For example, there have been experiments with GPS-equipped bags, RFID bracelets, school uniforms with tracking technologies, and (rarely, at least for now) implanted devices. Location services based on such devices could allow parents to check in on children occasionally or gather and store real-time information (Michael & Michael 2011b). In 2010 there was some controversy surrounding the decision by a U.S. school district to have pre-school children wear jerseys equipped with a device allowing them to be tracked. The system was designed such that any device noticed outside of normal areas could be easily detected, and whether a child attended school, obtained lunch, and went to proper locations could be easily monitored (Johnson, 2010).

ASSESSING PRIVACY LOSS

Privacy

Although "privacy" is a widely-deployed term, the nature and proper scope of privacy is disputed. That makes the task of evaluating privacy losses difficult. I will sketch some of the disputes about the nature and scope of privacy, drawing on examples from the previous section. The account of privacy offered is intended to be compatible with the widest array of philosophical views regarding the nature of privacy.

To begin, it is important to use the term "privacy" carefully in order to avoid vagueness and equivocation. Privacy is best understood as a relational concept, involving three parts: some person or persons (P) who has privacy, some domain (O) regarding which P has privacy, and some other person or persons (Q) with respect to whom P has, or lacks, privacy regarding O. Hence, merely stating "P has privacy" is incomplete and difficult to analyze, but we can specify such a proposition by specifying some domain (e.g., location) and some other person or persons (for example P's work supervisor or colleagues). Suppose that P informs her coworkers, but not her supervisor, that she will be going to meet someone involved in a project she is working on, or that she is leaving for the day. In that case, P may have privacy regarding her location with respect to her supervisor, but not with respect to her colleagues. And the mere fact that P lacks privacy regarding her location with respect to her colleagues tells us very little about her privacy overall, and nothing at all about her privacy with respect to her supervisor (Rubel, 2011).

This may seem a small point, but it is crucial. First, it avoids the inference that, because a person has disclosed (or someone else has otherwise obtained) some bit of information about herself, she has no privacy regarding that information. She has diminished privacy regarding that information with respect to that other person, and the possibility that the other person might disclose it to others could be considered a further diminution, but there would remain any number of others with respect to whom the first person still has privacy regarding the information. The second reason that the three-part relationship matters is that it explains how a person with comparatively little privacy can continue to lose privacy when information gathering extends to more domains. So, incarcerated persons (P) have very little privacy regarding their locations, activities, reading, relationships, habits, and so forth (O) with respect to prison guards. But a decision to implant prisoners with devices to sense their physiological states would decrease their privacy still further by extending official reach into a further domain. Hence, it would not be an adequate reply to objections regarding implantable devices that prisoners have no privacy. One might argue that the comparative lack of privacy makes the further intrusion less objectionable, or that prisoners have no claims to privacy in any domain with respect to prison officials (precluding even privacy in communications with counsel). But such claims would require argument, and do not simply follow from the fact that prisoners lack privacy in other respects.

The next issue to address is the precise composition of privacy. Privacy certainly concerns information about, and observations of, persons, but there is a significant dispute as to whether privacy properly includes a person's abilities to make certain kinds of decisions (Allen, 1988; DeCew, 1997). We need not address that issue here, as the focus is on surveillance. Even so, there is a question of the conditions under which a person's informational or observational privacy has diminished. Sometimes, the key issue is whether some Q can access P or information about P; where Q is able to learn information about P regarding O, P's privacy diminishes (Allen, 1988; Gavison, 1984; Parent, 1983). Others maintain that privacy consists not in Q's ability to gain information, but P's *control* over such information, regardless of whether any Q actually accesses the information (Inness, 1992; Moore, 2010; Rachels, 1975). There are important limitations to each account. One is that they do not cover falsehoods. Where Q receives information that appears to concern P in domain O, but which turns out to be false, Q has not accessed information about P nor has P lost control of information about P, for the information is not about P at all. Another limitation concerns increasing ability to analyze information. Q may have a great deal of information about P, and be able to make increasingly sophisticated inferences about P based on gathering further information about *other* people. That would appear to decrease P's privacy, even though P had no control in the first instance and even though Q had no more access to particular facts about P. We can accommodate these concerns by focusing on the inferences that Q can make about P and O. That

is, P's privacy diminishes as Q's ability to make particular inferences about P regarding O increases (Rubel, 2011). That ability will increase where Q has access to information, where Q's ability to analyze data increases, and where Q reasonably relies on false information.

Applying this framework to the cases outlined in the previous section, we can see that pervasive surveillance may both increase and decrease privacy, depending on the particular persons and domains involved. First consider persons under penal supervision. As a number of commentators have pointed out, tracking devices (including implantable RFID devices) could be an alternative to devices already worn by supervisees. These could either be passive devices that record a person's location or devices that record information such as whether the subject has used illicit drugs. To the degree that such devices allow more extensive monitoring of a supervisee's (P) location, activities, or physiological states (O) with respect to supervisors (Q), they would decrease P's privacy. But notice that implanted devices are less obtrusive than bracelets worn on the ankle. Hence, the implantable device would *increase* P's privacy regarding the fact that she is a convicted criminal (O) with respect to the public at large. Where P, who is under supervision, wears a band on his wrist or ankle, members of the public observing P can readily determine that P is under state supervision. This means that P's privacy regarding the fact that he is under supervision (O) decreases with respect to the public at large (Q). A less obtrusive device, such as an implantable RFID, would therefore increase P's privacy in this respect. Similarly, devices that monitor physiological states could afford P increased privacy to the extent that P could avoid drug tests. That is, where a device monitors P, P could potentially avoid having to travel to a particular location to take a test and be watched while taking the test. P's privacy regarding his location and his body (O) would increase with respect to test takers (Q), even if his privacy regarding whether he has used proscribed substances decreases with respect to his supervisors. Of course, devices less obtrusive than bracelets need not be implanted. But the privacy implications of implantation are less to do with implantation itself than difficulty of removal, and hence continuing susceptibility to surveillance.

A similar analysis can be applied in the other cases outlined above. The increased security of consumers' financial information better secures consumers' (P) privacy regarding the financial information (O) with respect to non-authorized persons (Q), even if providing greater information to credit providers or retailers. And the ability to track school children while on school grounds might decrease their privacy regarding location, having eaten, and so forth with respect to administrators. However, such an automated system could reduce the degree to which direct observation is necessary, thereby reducing the numbers of other persons with respect to whom children have some degree of privacy loss. Not every use of pervasive surveillance techniques is such a mixed bag; some collection is novel. The use of RFIDs to track driving behavior appears to be collection of information that would be difficult to collect systematically without the technology. Hence, such use only allows new information to be collected and only diminishes privacy, without a corresponding increase in some other aspect of privacy.

A further issue is whether devices actually allow more individuals to be monitored. So, the type of device used may make it the case that P has less privacy in some domain with respect to Q, and more privacy in some other domain with respect to Q, or more privacy in any domain with respect to yet another party. But if monitoring devices allow more persons to be surveilled, then there will be a different sort of privacy loss. If, for example, RFID use by employees decreases the need for security personnel to check badges, employees who would otherwise have to show a badge might have increased privacy regarding their whereabouts with respect to badge-checkers

but less privacy regarding their movements with respect to security personnel overall. Further, if RFIDs are inexpensive enough, it may be that a greater percentage of employees would be subject to some kind of monitoring; that is, more employees (more Ps) might have less privacy regarding their movements.

There are two things to note. The first is that privacy loss occurs in utterly mundane ways. There are innumerable cases in which P has no control over which information Q has access, or regarding which information Q can make inferences in some domain. Q's fortuitous observation of P shopping for groceries diminishes P's privacy regarding his whereabouts, his choice of food, his choice of stores, and so forth with respect to Q. However, such mundane privacy losses are in no way morally important, which is to say that for privacy loss to be morally problematic, there must be some further explanation. A corollary is that in considering privacy loss associated with technologies or practices we have to consider that loss in conjunction with privacy gains in other domains or with respect to different persons. Thus, what matters is not privacy loss per se, but *morally salient* privacy loss *and gain*. It may, then, be of greater moral importance to retain privacy regarding whether one is under state supervision with respect to the public at large than it is to have reduced privacy regarding one's location, one's attending meetings, one's location, etc. with respect to corrections workers.

PRIVACY AND VALUE

The discussion in the previous section addresses privacy itself, or what it means to say that a person's privacy has diminished. But the answer to that question only provides a partial understanding whether any such loss is morally weighty or whether any loss impinges a right. This is because privacy losses may occur in ways that are utterly insignificant and not at all morally problematic.

Morally important privacy losses will depend on whether there are values underwriting particular sorts of privacy, and whether privacy loss rises to the level of an impingement of rights will depend on the type and weight of the value underwriting the privacy.

Determining whether privacy loss is morally weighty, or of sufficient moral weight to give rise to a right to privacy, requires an examination of reasons why privacy might be valuable. As a preliminary matter, we should distinguish different ways in which something can have value. Something can be instrumentally valuable if it makes other good things happen, or more likely to happen. Instrumental value is a function of consequences. So, privacy might be instrumentally valuable insofar as it brings about other goods. In contrast, something can be intrinsically valuable if its value is independent of whether there are other goods it makes more likely. Pleasure might be like this: the value of pleasure is independent of whether pleasure leads to other good consequences. It is difficult to see how privacy could be intrinsically valuable in that sense. However, something can be valuable as a constitutive part of something else that is itself intrinsically valuable. A life in which one has rich aesthetic experiences may be intrinsically valuable. Art or music is a constitutive part of such an intrinsically valuable life, and is a feature of the life in virtue of which it has value. Hence, art and music would be constitutively valuable (Raz, 1988). Accounts of privacy include both instrumental and constitutive value accounts.

Instrumental Value

There are several ways in which privacy might be instrumentally valuable. Adam Moore argues, for example, that privacy is important for people to flourish. Where people are observed too closely or too often, they tend to suffer. Though the particular features of privacy necessary for flourishing may vary by culture, Moore argues that in all cases

there are some boundaries the transgression of which makes it less likely that persons will thrive (2010). Another prominent view of privacy's instrumental value is its importance in fostering personal relationships. This can occur in several different ways. On one view, the fact (if it is) that a great deal of information about oneself is available publicly diminishes one's capacity to share such information selectively, which is in turn an important feature of intimate relationships (Fried, 1984). A similar view is that privacy facilitates not merely intimate relationships, but also the broad range of relationships necessary to live a rich life in contemporary society. Just as P needs to share a substantial amount of information about many facets of her life with Q in order to be a close friend or other intimate relation with Q, P needs to withhold a great deal of information from others in order to foster different relationships. So, for P to have an appropriate distance from Q in order to be Q's teacher, physician, or arms-length business associate demands that P have privacy regarding some information with respect to Q, and vice versa (Rachels, 1975). It might be difficult for P to teach Q objectively, for example, if Q had no privacy regarding his dislike of the subject matter or regarding his propensity to play video games rather than study (O) with respect to P. Similarly, where people learn salacious facts about others—business associates, neighbors, coworkers—it may become difficult to think of much else, and those valuable relationships might suffer (Nagel, 1998).

Another good that privacy might further is democratic institutions and processes. Lack of privacy may, for example, dissuade potential office-holders from entering public service (either because of information that might be revealed about political candidates or due to background checks for civil service positions) (Gavison, 1984). Likewise, groups may need distance from outside scrutiny in order to develop dissenting political views. If premature scrutiny of such views impedes their development, and if the development of such views is an important facet of democracy, then privacy for such groups will be instrumentally valuable (Gavison, 1984).

Intrinsic Value

As noted above, privacy may be valuable as an element of an intrinsically valuable good or state of affairs. One version of intrinsic value account is that privacy is an important element of human dignity or personhood. Jeffrey Reiman, for example, argues that privacy confers "[moral] title to one's existence" (1976, p. 39). The idea behind such a view is that privacy allows the exercise of one's will to shape one's own personality, rather than acting in accord with others' views. Being free from constant observation (which is to say, having privacy in certain domains) allows one to act for one's own reasons rather than acting to please, preclude the criticism of, or avoid close scrutiny by others. This, in turn, affords one the opportunity to act with greater freedom and to be assured that the reasons for acting are one's own. Another respect for persons view is found in Bloustein (1964). If it is true that acting on the basis of one's own reasons is valuable in its own right, observation that causes a person to act for different reasons (fear of disapprobation, ease of acquiescing with others rather than deliberating and taking one's own path) will conflict with something of value. Privacy will be a constitutive part of an intrinsically valuable state of affairs.

Another type of intrinsic value account is based explicitly on respect for persons. Persons are, in most cases, moral agents. They have the ability to act according to reasons, they can understand how their actions affect others and tailor their actions accordingly, and they are able to make determinations for themselves about what matters to them, what confers meaning to their actions and lives, and what is of moral value. This ability may demand that each of us confer substantial respect to others as choosers. Stanley Benn, by way of

explanation, writes that respecting others at root requires that we understand ourselves as "capable of having projects, and assessing [our] achievements in relation to those projects" and understanding others as likewise choosers, "attempting to steer his own course through the world ... and correcting course as he perceives errors" (Benn, 1971, pp. 228–29). We take measure of our own lives according to the terms, values, and projects that we choose, and respecting others demands affording them the opportunity to instantiate the terms, values, and projects *they* choose. In many cases that will involve a significant degree of privacy. That may be because the person's values include privacy—indeed, many of us have strong preferences for privacy in lots of domains. But more importantly, it may be because a person's ability to act according to her own terms, values, and projects, demands *not* acting according to others' conceptions, which may occur when subject to extensive surveillance.

There is a further way in which privacy loss may implicate a failure of respect for persons. Deceit is a particularly important affront to the autonomy of individuals. Autonomous agents, as described above, are capable of living according to the values they choose, as they see fit. However, where such agents are deceived in important ways, their ability to act autonomously is diminished. They are unable to act according to what matters to them if they are deceived about important facts. Hence, consent for medical procedures or agreement to terms of a contract demands that the patient or party to the contract be provided with all information necessary to make an informed agreement, regardless of whether the person is better off medically or materially as a result. Moreover, respect for a person's autonomy demands that they have information enough to make sense of the world. Where a person lacks facts regarding important aspects of her life, she is unable to rationally assess those aspects of her life (Hill, 1984). Now, surveillance alone need not conflict with a person's autonomy; after all, we may be perfectly aware of the surveillance, or may not consider it important. But where surveillance, or the extent of surveillance, is not known, and where people consider it important (and many do), there is an important conflict with a person's autonomy interests (Rubel, 2007).

SURVEILLANCE AND THE VALUE OF PRIVACY

Turning back to the technologies discussed in the background section, we can see various ways in which pervasive surveillance can conflict with morally salient aspects of privacy. Consider first, instrumental value. One predominant view of privacy's instrumental value is its role in facilitating many and varied social relationships. The information disclosure necessary for people to establish intimate relationships—friends, lovers, family members, for example—is substantial. And it is important that such disclosure be voluntary. Finding out about a person based on other sources, after all, does not establish a relationship. Moreover, a degree of privacy is important in maintaining other relationships. The teacher-student relationship, for example, requires that neither party has too much information about the other; it might be difficult for a teacher to evaluate fairly and help enthusiastically the student whose habits he dislikes. But if students are tracked excessively, such information may become available. Likewise arms-length business associations depend on a degree of privacy; excessive information about people in a contractual relationship or employees may actually undermine those relationships. So, while being able to monitor workers to some degree may increase efficiency, it may undermine the relationships involved if supervisors know too much about the minutia of employee habits. Finally, family relationships, which like any intimate relationship, flourish where disclosure of information is voluntary may be affected if tracking is persistent.

Intrinsic value accounts are also implicated with some of the pervasive technologies discussed. To the extent that it is valuable that one acts for one's own reasons, where surveillance makes it the case that people act rightly just by virtue of being monitored, in an important sense those actions are not based on a person's own judgment of right. That is, the actions are not undertaken according to a person's values as she sees fit, but because she believes others watch. In the context of employee monitoring, persons thereby lose the opportunity to demonstrate that they are good, or virtuous, workers, for the reason behind performance is explicable by surveillance, not the will to do right. Those same surveillance activities can be seen as failures of respect for individuals as choosers, that they may be denied the opportunity to work in such a way as to instantiate the values that they choose—for example, doing good work—rather than working so as not be fired.

One potential way to avoid the worry that persons act rightly just because of being monitored would be to monitor surreptitiously. Pervasive surveillance may make that much easier, especially insofar as one may know that she is subject to some sort of tracking, but not the extent of the tracking. So, taking the cases involving RFID in cars or deployed to monitor employees, we can imagine that the information gathered is quite extensive, but the subjects of the information are aware only of the existence of some monitoring. In the case of employee tracking, we can imagine readers being deployed ever more widely, unbeknownst to the persons subject to the surveillance, and we can imagine that information being coordinated with other surveillance mechanisms (e.g., locations of other employees, closed circuit televisions, computer logs), all without the employee being aware. In such cases there may be a greater worry than the employee acting for reasons that are not her own; rather, she is unaware of important information about herself and her situation, which is potentially available to her, and without which she labors under an important misconception. That kind of hidden information, as explained above, impinges her autonomy interest in understanding important facts about her world. Monitoring without disclosure of the fact or extent of the monitoring is hence a failure of respect for the person (Rubel, 2007).

So far privacy and the value of privacy have been discussed, but not rights to privacy. The move from the value of privacy to rights to privacy is straightforward, but perhaps unsatisfying. Rights, in their essence, are valid claims of one's moral due (Feinberg, 1980). They provide reasons for others to respect the rights, but they are not all-things-considered judgments about what should occur or how one should be treated. So, my agreement to pay Jones a hundred dollars to fix my car gives Jones a right to that money once Jones has fixed my car. However, it may not be the case all-things-considered that Jones should get the money. For example, that debt might be discharged in bankruptcy. The failure to pay, though, is a wrong because of the right created by the agreement and action of fixing the car. A valid claim of one's moral due with respect to privacy may arise from instrumental or intrinsic value, if the privacy is of the type that either affords substantial instrumental value or is of the type that is generally a constitutive part of an intrinsically valuable state of affairs. Monitoring so as to preclude the possibility of acting rightly for one's own reasons, at least in many cases, would seem to be such a case. Likewise, surreptitious surveillance may impinge a right. This minimal conception of rights may be unsatisfying, though, insofar as it leaves open what considerations suffice to justify impinging a right.

FURTHER RESEARCH DIRECTIONS

The purpose of this paper is to analyze privacy claims in the context of pervasive surveillance. That involves looking at conceptions of privacy and conceptions of the value of privacy and analyzing how the technologies of pervasive surveillance may implicate privacy and whether such implications are morally important. Working from the ground up in such a way is important; otherwise, we run the risk of missing aspects of privacy, brushing over important considerations, and making claims broader than can bear analysis. But it leaves important things out. Most obviously, it leaves out other technologies and applications. Here I look at just a few ways in which the tools of pervasive surveillance might affect privacy, but there are countless others. The analysis offered here can certainly be extended to other cases, but is an area for further research.

More important, though, is the *conjunction* of pervasive surveillance technologies with all of the other tools and practices that bear on privacy. Analyzing discrete technologies can disclose discrete privacy claims based on either instrumental or constitutive value. However, the instrumental effects of *all* surveillance technologies are important as well. Each tool or instance of observation and information gathering may not have negative consequences for relationships, political processes, individual flourishing, etc., but the totality may. If, for example, Moore is correct about the importance of privacy for well-being, and negative effects occur only above some threshold (or they are offset by the value of disclosure up to some threshold), then analyzing a narrow range of privacy effects will fail to account for something important (Moore, 2010). Likewise, if our concern is the intrinsic value of actions being our own, and we tend to act based on something else only above some threshold, then looking at the narrow range will be inadequate. This, though, may be difficult to analyze in terms of rights to privacy. If the primary concern about privacy is loss from widespread sources, but those losses are not in violation of claims that individuals have against others, then it would appear to be a lamentable loss even without impingement of rights. In any case, the relation between widespread, incremental privacy diminishment and claims of persons' moral due warrants further work.

CONCLUSION

The purpose of this paper is to place controversies regarding pervasive surveillance or uberveillance into the context of philosophical analyses of the nature and value of privacy. Those analyses seek to clarify the concept of privacy, whether privacy has value, and what sort of value privacy has (if any). To that end, the paper described several potential and current uses of pervasive surveillance technologies, explicated some of the important threads of the literature regarding the nature and value of privacy, and described how the technologies might implicate some of those views. But of course there remain many open questions, including just how the technologies will develop, whether the use of the technologies will be such that they actually do conflict with privacy's value, and whether the goods realized by pervasive surveillance outweigh claims to privacy. Nonetheless, by being clear about the nature and value of privacy, we can better assess those issues.

REFERENCES

Allen, A. L. (1988). *Uneasy access: Privacy for women in a free society*. Totowa, NJ: Rowman & Littlefield.

Angeles, R. (2005). RFID technologies: Supply-chain applications and implementation issues. *Information Systems Management, 22*, 51–65. doi:10.1201/1078/44912.22.1.20051201/85739.7

Asif, Z., & Mandviwalla, M. (2005). Integrating the supply chain with RFID: A technical and business analysis. *Communications of the Association for Information Systems, 15*(1).

Benn, S. I. (1971). Privacy, freedom, and respect for persons. In J. R. Pennock, & J. W. Chapman (Eds.), *NOMOS XIII: Privacy* (pp. 1–26). New York: Atherton Press.

Bloustein, E. (1964). Privacy as an aspect of human dignity: An answer to Dean Prosser. *New York University Law Review, 39*, 962–1007.

Brady, B. (2008, January 13). Prisoners to be chipped like dogs. *The Independent*. Retrieved August 22, 2011 from http://www.independent.co.uk/news/uk/politics/prisoners-to-be-chipped-like-dogs-769977.html?pagewanted=all

Business Wire. (2007, November 28). VeriChip corporation to unveil plans for self-contained implantable RFID glucose-sensing microchip at Grand Hyatt in New York on December 4. *Business Wire*. Retrieved August 19, 2011, from http://www.businesswire.com/news/home/20071128005305/en/VeriChip-Corporation-Unveil-Plans-Self-Contained-Implantable-RFID

Clarke, R. (2007). Appendix to what überveillance is and what to do about it: Surveillance vignettes. In K. Michael, & M. G. Michael (Eds.), *From dataveillance to überveillance (uberveillance) and the realpolitik of the transparent society*. Wollongong, Australia: University of Wollongong.

DeCew, J. W. (1997). *In pursuit of privacy: Law, ethics, and the rise of technology*. Ithaca, NY: Cornell University Press.

Feinberg, J. (1980). The nature and value of rights. In *Rights, Justice, and the Bounds of Liberty: Essays in Social Philosophy* (pp. 143–158). Princeton, NJ: Princeton University Press.

Filipova-Neumann, L., & Welzel, P. (2010). Reducing asymmetric information in insurance markets: Cars with black boxes. *Telematics and Informatics, 27*, 394–403. doi:10.1016/j.tele.2010.03.003

Fried, C. (1984). Privacy: A moral analysis. In F. Schoeman (Ed.), *Philosophical dimensions of privacy* (pp. 203–222). Cambridge, UK: Cambridge University Press. doi:10.1017/CBO9780511625138.008

Gavison, R. (1984). Privacy and the limits of law. In F. Schoeman (Ed.), *Philosophical dimensions of privacy* (pp. 346–402). Cambridge, UK: Cambridge University Press. doi:10.1017/CBO9780511625138.017

Hill, T. J. (1984). Autonomy and benevolent lies. *The Journal of Value Inquiry*, *18*, 251–297. doi:10.1007/BF00144766

Inness, J. C. (1992). *Privacy, intimacy, and isolation*. New York, NY: Oxford University Press.

Iqbal, M. U., & Lim, S. (2007). Privacy implications of automated GPS tracking and profiling. In K. Michael, & M. G. Michael (Eds.), *From dataveillance to überveillance (uberveillance) and the realpolitik of the transparent society* (pp. 225–240). Wollongong, Australia: University of Wollongong.

Johnson, C. G. (2010, September 7). ACLU raises questions about microchip tracking of preschoolers. *California Watch*. Retrieved August 16, 2011, from http://californiawatch.org/dailyreport/aclu-raises-questions-about-microchip-tracking-preschoolers-4476

Kosta, E., & Bowman, D. M. (2011). Treating or tracking? Regulatory challenges of nano-enabled ICT implants. *Law & Policy*, *33*(2), 256–275. doi:10.1111/j.1467-9930.2010.00338.x

Kurkovsky, S., Syta, E., & Casano, B. (2011). Continuous RFID-enabled authentication: Privacy implications. *IEEE Technology and Society Magazine*, *30*(3), 34–41. doi:10.1109/MTS.2011.942306

Michael, K., & McCathie, L. (2005). The pros and cons of RFID in supply chain management. In *Proceedings of International Conference on Mobile Business* (pp. 623-629). Los Alamitos, CA: IEEE Computer Society.

Michael, K., & Michael, M. G. (2007). A note on uberveillance. In K. Michael & M. G. Michael (Eds.), *From dataveillance to überveillance (uberveillance) and the realpolitik of the transparent society*. Wollongong, Australia: University of Wollongong. Retrieved June 3, 2011 from http://works.bepress.com/mgmichael/24

Michael, K., & Michael, M. G. (2011a). Implementing namebers using microchip implants: The black box beneath the skin. In J. Pitt (Ed.), *This pervasive day: The potential and perils of pervasive computing* (pp. 1–51). London: Imperial College Press.

Michael, K., & Michael, M. G. (2011b). The social and behavioral implications of location-based services. *Journal of Location-Based Services*, *5*(3-4), 121–137. doi:10.1080/17489725.2011.642820

Michael, K., Roussos, G., Huang, G. Q., Chattopadhyay, A., Gadh, R., Prabhu, B. S., & Chu, P. (2010). Planetary-scale RFID services in an age of uberveillance. *Proceedings of the IEEE*, *98*, 1663–1671. doi:10.1109/JPROC.2010.2050850

Moore, A. D. (2010). *Privacy rights: Moral and legal foundations*. University Park, PA: Pennsylvania State University Press.

Nagel, T. (1998). Concealment and exposure. *Philosophy & Public Affairs*, *27*(1), 3–30. doi:10.1111/j.1088-4963.1998.tb00057.x

Nissenbaum, H. (2010). *Privacy in context: Technology, policy, and the integrity of social life*. Stanford, CA: Stanford Law Books.

Parent, W. A. (1983). Privacy, morality, and the law. *Philosophy & Public Affairs*, *12*(4), 269–288.

Rachels, J. (1975). Why privacy is important. *Philosophy & Public Affairs*, *4*(4), 323–333.

Raz, J. (1988). *The morality of freedom*. Oxford, UK: Clarendon Press, Oxford University Press. doi:10.1093/0198248075.001.0001

Reiman, J. H. (1976). Privacy, intimacy, and personhood. *Philosophy & Public Affairs*, 6(1), 26–44.

Rosenberg, I. B. (2007). Involuntary endogenous RFID compliance monitoring as a condition of federal supervised release - Chips ahoy. *Yale Journal of Law and Technology*, 10, 331–359.

Rotter, P., Daskala, B., & Compano, R. (2008). RFID implants: Opportunities and challenges for identifying people. *IEEE Technology and Society Magazine*, 27(2), 24–32. doi:10.1109/MTS.2008.924862

Rubel, A. (2007). Privacy and the USA patriot act: Rights, the value of rights, and autonomy. *Law and Philosophy*, 26(2), 119–159. doi:10.1007/s10982-005-5970-x

Rubel, A. (2011). The particularized judgment account of privacy. *Res Publica (Liverpool, England)*, 17, 275–290. doi:10.1007/s11158-011-9160-4

Stephan, K. D., Michael, K., Michael, M. G., Jacob, L., & Anesta, E. (2012). Social implications of technology: Past, present, and future. *Proceedings of the IEEE*, 100(13), 1752–1781. doi:10.1109/JPROC.2012.2189919

ADDITIONAL READING

Etzioni, A. (1999). *The Limits of Privacy*. New York: Basic Books.

Levmore, S. X., & Nussbaum, M. C. (Eds.). (2010). *The Offensive Internet: Privacy, Speech, and Reputation*. Cambridge, Mass: Harvard University Press.

Mill, J. S. (1863). *On Liberty* (2nd ed.). Boston: Ticknor and Fields.

Pennock, J. R., & Chapman, J. W. (Eds.). (1971). *Privacy. Nomos* (1st ed.). New York: Atherton Press.

Philosophical Dimensions of Privacy: An Anthology. (1984). Cambridge [Cambridgeshire], New York: Cambridge University Press.

Posner, R. A. (1977). Right of Privacy, The. *Georgia Law Review (Athens, Ga.)*, 12, 393.

Powers, M. (1996). A Cognitive Access Definition of Privacy. *Law and Philosophy*, 15(3), 369–386. doi:10.1007/BF00127211

Scanlon, T. (1975). Thomson on Privacy. *Philosophy & Public Affairs*, 4(4), 315–322.

Schoeman, F. D. (1992). *Privacy and Social Freedom. Cambridge studies in philosophy and public policy*. Cambridge, New York, NY, USA: Cambridge University Press. doi:10.1017/CBO9780511527401

Schoeman, F. D. (1992). *Privacy and Social Freedom. Cambridge studies in philosophy and public policy*. Cambridge, New York, NY, USA: Cambridge University Press. doi:10.1017/CBO9780511527401

Solove, D. J. (2002). Conceptualizing Privacy. *California Law Review*, 90, 1087–1155. doi:10.2307/3481326

Solove, D. J. (2004). *The Digital Person: Technology and Privacy in the Information Age. Ex machina*. New York: New York University Press.

Thomson, J. J. (1975). The Right to Privacy. *Philosophy & Public Affairs*, 4(4), 295–314.

Van Den Haag, E. (1971). On Privacy. In J. R. Pennock, & J. W. Chapman (Eds.), *NOMOS XIII: Privacy* (pp. 149–168). New York: Atherton Press.

van den Hoven, J., & Weckert, J. (2008). *Information Technology and Moral Philosophy* (1st ed.). Cambridge University Press.

Wenar, L. (2005). The Nature of Rights. *Philosophy & Public Affairs*, *33*(3), 223–252. doi:10.1111/j.1088-4963.2005.00032.x

Westin, A. F. (1967). *Privacy and Freedom*. New York: Athenaeum.

Zimmerman, M. J. (2010). Intrinsic vs. Extrinsic Value. In Edward N. Zalta (ed.), *The Stanford Encyclopedia of Philosophy (Winter 2010 Ed.)*, Retrieved September 16, 2011 from http://plato.stanford.edu/archives/win2010/entries/value-intrinsic-extrinsic/

KEY TERMS AND DEFINITIONS

Autonomy: The capacity of making determinations and decisions for oneself, or to govern oneself, including the ability to determine what one values and to act accordingly.

Constitutive Value: An element of something intrinsically valuable, and something in virtue of which the intrinsically valuable thing is valuable. By way of example, music may be a constitutive element of a life with rich aesthetic experience. If the life with rich aesthetic experience is intrinsically valuable, then music is constitutively valuable insofar as some of that life's value is in virtue of music.

Instrumental Value: Value based on consequences. Something is instrumentally valuable insofar as it helps cause, or make more likely, some other thing or state of affairs which is itself valuable. Money, for example, is instrumentally valuable because its value stems solely from being useful in doing other things that are valuable.

Intrinsic Value: Having value in and of itself, having value in its own right, or having value that does not depend on consequences. By way of example, a hedonist might say that pleasure is intrinsically valuable because the value of pleasure does not depend on whether pleasure has good consequences or any other good.

Privacy: The precise nature of privacy is disputed. As deployed in this paper, privacy is the condition of others having limited ability to make inferences about a person. Other commentators describe privacy as the condition of having control over information about oneself or the condition of others having limited access to information about oneself.

Rights: Moral rights are valid claims of one's moral due. They can be distinguished from legal rights, which depend on a legal system to be coherent, and from all-things-considered judgments of what must happen. The nature of rights is the subject of significant philosophical dispute.

Panel 14.1: The Social Implications of Radio–Frequency Identification
Moderator, Mr William Herbert, USA
Dr Katherine Albrecht, CASPIAN, USA
Professor Roger Clarke, Xamax Consultancy, Australia
Professor Rafael Capurro, IRIE, Germany
Dr Mark Gasson, University of Reading, England
Mr Amal Graafstra, RFIDToys, USA

Expert panel hosted at IEEE ISTAS13 on the 8 June 2013 at the University of Wollongong
http://www.youtube.com/watch?v=dI3Rps-VFdo

PANEL

William Herbert: Good afternoon everyone. My name is William Herbert, and for identification purposes only I am the Deputy Chair of the New York State Public Employment Relations Board. You may be wondering why am I here. In fact, my scholarship has been involved with issues involving RFID, GPS and other forms of technology, from a legal perspective. I was asked to moderate, I think partially, this panel because of my background in labour relations, in which we have conflicting views frequently in labour, and my agency's role is frequently brought in to try to bring some kind of bridges between varying positions on issues, at least in the workplace. We have over the past two days been very fortunate to hear very diverse viewpoints on the issue of RFID. And I thought it was appropriate that we try to bring those diverging voices together in seeking to bring some degree of bridging of these different ideas to try to aim towards bringing some degree of harmony about a perspective, or at least the first steps towards that perspective. As Roger Clarke mentioned earlier in his talk, there is a need for this kind of dialogue and I think this panel will be a very good first step or second step in that process.

So the question I'm going to be asking for the panellists today is: can societies develop a balanced response to radio-frequency identification (RFID)? And when I use the word RFID, I'm discussing both the technology, not limited to implants, but just the technology itself. So with that question, I'm going to first ask Roger to discuss whether societies can develop a balanced response to RFID technology.

Roger Clarke: Yes. What I normally do when I start a presentation is to say, "I'm actually an e-business consultant. Oh, and I also do some research and I'm also an advocate." And, so it's easy for me to say I've been in the IT industry 40 years. While there are technologies that are designed for evil purposes, to blow people up for example, the vast majority of technologies in the information technology arena are not inherently evil, or indeed inherently good. It's what we do with technology that matters- it's the framework, the context, the value systems. So, the word "balance" is a good one. There are good applications of RFID tags. The one that I'm not sure whether everyone will completely agree with me, but the example I frequently use is of RFID in the supply chain, up to the retail shelf... I have a lot of trouble thinking of evil or bad if they can make that work. But from there onwards, I get very, very concerned about it.

William Herbert: Well, let me follow up with you on the question of the accumulation of data and ways in which you can have a balanced approach on the issues of who controls the accumulation of data and the means by which that data can be then utilised for potentially evil purposes.

Roger Clarke: Well, I've actually gone beyond that now, because I spent the 80s and 90s on working on data surveillance, and now I believe we've reached a point where we have to start saying "no". We actually have to build in *forgetting* into our systems and I think that's a new departure that I'd never argued in the 1980s and 1990s. We are now collecting a huge amount of data just in case and we have got to get out of that mentality. We have got to talk about data destruction, data destruction at the earliest available point, and non-collection. That is to say, in sensing of data, extraction of that which we might happen to need is part of the processing but the rest we should let go of in the buffer. And so there out it's gone, been flushed by the next transaction that flowed in and we've retained that, which we have demonstrated we have a justification for collecting, and we're only holding that as long as we've got the clear justifications. We've got to get rid of this "just in case" mentality. We've got to teach our technologies to forget.

William Herbert: Raphael, you discussed yesterday the Article 8 of the European Convention of Human Rights and developments in Europe on the issue of data collection. I was just wondering what your response is to Roger's point about the idea of having data that would then essentially disappear and as a means of developing architecture aimed at avoidance of collection of data from RFID sources.

Raphael Capurro: May I say first that I agree completely with Roger- that good or bad are not properties of things; they are second order categories that we apply to things with properties; not properties of things. So they are a product of relations between humans and the world and so on. So they depend on the context. This is my first answer to your question. The second thing is the number or the amount or the quantity, of these products we are now putting into the market. I remember the case of cars 100 years ago. If you have one car, okay, then you don't need car regulation, but if you have thousands or millions of cars… it happened very soon in Europe that we had to regulate cars, and streets, and so on. So I agree, again, with Roger that we need probably very soon legal regulations with respect to RFID. Every technology changes the relationship between humans and between humans and the world. So no technology is neutral…so something is changing when we create a new technology, the way we are. So no technology is neutral. It is not just a question of bad or good use; it is a question that it changes something of our self-understanding, for good or for bad. So the question of dual use is a secondary question.

And my last remark is about the opinion of the EGE concerning the use of these kind of devices with regard to acceptance by a court. This is a speculative case because it is only related to implants as far as they theoretically could be part of an honour system used for surveillance in non-medical settings. So you know there are a lot of "ifs", and one important if is that, as far as I know, it is not possible, but please correct me if I am wrong, physically possible, to have implanted devices that can be part of an honour system, say, of hundreds of thousands of kilometres, because the signals going out and into your body will destroy your body. So this is what I heard some years ago when we discussed this in Brussels, and so this is why this is science fiction, if you want. But we wanted when we wrote this, we wanted to include this possibility, just because you never know if an engineer finds some way of, you know, in which this kind of system would be possible. So in this case, only in this case, when you have implants, part of an honour system, for surveillance purposes in non-medical settings, in this and only in this case we said, we didn't say "no" at all, we said in this case then with the approval of a court. Okay so this is just an extreme possibility, okay? In all other cases, we were a little bit, how could I say?... reluctant with regard to surveillance purposes. But I think that in the case of medical situations, this can be useful and even in the case of non-medical uses. But the main point we made from an ethical point of view was the question of how invasive the systems are, and if there are less invasive methodologies, for instance, a cell phone, then you'd use that, and so on. So you see, this is, I still think this is good thinking. I still think we are not necessarily obliged to use, you know, the whole society, these kinds of invasive technologies.

William Herbert: Mark, I know you presented the work that you're doing at the University of Reading. I'm just curious: in terms of the work that's being done there, is your work being done in conjunction with ethicists or people who are studying the issue of RFID and that type of technology, as well as implants, and sort of joint work that is being done, for example, in bioethics?

Mark Gasson: Yeah. I mean, we've been involved with several large-scale European projects which are interdisciplinary, because we've appreciated for quite some time the benefits of talking to other people outside our own discipline. Thinking in a very blinkered way is extremely limiting. So yeah, I mean, certainly we work very closely with legal, ethical, social scientists, but I think the problem with RFID is that it tends to get enormously bad press because the fundamental limitations of the technology are exaggerated. The idea that you can globally track someone through an implanted RFID tag is fiction. You could put a reader by a doorway, and if I had an RFID tag and I walked through that doorway, you could read a number. Well, okay. You could potentially collate information about when that tag walked through that door, passed through that door, at certain periods during the day. You could collate that information over months, but actually what use is that data? It certainly doesn't necessarily link explicitly to me. You may be able to data-mine to some degree in order to work out it's me, but actual, what actual value is there to that?

William Herbert: In terms of the work you're doing with these, with other disciplines, how is that, the approach from the legal perspective or the ethical perspective, affected your testing and your modelling in terms of the programs that you're doing at the University?

Mark Gasson: Well, we're particularly interested in the limitations of the technology. So what we tend to do is explore the potential applications that they have and then feed this information into the other groups. So it's through this mechanism that that goes on to then potentially feed into legislators and other stakeholders. So through those mechanisms, we're exploring what is possible. We're not necessarily saying, "Well, this is possible and therefore we should commercialise it and there should be a technology that's used in this application."

William Herbert: What about Roger's point about there being always the potential for misuse? And have you structured any kind of way of examining that?

Mark Gasson: The problem is, with a lot of technologies, that there's the possibility of function creep. So if something like RFID that, if you're using it in a warehouse tracking scenario, as Roger was saying, then it has a perfectly valid and safe application. It's when the RFID tag isn't disabled and then it goes out and is used by the people, it may be in their clothing or in their shoes, because it's then a legacy. It's left in there from its previous use. So that sort of function creep phenomenon isn't unique to RFID, but certainly there are valid privacy concerns which do need to be addressed, but they need to be addressed in the context of what the reality really is.

William Herbert: Well, a question I have is then how you expressed concern about the press that the technology is getting. Do you agree with Raphael that technology is not neutral? Roger expressed that technology is inherently neutral; it's just a matter of, the manner in which it's utilised. I think Raphael takes a different perspective on that. What's your perspective?

Mark Gasson: I take the perspective really, certainly from our research perspective, that the technology is neither inherently evil, and it's not inherently good. It's how you then go on to apply that technology.

William Herbert: Now, in terms of your point about the concerns about the way RFID technology is perceived through the media, what steps would you conceive would provide a more balanced presentation of information? What would be the means to get that information out to the general public to have a more balanced perspective?

Mark Gasson: The problem is, in part, is that it's the sensational stories that really capture the imagination, and to say, "Well, basically there are these implants, these RFID chips that you may end up having implanted or you may end up carrying around. And there's all these problems and it's going to cause you an enormous amount of privacy, invasive security issues." Those are the stories that actually the media is more likely to pick up and perpetuate. So actually I think it's a very difficult issue. But certainly if you look in the UK, we've got an enormous problem with CCTV cameras. We've probably got more CCTV cameras in the UK than just about any other country on the planet. Now, this has been discussed quite readily in the press and it really is a problem. If it's surveillance on a 24/7 basis and you've got automated technologies which can capture your image, there's facial recognition or gait recognition. It can work out it's you, where you've been, what you've done. There's an enormous amount of data-mining that you can do there. In comparison with the problems that we currently have with RFID, that far outweighs the privacy invasiveness of the RFID technologies that we have at the moment.

William Herbert: Katherine, I know you've expressed concerns not just simply about implants but also about the use of RFID technology in general. So the question I have for you is what means can you see society applying a balanced approach to that technology?

Katherine Albrecht: I think it's interesting, because I've been working on this issue since 2002, when I first discovered the Auto-ID Centre at MIT, and it was in 2003 that a group of civil liberties organisations, over 40 of the world's leading privacy advocates and civil liberties groups, got together and we drafted a document, the *Position Paper on the Use of RFID in Consumer Products*. And one of the things we did, and this was the ACLU, this was EPIC, EFF, Privacy International in the UK etc …What we looked at were the applications that we considered to be acceptable uses of RFID, and those that we considered to not be acceptable. And it's very interesting that since that time, we have seen RFID where we said, "Go ahead, use it in the back room. Use it in the warehouse. Use it in a supply chain. But when it gets to the point of human beings, that's where the line needs to be drawn."

William Herbert: So the question …

Katherine Albrecht: There was a pretty universal agreement among those 40 different organisations and individuals that the use of RFID on human beings, to track people, was inappropriate.

William Herbert: So you're saying then that this group of civil liberties groups took the position that all forms of using RFID with humans should be prohibited?

Katherine Albrecht: That was the position that was taken in 2004. And that-

William Herbert: And is that your position today?

Katherine Albrecht: Well, that was the position not only of CASPIAN, but that was the position of-

William Herbert: Is that your position today?

Katherine Albrecht: It is, but if I could finish my point, that was not only our position; that was also the position of the RFID industry itself. So EPC Global's, Kevin Ashton, one of the primary developers of this technology, pretty much everybody across the board who was promoting RFID as a technology at that point, said that they were in absolute agreement. And in fact, they publicly stated, and you can see this on television and news articles, that they said, "Absolutely, this is only for product tagging. We'll never use it for tracking people."

William Herbert: So let me just ask you, in terms of this group that you're describing, so for six years, they've been advocating prohibiting the use of RFID for entry and exit monitoring in workplaces?

Katherine Albrecht: I guess I'm trying to make a slightly different point. And the point that I want to make is maybe the function-creep point that's been made several times, which is that you can have a point at which even the proponents of the technology say, "Here's where we draw the line," and then a year or two goes by, three or four years go by, that line moves. So I think it's a moving target. I don't know that we can reasonably say at this point whatever restrictions this group or any other group agrees to put on RFID. I don't believe that it's going to stop there because I've watched it not stop there before.

William Herbert: Okay. So I just, what would be, right today, what would be your position about a balanced approach to the use of RFID in dealing with humans?

Katherine Albrecht: I'd probably stand by my earlier position, that RFID is dangerous for use on human beings.

William Herbert: Period?

Katherine Albrecht: Should not be used on human beings, correct.

William Herbert: So that would be inclusive of RFID on tags in entering and exiting …

Katherine Albrecht: Correct.

William Herbert: … premises?

Katherine Albrecht: That's my position.

William Herbert: Okay. Amal, you take, I think, a different perspective on that, and I'm just curious… you're someone who's taking sort of individualism to a new stage in terms of your approach to RFID. What would you see as being an appropriate means of balancing your personal desire with a societal approach to developing public policy in the area of the use of RFID with respect to humans?

Amal Graafstra: Well, actually, I've been talking with Mark Gasson a little bit about the body integrity laws in the EU, and I think they're interesting, that a human has a right to do whatever they want with their own body… that premise. That concept is really interesting. I'm kind of sad to say that we don't have that, not that I know of anyway, in any capacity in the US. I think that beyond that, when you start talking about, you know, technology, a balanced approach to the technology, I think there is a lot of great uses that it can be used for, even with humans.

William Herbert: I'm just curious, do you see any rule in regulating limits about the use of RFID by individuals like yourself, or anyone else?

Amal Graafstra: I would say in a do-it-yourself context, in a non-commercial context, I don't think there should be any limitations. I think people should be free to experiment with their bodies in much the same way there are piercing artists or tattoo artists are, you know, applying their craft, so …

William Herbert: But in most states in the United States, body tattooing is regulated.

Amal Graafstra: Yeah.

William Herbert: So I'm just curious whether or not you would think that, just like body tattooing is regulated, that piercing your body to take an implant should be subject to regulation as well?

Amal Graafstra: Yes, I think the medical, the procedure of receiving an implant should be regulated, just like any medical procedure. But the technology itself, I think for the individual and individual's use, I don't think that should be prohibited.

William Herbert: So would you be supportive of licensing of people, doctors, to, in terms of people who are going to be doing the implants? Just like tattoo artists can be licensed in some states?

Amal Graafstra: I think the concept of licensing to perform the procedure is interesting. I think that probably doctors are already licensed to do that. They, you know, perform different various, you know, birth control implants, things like that. I think their licensing question could be extended to piercing parlours and things, because there has been a lot of confusion in that arena as to whether or not they're legally able to do this procedure.

William Herbert: So you support that form of regulation? With respect to who is the one providing the implant?

Amal Graafstra: Right. I mean, in today's world, it's a confusing time for people that do want to experiment with this technology in themselves. So yeah, regulating it I think is a good idea.

William Herbert: Katherine, I think you would agree with that, that at least the minimum, that would be something that you would support, which-

Katherine Albrecht: I would actually take the opposite position. As a libertarian, I believe you should have the right to do whatever you want and no-one has the right to regulate or get involved or legislate about that.

William Herbert: See I thought this was one area where you would agree, but okay. So from a libertarian perspective: you support the notion of free-

Katherine Albrecht: Amal, can do whatever he wants.

William Herbert: … he wants to do.

Katherine Albrecht: He can take a tattoo, he can cut off an arm, he can do whatever he chooses with his own body.

William Herbert: Okay. Now, the next question I want to move towards is the question about developing a knowledge base to examine the means of regulation and how this society can approach this problem of conflicting information in the society to develop a framework for having a discussion about regulation on some form with respect to RFID, and I'd like to start with Roger, because I think Roger started talking about that towards the end of his presentation.

Roger Clarke: Yeah. It's certainly essential that professionals take the responsibility to inform the public, so we've each got to be writing the papers that are for this kind of a conference, but also for the next level, so the intelligent, interested, educated public, and we've got to be prepared to go that bit further and reach out to the great unwashed public who aren't going to understand the long words and get some simple 15-second grabs out there on television. But we've got to go further than that. We have got to not just look at each technology in isolation. Now, that's my difficulty with Mark and Amal's position in relation to this proposition that global tracking with RFID is nonsense. Well, by itself, the technology doesn't achieve global tracking. That's quite clear. For starters, you've got to have readers. But there's a very simple structure whereby RFID tags do provide a global, a widespread tracking mechanism. You have got to look at the technology within a context of other technologies, within a context of other social institutions, within a context of other existing databases.

William Herbert: So you're now moving more into a question of data collection, going back to your position about the need for the data to disappear through the architecture of design?

Roger Clarke: That's going to be one measure that can be taken that will be a big contributor to overcoming some of these problems, but the point I'm making here is that in informing the public, it's not just the technology, on what range, with how many megahertz and what size is that thing and does it actually hook itself into a location within the body or does it wander? We need all of those things about the technology, but we need to also put the technology in a context and say that, "You do realise that this intersects with all of the registries and all of the databases, and all of the multiple systems that are capable of picking up signals from this highly promiscuous device?"

William Herbert: When you referred to professionals getting together, just curious; what professions do you view as being appropriate to be at the table, to have this discussion?

Roger Clarke: I'll answer it from an Australian perspective, because it's easiest that way. In Australia, there's a couple of engineering organisations, IEEE and IEAUST, and the Australian Computer Society (ACS). And they're the front three that come to mind in our particular context. Now, each of the countries has a rather different structure of professional bodies, so in the United States, the ACM isn't quite the same as the ACS and the BCS. But it's those sorts of organisations, plus the engineers.

William Herbert: Okay. So, in addition to those professionals, what about advocacy groups like Katherine's organisation or something similar in Australia to get together and work with the engineers to hear from advocacy groups, concerns about the technology, as well as lawyers, ethicists, people who may not know the technology per se but can add value to the discussion?

Roger Clarke: Absolutely. I'm currently chair of the Australian Privacy Foundation. It is absolutely crucial because there are people in the Australian Privacy Foundation who bring a very different perspective to the one I bring, which is an e-business, IT, 40-year professional perspective. And there are other people who come to it quite differently, and their voices must be heard. And they sometimes disagree with mine and I have to crawl away and let somebody else do that bit, it's important.

William Herbert: Katherine, your organisation, you explained the proposed legislation that you're advocating for, but I'm just wondering, what other professionals do you bring into the discussion as a part of your advocacy to educate the public in terms of discussing with people with legal backgrounds, with philosophical backgrounds? Is that a part of your organisational structure, to have those different voices spoken before the organisation comes out with a position?

Katherine Albrecht: Well, as I mentioned yesterday, we're putting together an organisation to deal with a very specific issue, which is microchip implants in pets. For that, we're looking to bring together pathologists, veterinarians, policymakers, people concerned with animal health, hopefully some folks from IEEE. So obviously when you're creating a new venture, you're looking at all of these kinds of voices. I would say something like what you're discussing. I'd like to see historians, philosophers, politicians, lawyers, advocates, I mean, a whole group across the board of people involved.

William Herbert: And Raphael, I know that in Europe there's the, under the privacy directive, there's the the Article 29 Working Party. Are you familiar with that?

Raphael Capurro: A little bit.

William Herbert: Okay. Well, in terms of the work that you've done with the European Commission, what's the interaction between, you described yourself working with someone from Italy who has a legal background. What other voices were heard as you were developing your program, your Opinion?

Raphael Capurro: Well, we are a multi-disciplinary group, a European group on ethics, so biologists and philosophers and lawyers, theologians and so on. So there were many voices, but we invited also many experts and so on. And so we had workshops and open forums. But it was five years ago, so things change very quickly in this field. And I think there is no absolute liberty; there's always a question of where to draw the line in specific settings and cultures. I don't think this is a, just a universal technology that can be applied, should be applied equally in all cultures, just because people are different and the settings are different and the risks are different, of misuse and so on. So I think it's a complex issue. And also because the question of regulation is not just a question of legal, it's also a question of moral, it's a question that I decide for me to use it or not, if law allows this. And so it's a question of how much freedom we can allow in our societies, because there is no absolute freedom from nothing. So it's just a question of how free individuals in society want to be. In consideration of the risks, of the issues of control, of surveillance, of bad guys, or bad guys events, and whatever it is we are confronted with. It's a complex world, so-

William Herbert: But in light of globalisation and the fact that we're all here, many of us travelled very far, we know each other's works through the Internet, isn't there going to be, at some point, a need for more of a uniform approach to these questions? Because for example, if there's a country which prohibits someone voluntarily putting in an implant, he may not be able to get into that country. This is one example.

Raphael Capurro: Yes, interesting because we're discussing this now in Europe, for instance, this body scanning in airports. And for instance, okay, you're going to some place and want to go to the airplane and they say, "Uh uh, sir, you have something inside. Either you take it out or you're not allowed to go there." So I think there should be an international discussion about this, because if you say, "Okay, we in the States do this and do that, and you in Europe do this and do that." And these are not in some way connected, then we'll have chaos. So we need some kind of standards about this, basic standards. And then we need a deep discussion on the application, free application, also free in the sense of legal regulations, dependent on different cultures and different, how could I say, feelings of people with regard to this... because I am half as European, and Latin American European... So I think it's a very, you know, the body is something extremely personal, right? I am probably more conservative. But others are different. So it's really a very complex thing and I don't see a possibility of taking a general, you know, kind of standard for everybody and decide it.

But on the other hand, we have to do something if in case this technology, whatever this technology is, because it's so complex, it's not just the implants, but everything we're using. And we are seeing this now with the Internet and with all these kinds of technologies we have, cell phones and all these kind of things, robots and bionics and so on. And I have no simple solution for all these questions, and I think nobody has these. And this is why I think the most important thing is to keep tracking this discussion internationally and inter-culturally, and this is a start, this conference, I think. And not fixing positions dogmatically. That's the worst thing we can do. And don't be afraid. I think the only thing to be afraid of is bad counselling. But don't be naive, of course, again. So sorry, I have no more solution than that, but I think this can help us to open, to be open to different solutions.

William Herbert: Mark, and you were describing before you're working with multi-disciplinary structure. Just wondering whether or not you can envision, in light of your background lab work that Professor Warwick and you are doing, that, for bringing in other disciplines to work together… What perspective do you have on that?

Mark Gasson: Yeah, I mean, I think there's great value in doing that, but I think when you start a discussion like that, you can't start from the perspective of, "this technology is fundamentally wrong, implanting it in the body is fundamentally unacceptable, and therefore we can't do that." That's not a tenable position to start from. So we have to be realistic. I mean, certainly the implants that Amal has, and there are probably 200 or 300 other people around the planet that have this sort of implant …

William Herbert: What, when you say, you use the word realistic, so I just wanted to follow up; what do you mean by realistic?

Mark Gasson: Well, I think you have to start from a sensible position. It's like saying, "Okay, mobile phones. You can track someone using the mobile phone, therefore, we should all get rid of our mobile phones." It's completely unrealistic.

William Herbert: I'm not asking the position where someone comes from asking sort of the professional groupings that would be appropriate at the table, to try to develop a consensus about an approach to moving forward. How you would envision that kind of dialogue?

Mark Gasson: I think that sort of discussion has to happen between the people developing the technologies, the people that are looking to commercialise the technologies, certainly a lot of the research is driven by commercialisation. The legislators, we tend to find that there needs to be a flow of information to the policymakers because largely, they don't understand the technologies and the implications of using and developing the technologies, and the ethicists who are active in the area of technology as well. I mean, this is a good base for sensible discussion.

William Herbert: Well, one, I mean, one of the balances that are inherent in this is a question between marketing and regulation, between commerce and regulation. And you had mentioned that commercial ventures are sort of supporting some of this research. In light of that, and that's also turning to the bio-ethics field as well, but the difference is in bioethics, it seems that there has been far greater concerns and more proactive steps taken in that field. And I'd like to know whether you see an ability to move from that model in which bioethicists are raising the issues simultaneously with the development of technology. Do you see that as being something that could work in terms of the research that you're conducting right now?

Mark Gasson: To some degree, but the type of research that you're talking about there is stemming from some very fundamental research paradigms which people are finding very difficult to cope with anyway. So for example, stem cell research. There's a lot that feeds from that fundamental research, so to say, "Well okay, we need to question whether we should allow that," when we're looking at engineering, we've got a broad range of technologies that are being developed and a broad range of contexts using a broad range of different research agendas. The problem is that it seems to be a continual chase to point the finger at the one that actually we should be concerned about now and RFID seems to be the one that currently people are finger-pointing and saying, "Right, well okay, this is the one that's going to cause us privacy concerns and security concerns, and therefore the one that we need to legislate."

William Herbert: Well, you know, the other area where there's been ethical analysis given before the research was conducted is the area of genetics. The question in the area of RFID is how comfortable you would be with that approach being done now that we know that there is an issue out there in terms of the public, whether it would improve or aid your research to have that kind of dialogue going on simultaneously with your research. And the understanding that, of course, the ethicists may have a very different perspective from your research and your approach.

Mark Gasson: Yeah. I think there is a lot of value in that, but the commercialisation of these technologies is a long, long way down, a lot further down the line than the other technologies that you're talking about- the genetic engineering technologies. So we're looking at a technology which is already commercialised, which is already in this room in a variety of contexts. So to say, "Well, okay, actually we may need to wind back a bit and we shouldn't be using it in this context," I think that's a very difficult position to take.

William Herbert: Well, we're discussing RFID generally, but I guess the issue then comes with respect to implants and ethics. And Katherine, what was the number of people who had implants in the United States? You gave that number yesterday.

Katherine Albrecht: Well, it depends. If you trust the VeriChip Corporation's numbers, they claim that about 2,000 people worldwide have been implanted. I know that they tried to implant 200 people at an Alzheimer's centre and they had about 50 people enrolled in a clinical trial.

William Herbert: So maximum as far as the information you have is 2,000 people worldwide?

Katherine Albrecht: Yes, that's the maximum that I know.

William Herbert: So it's relatively fresh issue in terms of marketing. So the question then comes with respect to RFID implants, whether or not there's a way of developing this through the IEEE SSIT society or in other forms for there to be discussions with lawyers, ethicists, tied to questions about what happens when it gets rolled out in the commercial field.

Mark Gasson: Yeah, but we also go back to the idea that, as Amal was saying, that the right to do what you want with your own body, if you want to implant an RFID tag, then I don't see why there should be legislation that stops you from doing that. I think that's what you're talking about …

William Herbert: I was asking you more about ways in which to have a dialogue over those issues to hear from various experts in the field, rather than allowing for regulation to be developed out of context.

Mark Gasson: Well sure, but I think these discussions are happening. Certainly the European projects that we've been involved with, the topic of RFID in general is heavily discussed, and the specific applications to implantable RFID tags is largely discussed as well. So I don't think that there's a void of discussion from these disciplines.

William Herbert: Well, could you just give us some examples in terms of organisations that are working in that field?

Mark Gasson: Well, largely we're looking at academic institutions involved in these European projects. So there are a range of, there were 24 institutions across Europe involved in the last European project that we were involved with, and that includes some enterprises, some small and medium-size enterprises as well.

William Herbert: Are you involved with the Internet of Things, the project tied with the questions of applying RFID generally?

Mark Gasson: We're involved with it in the context that our research falls under that category, but there are some very specific groups which are discussing that specifically. But I think the Internet of Things is just another label that's really applied to a technology which can become ubiquitous. And RFID being cheap enough, it really is that technology.

William Herbert: With respect to the EU and the EU has the data privacy directive, and also it has the Article 29 working party. Have they issued reports on the question of privacy in RFID? How much of their work has influenced your work?

Mark Gasson: Well actually, I can turn that on its head. I know that our work has directly influenced what they've been doing, so we've been directly feeding into those groups. So on those levels, there's already interaction.

William Herbert: Amal, in the United States, there isn't very much dialogue on RFID, as hard as you try and Katherine tries.

Amal Graafstra: Oh, probably not as much as you'd hope to have.

William Herbert: Right. Now, as an advocate for RFID implants, how would you see as being a means of developing a dialogue within the States on this issue, factoring in Katherine's organisation which takes a very strong position against implants, in trying to develop some kind of a balanced viewpoint towards this technology?

Amal Graafstra: Actually, I don't know that I would call myself an advocate of RFID implants. I would definitely call myself an advocate of personal use of RFID implants. I do have concerns about commercial use and commercialisation of implantable technology-

William Herbert: What are those concerns?

Amal Graafstra: Much the same concerns that Katherine has. How do you control the data? How do you control consent? What happens to the data after it's been collected? You know, that's why I think that any balanced approached to the technology should include legislation or a legal means of recourse for controlling, you know, what happens to your data after it's collected? How is it used? Is it sold or migrated? And then, you know, Roger's case, he was talking about building in forgetfulness into systems, and I think that has validity.

William Herbert: Katherine, technology moves rather rapidly and the law moves much slower... can you think of a means by which societies could develop an approach to regulation with would allow for the regulatory process to move at a faster rate to match the speed by which people like Kevin and Mark do their research?

Katherine Albrecht: Well, part of the problem with that approach is the only people who really know about RFID are the people who stand to profit from it, the people who are developing it, the people who are the proponents of RFID, outnumber probably 10,000 to one the number of people in the general public who might oppose the technology if they understood more about it. So I think it's difficult, whenever you get into a regulatory framework environment, particularly in the US and probably the same in Australia and elsewhere, that the people who come to the table to have the discussions tend to be the people with the very strong agenda in favour of whatever that is.

William Herbert: Well, how could you change the regulatory paradigm then?

Katherine Albrecht: Well, I think it would be very challenging to try to do that because, you know, as I've been trying to educate people about this issue for eight years now, people don't know about it. And again, you know, I go to testify before a state level body, for example, who might be considering legislation about this. I walk into the room and there are 15 lobbyists who have been flown in from all over the country at great expense, and then there's me. So, you know, I would be concerned about any kind of a framework like that being put together, developing policy right now, because the funding is all on the side of the people who want to see less restriction, more, even government funding. We had this come up back in 2004 where, within the United States, there actually was the General Accounting Office (GAO) that actually sent out a letter to government agencies asking them to spend taxpayer money on ways to enhance, increase and further the RFID industry. That was within my own government. So how can I trust those folks to have an honest discussion?

William Herbert: When you were testifying at various state legislatures, were you testifying in favour of regulation or against regulation?

Katherine Albrecht: We, well at that point, we were testifying in favour of simply labelling on products containing RFID: shoes, shirts, etc. That-

William Herbert: So that's a form of regulation that you would support?

Katherine Albrecht: That's pretty much the only form of regulation that we've ever called for is labelling so that people have, well that and then for implants, the ability to say "no," and not been penalised for saying "no".

William Herbert: Roger?

Roger Clarke: In order to get a process to work and to get institutions to reflect information, we need just a couple of simple laws, and after that, things flow. And the simple law is thou shalt assess, and in the process of doing assessment, thou shall provide information and thou shalt consult, and thou shalt consult all the more widely the more complicated and difficult the technology and its implications become. Now, we have lots of precedence for this. Environmental impact assessments have been around for a long time. The US and Europe used to do technology assessment quite well in the 70s. It died. The Office of Technology Assessment was one of the great institutions of the United States, and it died. Now-

William Herbert: And let me guess which decade it died in.

Roger Clarke: Yes. Look, switch hats again. I'm an e-business consultant, but about 20% to 30% of my work has been in privacy. Privacy strategy is hard to get organisations to do. I try to do that. But what I can get organisations to do is privacy impact assessment built on all those principles. Now, take a simple example. Let's get it away from implants, because that's down at the bottom right-hand corner of my diagram. Let's get a simple one. RFID tags applied to road-tolling. What have we got here in Sydney? What have got here in Melbourne? What have we got here in a range of other places? As a result of the way it's been done, we've got the denial of anonymous travel because of the way they implemented RFID, active RFID tags, rather than simple passive ones, but they just implemented it. They wanted to get rid of the cash booths and they didn't want to do anything complicated or think hard about the way in which people would load up or settle up on the bill or pre-pay the chip. They didn't want to think about that. They did no assessment, they did no consultation, and as a result, we're, well, I pay by cash because I send cash through the mail and break the law and they've never called my bluff because they know who I am. I'm waiting for that court case, because they deny anonymity. Now, what, roll it back, what should we have done? We should've had a requirement that they do an assessment, provide information to the public about what's actually going to be going on here, invite and bring in consultation from multiple perspectives, and then it would've been apparent to all concerned. And at the beginning, it would've been really cheap to have designed an anonymous option for road transport.

William Herbert: Well, the question is, in the United States, we do not have privacy commissioners. In Australia, you've got privacy commissioners, in the EU, and in Canada, there are privacy commissioners. Isn't a privacy commissioner's office the appropriate place for those kind of privacy assessments? Without getting personal?

Roger Clarke: Without getting personal, we have ample empirical evidence now that privacy commissioners have not been an effective mechanism to achieve that consultation. They have been used as shields by government agencies and in those jurisdictions where they relate to the private sector, they've been used as shields by the private sector. They're part of the bureaucracy. They should exist and they should have more powers than they've got, particularly in Australia, but they are not a substitute for that process. The privacy impact assessment process is needed and the privacy commissioner is merely one of the players. They're the bureaucrat that knows the Act and, to some extent, reflect the other bureaucrats' interests and, to some extent, reflect the interests of the public, but not usually much.

William Herbert: So have I heard you correct? Do you agree with Katherine's perspective that trying to get regulation through a governmental process is filled with problems because of the influence of lobbyists from the commercial sector?

Roger Clarke: That I agree with. I'm not a libertarian. So I have a stronger belief in regulation and the responsibility of governments to regulate. I'm also a professional and I also like technology, so I don't want stupid regulation and I don't want too much regulation and I don't want inefficient regulation, but I want more regulation than Katherine does. So our positions aren't quite the same on that, but there's quite a few theories in which we have agreement, yes.

William Herbert: Raphael, we're just discussing privacy assessments and I seem to me, I need to go back. When you were talking about privacy assessments who does the assessments in terms of your vision of going back to the pre-80s?

Roger Clarke: The organisation that is the sponsor of the project, and obviously it's now situational, but there's going to be a tolling company or a public private partnership between, say, the New South Wales Government and Toll Roads Inc. That sponsor has the responsibility, in my perfect legal situation, has the responsibility to undertake that assessment subject to some constraints, of course, on how they go about it.

William Herbert: So they would view, there would be like an environmental impact statement, they would have to prepare a statement before they implement anything and presumably someone can sue if that privacy and impact statement has not been prepared?

Roger Clarke: It's actually beyond a statement, and that's one of the things that we moved on from. Environmental impact statements, the history was organisations were required to do that, and of course it became internal and manipulated and supressed the important information, and it became non-credible, the public got upset, so they had to change it. It was opened out much more to an assessment process. Sure, there's a report at the end of it, but it's not the statement that matters, it's the process. The product is secondary. And there's quite well-established documented techniques for doing that.

William Herbert: Katherine, what's your view of that process of requiring entities to create privacy assessments that would then be presumably available to the public?

Katherine Albrecht: I think it'd be terrific. I think it'd be terrific.

William Herbert: And Raphael, what's your perspective on not moving away from the bureaucratic approach to regulation but more towards a self-regulatory process involving, requiring privacy assessments?

Raphael Capurro: Well, we have in Germany last month, our two or three hot topics. One topic was about the government allowing to go into your computer, which is sometimes more dangerous than allowing the government to go into your body, because we externalise a lot of our data in our laptops. So we called it a Trojan horse. The second hot point was about censorship and about, well, we have a big discussion on child pornography and all these things. What I want to say, is that I think you agree with me, that with this technology, the questions are all connected, more or less. And so the fear is particularly in Europe and particularly in Germany, if you allow the State to have too much transparency into your privacy, then we learn from our history, okay? So this is where I say again, there is a context and is history dependent. And the wounds of the past are still there in Europe. So, and it is different if we think and talk about this, than about leadership, which has translated into German as "der Führer", which is a German word for "leader". So we can't use this word just as you use the word "leadership", okay? I mean this because the language is very specific and it is loaded with wounds and with experiences of the past. So we use the word "leadership" in English but in Germany when we talk about leaders, we use the word "leaders" and not "Führers", for good reasons.

So what I mean is that if we want to have a discussion on this, we have to have broad decisions, not just about my body, because my body is just, you know, a point of connection. I was talking about the body as data. It is not just that I have decided on this body, but this body is a connection point of a complex system of relationships. So it is not just that I decide in a Cartesian body, "It is my body, separated from everything else." It was a nice Cartesian metaphor, but it doesn't exist. The body is a connection of things. And now, it is a connection with the information technology. And this is what I count as your point, and I say, "This is my body. I can do what I want." This is an abstraction, okay? My body is my body and it's your body and everything is more or less related, particularly through all kinds of devices.

So, and the problem is that on the one hand, we have the State and we have a more or less paternalist State philosophy in Europe, differently to the United States, which is good and bad, it depends, in which we trust the State to take care of some official things. But we don't want the State to go too much into our privacy. And on the other hand, we cannot say we are libertarians, because of our tradition partly, we have libertarians too but our tradition is more that the society's organised. So there are some rules that everybody accepts and so on, and so the range of liberty changes, also within Europe, the UK and Germany and Italy and so on.

And this is what makes this activism and so on so complex in Europe. As I told you, there's just these two cases about child pornography and about the State. And we don't like the State to have too much transparency, so when you speak about open and transparent society, some of us say, "Please, not too much transparency. Don't allow the State to mix too many databases and so no, because this transparent society can be the reverse." So this is a way we now try in some way to have some intransparent society, opaque, we can say, opaque society, so that it's not so easy to look through. And I think this is a challenge- how to have a kind of parts of our society that remain opaque for each other too, so that you don't have, I don't want you to look to me too much, okay? Better if I remain a little bit secret for other people, okay? Also, for the State, I think that this is the situation we have. And it is changing so quickly because of this technology is about transparency. And so we are putting walls and trying to make opaque some things, and sometimes walls are seen as something bad, but from the other perspective, something good. So this is, I have no solution again, for that, but the discussion is about this and every month we have a new discussion about new technologies and new things.

William Herbert: We have to end this here unfortunately because the time is running out, but I want to say that, Raphael, you did a perfect conclusion in terms of laying the framework for how we can step forward towards a balanced approach to RFID. I want to thank each and every one of you for participating.

Chapter 15
Neuroethics and Implanted Brain Machine Interfaces

Ellen M. McGee
Independent Researcher, USA

ABSTRACT

Transformations of humans through advances in bioelectronics, nanotechnologies, and computer science are leading to hybrids of humans and machines. Future brain-machine interfaces will enable humans not only to be constantly linked to the Internet, and to cyber think, but will also enable technology to take information directly from the brain. Brain-computer interfaces, where a chip is implanted in the brain, will facilitate a tremendous augmentation of human capacities, including the radical enhancement of the human ability to remember and to reason, and to achieve immortality through cloning and brain downloading, or existence in virtual reality. The ethical and legal issues raised by these possibilities represent global challenges. The most pressing concerns are those raised by privacy and autonomy. The potential exists for control of persons, through global tracking, by actually "seeing" and "hearing" what the individual is experiencing, and by controlling and directing an individual's thoughts, emotions, moods, and motivations. Public dialogue must be initiated. New principles, agencies, and regulations need to be formulated and scientific organizations, states, countries, and the United Nations must all be involved.

INTRODUCTION

Hybrids of humans and machines, facilitated by advances in bioelectronics, computer science and nanotechnologies promise to transform the nature of humankind. In the future, brain-machine interfaces will permit the emergence of humans who are essentially connected to bioelectronic devices. Brain machine interfaces (BMI), those technological interventions that establish direct communication pathways between the brain and an external device, are also referred to as brain

DOI: 10.4018/978-1-4666-4582-0.ch015

computer interfaces (BCI), or neuromotor prostheses (NMP). The interface interprets signals from an array of neurons and uses computer chips and programs to translate the signals into a desired action. Thus, increasingly, in the future man will have an intimate relationship with machines and become cybernetic organisms, science fiction's "cyborgs", humans who are intrinsically coupled to bioelectronic devices.

The purpose of this paper is to examine the history of the development of brain-machine interfaces in order to demonstrate the feasibility of these brain implants, to present some examples of the state of the art in this field, and lastly to elucidate the ethical and social challenges arising from this technology. The ethical issues delineated here fall within the growing field of neuroethics, "a term used to describe the study of the ethical, legal, and social implications of new technologies from neuroscience Inasmuch as neuroethics is a subfield of bioethics, it has adopted the principles and rules commonly utilized in the field of bioethics and applied them to the issues arising in the neuroethical domain. Thus, as in bioethics, the primary working principles employed are nonmaleficence, beneficence, justice and autonomy (Beauchamp and Childress, 1979). Nonmaleficence is concerned with the responsibility not to intentionally harm another, beneficence is the requirement that if one can do good one has an obligation to do so, justice refers to the fair allocation of scarce resources and autonomy emphasizes the duty to respect the self-direction of persons. Thus, anxieties arising from safety and efficacy, which fall under both maleficence and beneficence, will be addressed, as well as those of fairness and justice. In particular, the ramifications of brain-computer interfaces for privacy and autonomy, for ubersurveillance, will be explored. Finally, the chapter will conclude with suggestions for means to address the concerns raised, including principles, standards, a regulatory framework and a forum for discussion.

BRAIN COMPUTER INTERFACES

Two kinds of interfaces can be identified – those that input to the neural base and those that output or record electrical brain signals. Interfaces that input to the neural base include clinical devices that aim to restore function to body systems.

This type of interface is comprised of three varieties that are presently undergoing research: non-invasive, partially invasive and invasive. Non-invasive neural interfaces record brain activity from an external device mounted on the scalp. Recording of electrocorticographic activity from the cortical surface has been used to create games that read alpha and beta waves, and to allow patients, after extensive training, to detect, modify and use a computer to direct a cursor on a screen or to control lights, TV and stereo sound (Donoghue, 2006). EEG recording is the most widely studied non-invasive Brain Computer Interface; in addition, magnetoencephalography and functional magnetic resonance imaging are employed. These non-invasive methods suffer from poor signaling resolution due to interference from the skull, and the intensive and demanding training needed to operate the technology. Nevertheless, several commercial models are available to control gaming systems, educational applications and investigative medical applications (www.emotiv.com, www.neurosky.com). The company InteraXon has created a suite of brain training games, and has introduced a device called MUSE which measures brainwaves, and "allows you to control games, reduce stress, improve memory and concentration, and eventually to control devices directly with your mind." (Interexon) The first commercial effort of a computer interface designed for patients with locked-in syndrome, the *Intendix* lets users input text using only their brains; another application lets users create paintings. (Intendix)

Somewhat better results have been achieved using partially invasive brain computer interfaces. In this type, the device is implanted inside the head, but on the surface of the brain rather than inserted into the brain. These have the advantage of producing better signals than non-invasive devices, but, unfortunately, also have the potential to form scar tissue in the brain.

The third type of brain computer interface is surgically implanted in the brain; these are implanted directly into the grey matter of the brain. These devices produce the highest quality signals, and as materials are developed which permit long-term implantation without scar tissue buildup and without degradation of the signal, they will become the preferred method of enhancement.

A report issued in 2006 noted that there are 175 American, 69 European and 54 Asian laboratories pursuing brain computer interfaces (WTEC, 2006). In North America, the majority of these laboratories are pursuing invasive brain computer interfaces.., while those pursued in Europe emphasize noninvasive modalities, and the development of biologically inspired robots.

Development of Brain Computer Interfaces

Devices produced to restore functions to disabled limbs, hearts, and brain are moving towards facilitating a close interface between brains and microcomputers. The field has progressed forward in stages, marked by advances in prosthetic technology and computer science. Initial use of prosthetics, used to restore function to disabled limbs, has progressed from the use of crutches and peg legs to the development of devices which employ microprocessors and turn out to be "bionic" joints, artificial knees, and smart legs. Research has progressed to a "biohybrid" limb actually using brain signals to directly control a prosthesis (Lawton, 2004).

At the same time a revolution in bioengineering and implant technology has resulted in millions of people using implants (pectoral, testicular, chin, calf, hair, hormonal, dental and breast) and millions more using cardiac pacemaker and cardiac assist devices. In developing these implants safer interfaces between neural tissues and the substrate micro probes have been produced.

Cochlear implants, which successfully make hearing possible for totally deaf individuals, have been in use since the 1980s, and have over 52,000 users worldwide. Prosthetic vision, studied since the 1960s, is in an earlier stage of development; it employs a diversity of visual stimulating implants: retinal, cortical, optic nerve, and biohybrid. One type, the retinal visual prosthesis delivers direct electrical stimulation to those cells that carry visual information to the brain from a tiny camera; data is then wirelessly transmitted to a microelectronic prosthesis. Another type of vision prosthesis, the cortical implant, directly stimulates the visual cortex, and can be either surface type implants, such as the Dobelle implant, or penetrating (The Dobelle Institute, 2000).

Applied neural control is also used for the bladder, to help Parkinson's patients control tremors, to treat epilepsy and to mitigate intractable depression. Deep brain stimulation (DBS) utilizes 'brain-pacers' that connect wires from a device implanted in the chest to targeted areas of the brain to control unwanted symptoms. There are more than 80,000 people already implanted with these devices, treating, in addition to Parkinson's disease, depression and epilepsy, dystonia, essential tremor, pain conditions and obsessive-compulsive disorders. A second generation of these devices will be capable of responding to brain activity, shutting off when not needed, and increasing action when required. The Rehabilitation Nano Chip (ReNaChip) is bidirectional – it is involved in monitoring and regulating both electrical input and output (Halley, 2010). The development of this capacity, where output interfaces record electrical signals from the brain, foreshadows the development of new forms of connection to humans. Neural activity can and will be monitored, interpreted and directed. These devices will permit the decoding of human intentions. Initially the intentions studied are those that presage limb movement, allowing actions to be initiated by paralyzed patients. The Modular Prosthesis Limb (MPL) project is planning to test a mind-controlled arm (Drummond, 2010). In June of 2011 scientists published results of experiments that demonstrated creation of an implant for the hippocampal system that restored lost memory function in rats. In addition, they were able to strengthen memory capacity in normal rats (Berger et al., 2011; University of Southern California, 2011). These advances and their probable future developments raise issues of privacy, autonomy and control.

These neuroprosthetics, many of which simply connect the nervous system to a device, represent a step towards the development of more comprehensive brain computer interfaces, which involve connection of the brain with a computer system. Brain computer interfaces, where the interface is surgically implanted in the brain, provide for greater energy efficiency and will eliminate the need for TVs, newspapers, GPS units, and cell phones or other separate devices (Maguire, 1999). Because they are more energy efficient, have greater bandwidth, and are invisible, these devices will be preferred to partially invasive or non-invasive interfaces.

Kevin Warwick, through projects entitled Cyborg 1 and Cyborg 2. has led experiments involving the insertion of an active microchip into the nerves of his left arm in order to link his nervous system directly to a computer. Initially in 1998, the device served to simply open doors, turn on lights, and heaters. In 2002, with the implantation of a more complex neural system into both Kevin and his wife, the first purely electronic communication experiment between the nervous systems of two humans was achieved. (Warwick, 2004)

For the most part, however, and because of the strictures for research on humans, animals are the research subjects, closely followed by trials on human patients. Preliminary work on linking the brain directly, with both local and remote manipulators, has been verified by neuroscientists at Duke University; they have trained a monkey to manipulate a motorized arm just by thinking (Lemonick, 2003). Additional work in this area has been led by Yang Dan at the University of California, who in 1999 implanted electrodes in cats (Whitehouse, 1999), and by Miguel Nicolelis (2011) whose research has focused on decoding the brain activity of rats and owl monkeys. In October of 2011, monkeys were enabled "to interpret the signals fed to their brains as a kind of artificial tactile sensation that allowed them to identify the 'texture' of virtual objects." (Guizzo, n.d.)

The first research to treat a human with a brain implant was led by investigators at Emory University in 1998. A "locked in" patient was enabled to communicate by using his thoughts to move a cursor (Headlam, 2000). This work resulted in the formation of a company to develop these interfaces for clinical use in patients. A device, called *Braingate*, allows the paralyzed to control a computer through a neural interface, and has been tested successfully on severely paralyzed patients since 2004 (Hooper, 2004). The first subject, a quadriplegic 25 year old, was effectively implanted with a brain chip which enabled him to check e-mail, play computer games, control a television, and turn lights on and off by thought alone; he succeeded in employing an artificial hand directed by his thinking alone. *Braingate2*, now in clinical trials, is expected to operate with a wireless interface, which has already been tested on non-human primates: "Sensors attached to the neurons in your brain would be implanted as with the original Braingate technology. Now, however, power, and control would be supplied by a radio frequency signal (RF) into the brain" (Saenz, 2009).

Further support is provided for brain-computer interface research by the United States Defense Advanced Research Projects Agency (DARPA), which has allocated over $24 million to support projects of six laboratories for brain-machine systems (Zimmer, 2004). These proposals seek to manipulate airplanes and robots through the mind alone. The United States National Aeronautic and

Space Administration's (NASA) Extension of the Human Senses Group (EHS) "focuses on developing alternative human-machine interfaces by replacing traditional interfaces (keyboards, mice, joysticks, microphones) with bio-electric control and augmentation technologies" (Dino, 2008). The goal of these efforts is the Cyber Soldier.

Using similar technologies, a computer directed rat was unveiled at the State University of New York, amid suggestions that it could search for people after an earthquake, and detect explosives or fulfill other dangerous tasks (Suny, 2002). In further research on monkeys, a primate at Duke University was implanted with a brain computer interface, and it succeeded in controlling, with its thoughts, a robot walking on a treadmill, continents away in Japan (Greenemeir, 2008). Brain implants have been used to detect activity in the parietal reach region of monkey's brains, the area where higher-level thoughts are initiated. This enables researchers to actually assess the degree of enthusiasm of monkeys and involves decoding higher brain functions. Such work enables understanding not only the goals and intentions of individuals, but also their moods and motivation (Begley, 2004).

In 2010, scientists demonstrated that an implant could read the thoughts of an individual and move a cursor, demonstrating that it will be possible to know what individuals are thinking. (Science Daily, April 2011). Both government-sponsored entities in Europe, Asia and the United States, and commercial projects, are involved in furthering these efforts. The commercial promise of this technology has stimulated more than three hundred private companies to investigate such devices, seeking cures for conditions caused by defective nerves, brains and spinal columns. Once clinical uses are implemented, non-clinical enhancement uses will follow. The developmental mid-point of these projects will involve humans, implanted with these devices, who will be constantly linked to each other, to their own perfect memory systems, and to the Internet. The boundaries between self and the other will be undermined; the community will be thoroughly part of the individual, the self will no longer have the possibility of being an isolated entity, nor will traditional concepts of privacy and autonomy be maintained. Brain-machine interfaces will permit humans not only to be constantly linked to the Internet and to cyber think, that is allow for invisible communication with others, but will also permit technology to lift information directly from the brain, even to direct the thoughts and actions of technology-enabled entities. Brain-computer interfaces, where a chip is implanted in the brain, will facilitate a tremendous augmentation of human capacities, and the radical enhancement of the human ability to remember and reason. Upcoming brain-machine interfaces may also enable the individual to achieve immortality through cloning and brain downloading; the thoughts, memories and emotions of an individual may be able to be downloaded and stored. Humans are already familiar with a life that is constantly online and connected to the Internet. The wearable, personal information structures used by Thad Starner (Boran, 2011) and the "Wear Cam" of Steve Mann (Mann, 1997) are early harbingers of technologies that combine wireless communication with information systems and allow the augmentation and enhancement of experiences, memories and networking. Gordon Bell and a team at Microsoft Research, for example, have been working on creating a digital archive of his entire existence, of every aspect of his life, including all images, sounds, memories and experiences (Bell and Gemmell, 2007). Memoto is being marketed as the world's smallest wearable camera; it can record every instance of life and have it stored and organized. (Memoto)

In September of 2012, The Journal of Neural Engineering reported on "a device that improves brain function internally, by fine-tuning communication among neurons." (Carey) The device, a brain prosthesis, restored memory functions in monkeys, and improved decision-making.

At Johns Hopkins, surgeons have implanted a deep brain stimulation device that has stimulated the growth of the hippocampus in several Alzheimer's patients. (Rubio) The Intel Corporation is proposing that consumers will adopt brain implants to control a myriad of gadgets by 2020, dispensing with the need for I phones, and keyboards to access the web. (Hsu)

The development of a wireless interface scheme, which was developed by Brown University's ArtoNurmikko, promises to facilitate the adoption of brain computer interfaces, since multiple chips could be implanted in the brain, allowing access to more neurons and permitting complicated thoughts to progress to action. (Patel)

Ethical Issues

Ethical issues surrounding brain computer interfaces are neither unique to this technology nor premature. Waiting until the full development, deployment and adoption of the technology is risky inasmuch as it is far more difficult to impose standards and regulations after the use of technologies than before. The proposal that scientific inquiry should be free and unfettered is itself a moral stance. It relies on the common tendency of computer specialists and scientists to compartmentalize and focus solely on technological challenges. There is a responsibility to evaluate the broader implications of scientific work; ethical deliberation on technological advances should precede not follow development, and engage the worldwide community, not just researchers. Otherwise technology drives society, and reinforces the stance that technology creates and operates within a deterministic system. Various scientific and computer specialist organizations have adopted and promulgated codes of professional responsibility. Invariably, these codes require professionals to consider the welfare of the public. The code of Computer Professionals for Social Responsibility states that professionals

"shalt think about the social consequences of the program you are writing or the system you are designing." (CPSR) The IEEE-CS Code of Ethics makes the public a priority in its code and agrees "to accept responsibility in making decisions consistent with the safety, health, and welfare of the public, and to disclose promptly factors that might endanger the public or the environment". (IEEE Code)

The trajectory of technology introduction will undoubtedly proceed from methods to restore species typical function to enhancement of sensory perceptions, and intelligence. The time for consideration, proposal and adoption of regulations to guide the development of this humanity altering technology is in the present, not after dissemination. In this instance regulation is called for, based on a preventive ethics, similar to the proposals for implementing the precautionary principle. Since the risks of harm are uncertain, but momentous, scientific research should be guided by public accountability. Frequently cited in regard to environmental concerns, the tenor of the precautionary principle should guide consideration of the benefits and burdens of this technology. This principle, of more common usage in Europe than the United States, holds, according to the Wingspread Statement, that "when an activity raises threats of harm to human health or the environment, precautionary measures should be taken even if some cause and effect relationships are not fully established scientifically." (Wingspread Statement)

Since brain computer interfaces will allow for 1) the enhancement and augmentation of human capacities, 2) the possibility of human immortality through cloning and implantation of bioelectronic chips with the uploaded emotions, memories and knowledge of the source human (McGee and Maguire, 2007), and 3) the prospect that homo sapiens may be superseded by the next stage in guided evolution, it is reasonable to purpose that regulation be debated, considered and adopted.

Thus a precautionary ethic in this instance would call for adoption of standards and regulation, not outright bans.

The ethical issues arising from this technology are myriad. Among them are safety, equity, costs, privacy, autonomy and justice. The ethical concerns are magnified when the devices are used for enhancement, rather than therapy. Devices that enable those with sensory, motor or cognitive disabilities to see, hear, move and remember simply raise the normal ethical concerns involving safety, efficacy, informed consent, and fairness. Such devices, when used for enhancement, however, raise, in addition, concerns that include: trepidation about safety, risk, and informed consent, issues surrounding manufacturing, upgrading, and scientific responsibility, apprehension about the psychological shock of enhancing human nature, worries about possible usage in children, concerns about increasing the divide between the rich and the poor, and most troublesome, issues of privacy and autonomy (McGee & Maguire, 1999). It is virtually impossible to predict all the effects of this technology, especially when mind uploading and cloning are factored in, yet a preventive ethic requires an attempt to fully explore the issues.

Of great complexity, is the question of principled standards for use of this technology in enhancement. Using the technology to augment human capacities, to radically enhance humans, to even assist humans in a further stage of evolution, which will involve phasing out of the embodied self, changes the kind of ethical issues raised. Already, the stage is set for allowing the unimpaired to acquire new sensory perceptions. As a sensory amplifier, the implantable chip will enable the user to augment sensory abilities enabling, for example, night vision, x-ray vision, seeing currently invisible wavelengths, and adding the ability to zoom. Humans might become able to hear sounds previously accessible only to animals, to smell with the precision of dogs, even to detect the earth's magnetic field. The achieve-ment of greater memory and intelligence will render obsolete strategies presently promoted to enhance memory, and pills marketed to promote improved recollection. Individuals will be able to access information at any time and any place, as it is needed. No longer will the individual operate within a zone of privacy; each person will be in constant and intimate contact with others. "The emergence of a Borg type collectivity or hive mind, where personal identity is lost and assimilation is the preeminent value is a real possibility" (McGee, 2008).

Further ethical implications arise from the possibilities of combining brain chips and cloning, or even creating nonbiological conscious selves (McGee and Maguire, 2007). If an individual's physical reality could be cloned and its narrative identity replicated, uploaded, and stored on a chip, then that individual could achieve immortality. This data could be collected by biological probes receiving electrical impulses and would enable a user to recreate experiences or even to transfer (transplant) memories from one brain to another. Whenever it is possible "to make a full duplex mind link between man and machine," (Pearson,) thought transmission between humans could be achievable, and backup copies of our brains could be made. One result of uploading minds is that immortality could be assured because uploaded minds would not age. Combining cloning with techniques for uploading and implanting the narrative identity, which makes a self a self, would then also become feasible. Very real concerns thus arise regarding the loss of an open future for the clone, and the impact on autonomy, uniqueness, and individuality.

Even more disquieting is the prospect of a post-carbon future where our minds would be copied to another medium, or exist in virtual environments. Predictions forecast that the brain will be successfully reverse-engineered by the 2020s, and that by 2045 artificial intelligence will vastly exceed the sum total of human intelligence. This event is termed the Singularity by Ray Kurzweil: "The Singularity will allow us to transcend these limitations of our biological bodies and brains ... There will be no distinction, post-Singularity, between human and machine" (Kurzweil, 2005). When a brain is scanned with all of its memories, emotions and intentions, it can be uploaded to a computer, or a robot. An individual could exist without a body; minds will survive in virtual environments and have experiences in virtual reality. This prospect is not necessarily the hope of all men; many prefer an embodied existence with all its vulnerabilities. Nor is it clear that this future vision for mankind is one that should be embraced.

Safety Issues

It is probable that safety concerns will be addressed in clinical trials arising from clinical uses of the technology to overcome disability. Before proceeding, scientists need to address the requirements for non-toxic materials that can meld with the brain without causing inflammation, or the accumulation of scar tissue and/or rejection. Safety and efficacy issues should also include guarantees, the responsibilities of manufacturers for failures and upgrades, standards for industry-wide implementation, and parameters for acceptance of candidates for implants. Judging from research results in implanting devices for clinical diseases such as Parkinson's, there is also the potential for changes in personality, mood and cognition, and for mistakes in placement of the electrodes (Ford and Kuba, 2006). Moreover, it will be imperative to safeguard the implants from computer viruses, as Dr Mark Gasson of the University of Reading found when the high-end Radio Frequency Identification (RFID) chip implanted in his hand both to provide access to the University building and his mobile phone, and to enable him to be tracked and profiled, acquired a computer virus (Science Daily, June 2010).

Justice Issues

In an age of increasing medical costs and diminishing funds, the question of how to pay for brain implants is a serious issue. Neither insurance nor government programs will be able to easily add cost free access to this technology. Where the technology is used to remediate disability the need for coverage will be imperative since the technology will enable normalcy. How to fairly distribute this scarce and, presumably, expensive care is highly problematical. But, as the technology moves from treatment to enhancement the problem will become more complex. The divide between haves and have nots will be fueled by a technology that essentially changes the type of human that one is; those who already have more financial capital, will be able to use it to generate even more advantages. In a capitalist system free trade will result in some enhanced humans and many more unenhanced. Certainly, the poor of the developing world will have little access to the technology.

Privacy and Autonomy Issues

A significant ethical issue with this emerging technology is the prospect of loss of privacy, autonomy and control by individuals. Presently, it is possible to track others. In the future, when an implanted device allows persons to access the feelings and thoughts of others, not only will awareness be enhanced, but also communication. But, with this will come a significant possibility of a loss of privacy and individuality. "A person with a suitably wired brain could be aware of other people as if they were part of her own body, the same way she knows where her own fingers are" (Chorost, 2011, p.11). When one brain can communicate with another seamlessly and instantly, the individual's inner self will be open to the "other". The borders between self and others and even groups will be gradually destroyed until they are in effect eliminated. Potential commercial uses of the technology include using it to influence and understand buyers' desires, or to detect lies and validate testimony in court cases (Greeley, 2004). One company *Applied Digital Solutions*, has developed global positioning and radio frequency identification products to track pets and livestock. Its technology could easily be used for children, the elderly and the medically compromised. It should then prove easy to know where and with whom anyone is associated, and to monitor, or even control and direct the actions of anyone. Utilization of microchip implants has been suggested for medial records, financial transactions, criminal status and national security reasons. Darpa is funding a project called "Silent Talk" with the goal of promoting "user-to-user communication on the battlefield without the use of vocalized speech through analysis of neural signals." (Drummond) Initially, it is easy to foresee uses for children, the demented, prisoners, and certainly the military. The efficiency of armed forces would be greatly improved with brain computer interfaces which would facilitate warfare, and communication, but also open the path to control by government of thousands of soldiers, who will often later return to being private citizens. Once implanted individuals could always and everywhere be tracked, and if there were untrackable areas, the individual could be trained to avoid them. "With implantable devices, messages and information could be transmitted to the brain, actions could be initiated by remote control, and information could be transmitted both to and from the brain" (McGee & Maguire, 2007).

Once implanted, the boundaries between the real and virtual world will blur, and a self constantly wired to the collective will be transformed. The emergence of a Borg type collectivity or hive mind, where personal identity is lost and assimilation is the preeminent value is a real possibility. Selves will be able to have relationships and interact in highly realistic virtual reality environments,

transforming the sense of both the individual and reality. This transformation in the relationship of the self to the other and to the virtual will completely change the ways of mankind.

Certainly, safeguards need to be put in place. A multitude of threats to personal autonomy will need to be addressed. First and foremost, access to one's brain will need to be controlled to prevent the hijacking of perception and memories or the imposition of a mind virus. The individual needs to have rights to control access to the self, both in terms of what the self will reveal and what the self can be constrained to carry out. Without safeguards,

there would be no end of nefarious things to do to a brain. Create sensations of pain … Create aversive sensations when certain thoughts come up, such as a desire to vote for an opposing candidate. In the worst possible case, it might even be possible to "crash" a brain by generating so much conflicting input that the person is incapacitated. (Chorost, 2011, p.197)

At the 21rst Usinex Security Symposium, a paper explored the feasibility of extracting private information from electrical signals when users employ neuro-tech headsets such as those made by Emotiv or Neurosky. " The captured EEG signal could reveal the user's private information about, bank cards, PIN numbers, area of living, the knowledge of the known persons."(Martinovic et al.)

Concerns about privacy have fueled a growing litany of protests, including cases about the right to share and sell private pharmaceutical information, about Facebook's use of an individual's pictures and private information, and about online data collection and tracking. Data mining and storage on the part of companies and governments are clear threats to individual privacy and render obsolete the notion of anonymity and privacy. Troublesome revelations of the overreaching of companies have forced Apple and Google to cease collection of location data. Nevertheless, suspicions have arisen that United States spy agencies are collecting geolocation information clandestinely (Akerman, 2011). Murdoch's news empire has recently been embroiled in a scandal affecting the British government, and hackers and whistleblowers like Wikileaks are violating traditional privacy norms, even redefining what norms can be expected (Greenemeirer, 2011). No clear rules, regulations or laws exist for the collection and dissemination of information gleaned from what individuals do on the web. European agencies tend to hold to a stricter view of privacy rights than those in the United States, but where governments are prohibited from collecting data, they are turning to private companies and buying the information from them (Bloomberg Business Week, 2006). Clearly, some governments would have no qualms about using the technology to monitor, understand and control its citizens, certainly, any citizens who pose a threat to the ruler's dominance.

There are often conflicts between individual autonomy and the common good. In the case of brain-machine implants it is conceivable that the goods achieved by individuals through the use of these enhancement technologies – greater knowledge, communication, memory, sensory abilities – might conflict with the good of society in general. Where now there is variation in intelligence within a range, after the implants the magnitude of change will result in almost a new species. Mankind constantly seeks the means to improve. The enhancements achieved through brain computer interface implants will result in a significant self-directed transformation of humanity. These radical human enhancements may reasonably be viewed with caution, even foreboding. As humans are faced with the decision of what kind of human to be – naturals, cloned immortals or beings that exist solely in virtual reality, mankind is faced with deciding whether to adopt a hands-off policy towards this future, or to ban such progress, or to adopt regulations.

Raising this problem reveals that the basic issue involves the philosophical query of what is the good for man. It may well be that some conceptions of human flourishing, and of the good, are superior to others (Wall, 2007).

There is no way to consider whether or not to regulate, ban, or adopt a permissive attitude toward radical human enhancement without considering the question of what is the human. How do humans differ from nonhumans? How do humans differ from the future's 'post humans'? What does it mean to be human? Would it be better to be more than human? Or more precisely, other than human?" (McGee, 2010, p.50)

A response to this question can only be generated by mankind itself. It should not be left to happenstance, or the vagaries of the market, or the actions of scientists, researchers, or even isolated governments.

NEED FOR DISCOURSE AND REGULATION

There is an urgent need for debate on these issues. World governments have succeeded in regulating the environment, nuclear materials, research on humans, and investigations of active infectious agents. Similar regulation of brain computer interface implants needs to occur. It is imperative that mankind debate, and regulate these technologies. Questions that an open debate should address are: Is it wise to pursue these developments? What parameters should guide these advances? What standards and agencies need to be created to monitor and regulate these technologies?

The enhancements achievable with brain computer interfaces require reflection and policy development. Ethics suggests that new systems of review must be instituted. Many forums for deliberation should be utilized, including those of scientific societies, of state and federal legislatures, of individual nations and organizations of nation states, and of the United Nations. Public hearings must work to establish research guidelines for this transformative technology. In so doing there must be substantive discourse on the nature of the good for man. It will be necessary to create international regulatory agencies for biotechnological developments because public discussion engaging individuals from all over the globe is required; deliberations need to be highly visible and be preceded by debate within scientific and professional societies, states and countries. Inasmuch as public funds are being expended in the United States, Europe and Asia, the goals and results of these investments need to be openly debated; decision makers need to be publically accountable. Procedures for regulation need to be devised, and promulgated. Presently there is no system in place for review of enhancement technologies; the safety and efficacy norms of FDA type approval are inadequate for the complexities of cybernetic technologies that can change the nature of man. Standards need to be established to ensure that enhancement uses of interface technology include provisions that allow for informed consent, reversibility, and initial evaluation in limited trials (McGee and Maguire, 2007). International regulation is required since these technologies will cross national boundaries. This enterprise requires the agreement of established governments, and should be facilitated by the United Nations.

REFERENCES

Ackerman, S. (2011). Senators ask spy chief: Are you tracking us through our iPhones? *Wired*. Retrieved on July 18, 2011 from http://www.wired.com/dangerroom/2011/07/senators-ask-spy-chief-are-you-tracking-us-through-our-iphones/

American Friends of Tel Aviv University. (2010, June 29). Pacemaker for your brain: Brain-to-computer chip revolutionizes neurological therapy. *ScienceDaily*. Retrieved July 20, 2011, from http://www.sciencedaily.com/releases/2010/06/100628152645.htm

Beauchamp, T., & Childress, J. (1979). *Principles of biomedical ethics*. New York, NY: Oxford University Press.

Begley, S. (2004, July 9). Prosthetics operated by brain activity move closer to reality. *Wall Street Journal*. Retrieved July 22, 2011 from http://online.wsj.com/article/0,SB108931910809458926,00.html

Bell, G., & Gemmell, J. (2007). A digital life. *Scientific American*. Retrieved on July 18, 2011 from http://www.scientificamerican.com/article.cfm?id=a-digital-life

Berger, T. et al. (2008). *Brain computer interfaces: An international assessment of research and development trends*. New York, NY: Springer. doi:10.1007/978-1-4020-8705-9

Berger, T. et al. (2011). A cortical neural prosthesis for restoring and enhancing memory. *Journal of Neural Engineering*, *8*(11). PMID:21677369

Bloomberg Business Week. (2006). The snooping goes beyond phone calls. *Bloomberg Business Week*. Retrieved on July 21, 2011 from http://www.businessweek.com/magazine/content/06_22/b3986068.htm

Boran, M. (2011). *Thad Starner: Wearable computing for smarter living*. Retrieved on July 21, 2011 from http://newtechpost.com/2011/03/16/thad-starner-wearable-computing-for-smarter-living

Carey, B. (2012, September 14). Brain implant improves thinking in monkeys, first such demonstration in primates. *The New York Times*, pp. A19.

Chorost, M. (2011). *World wide mind, the coming integration of humanity, machines and the internet.* New York, NY: Free Press.

Computer Professionals for Social Responsibility. (n.d.). *Ten commandments of computer ethics.* Retrieved from http://cpsr.org/issues/ethics/cei/

Dino, J. (2008). Extension of the human senses. *Ames Technology Capability and Facilities.* Retrieved July 22, 2011 from http://www.nasa.gov/centers/ames/research/technology-onepagers/human_senses.html

Dobelle Institute. (2000, January 18). Artificial vision system for the blind announced by the Dobelle Institute. *ScienceDaily.* Retrieved July 20, 2011, from http://www.sciencedaily.com/releases/2000/01/000118065202.htm

Donoghue, J. (2006). A link between mind and machine that can turn thought into movement. *Nature*, 442.

Drummond, K. (2009, June 14). Pentagon preps soldier telepathy push. *Wired.* Retrieved January 19, 2013 from http://www.wired.com/dangerroom/2009/05/pentagon-preps-soldier-telepathy-push/

Drummond, K. (2010, July 15). Human trials next for darpa's mind-controlled artificial arm. *Wired.* Retrieved July 20, 2011 from http://www.wired.com/dangerroom/2010/07/human-trials-ahead-for-darpas-mind-controlled-artificial-arm/

Ford, P., & Kuba, C. (2006). Stimulating debate: Ethics in multidisciplinary functional neurosurgery committee. *Journal of Medical Ethics*, 32(2), 106–109. doi:10.1136/jme.200X.013151 PMID:16446416

Greely, H. (2004). Remarks at the Regan lecture. In *Prediction, litigation, privacy, and property: Some possible legal and social implications of advances in neuroscience.* Retrieved on July 24, 2011 from http://www.scu.edu/ethics/publications/submitted/greely/neuroscience_ethics_law.html

Greely, H. (2006). Neuroethics and ELSI: Similarities and differences. *Minn. J. LSCI. & Tech., 599.*

Greenemeier, L. (2008). Monkey think, robot do. *Scientific American.* Retrieved on July 21, 2011 at http://www.scientificamerican.com/article.cfm?id=monkey-think-robot-do

Greenemeier, L. (2011). Does the Murdoch hacking scandal signify the end of privacy? *Scientific American.* Retrieved on July 21, 2011 from http://www.scientificamerican.com/article.cfm?id=murdoch-phone-hack-scandal-privacy

Guizzo, E. (n.d.). Monkeys use brain interface to move and feel virtual objects. *IEEE Spectrum.* Retrieved from http://spectrum.ieee.org/automaton/robotics/medical-robots/monkeys-use-bidirectional-brain-machine-interface-to-feel-virtual-objects

Halley, D. (2010). Computer chip implant to program brain activity, treat Parkinson's. *Singularity Hub.* Retrieved July 20, 2011, from http://singularityhub.com/2010/07/21/computer-chip-implant-to-program-brain-activity-treat-parkinsons/

Headlam, B. (2000, June 11). The mind that moves objects. *NY Times,* 63-64.

Hooper, S. (2004). Brain chip offers hope for paralyzed. *CNN.com.* Retrieved July 20, 2011 from http://ed.cnn.com/2004/TECH/10/20/explorers.braingate/

Hsu, J. (n.d.). Intel wants brain implants in its customers' heads by 2020. *POPSI.* Retrieved from http://www.popsci.com/technology/article/2009-11/intel-wants-brain-implants-consumers-heads-2020

IEEE Code of Ethics. (n.d.). Retrieved from http://www.ieee.org/about/corporate/governance/p7-8.html

Intendix. (n.d.). Retrieved from http://www.intendix.com

Kurzweil, R. (2005). *The singularity is near.* New York, NY: Viking.

Lawton, W. (2004). *Research group exploring limb loss hopes biohybrid will bridge gap between human and machine.* Retrieved from http://www. brown.edu/Administration/George_Street_Journal/vol29/29GSJ06e.html

Lemonick, M. (2003, October 21). Robo-monkey's reward. *Time.* Retrieved July 20, 2011 from http://www.time.com/time/magazine/article/0,9171,524491,00.html

Maguire, G. (1999). *Transforming humanity: Toward the implantable walkstation.* Retrieved July 20, 2011 from http://www.scribd.com/doc/27151750/1/Transforming-Humanity-Toward-the-Implantable-Walkstation

Mann, S. (1997). Wearable computing: A first step toward personal imaging. *Computer, 30*(2). doi:10.1109/2.566147

Martinovic, I., Davies, D., Frank, M., Perito, D., Ros, T. D., & Song, D. (n.d.). *On the feasibility of side-channel attacks with brain-computer interfaces conference: USENIX security '12.* Retrieved from https://www.usenix.org/conference/usenixsecurity12/feasibility-side-channel-attacks-brain-computer-interfaces

McGee, E. (2008). Bioelectronics and implanted devices. In B. Gordijn & R. Chadwick (Eds.), *Medical enhancement and posthumanity* (pp. 207-224). Berlin: Springer Science + Business media B.V.

McGee, E. (2010). Toward regulating human enhancement technologies. *AJOB Neuroscience, 1*(2), 49–50. doi:10.1080/21507741003699405

McGee, E., & Maguire, G. (1999). Implantable brain chips? Time for debate. *The Hastings Center Report, 29*(1), 7–13. doi:10.2307/3528533 PMID:10052005

McGee, E., & Maguire, G. (2007). Becoming borg to become immortal: Regulating brain implant technologies. *Cambridge Quarterly of Healthcare Ethics, 16*(3), 291–302. doi:10.1017/S0963180107070326 PMID:17695620

Memoto. (n.d.). Retrieved from http://memoto. com

Nicolelis, M. (n.d.). *Laboratory of Miguel Nicolelis.* Retrieved, July 20, 2011 from http://www. nicolelislab.net/

Patel, P. (2009). The brain-machine interface, unplugged. *IEEE Spectrum.* Retrieved from http://spectrum.ieee.org/biomedical/devices/the-brainmachine-interface-unplugged

Pearson, I. (2012). Future visions. *Future Human Evolution.* Retrieved July 23, 2011 from http://www.humansfuture.org/future_post_human_futures.php.htm

Rubio, J. (2012, December 7). US begins testing brain implants with hopes of slowing Alzheimer's. *The Verge.* Retrieved from http://mobile.theverge.com/2012/12/7/3740988/us-deep-brain-stimulation-implant

Saenz, A. (2009). Braingate2: Your mind just went wireless. *Singularity Hub.* Retrieved on July 22, 2011 from http://singularityhub.com/2009/06/17/braingate2-your-mind-just-went-wireless/

University of Reading. (2010, May 26). Could humans be infected by computer viruses? *ScienceDaily.* Retrieved July 23, 2011, from http://www.sciencedaily.com/releases/2010/05/100526095830. htm

University of Southern California. (2011, June 17). Scientists turn memories off and on with flip of switch. *ScienceDaily.* Retrieved July 20, 2011, from http://www.sciencedaily.com/releases/2011/06/110617081543.htm

Wall, S. (2007). Perfectionism in moral and political philosophy. In E. Zalta (Ed.), *Stanford encyclopedia of philosophy*. Retrieved July 21, 2011 from http://plato.stanford.edu/entries/perfectionism-moral/

Warwick, K. (2004). *I cyborg*. Chicago: University of Illinois Press.

Warwick, K. (n.d.). *Te next step towards true cyborgs*. Retrieved from http://www.kevinwarwick.com/Cyborg2.htm

Whitehouse, D. (1999). Looking through cats' eyes. *BBC Sci/Tech News*. Retrieved July 20, 2011 from http://news.bbc.co.uk/2/hi/science/nature/471786.stm

World Health Organization. (n.d.). *Wingspread statement on the precautionary principle*. Retrieved from www.who.int/ifcs/documents/forums/forum5/wingspread.doc

Zimmer, C. (2004, February 2). Mind over machine. *Popular Science*, 47-48.

KEY TERMS AND DEFINITIONS

Autonomy: The state of functioning independently, without extraneous influence.

Bioelectronics: The application of the principles of electronics to biology and medicine.

Brain Computer Interfaces: Also known as mind-machine interface (MMI) or a brain–machine interface (BMI) is a direct communication pathway between the brain and an external device. BCIs are often directed at assisting, augmenting, or repairing human cognitive or sensory-motor functions.

Braingate: Is a brain implant system designed to help those who have lost control of their limbs, or other bodily functions, such as patients with amyotrophic lateral sclerosis (ALS) or spinal cord injury. A sensor which is implanted into the brain, monitors brain activity in the patient and converts the intention of the user into computer commands.

Brain Machine Interfaces: Is a direct communication pathway between the brain and an external device. Brain-Computer Interfaces are often directed at assisting, augmenting, or repairing human cognitive or sensory-motor functions.

Cochlear Implants: Is a surgically implanted electronic device that provides a sense of sound to a person who is profoundly deaf or severely hard of hearing.

Cloning: In biotechnology refers to processes used to create copies of DNA fragments (molecular cloning), cells (cell cloning), or organisms.

Cyborgs: Is short for "*cyb*ernetic *org*anism" and is a being with both organic and artificial parts.

Deep Brain Stimulation: Is a surgical treatment involving the implantation of a medical device called a brain pacemaker, which sends electrical impulses to specific parts of the brain. DBS in select brain regions has provided therapeutic benefits for otherwise-treatment-resistant movement and affective disorders such as chronic pain, Parkinson's disease, tremor, and dystonia.

Disability: Is the consequence of an impairment that may be physical, cognitive, mental, sensory, emotional, developmental, or some combination of these. A disability may be present from birth, or occur during a person's lifetime.

Enhancement Technologies: Are techniques that can be used not simply for treating illness and disability, but also for enhancing human characteristics and capacities.

Forecasting: Is the process of making statements about events whose actual outcomes have not yet been observed.

Neuroethics: Concerns the ethical, legal and social impact of neuroscience, including the ways in which neurotechnology can be used to predict or alter human behavior.

Neurons: Is an electrically excitable cell that processes and transmits information through electrical and chemical signals. A chemical signal occurs via a synapse, a specialized connection with other cells. Neurons connect to each other to form neural networks. Neurons are the core components of the nervous system which includes the brain, spinal cord, and peripheral ganglia. Sensory neurons respond to touch, sound, light; motor neurons cause muscle contractions; interneurons connect neurons to other neurons within the same region of the brain or spinal cord.

Neuroprosthetics: Are a series of devices that can substitute a motor, sensory or cognitive modality that might have been damaged as a result of an injury or a disease.

Obsessive Compulsive Disorder: Is an anxiety disorder characterized by intrusive thoughts that produce uneasiness, apprehension, fear, or worry; by repetitive behaviors aimed at reducing the associated anxiety; or by a combination of such obsessions and compulsions.

Chapter 16
We Are the Borg! Human Assimilation into Cellular Society

Ronnie D. Lipschutz
University of California - Santa Cruz, USA

Rebecca J. Hester
University of Texas Medical Branch, USA

ABSTRACT

As cybersurveillance, datamining, and social networking for security, transparency, and commercial purposes become more ubiquitous, individuals who use and rely on various forms of electronic communications are being absorbed into a new type of cellular society. The eventual end of this project might be a world in which each individual, each cell in the electronic "body politic," can be monitored, managed, and, if dangerous to the social organism, eliminated. This chapter examines the objectives, desires, and designs associated with such a cellular biopolitics. Are individuals being incorporated into a Borg-like cyber-organism in which they no longer "own" their substance, preferences, desires, and thoughts and in which they are told what they should be doing next?

INTRODUCTION

We know roughly who you are, roughly what you care about, roughly who your friends are.

The power of individual targeting—the technology will be so good it will be very hard for people to watch or consume something that has not in some sense been tailored for them.

I actually think most people don't want Google to answer their questions … They want Google to tell them what they should be doing next. (Eric Schmidt, CEO of Google (quoted in Jenkins, 2010)

It is a dream of power to control human minds and bodies. Obstacles to this end are the materiality of the latter and the noncorporeality of

DOI: 10.4018/978-1-4666-4582-0.ch016

human thought. In this chapter, we explore the technological and social potential for creation of a cybernetic collective, not terribly dissimilar from *Star Trek's* "Borg". We propose that such a socio-technological formation might not be quite the science fiction fantasy it is generally thought to be. Although we do not anticipate the full fusion of minds, as among the Borg, the combination of RFID-type brain implants, neuropsychological research and changes in individual subjectivities point toward a "cellular society", in which individual identities and autonomy are submerged in a greater whole. Our goal here is to assess the current state of technology, politics, and social control where minds and bodies are concerned and to suggest how new developments, yoked together, could lead to a re(B)organized cellular society, in which the individual members are linked to each other, in real time, via centralized data bases and surveillance systems available to state authorities.

Our conceptualization of the Borg centers on the collective ontological and cybernetic formation that results from being connected to other brains and bodies through embodied technology. Because of its connectedness, the Borg is more than a cyborg. That is, it is not just a fusion of biology and technology such that a new, bionic man or woman results. It is more akin to Michael and Michael's (2007) notion of the electrophorus in which a "bearer of electricity" acts like a network element or node in a larger electromagnetic field. The novelty of this networked being is not only that technology and society are fused (Stephan et al., 2012) such that human capacities are expanded and improved, but also that mechanisms for surveillance and social control are internalized, opening up the possibility that the Borg can be externally manipulated (Duhigg, 2012; Singer & Duhigg, 2012). The concern here expands beyond whether this kind of networked society is threatening long-standing notions of what it means to be human (pace the debate between Bostrom (2003) and Fukuyama (2004) on transhumanism). Certainly

new neurotechnologies are questioning and even threatening the primacy of traditional humanistic "mind over matter" world views (Benedikter et al., 2010). There is no doubt that humans continue to evolve in relationship to the technologies they develop. What is at stake, however, is the extent to which a re(B)organized society can exercise political and moral agency if its thoughts are tracked and controlled from without and if those in such a society feel confused, naked and lost (Mann, 1997) when they are not "jacked in" to the network (Gibson, 1984).

We begin with a discussion of the political motivations for extended electronic monitoring of "unruly bodies" that pose risks and dangers to the self-discipline and social order underpinning advanced liberal society. Preventing and pre-empting risks to minds and bodies is a central logic driving what we call the "re(B)organization" of society. The following two sections examine, first, recent technological and neurological efforts to measure and collect *in vivo* data on biochemical and neurotransmitter levels in brains and bodies, body temperature, toxins, and viruses, and to communicate real time data to remote electronic databases for assessment of risk potential; and second, recent developments in neuropsychology, mindreading and synthetic telepathy. In the fourth part of the chapter, we review recent research on changes in individual subjectivities following from instant and continuous communication with friends and families afforded by near-ubiquitous cell phones. Recent experiments with implantable RFID chips point toward more sophisticated, brain-implanted receiver-transmitters offering access to the world's communication networks while sending out streams of biodata. Already, the current mix of security, technology and subjectivity is transforming both society and individuality; we should not be surprised if future developments are welcomed with open arms and minds (Collins, 2002). We conclude with a discussion of the implications of a Borg-like cellular society.

THEORY OF CELLULAR SOCIETY

The diffusion of advanced liberalism (Rose, 1993) throughout the Global North and many parts of the developing world has heightened a broadly-held sense of imminent and incipient risk and danger. This is manifest not only in what Ulrich Beck (1992) has called "risk society", based on fears of the effects of, and threats to, advanced technologies, but also in state and popular anxieties about dangerous people and unruly bodies, including terrorists, hackers, lone shooters, homegrown radicals, immigrants and xenophobes, the obese, the diseased and the invisible (Lupton, 1993; Lupton and Petersen, 1996; Inda, 2006; Lipschutz, 2008). Yet, risk and risky individuals are ubiquitous—indeed, market society is premised on risk—and, therefore, strategies for preempting risk and rendering society more secure must also be all-pervading. Further, everyone must participate in the detection and prevention of risk if such a project is to be effective. Failure to do one's part puts others at risk and can increase speculation and, perhaps, suspicion that one is a risky "person of interest".

Elsewhere, Lipschutz (2008) has argued that such "risk" is a concomitant of individual "freedom" to operate within advanced liberal society and to consume goods, beliefs, and behaviors without hard coercive limits, and that the expansion of normative restrictions is "good for the economy" even if risky to society. The assumption that lightly-regulated consumerism makes people happy, obedient, and placid came undone on September 11, 2001, impressing upon governments the need to monitor actions, movements, and plans so as to detect and apprehend dangerous individuals before they can harm themselves or others. How is this to be done?

Over human history, governments have tried many techniques to instill obedience and docility in their subjects. They have imprisoned violators of the law, judged insane those who rejected social norms, and killed any found to be unredeemable (Orwell, 1949). Even so, total control over its subjects remains beyond the reach of the state. In advanced liberal society, obedience and order rest on individual self-control and discipline, itself an unreliable strategy in an era of risky persons. If liberal society is to thrive, it must ensure that its subjects will not face deadly dangers in living their liberal lives and not pose such dangers to others; because self-discipline is a thin reed on which to base security, liberalism must somehow come to control or eliminate those who behave badly. Although coercion is always available, safety is much more effective and sustainable if it is achieved through consent.

Michel Foucault (2003) coined the term "governmentality" to describe the way that order and obedience are produced in individual subjects through the internalization of normalized discourses, a process that guides behavior from afar yet in which individuals are also active participants. Importantly, governmentality does not express what States, corporations or other forces and institutions do *to* people; it is not a practice of oppression or domination. Rather, it is "the conduct of conduct", a form of *productive power* that enables individuals to govern themselves without obvious external coercion, albeit not always as they please or desire. Discussions and discourses of risk and danger posed to society by unruly bodies, and associated practices, such as airport security, anti-virus programs and identity theft protection, all heighten individuals' sense of threat and enroll them as participants in governmentality. Risk discourses and experts draw on people's fear of death and of cultural "others" while at the same time offering avenues for ensuring and insuring more security. "Risk management" has thus become a technique to conduct the conduct of individuals and collectives through transformation of mentalities and practices. It is this governmental process that, we argue, facilitates individual desire for constant contact with others, real-time awareness of risk and danger and, ultimately, assimilation into cellular society.

Foucault (2003) pointed out that governmentality included "as its essential technical means apparatuses of security" (p. 244). Here, "security" is to be understood as encompassing much more than national defense or war. The broad social demand for "security" resulting from 9/11 and the Global War on Terror, as well as the state's concern for its own integrity and the continued operation of economies and social systems, have led to an enormous expansion of the term's remit. Not only have the budgets of the United States Defense Department and the country's 40-odd intelligence agencies exploded, the establishment of the Department of Homeland Security, passage of numerous anti-terrorist laws, and incorporation of counter-terrorist provisions in state contracts have made personal and social security in the U.S. all-encompassing (Lipschutz & Turcotte, 2005; Priest & Arkin, 2011).

The expansion of "technical apparatuses" is especially evident in the doctrine of pre-emption. In security studies, prevention is distinct from preemption insofar as the latter is based on the use of force in self-defense against an enemy who is deemed to constitute an imminent danger (Keller & Mitchell, 2006, p. 4). Prevention, by contrast, involves measures to forestall a future, and possibly imaginary, threat of danger. Since 9/11, distinctions between preemption and prevention have become increasingly blurred, both rhetorically and in practice. As Keller and Mitchell (2006) explain, "the resulting fog of semantic confusion facilitates a mix-and-match rhetorical strategy that defends preventive military force by linking it to the more legitimate aspects of preemptive action" (p. 10). How far such confusion has gone can be seen in the extension of prevention into the national health agenda, where it is deployed in detection and prevention of bioterrorism, sometimes in connection with individuals' health (Fauci, 2002; Fidler, 2003).

Because information is key to discovery and prevention, governments have been motivated to increase SIGINT (signals intelligence) activities,

a task made easier, but also more daunting, by the rise of both hardwired and wireless electronic communications technologies and networks. SIGINT has, traditionally, signified communication emissions by states, but detectable electronic signals have become pervasive across the world courtesy of the semiconductor revolution and its role in the global economy. In order to buy, sell, travel, read, write, blog, listen and speak, people are asked or required to provide various forms of personal information, while the very practices of surfing the Internet and accessing electronic networks, such as ATMs, credit card swipers and self-serve gasoline pumps, make available all sorts of electronic data and "pocket litter" which are collected and stored in government and corporate data bases (Lipschutz, forthcoming 2014). Such information is of especial interest to those agencies and institutions responsible for addressing risks and ensuring security. As Clarke (1988) has argued, this form of mass dataveillance has been used by U.S. federal agencies "to develop a wide variety of profiles including drug dealers, taxpayers who underreport their income, likely violent offenders, arsonists, rapists, child molesters, and sexually exploited children."

Acquisition of this data is captured in the concept of "Total Information Awareness", the name of a 2002 data mining program launched out of the Defense Advanced Research Projects Agency (DARPA) and its "Information Awareness Office" by Admiral John Poindexter. According to one law journal article on data mining,

The TIA program itself was the "systems-level" program of the IAO [Information Awareness Office] that "aim[ed] to integrate information technologies into a prototype to provide tools to better detect, classify, and identify potential foreign terrorists [with the goal] to increase the probability that authorized agencies of the United States [could] preempt adverse actions". As a systems-level program, "TIA [was] a program of programs whose goal [was] the creation of a counterterrorism information architecture" by integrating technologies from other IAO programs (and elsewhere, as appropriate) (IAO 2003, cited in Taipale, 2003, p. 46).

The TIA moniker was quickly dropped after widespread public and Congressional criticism, and the widely-held concern that TIA might be used to spy and inform on civilians. Indeed, in 2002, then-Attorney General John Ashcroft announced the launch of the Terrorism Information and Prevention System (TIPS), which would "form a corps of truck and bus drivers, port workers, meter readers, letter carriers and others to report suspicious activities around the nation" (Clymer, 2002). Nevertheless, many elements of the program have continued under other names. For example,

According to current and former intelligence officials, the spy agency [NSA] now monitors huge volumes of records of domestic emails and Internet searches as well as bank transfers, credit-card transactions, travel and telephone records. The NSA receives this so-called "transactional" data from other agencies or private companies, and its sophisticated software programs analyze the various transactions for suspicious patterns (Gorman, 2008).

Darpa's short-lived Lifelog program aimed to collect all of this information and more. As one analyst put it, Lifelog had the potential to become TIA cubed in its attempt "to collect all the threads of an individual's life" (Schactman, 2003; the recent outing and resignation of CIA Director David Petraeus illustrates how such webs are weaved from disparate threads).

For the most part, however, data mining has not resulted in the apprehension of terrorists or the prevention of attacks, whether failed or successful. In the United States, at least, a growing number of indictments on charges of plotting terrorism involve entrapment by FBI informants or the for-

tuitous inspection of electronic media belonging to individuals entering or transiting the country. Recently, a Saudi Arabian student at a college in Texas was arrested for ordering and possessing restricted chemicals, researching potential targets via the Internet, and keeping an incriminating journal. But this occurred only because "He came to the government's attention on Feb. 1, when a North Carolina supply company reported that he had tried to order five liters of a chemical that can be used to make an explosive" (Savage & Shane, 2011). Other cases seem to have rested as much on accidental or coincidental contacts as concrete intelligence gathering.

The upshot is that "total information awareness" has proved an inefficient means of detecting and pre-empting threats and risk, if only because it seeks to infer intentions from the pocket litter generated by people as they proceed through their daily routines. More effective would be methods for detecting individual intentions, whether from appearances, behaviors, or thoughts, as in the following examples:

Select TSA [US Transportation Security Agency] employees will be trained to identify suspicious individuals who raise red flags by exhibiting unusual or anxious behavior, which can be as simple as changes in mannerisms, excessive sweating on a cool day, or changes in the pitch of a person's voice. (Donnelly, 2006)

Tiny cameras the size of a fingernail linked to specialist computers will be used to monitor the behavior of airline passengers as part of the war on terrorism ... Fitted to seat-backs, the cameras will record every twitch or suspicious movement before sending the data to onboard software that will check it against individual passenger profiles ... Scientists from Britain and Germany ... say rapid eye movements, blinking excessively, licking lips or ways of stroking hair or ears are classic symptoms of somebody trying to conceal something ... A separate microphone will record speech, including whispers; Islamic suicide bombers whisper texts from the Koran in the moments before they explode bombs. (IOL, 2007)

As an August 2007 report by the New York Police Department's Intelligence Division, entitled "Radicalization in the West: The Homegrown Threat", put it

The challenge to intelligence and law enforcement agencies in the West in general, and the United States in particular, is how to identify, pre-empt and thus prevent homegrown terrorist attacks given the noncriminal element of its indicators, the high growth rate of the process that underpins it and the increasing numbers of its citizens that are exposed to it. (Silber & Bhatt, 2007, p. 85)

One response to this challenge has been the recently-funded $1 billion Next Generation Identification System (NGIS) which uses biometrics to detect "risky" subjects (Webster, 2011a). While narrower in scope than the Total Information Awareness program, which targeted private individuals all over the world instead of just suspected criminals and terrorists, the NGIS constitutes a "revolution in law enforcement technology" because it will have the capacity to scan fingerprints and retinas, map faces and voices, take palm prints, and analyze handwriting (Webster, 2011b). Through NGIS, state and local police agencies will receive electronic fingerprint scanners with which they can collect biometric data from "suspects" (as opposed to those convicted of crimes). That data will be accessible from anywhere in the world through a Department of Justice (DOJ) database. A sub-database, called "the Repository for Individuals of Special Concern"' will also be created in conjunction with the NGIS, reportedly to track wanted criminals, registered sex offenders and "suspected terrorists". This system will work in conjunction with the Secure Communities Initiative which requires local law enforcement to run biometric data through the Department of Homeland Security's immigration records, in addition to running it through the DOJ. Despite concerns from law enforcement officials and the refusal of some counties to participate in the Secure Communities Initiative, U.S. Immigration and Customs Enforcement has made it clear that no one is allowed to opt out of participation.

READING BODIES

Such efforts point beyond mere collection of electronic data whose analysis cannot provide reliable indications of individual mentalities to the search for a reliable means of "reading" thoughts, intentions and plans. While mindreading continues to be regarded as somewhat farfetched, and the provenance of magicians and fortune tellers, technology is rapidly catching up. The transhumanism movement is at the forefront of this effort. Tranhumanism, according to its leading proponent Nick Bostrom, "is the name for a new way of thinking that challenges the premise that the human condition is and will remain essentially unalterable."[1] It thus seeks to transcend human limits through the use of technology. Transhumanists hold that everyone should have access to the means to enhance various dimensions of their cognitive, emotional and physical well-being (Bostrom, 2004). Spearheaded by computer scientists, neuroscientists, nanotechnologists and researchers at the forefront of technological development, this movement challenges the idea that human enhancements will deprive people of the capacity for moral agency by changing what it means to be human (Bostrom, 2004). The use of implants for control purposes (Michael & Michael, 2011) should give us reason to question that claim. While "homebrew biohackers" obsessed with the idea of human enhancement (Smartnews, 2012) and those who use implants for convenience may be excited by transhumanist possibilities, the black box beneath the skin (Michael & Michael, 2011) does not just offer physical transcendence but also assimilation and social control. Implants for care-related applications (Michael & Michael, 2011) are the entry point into the re(B)organization of society.

In anticipation of the unveiling of implantable microchips for use in humans, at an October 2000 event in New York City, Dr. Peter Zhou remarked: "We will be a hybrid of electronic intelligence and our own soul" (Kohlbrand & Foster, 2000). When Zhou made this comment, he was chief scientist at Digital Angel.net, the company that mainstreamed implantable chips for identifying lost animals. The Digital Angel micro-chip, designed to be inserted just under the skin and to communicate with an external watch-like transmitting device, is a unique and powerful technology uniting wireless telecommunications, bio-sensors that monitor critical body functions in real time, and GPS technology into one functioning system that sends biodata to a variety of computers and third parties. The company's prototype, about the size of a dime, was powered electromechanically through the movement of muscles, and activated either by the "wearer" or a monitoring facility. In the article announcing the event, Zhou was enthusiastic about the possibilities of the implantable chips designed to monitor the physiology and whereabouts of human bearers. Digital Angel, he said, "will be a connection from yourself to the electronic world. It will be your guardian, protector. It will bring good things to you "(quoted in Kohlbrand & Foster, 2000). Those "good things" include the chip's ability to save lives by remotely monitoring the medical conditions of at-risk patients while providing emergency medical units with a patient's exact location. Other "good" things are the commercial potential for businesses interested in pinpointing a consumer's location and agri-industry monitoring of livestock (and meat) trajectories from feeding pen to supermarket.

The medical and commercial benefits of Digital Angel are, indeed, remarkable. But perhaps more remarkable still are the ways this technology has opened up avenues for "conducting the conduct"

of individuals and populations through the use of biodata and feedback mechanisms to warn bearers when they are at risk. While Zhou's comments address one aspect of the connection that occurs through the implantable chip technology—between an individual and the electronic world—he neglects to point out that this "electronic world" is, in turn, connected to an unknown number of people and systems who can constantly monitor and surveil the embodied states and movements of the chip bearers. Who are those third parties? And what kind of access might they have into the minds and bodies of the chipped? Such questions are as yet unanswered.

Even a decade later, although uncertainties remain about how such chips might be used, other companies are developing new uses for implantable microchips. For example, PositiveID (www.positiveidcorp.com), a company with a long-standing relationship to Digital Angel, holds a patent for an "embedded biosensor system", based on the use of an implantable, bio-sensing RFID microchip, now smaller than a grain of rice, to measure and communicate glucose levels in the body to remote terminals in real time (PositiveID, n.d.; Carlson, Silverman & Mejia, 2007). More recently, the company has entered into an agreement with Raytheon Microelectronics to produce a microchip for use in Medcomp's vascular ports (StreetInsider.com, 2011) and partnered with Siemens to develop the "Wireless Body" system, an implantable technology that allows communication from within the body to outside of the body on an integrated platform (PositiveID, 2010).

A press release by the company touts the integrated platform, announcing that it will "enhance the management of diabetes by allowing disease management systems to communicate with each other and deliver solutions to patients seamlessly, enhancing the ability to deliver personalized medical solutions wirelessly" (id.). While specifically used for diabetes management, the wireless body platform technology will be available for "other disease management applications in the future" (id). For example, wireless body sensor networks have been presented as an "opportunity to sense acute disease processes and monitor chronic illness quickly and efficiently" (Aziz, et al., 2005, p. 131). Indeed, a next-generation technology could move away from static sensors to those that can move through blood vessels, the urinary tract, ventricles of the brain, spinal canal, lymphatic and venous systems (Drummond, 2012). As Aziz, et al. (2005) note, "Recent advances in nano-technology have meant that delivering sensors within these luminal cavities is for the first time a real possibility" (id.). These platforms point not only to monitoring but also to two-way data streams linking individuals and technologies into seamless, wireless networks without the need for an external transmitting device. While there has been much public outcry against micro-chips, a number of implantable devices with wireless capabilities are already being used for medical purposes (Yang, 2006) without much public debate. For example, "The European Commission project 'Healthy Aims' has been focusing on specific sensor applications, namely, for hearing aids (cochlear implant), vision aids (retinal implant), detecting raised orbital pressure (glaucoma sensor), and intracranial pressure sensing (implantable pressure sensor)" (Aziz et al., 2005, p. 131).

The mix of military and medical partnerships, in particular, that have developed around the technology are revealed through PositiveID's research and its agreements with Raytheon and Siemens, two companies that work in the area of defense technology. PositiveID has been working to engineer its micro-chip to detect and communicate the presence of biothreats in an individual's body. Although only one strain of one virus can be detected at the moment, research on detection of multiple threats is underway. The imbrication of military and medical have been extended further through the 2009 merger of PositiveID's predecessor company, VeriChip, with SteelVault, a consumer credit reporting company, in order to

"focus on securing consumers' financial information and addressing the critical need for secure, online personal health records…" (PositiveID, 2009). Mergers such as that between VeriChip and Steel Vault permit personal records, and real time bio- and credit-data, to be accessed and cross-referenced by public authorities for security purposes. Evidently, such information will be of value to hospitals and other medical facilities, which will be able to determine whether patients seeking admission can pay for services. It would not take much to further connect these data to other on-line personal information, with far reaching consequences for the surveillance and disciplining of populations. PositiveID is not the only entity developing such systems, nor are medical and financial purposes the only ones for which such chips are being designed. For example, a 2009 report by the Telemedicine and Advanced Technology Research Center, a United States Army research center, highlights the use of implantable biosensors for the continuous monitoring of the glucose, lactose, and oxygen levels of fighter pilots (Telemedicine, 2009). Another study outlines the use of this technology for monitoring pressure and strain in the human body (Tan et al., 2009).

For many, there are real consequences from the mainstreaming of microchip implants. For example, many American diabetics are minorities, especially Latinos. If medical chipping becomes widespread, it is possible that more minorities will be surveilled 24/7/365, a matter of some concern given the fact that in the United States, as well as in some European countries, chipping has been proposed as a means of keeping track of immigrants (Hancock, 2010). Moreover, people wanting to qualify for public or private health insurance might be required to undergo chipping, while soldiers might be chipped in order to monitor their well-being and location. Finally, as the science and technology become more sophisticated, biochemical indicators of autonomic arousal (anxiety, anger, fear) could be measured, assessed for and cross-referenced with other biodata (heart rate and breathing), biometrics (gait and facial expression) external behaviors (praying, looking nervous, sweating) and consumption patterns in order to identify "risky" or suspicious people for followup (see below). While such technology is not yet in hand, it is only a matter of time before the mix of technology and biochemistry will permit those with access to one of the privileged databases to "read" the internal status as well as the external bodies of "chipped" individuals.

To some, such monitoring measures might seem harmless as long as one has nothing to hide. Yet, in an environment where risk is perceived to be ubiquitous, abnormalities in an individual's embodied condition could become as suspicious or dangerous as openly advocating public violence. More importantly, the fact that *in vivo* biodata can work within a feedback loop that generates real time *in vivo* responses, as is the case of the implantable insulin pump, suggests that control from a distance is not far off. Because in-body wireless glucose monitoring is being linked to health, the security implications of the technology are obscured. Furthermore, the ability to transmit data from bio-sensor systems to digital medical records will make it appealing to doctors, insurance companies, credit companies and others who seek access to people's health histories. Such developments mean that we now have a two-way flow of information that can both "read" bodies and control them from afar (Singer, 2010). Because of its potential to reach a variety of audiences, the hybrid described earlier by Peter Zhou goes beyond a relationship between one man and one machine; indeed, the implantable chip technology can connect multiple humans with multiple electronic databases into a networked, increasingly-singular and globalized entity that will expand and become increasingly powerful as the technology improves and its purposes diversify. And this, as previously suggested, could extend to reading minds.

READING MINDS

At the same time as two-way chips communicating with remote databases and observers are being developed, neurobiological and neuropsychological researchers are studying activation of portions of the brain in response to various forms and types of visual and other sensory stimuli. Of especial interest in this regard is recent research using functional magnetic resonance imaging (fMRI) to study the

functioning of the human brain in real time … and thereby to access both sides of the mind–brain interface — subjective experience (that is, one's mind) and objective observations (that is, external, quantitative measurements of one's brain activity) — simultaneously (deCharms, 2008, p. 720).

Of even greater interest and, perhaps, concern, is the burgeoning scope of research in neuroimaging and what Charles Jennings (2006) has called the "battlefield between the ears". R. Christopher deCharms (2008) explains that fMRI has developed to the point that

specialized antennas … positioned around the subject's head [can] measure the signals that are emitted from many points all at once, and the spatial origin of each signal component is then separated mathematically from the total signal on the basis of each signal component's time, frequency and phase (p. 721).

As analytical algorithms become increasingly sophisticated, "mind-reading" may become feasible, and "in some cases provide enough information to allow a good prediction of what a person is experiencing or doing" (deCharms, 2008, p. 723). Utilized as a sophisticated form of "lie detection", deCharms writes that

Theoretically, such a system could put criminals in prison and keep innocent people out on the basis of their accounts of what happened. In national security it could be used for screening out foreign countries' agents during security clearance, and for counter-terrorism' (p. 724).

And, he warns:

The potential power to read information from a person's brain also leads to a new frontier of personal confidentiality: mind privacy. The foreseeable ability to read the state of a person's brain and thereby to know aspects of their private mental experience, as well as the potential to incorrectly interpret brain signals and draw spurious conclusions, raises important new questions for the nascent field of neuroethics. One can imagine the scenario of someone having their private thoughts read against their will, or having the contents of their mind used in ways of which they do not approve. Many of these concerns also apply to non-real-time applications (p. 728).

Other have pondered such applications and expressed similar concerns (Simpson, 2008; Denning, Matsuoka, & Kohno, 2009).

In 2008, the U.S. Army Research Office awarded $4 million to a group at the University of California, Irvine to "study the neuroscientific and signal-processing foundations of synthetic telepathy" (*UC Irvine Today*, 2008). The project uses "non-invasive brain-imaging techniques like EEG, MEG and fMRI to learn more about how the brain produces imagined speech when one thinks" (MURI, 2008; interestingly, there is no information on this web site dated later than Oct. 13, 2008). The ultimate objective of the project is to "capture those brain waves with incredibly sophisticated software that then translates the waves into audible radio messages for other troops in the field" (Thompson, 2008). Although researchers have claimed that reading the thoughts

of those not trained in the eventual technology will be impossible, this cannot be wholly ruled out. Indeed, the "Future BNCI" (Brain/Neuronal Computer Interaction) website, based in Europe and partially-funded by the European Commission, reports on various efforts to develop such systems:

No matter what the future holds for BNCI, we know that it will involve signals from the brain or body, that these signals will be captured with sensors of some kind, and that those signals will be processed to extract useful information and features (BCI Basics, 2010; Graimann, Allison & Pfurtscheller, 2010).

For the moment, efforts to explore brain neural activity and develop means for extracting signals are directed toward dealing with pain, using thoughts to manipulate devices, and toward soundless communication on the battlefield, whereby soldiers can control weapons remotely. Ultimately, however, more sophisticated microchip implants and signal detection systems will be fused into implantable devices that can detect and broadcast brain activity and even thoughts (MsGee & Maguire, 2007). For the moment, this possibility remains relegated to the sphere of the paranoid and psychotic, although a great deal of the open-source research underway points toward such devices. As McGee and Maguire report, "remote control of rats has already been demonstrated, and remote control of humans is equally feasible"(2007). It is not our intention here to judge their feasibility—we are manifestly unqualified to do so—but, rather, to pursue the longer-term social implications of "being watched" all the time (*Homeland Security News Wire*, 2012)

Insofar as microchips are regarded as a desirable medical technology, there is no need to for coercion to incite adoption and use. Like safeguards against identity theft and cell phones, medical surveillance will be presented and received, first, as a modern convenience and, later, a necessity. Indeed, the first family ever to be "chipped" dismissed the concerns of privacy advocates expressing the idea that "losing a little personal privacy is a small price to pay for what one day may save their lives" (Collins, 2002). Among those developing the devices, chipping is regarded as a less obvious and gentler form of social control than earlier methods of surveillance and policing. It is in this context that the health and practices of individuals, especially those judged to be "high risk" (and potentially costly), are being followed.

This sort of prophylactic approach to risk has become a driving force in other aspects of contemporary liberal life, as consumers are enrolled in the globalized electronic matrix of consumer capitalism and are warned repeatedly to safeguard personal information lest it be stolen or contaminated by unauthorized hackers. As Haggerty and Ericson (2000) write (citing Zygmunt Bauman),

Instead of being subject to disciplinary surveillance or simple repression, the population is increasingly constituted as consumers and seduced into the market economy ... monitoring for market consumption is more concerned with attempts to limit access to places and information, or to allow for the production of consumer profiles through the ex post facto reconstructions of a person's behaviour, habits and actions. In those situations where individuals monitor their behaviour in light of the thresholds established by such surveillance systems, they are often involved in efforts to maintain or augment various social perks such as preferential credit ratings, computer services, or rapid movement through customs (p. 615, citing Bauman, 1992, p. 51).

The result is the securitization of daily life, as the private and personal become part of the "public" record and as producer-consumers participate not only in sustaining the mechanisms ensuring discipline and obedience but also the very op-

eration of a "surveillant assemblage" (Haggerty and Ericson, 2000, p. 608, citing Patton, 1994, p. 158; the term comes from Deluze and Guattari, 1987; "'Assemblages' consist of a 'multiplicity of heterogeneous objects, whose unity comes solely from the fact that these items function together, that they 'work' together as a functional entity", Patton, 1994, p. 158). All of this might sound like Bentham's panopticon, but it reflects a more complex environment. The prisoner in the panopticon self-regulates in the knowledge that s/he is being watched; the prisoner does not, however, reveal inner thoughts or intentions that might be exploited to pre-empt future behaviors (much like the "pre-cogs" in the film *Minority Report*).

Within this surveillant assemblage, and exposed to it on a daily basis, each individual must be responsible for the "security" of all, in terms of civil behavior and surveillance. Those in both physical and virtual motion within this assemblage are told and warned repeatedly to be watchful and cautious, to report suspicious items, people and events (without those ever being fully defined), and to fear those who do not comport themselves in a "proper" fashion (again without specific definition). To this end, travel, transactions, telecommunications and tendencies can be scrutinized for signs of deviance, hostile intentions and disruptive potential even as each individual is expected to comport himself or herself according to norms that maintain security. Each of us is thus engaged in self-regulation, self-surveillance, self-discipline and self-garrisoning, even as our individual behaviors are subject to constant surveillance, scrutiny and assessment by others, mechanical, electronic and flesh (Lipschutz, 2008; Lipschutz & Turcotte, 2005).

At this point, the full insidiousness of chipping becomes evident, as the body politic maintains eternal vigilance against "infection" from within and without. At the heart of notions of preventive security and medical strategies is the idea of building societal "immunity" from all sorts of risks. Rather than protecting the body (individual and social) from bacteria, injury and aggressors, the risk is from the body to the body, manifest in what might be called "social auto-immune disease" (SAID). What happens when an individual, against her will and possibly without her knowledge, becomes the risk rather than being exposed to risk? What if one's biological, anatomical and psychological make-up poses a risk and is being monitored and detected by security agencies for security purposes?

RE(B)ORGANIZATION

The final piece of the puzzle is somewhat more speculative than preceding ones, but it expands on the notion that, once a population is "wired" into a networked electronic assemblage, individual subjectivities will also be transformed, from the notionally autonomous individuality of modern liberalism to a member of a cellular society in constant communication and under constant watch. Two questions follow: first, will people willingly accept assimilation? Second, is assimilation already underway? Unbeknownst, perhaps, to those of an age to recall public payphones, such a process can be observed among those for whom cell phones are part of life's everyday baggage and for whom the external world is mediated primarily through smartphones and social media.

There is a considerable body of research examining the effects of new communications technologies on society and individuals. As far as we know, however, there has been no research on how "being followed" affects individuals and their social relations, but cell phone use among adolescents and young adults may provide a close analogue. García-Montes, Caballero-Muñoz and Pérez-Álvarez (2006) argue that "the use of this and other technologies *favours*, *promotes* or *foments* a particular way of behaving and of understanding one's own identity" (p. 68). Moreover, they claim that

the mobile phone promotes the development of an individual uncoupled from traditional institutional forms. This new kind of subject would be a sort of node in a web of social relationships that is woven and unwoven according to quite variable and diverse circumstances (p. 78).

Other researchers (Walsh et al., 2011) find that cell phones have "become integrated into many young people's self-identity ... [and have] become a materialistic representation of the self" (p. 334).

What is most interesting, however, is anecdotal evidence that younger users feel isolated, disconnected and even at risk when deprived of their cell phones (Major, 2011). This may be due to several factors. As a recent PhD dissertation points out,

In the context of family life, the mobile phone has come to play a significant role within the existing dynamic of struggle and control between parents and young people. While it potentially gives young people a greater amount of personal freedom, the mobile phone also exposes them to the prospect of increased and intensified parental monitoring and surveillance. It can be a means through which parents aim to maintain some knowledge and control over the activities of their children, while equally allowing young people themselves a greater amount of private communication. However, the mobile phone can provide young people with a means to elude parental surveillance and carry out social activities independent of parental control and supervision (O'Brien, 2010, pp. 26-27).

There is also a pervasive anxiety connected with being "out of touch", especially since "something might happen". As MIT Professor, Sherry Turkle (2009), explains about her research with teenagers,

One of the things I've found with continual connectivity is there's an anxiety of disconnection; that these teens have a kind of panic. They say

things like: "I lost my iPhone; it felt like somebody died, as though I'd lost my mind. If I don't have my iPhone with me, I continue to feel it vibrating. I think about it in my locker". The technology is already part of themselves.

Turkle links the need for continual connectivity to 9/11 which was "a moment of trauma for parents, where they wanted that connection with their children". The constant presence and use of cell phones, and their ubiquity, point to a changing social subjectivity embedded in an increasingly networked world. Turkle is insightful on this point, "with the constant possibility of connectivity, one of the things that I see is ... a very subtle movement from 'I have a feeling I want to make a call' to 'I want to have a feeling I need to make a call' – in other words, people almost feeling as if they can't feel their feeling unless they're connected" (Turkle, 2009).

We suggest that, although the cell phone gives rise to certain forms of autonomy and independence, it must also be regarded as a form of constraint and even bondage, insofar as "cellular society" can lay claim to the individual and her/his attention at all times and in all places. To be sure, the cell phone is not always to hand and it is not always switched on: the user has a "choice" in this regard. But, if risk anxiety becomes pervasive, the off switch is not really an option. As Hans Geser (2006) points out, "cell phones support the maintenance of highly pervasive social roles that bind individuals wholly into particular groups, communities or occupational functions. This diminishes their capacity for keeping a separate private life or maintaining any other commitments" (pp. 15-16). He goes on to argue that

In contrast to many earlier negative visions of an emerging "surveillance society" (Marx), it is less likely to be some sort of "Big Brother" wishing to trace our whereabouts than it is our own "little brother", sister, parent or child. In other words: the Orwellian visions of "totalitarian control"

emanating from unlimited governmental and mass media power have given way to a sort of "neocommunitarian" control emerging from a denser horizontal cohesion of informal groupings facilitated by the ubiquity of mobile digital communication (p. 15).

All of this leads to the end of solitude, of being alone with one's thoughts, and renders individual mental activity more and more subject to interruption and even inspection.

WHAT IS TO BE DONE?

The possibility that, in the not-too-distant future, people will willingly seek (and perhaps be required) to be permanently "jacked into the matrix" also opens up a host of questions and avenues for research. For example, a number of ethical concerns follow: Are there data that should not be extracted from people's minds and bodies? How can society decide what is fair game and what is not? What are the risks of trying to preempt threats through mind and body "reading"? Is there a balance between right and wrong in deploying such technologies (Escoffier, 2010)? If so, what is it? Are there other, better ways to keep individuals in society safe and secure? These questions become all the more pressing as a variety of social and political uses emerge for chipping technologies (see, e.g., Baldwin, 2010). Will these technologies create new forms of stratification in society? We are already seeing disparities occur around access to technology and the internet. Will there be technological have's and have-not's? Will the military utilize these technologies to outfit super or trans-human soldiers who will reinforce the re(B)organization of society? Will we be able to opt out of being "jacked in?"

The use of implantable, wireless technologies also opens up a series of legal issues to do with cybercrimes. We need a legal framework to deal with hackers who penetrate biodata systems and alter individual's minds and bodies, or who may even kill a person by tampering with or reprogramming her medical device from afar (Maisel & Tadayoshi, 2010). While such concerns may seem far-fetched, one case has already been reported in which someone hacked into an epilepsy support website and, using a combination of Javascript coding and flashing animations, was apparently able to trigger epileptic fits in the site's users (O'Neill, 2008). In another incident, a team of researchers was able to conduct software radio-based attacks on an implantable cardioverter defibrillator utilizing an external programmer that communicated with the device (Halperin et al., 2008; Takahashi, 2008) These are examples of intentional, malicious attacks. What happens when two-way communication between the implanted chip and the database occurs without an individual's knowledge, or worse, with knowledge but without any control? What if these technologies are implanted in some delicate location, such as beside the heart or inside the gum, so as to discourage uncooperative subjects from removing them (Michael & Michael, 2011)? The very real potential creates pressing questions for a variety of fields including linguistics, ethics, politics, law, medicine and sociology.

Finally, there are serious ontological questions to consider. What will become of our humanity in a re(B)organized society? Will this society be stratified such that a super or trans-human species emerges and is able to subjugate and oppress those who are physically and intellectually inferior? What will become of liberal individualist notions of human agency in this networked collective?

CONCLUSION

At this point, we return to the neural future. We cannot predict whether chipping will be required for the citizen of the electronic security state or simply a sociocultural custom that, like piercing and tattoos, becomes a norm rather than an exception. We can expect neural devices not only to monitor medical status and permit "hands-free" wireless communication with others similarly kitted out, but also to provide constant GPS positioning for one's self as well as one's "friends". Moreover, should synthetic telepathy and similar "mind reading" technologies prove feasible, we may also see real-time monitoring of speech, thoughts and communications. Ultimately, any remaining individual subjectivity will be subsumed into a stream of constant input and output, diminishing the distinction between self and society. It might not be The Borg, but neither will it be that distant from It/Them.

REFERENCES

Agar, N. (2007). Where to transhumanism? The literature reaches a critical mass. *The Hastings Center Report*, *37*(3), 12–17. doi:10.1353/hcr.2007.0034 PMID:17649897

Aziz, O., Lo, B., Yang, G., & Darzi, A. (2005). Pervasive healthcare: Clinical drive, technological innovations, and socio-economic benefits. In *Proceedings of the Perspective in Pervasive Computing*. Retrieved April 29, 2011, from http://ieeexplore.ieee.org/stamp/stamp.jsp?tp=&arnumber=5678079&isnumber=5678066

Baldwin, T. (2010, March 25). National healthcare will require RFID chips. *The New American*. Retrieved April 17, 2011, from http://beforeitsnews.com/story/28/042/National_Healthcare_Will_Require_National_RFID_Chips.html

Basics, B. C. I. (2010). *Future BNCI (brain neural computer interface)*. Retrieved February 28, 2011, from http://future-bnci.org/index.php?option=com_content&view=article&id=58&Itemid=59

Bauman, Z. (1992). *Intimations of postmodernity*. London: Routledge.

Beck, U. (1992). *Risk society: Towards a new modernity*. London: Sage.

Benedikter, R., Giordano, & Fitzgerald. (2010). The future of the self-image of the human being in the age of transhumanism, neurotechnology and global transition. *Futures, 42*, 1102–1109. doi:10.1016/j.futures.2010.08.010

Bostrom, N. (2004). Is transhumanism the world's most dangerous idea? *Betterhumans*. Retrieved from http://www.nickbostrom.com/papers/dangerous.html

Bostrom, N. (2005). In defense of posthuman dignity. *Bioethics, 19*(3), 202–214. doi:10.1111/j.1467-8519.2005.00437.x PMID:16167401

Carlson, R. E., Silverman, S. R., & Mejia, Z. (2007). *Development of an implantable glucose sensor*. PositiveID Corp. Retrieved April 10, 2011 from http://www.positiveidcorp.com/files/Glucose-Sensor.pdf

Clymer, A. (2002, July 26). Worker corps to be formed to report odd activity. *New York Times*. Retrieved May 10, 2010 from http://www.nytimes.com/2002/07/26/us/traces-terror-security-liberty-worker-corps-be-formed-report-odd-activity.html

Collins, D. (2002, May 10). Florida family takes computer chip trip: Tiny implants contain phone numbers, information on medications. *CBS News*. Retrieved April 11, 2011 from http://www.cbsnews.com/stories/2002/05/10/tech/main508641.shtml

deCharms, R. C. (2008). Applications of real-time fMRI. *Nature Reviews. Neuroscience, 9*, 720–729. doi:10.1038/nrn2414 PMID:18714327

Deleuze, G., & Felix, G. (1987). *A thousand plateaus*. Minneapolis, MN: University of Minnesota Press.

Denning, T., Matsuoka, Y., & Kohno, T. (2009). Neurosecurity: Security and privacy for neural devices. *Neurosurgical Focus, 27*(1), 1–4. doi:10.3171/2009.4.FOCUS0985 PMID:19569895

Donnelly, S. B. (2006, May 17). A new tack for airport screening: Behave yourself. *Time*. Retrieved February 1, 2008 from http://www.time.com/time/nation/article/0,8599,1195330,00.html

Drummond, K. (2012, September 27). DARPA's 'transient electronics' will disappear anywhere. *Forbes*. Retrieved November 19, 2012 from http://www.forbes.com/sites/katiedrummond/2012/09/27/darpa-transient-electronics/

Duhigg, C. (2012). *The power of habit—Why we do what we do in life and business*. New York: Random House.

Escoffier, L. (2010). Radio frequency identification tags, memory spots, and the processing of personally identifiable information, and sensitive data: When there is no balance between right and wrong. In *Proceedings of 2010 Association for the Advancement of Artificial Intelligence Spring Symposium Series*. Retrieved April 28, 2011 from http://www.aaai.org/ocs/index.php/SSS/SSS10/paper/viewFile/1050/1484

Fauci, A. S. (2002). Bioterrorism: Defining a research agenda. *Food and Drug Law Journal, 57*, 413–421. PMID:12703508

Fidler, D. P. (2003). Public health and national security in the global age: Infectious diseases, bioterrorism and realpolitik. *The George Washington International Law Review, 35*, 787–856.

Foucault, M. (2003). Governmentality. In P. Rabinow, & N. Rose (Eds.), *The essential Foucault* (pp. 229–245). New York: New Press.

Fukuyama, F. (2004, September 1). Transhumanism. *Foreign Policy*.

García-Montes, J. M., Caballero-Muñoz, D., & Pérez-Álvarez, M. (2006). Changes in the self resulting from the use of mobile phones. *Media Culture & Society, 28*(1), 67–82. doi:10.1177/0163443706059287

Geser, H. (2006). Is the cell phone undermining the social order? Understanding mobile technology from a sociological perspective. *Knowledge, Technology & Policy*, *19*(1), 8–18. doi:10.1007/s12130-006-1010-x

Gorman, S. (2008, March 10). NSA's domestic spying grows as agency sweeps up data. *The Wall Street Journal*. Retrieved June 14, 2010 from http://homepage.mac.com/imfalse/chapel_annex/NSAs_Domestic_Spying_Grows_As_Agency_Sweeps_Up_Data_WSJ.pdf

Graimann, B., Allison, B., & Pfurtscheller, G. (2010). Brain-computer interfaces: A gentle introduction. In B. Graimann, B.Z. Allison, & G. Pfurtscheller (Eds.), *Brain-computer interfaces: Revolutionizing human-computer interaction*. Berlin: Springer-Verlag. Retrieved April 29, 2011 from http://www.springerlink.com/content/um1k229348172m20/fulltext.pdf

Haggerty, K. D., & Ericson, R. V. (2000). The surveillant assemblage. *The British Journal of Sociology*, *51*(4), 605–622. doi:10.1080/00071310020015280 PMID:11140886

Halperin, D., et al. (2008). *Pacemakers and implantable cardiac defibrillators: Software radio attacks and zero-power defenses*. Paper presented at the 2008 IEEE Symposium on Security and Privacy. Oakland, CA. http://doi.ieeecomputersociety.org/10.1109/SP.2008.31

Hancock, J. (2010, April 27). Install microchips in illegal immigrants, GOP candidate says. *Iowa Independent*. Retrieved April 20, 2011 from http://iowaindependent.com/32926/install-microchips-in-illegal-immigrants-gop-candidate-says

Homeland Security News Wire. (2012, November 19). *DARPA seeking surveillance technology to predict future behavior*. Retrieved November 19, 2012 from http://www.homelandsecuritynewswire.com/dr20121119-darpa-seeking-surveillance-technology-to-predict-future-behavior

IAO (Information Awareness Office). (2003). *Report to congress regarding the terrorism information awareness program*. Retrieved June 17, 2010 from http://www.globalsecurity.org/security/library/report/2003/tia-di_report_20may2003.pdf

Inda, J. (2006). *Targeting immigrants: Government, technology and ethics*. Malden, MA: Blackwell Publishing. doi:10.1002/9780470776315

IOL. (2007). *Big brother will be watching you fly*. Retrieved February 1, 2008 from www.iol.co.za/index.php?set_id=14&click_id=116&art_id=iol1171352398668C560

Jenkins, H. W., Jr. (2010, August 14). Google and the search for the future. *The Wall Street Journal*. Retrieved August 20, 2010, from http://online.wsj.com/article/SB10001424052748704901104575423294099527212.html

Jennings, C. (2006). Battlefield between the ears. *Nature*, *443*(326), 911. doi:10.1038/443911a

Keller, W. W., & Mitchell, G. R. (Eds.). (2006). *Hitting first: Preventive force in U.S. security strategy*. Pittsburgh, PA: University of Pittsburgh Press.

Kohlbrand, J., & Foster, J. (2000, August 13). Human ID implant to be unveiled soon--'Wearers' of digital angel®' monitored by GPS, internet. *World Net Daily*. Retrieved April 8, 2011 from http://www.wnd.com/news/article.asp?ARTICLE_ID=17601

Lipschutz, R. D. (2008). Imperial warfare in the naked city—Sociality as critical infrastructure. *International Political Sociology*, *3*(3), 204–218. doi:10.1111/j.1749-5687.2008.00045.x

Lipschutz, R. D. (2014). World war infinity: Total information awareness in the global garrison. In S. Hurt, & R. D. Lipschutz (Eds.), *The public-private hybridization of the 21st century state*. Academic Press.

Lipschutz, R. D., & Turcotte, H. (2005). Duct tape or plastic? The political economy of threats and the production of fear. In B. Hartmann, B. Subramaniam, & C. Zerner (Eds.), *Making threats—Biofears and environmental anxieties* (pp. 25–46). Lanham, MD: Rowman & Littlefield.

Lupton, D. (1993). Risk as moral danger: The social and political functions of risk discourse in public health. *International Journal of Health Services*, *23*(3), 425–435. doi:10.2190/16AY-E2GC-DFLD-51X2 PMID:8375947

Lupton, D., & Petersen, A. (1996). *The new public health: Health and self in the age of risk.* London: Sage.

Maisel, W. H., & Kohno, T. (2010). Improving the security and privacy of implantable medical devices. *The New England Journal of Medicine*, *362*, 1164–1166. doi:10.1056/NEJMp1000745 PMID:20357279

McGee, E. M., & Maguire, G. Q. (2007). Becoming borg to become immortal: Regulating brain implant technologies. *Cambridge Quarterly of Healthcare Ethics*, *16*, 291–302. doi:10.1017/S0963180107070326 PMID:17695620

McNamee, M.J., & Edwards. (2006). Transhumanism, medical technology and slippery slopes. *Journal of Medical Ethics*, *32*, 513–518. doi:10.1136/jme.2005.013789 PMID:16943331

Michael, K., & Michael. (2007). Homo electricus and the continued speciation of humans. In M. Quigley (Ed.), *The encyclopedia of information ethics and security*. Hershey, PA: IGI Global.

Michael, K., & Michael. (2011). Implementing namebers using microchip implants: The black box beneath the skin. In *This pervasive day: The potential and perils of pervasive computing*. London: Imperial College Press.

MURI (Multi-Disciplinary University Research Initiative). (2008). *Synthetic telepathy.* Cognitive NeuroSystems Lab, University of California, Irvine. Retrieved February 28, 2011 from http://cnslab.ss.uci.edu/muri/research.html

O'Brien, M. (2010). *Consuming talk: Youth culture and the mobile phone.* (Unpublished doctoral dissertation). Department. of Sociology, National University of Ireland, Maynooth, Ireland. Retrieved March 2, 2011 from http://eprints.nuim.ie/2253/1/MOB_PhD.pdf

O'Neill, M. (2008, March 9). Hackers attack epileptic forum and make sufferers convulse. *Geeks are Sexy*. Retrieved April 27, 2011 from http://www.geeksaresexy.net/2008/03/29/hackers-attack-epileptic-forum-and-make-sufferers-convulse/

Orwell, G. (1949). *Nineteen-eighty four—A novel.* London: Secker & Warburg.

Patton, P. (1994). MetamorphoLogic: Bodies and powers in A Thousand Plateaus. *Journal of the British Society for Phenomenology*, *25*(2), 157–169.

Positive, I. D. (2009, September 8). *VeriChip corporation agrees to acquire steel vault corporation to form PositiveID corporation.* Retrieved April 15, 2011 from http://investors.positiveidcorp.com/releasedetail.cfm?ReleaseID=428367

Positive, I. D. (2010). *PositiveID corporation launches the wireless body at ID WORLD international congress in Milan, Italy, Nov. 17.* Retrieved February 28, 2010 from http://investors.positiveidcorp.com/releasedetail.cfm?ReleaseID=531278

Positive, I. D. (n.d.). *Glucose-sensing microchip.* Retrieved April 21, 2011 from http://www.positiveidcorp.com/glucose_sensing.html

Priest, D., & Arkin, W. (2011). *Top secret America: The rise of the new American security state*. Boston: Little, Brown.

Savage, C., & Shane, S. (2011, February 24). U.S. arrests Saudi student in bomb plot. *The New York Times*. Retrieved February 25, 2011 from http://www.nytimes.com/2011/02/25/us/25terror.html

Silber, M. D., & Bhatt, A. (2007). *Radicalization in the West: The homegrown threat*. New York: New York City Police Department Intelligence Division. Retrieved April 29, 2011 from http://msnbcmedia.msn.com/i/msnbc/Sections/NEWS/PDFs/nypd_radicalization_report.pdf

Simpson, J. R. (2008). Functional MRI lie detection: Too good to be true? *The Journal of the American Academy of Psychiatry and the Law*, *36*(4), 491–498. PMID:19092066

Singer, E. (2010, July 29). Glucose monitors get under the skin: Implantable devices work in diabetic pigs for over a year – Human tests could be next. *Technology Review*. Retrieved April 14, 2011 from http://www.technologyreview.com/biomedicine/25889/?a=f

Singer, N., & Duhigg, C. (2012, October 27). Tracking voters' clicks online to try to sway them. *The New York Times*. Retrieved November 19, 2012 from http://www.nytimes.com/2012/10/28/us/politics/tracking-clicks-online-to-try-to-sway-voters.html

Smithsonian. (2012, August 19). These people are turning themselves into cyborgs in their basement. *Smartnews*. Retrieved from http://blogs.smithsonianmag.com/smartnews/2012/08/these-people-are-turning-themselves-into-cyborgs-in-their-basement/

Taipale, K. A. (2003). Data mining and domestic security: Connecting the dots to make sense of data. *Columbia Science and Technology Law Review*, *5*(2). Retrieved June 17, 2010 from http://ssrn.com/abstract=546782

Takahashi, D. (2008, August 8). Defcon: Excuse me while I turn off your pacemaker. *VentureBeat*. Retrieved April 28, 2011 from http://venturebeat.com/2008/08/08/defcon-excuse-me-while-I-turn-off-your-pacemaker/

Tan, R., McClure, T., Lin, C. K., Jea, D., Dabiri, F., & Massey, T. et al. (2009). Development of a fully implantable wireless pressure monitoring system. *Biomedical Microdevices*, *11*(1), 259–264. doi:10.1007/s10544-008-9232-1 PMID:18836836

Telemedicine and Advanced Technology Research Center. (2009). *Annual report*. Retrieved April 29, 2011 from http://www.tatrc.org/docs/TATRC_report_2009.pdf

Thompson, M. (2008, September 18). The Army's totally serious mind control project. *Time*. Retrieved February 28, 2010 from http://www.time.com/time/nation/article/0,8599,1841108,00.html

Turkle, S. (2009, February 2). Interview. *Frontline*. Retrieved April 28, 2011, from http://www.pbs.org/wgbh/pages/frontline/digitalnation/interviews/turkle.html

UC Irvine Today. (2008, August 13). *Scientists to study synthetic telepathy*. Retrieved February 28, 2011 from http://today.uci.edu/iframe.php?p=/news/release_detail_iframe.asp?key=1808

Walsh, S. P., White, K. M., Cox, S., & Young, R. McD. (2011). Keeping in constant touch: The predictors of young Australians' mobile phone involvement. *Computers in Human Behavior*, *27*, 333–342. doi:10.1016/j.chb.2010.08.011

Webster, S. C. (2011a, March 26). $1 billion FBI biometrics identification programme. *The Nation*. Retrieved April 29, 2011 from http://www.nation.com.pk/pakistan-news-newspaper-daily-english-online/International/26-Mar-2011/1-billion-FBI-biometrics-identification-programme

Webster, S. C. (2011b, March 24). FBI dedicates $1 billion to massive biometrics identification program. *The Raw Story*. Retrieved April 7, 2011 from http://www.rawstory.com/rs/2011/03/24/fbi-dedicates-1-billion-to-massive-biometrics-identification-program/

William, M. (2011, January 16). Thoreau's cellphone experiment. *The Chronicle of Higher Education*. Retrieved February 5, 2011 from http://chronicle.com/article/Thoreaus-Cellphone-Experiment/125962/

Yang, G.-Z. (2006). *Body sensor networks*. London: Springer Verlag. doi:10.1007/1-84628-484-8

KEY TERMS AND DEFINITIONS

Biodata: Medical data about individual physiological functions, usually entered into a computerized database for recordkeeping, billing and analysis.

Biometrics: Measurement of individual physiological features (fingerprints, voiceprints, body temperature, iris) for either identification of specific individuals or detection of unusual behaviors by individuals.

Danger: Potential or actual presence of conditions that might lead or contribute to injury, death or destruction of property (e.g., from a live but unexploded bomb).

Governmentality: The "conduct of conduct" through norms, regulations and practices that lead individuals to comport themselves and behave according to appropriate conditions and in such a way as to not pose dangers or risks to the self or others, through mechanisms of security, economy and politics.

Pocket Litter: The "stuff" found in the pockets and possessions of a suspect; here referring to the signals, data and traces resulting from daily individual interactions with various electronic communication networks.

Risk: Probability of development of some set of conditions that could lead to individual injury, death or destruction of property (e.g., chances of coming into contact with an exploding bomb).

Surveillant Assemblage: The complex of electronic (SIGINT) and human (HUMINT) data surveillance and collection systems found in everyday life.

Synthetic Telepathy: Capture of electronic and biological signals produced by the brain to be transformed into radio waves for remote communication and control.

Threat: The potential existence of conditions or actions that might lead to injury, death or destruction of property (e.g., possession of a bomb that could be detonated).

Total Information Awareness: Name of a U.S. government program designed to collect a broad range of electronic and other data on individuals, for data mining in order to detect and surveil persons of interest, especially actual or potential terrorists.

ENDNOTES

[1] http://www.nickbostrom.com/old/transhumanism.html, accessed 11/14/2012

Interview 16.1
The Screen Bubble

Katina Michael, Faculty of Engineering and Information Sciences, University of Wollongong
Interview conducted by Jordan Brown on 14 June 2013 at the University of Wollongong

Jordan Brown: So give us a recap of how we've gotten to where we are today.

Katina Michael: Well, the ENIAC was a 1946 innovation, and is widely considered the "first" automated general-purpose computer. These things were really huge and they covered the wall to ceiling, and large floor spaces dedicated to just a single computer.. I think as the transistor was then invented post the vacuum tubes, that it was inevitable that, yes, chipsets would come next, and that lots and lots of these would fill computers and larger rooms, and that miniaturisation was inevitable. These computers got smaller; they didn't cover walls and floors of buildings, but actually got to the point where you could put them up against a single wall (e.g. mainframes and minicomputers and the like); and then toward the mid-1980s microcomputers entered the scene; and then we got to the point where we could carry these and lug these around with us in the 1990s, and now we can even wear them. Digital Glass for example, Google's latest product, enables us to have lots of sensors and lots of chips in our digital eye glasses. So what's the next step? We've gone from luggables to wearables. Are bearables the next phase of innovation? How much smaller are we going to get? What's the next quantum leap for computing and humankind?

Jordan Brown: So where do you think it sits at today in terms of the size? What do you see happening after that? We've had this progression from the size and capability, what's next?

Katina Michael: So the progression has been that we have observed the chip in other objects. We then have the chips in devices that we carry with us, like smartphones, and inevitably perhaps one can ponder that they'll be injected into our bodies. We've already seen this demonstrated in heart pacemakers since the 1960s for instance; and much later the innovation of the Cochlear implant (an implantable device for the deaf). The question is when we take these commodities and apply them to non-traditional areas of application. What you then have are typical biomedical devices applied to everyday contexts for convenience as opposed to need.

So what do I see? I see little tiny cameras in everyday objects, we've already been speaking about the Internet of Things—the web of things and people—and these individual objects will come alive once they have a place via IP on the Internet. So you will be able to speak to your fridge; know when there is energy being used in your home; your TV will automatically shut off when you leave the room. So all of these appliances will not only be talking with you, but also with the suppliers, the organisations that you bought these devices from. So you won't have to worry about warranty cards; the physical lifetime of your device will alert you to the fact that you've had this washing machine for two years, it requires service. So our everyday objects will become smart and alive, and we will be interacting with them. So it's no longer people-to-people communications or people-to-machine, but actually the juxtaposition of this where machines start to talk to people.

Jordan Brown: And so that's happening today?

Katina Michael: So we've had, for example, Auto ID Labs which was an MIT initiative which first began the notion of the Internet of Things, and the Internet of Things really coincided with the rise of Radio-Frequency Identification, and so we've taken this proposed way of interlinking RFID with everyday objects and everyday applications, and we're now at a point with the emergence of Smart Grid infrastructure and Smart Meters where yes, we have the capacity to interact with everyday objects. And so there are distinct relationship links between objects and subjects—and subjects being people, objects being things. So things now have come alive, and people relate to things. People relate to animals, they own animals with implants; people have relationships with other people in social networks, people use machines etc. So I want you to think about the Internet of Things as one big social network—it's just that your social network might actually have a list of devices you actually own and have purchased as well, as well as your family members, pets, car etc.

Jordan Brown: Can you explain what the Internet of Things is, or what you see it to be?

Katina Michael: The Internet of Things is when a convergence between various network levels- infrastructure, the core, the edge, application devices that users carry—begin to interact with one another and share data that's collected from the field. This to me has a lot to do with the field of telematics. So in a simple scenario, with fleet management, for example, you may have a fleet of vehicles. You may be transmitting to them information about where to go next or the shortest path to a delivery destination. But these fleets may also be interacting with you as a fleet manager and be telling you "we're here, we're currently experiencing unanticipated congestion, and the sensors on the street have alerted us to a better route to take so that we save on fuel and optimise our business processes". So it's these constant data flows back and forth from the field to the network operation centre, to the storage centre, to the hubs, in this large complex system of systems. So there's a lot of embeddedness. There's a lot of nestedness. There are lots of larger systems interacting with embedded systems within. There are data flows and protocols and handshakes occurring between subjects (as in people and objects as things) and so it's all about business process optimisation on the back-end. For example, how do you best optimise your business processes, say in order to save on operational expenditure, and ensure you how your money is being expended. Now on the subject side, on the human side, companies will know where their employees are every moment of the day and this will help in business process optimisation, the only thing is people are not machines. People work well under pressure but once pushed over the edge may feel like they've been driven too far and so break like elastic bands. You can quash the human spirit because humans are not robots.

Jordan Brown: Do you think that premise that people are machines or even the phrase "Human Resources" is a part of the ideology of the advancement of these things (say the Internet of Things, or technological progress in general)? Do you think there's an assumption that humans are just a part of the vast machine to be controlled in a further granularised way with all of this data coming back to the people that own and run the networks?

Katina Michael: Humans are a vital part of any business process. I think we can build really smart systems in order to do away with some of the human judgement that takes place in order to reduce risk; so then you have automated data-driven innovative processes that can help you in order to make the right decisions at the right time, for example, during emergency responses—in fires. So you have these sensors in the field, for example, that give you and feed back to you, the right and appropriate information, and automatically may give you the best way forward. My problem with removing humans from business processes however altogether, is that the machine will never know what is inside the brain of the human and so the "what if" exhaustive options used in an automated decision-making process may not be so exhaustive because they've been programmed by people with limitedness based on predefined options. But if you bring together a vast array of people and try to respond to a particular decision that is required to be made with a lot of intellect, a lot of cross-disciplinary knowledge, I don't think you can replicate that on any artificial machine, or with alleged artificial intelligence.

Of course what we've seen over the last few decades is this need to do away with large items of operational expenditure and any company—its labour force is usually in excess of 40% of its ongoing operational expenditure. So what do companies do at times of difficulty? They look at their financials and they think "How can I get rid of X percent of my labour force and replace this labour with machines that have a lower expenditure and a higher return on investment because they never get tired, and don't need to sleep, and don't need to go on annual leave?" If we're talking about a manufacturing industry, for example, where robotics has done very well, we have whole factories now being run by robots. Of course these are human-programmed robots where people have fed instructions and creatively produced software to enable these factories to run. But the question is: What do we do with these people who don't have jobs? Are we reskilling them in any way or just de-skilling them and bringing them out of the labour force?

Jordan Brown: I'm not sure if I should ask about drones on that point?

Katina Michael: Actually, I would rather talk about countries like Bangladesh, Nepal and India, and numerous hotspots in South America. For some time, we have been aware that there are many organisations that perhaps use human resources as if they were machines and that the amount of money that they get paid for their labours is not commensurate anywhere to the retail price of these garments, for example, being distributed and sold in more developed countries. Someone may be working for a dollar a month and I pay for a garment which is in excess of one hundred dollars, and don't know what kind of effort has gone into that particular garment. The business process optimisation in these organisations and factories are benchmarking in seconds. So if you don't complete a part of the garment within X number of seconds, then you haven't done your job appropriately; you're not working; and you're at risk of losing your job. So the pressure in these factories and I want to point to perhaps the example of the production of Information Technology components and devices, especially, is overwhelming to the individual manufacturing worker. For example, there have been multiple suicides in large factories which build everyday components for companies like Sony and Apple and many other well-known brands, whereby employees have wished to end their life from the pressure of engaging with difficult employers who drove them so hard that they've just thrown themselves off a 10th floor. I mean, how do you cope with 17 suicides in a single organisation over a 2 year period? That's a high number and quite concentrated when compared to the rest of the world's statistics on suicides.

Jordan Brown: They put nets up to stop them successfully jumping down. Send them back in.

Katina Michael: That's it. They did. So that's even worse. 'Die slowly.'

Jordan Brown: Yes. I want to go back to the Internet of Things. What I really want to discuss is: who owns that infrastructure? It's not the consumer—they're just a passive recipient of the technology existing in that space. I want to discuss possible manipulation. I want to examine the forces impacting on people existing in that space that they might not be aware of, and even if they are aware, there seems to be lots of excuses for 'no cause for concern.' I'd like to get your take on that. So if people have very little control over the infrastructure for the Internet of Things on the outset, what's happening to them?

Katina Michael: As consumers we buy products from stores. We buy computers, mobile phones, and other high-tech gadgetry. When we bring these into our homes and start using them—whether we're at work, for personal application, at university, or around with our friends—what we're doing is actually buying in to these new innovations and their externalities. What do I mean by that? We've become enslaved by our adoption of these devices and constantly feel like we have to upgrade to the next device. We constantly feel that we have to be using, to be seen to be up with the Joneses, and beyond that. Actually at times, I think people have no choice—we wish to maybe stop this cycle of usage but cannot. All of these devices which are online devices, they can be used wirelessly in any space, any geographic context. These allow us to be located, tracked and identified (because you've got passwords, you've got a location where you receive signal strength from your nearest Wi-Fi access point. All of these put us onto a grid of sorts—it's just called the Internet. So we can be tracked, our behaviour and our behavioural patterns of how often we log on, where we log on from, who we are logged on with; whether we are visiting a social networking site for example— all of these behaviours are logged and audited. Now shared, we can provide adequate infrastructure for these services. From the service provider point of view what you're doing is saying, "The more we know about you, the better we can service you." But in actual fact what's being done is your data is being collected (potentially anonymised) but collectively used to identify more of what you're interested. So your buying habits, where you visit online—we'll just throw a little bit more about this or that product at you because that's what you've been searching. So what is in that search box, that search engine that you use to look around for various likes and dislikes, and questions that you might have, actually depicts you as a person—it's a digital DNA footprint which distinguishes you from everyone else.

Jordan Brown: And that's called data mining?

Katina Michael: This is data analytics, predictive analytics; data mining in the traditional sense. But whereby data mining was very applicable to a data-surveillance world, we are now looking at über views—holistic views of the person about identity, location, video that your watching, images that your uploading, etc—all of these different contexts and sensory data are coming back to give us an über view, in an überveillance society. So we've moved on I think from data analytics to über-analytics—being able to define you not just by the transactions that you do at a store, but by absolutely every touch-point that you become engaged with in an online context.

Jordan Brown: And across all the devices you're using…

Katina Michael: So service providers have an über-context to work with. They can analyse your movements and analyse your behaviours across a vast array of devices. So in the morning you might be checking your email via your desktop but as you exit the house, you have your smartphone clipped on. And so your interactions and the different devices that you use and the different applications that you may interact with on a day-to-day basis, for example, social networking tools, define who you are and define how much activity you have with particular groups in society.

Jordan Brown: So taking that, I'm going to play devil's advocate for a moment. Imagine that I'm Joe Consumer. How would you convince me that all of these technologies that I'm using and all those processes that are happening, pose a risk to me in any way? Or even us collectively as a society? The line we often hear is "It's convenient. I love these things. I'm happy to make that trade-off (of mass surveillance)." So if you were to convince someone or to even just point out the risks, what would you say?

Katina Michael: The phrase that's often used: "I've got nothing to hide, so I've got nothing to fear", is something that's often said by people. To that I always say, "You've got nothing to fear and nothing to hide until somebody identifies that you have otherwise." The other thing that is very pronounced is that as consumers the very information we give over, gives rise to our own exploitation and manipulation. It is like luring consumers to admit to certain weaknesses in buying particular goods or services, because they have stated that they are considering buying product x on Facebook.

The other thing about the current reality is that attacks on consumer privacy are asymmetric. To the vast population, they go on about their daily workings and activities as normal, but to those individuals who become caught up—by accident—in further questioning about their particular physical or online behaviours, there is an asymmetric trade-off and that is, I am one person against perhaps an army of people being accused of X or Y, and all I've done is actually searched for a piece of information on the Internet. The asymmetry occurs just like in credit card fraud. You know, where a lot of money is stolen by hackers annually and the credit card companies pass this off as just a liability and another cost to their business (because the interest they're making on credit cards is so huge. They do not invest in more secure technologies because they can write off the losses). But if you are the victim of a credit card fraud crime and somebody has stolen your credit card, gone and visited an escort service, and then that particular line item appears on your credit card and creates a family furore, then you're actually on the other side. So it's an asymmetric relationship. You feel victimised, but the credit card company really doesn't care and what they do is just reimburse you the amount of money, but what you want to happen is that you want that line item removed, but you can't get it removed—it's a service that has been paid for. I'm not saying people should not use credit cards, what I am saying is that we only feel the full force of this current reality if we find ourselves as victims, otherwise we might be oblivious to the goings-on.

Jordan Brown: So that's the thing. What happens when we have a culture imbedded in the world of the screens, where information flows coming out of the screens are seen to be valuable, and objective, and indisputable—what happens when something like you've mentioned above happens and you have the fallout, the impact on relationships, the question of what is trusted, what is real and what isn't real, all tied up in the perception of 'computers never lie'?

Katina Michael: The culture of screens is a very misleading culture. Don't believe everything you see. That's what we're taught from a young age. Increasingly, I believe, people do believe what they see. We've seen examples of this where police, for example, have been accused of police brutality without provocation but more evidence from smartphones has indicated that the context has been missing and therefore identifies that the police were provoked perhaps into some force. On the other side, on the flipside, we now have police trialling wearable cameras on their bodies in order to decrease complaint handling. However, censorship is still possible through the screens—my point of eye, where I'm looking towards may not be where the crime is taking place. So if I'm a police officer taping a whole scene of an alleged crime, but don't wish for particular brutality to be shown on a screen, I simply look away. I'm recording this way, the activity is occurring the other way, and I've just done censorship. So don't believe everything you see on screens. It is often not the whole truth.

Jordan Brown: Do you see manipulation happening in screen culture today?

Katina Michael: I think manipulation on the Internet certainly occurs. We call this disinformation. This is nothing new. Propaganda is a historical element of this screen culture—if I tell my message to enough people out there, they'll believe it. If it comes from enough credible sources, the populace will believe it. If I look up on Wikipedia, a particular entry, "definitely it's correct" [*said ironically*], I believe it. There are administrators over 1500 now correcting and making edits to Wikipedia. So we believe what we read by nature but who are these administrators? Nobody knows you're a dog on the Internet, right?

So is there disinformation occurring? Sure. Are people of all walks of life engaging in disinformation? Certainly! It doesn't just mean online communities, or communities that are related to organisations. Just even the idea of brand awareness is a type of propaganda—"I'm pushing forward a particular brand, I'm advertising, I'm pushing this to your screen." Every time you go onto Google, for example, and do a search and it's related to this or that product, I will push more of this product to you. So we are subliminally being provided with messages whether we realise it or not that perhaps sway us to a particular brand, but also sway our intentions and motivations towards X or Y. This can be done by companies, by politicians, by government agencies etc.

Jordan Brown: And do you think that an element of that manipulation is what is driving technological advancement? Say with targeted advertising for example?

Katina Michael: Companies tend to defend their practices as being purely related to marketing. "We elicit this response from you; we use the behavioural tracking and cookies in order to perhaps sell you more of what you want to see. We're not doing anything bad; we're just giving you more of what you asked for." The question however becomes when you start to consider at a much broader level, at a higher level, when all of this advertising and affiliate advertising and affiliate sharing of data and partnership sharing of data becomes used to exploit the consumer. At what point do we say enough is enough? And at what point do we say, yes we would like more. I think people are sick of getting more of the same, but I think we are oblivious to the fact that actually push marketing or push advertising is occurring because we've become immune to the practice. If someone was to film a heavy user of a smartphone for a 24 hour period, and then replay back to that user what they looked like while using their smartphone, I am sure that user would be asking questions about their 'conditioning' to all consumer electronics.

Jordan Brown: Does that mean that screen culture is creating bubbles around people then too?

Katina Michael: The screen culture makes people look within and not to look outside. So when I'm using my smartphone, and I'm being sent instant messages, and I'm being communicated to—it's about me. It's about me and my interactions, and people can say that's great for personalisation, that's how I want it—I want to customise my whole life, but in fact, we're internalising a lot of things. If I think about "me" then most likely I will neglect my children, I will neglect my partner, I will neglect my workplace, because it's about me and my interactions and the instantaneous communications that take place. There's always a danger in that—in ignoring your neighbour, in a lack of collective awareness. It's about insular things, and in so doing, what you are doing is removing your ability to think, removing your ability to pray and be peaceful about things because you're constantly being bombarded by messages (which may be entirely irrelevant- for example, spam). You're constantly thinking that these are more urgent than the baby crying in the next room who requires milk or food, and the screen culture just propagates itself. So in order for me to internalise my communications and look down and keep texting and keep messaging back, I also impose the same culture on my children because I just tell them to go look at the TV for a little while longer, go onto the Internet, search some more things. So I am spreading this mimicry of sorts and I can't stop this cycle because I'm deeply engaged in it. And so when my senses are enveloped and it's about me and my communications, it's not about my children, it's not about my partner, is not about my workplace—it's about me, and I think there is a great danger in trust within society, in building relationships with one another or a lack of building, when we are concerned about the me.

Jordan Brown: Is there an addictive quality as well?

Katina Michael: I think our use of smartphones and our impact on our daily life by smartphone communications, for example, and the screen culture, is not only addictive but "obsessive-compulsive" addictive, like a cyber-drug of sorts. It's a health problem and we've yet to really master even to begin to ask the right questions. It's taken us 30 years to realise that fast foods cause obesity. This is a well-known fact now but was unknown for decades. Fast food advertising—even in the sports arena—causes obesity. How long is it going to take us to realise the addictive nature of smartphone usage to our being and our family units? 5 years, 10 years? Is that going to be too late by then, because the mimicry will have been so well entrenched in the next generation? What do you do about that then when all the Millennials are entrapped in a particular way of life, resembling what some would argue is a "zombie-like" state? The thing is you've got to do something about it today. We're not even coming to grips with the obsessive-compulsive disorders of young people suffering from anorexia because their online gaming too long; of young people wetting their pants because they forget to go to the toilet because they are almost at the next level of a multiplayer game with their friends; and of young people being stuck in the room because of the screen culture which has pervaded bedrooms that were traditionally used for study and sleep.

Jordan Brown: People dying in Korea in Internet cafés…

Katina Michael: Yes, people dying in Korea in Internet cafes. Babies being starved to death because their parents are raising online children in Korea, the most networked nation in the world. We've got to stop and think about our next phase as a humanity, and our next movements forward as a populace.

We got people starving in less developed countries, people being treated like slaves in newly industrialised countries, and here we are in the more developed countries saying "I've got Google glass". Well congratulations. New innovations, augmented reality, and perhaps augmented death at the same time.

Jordan Brown: So what are these companies building then? If it's not us building these things—we're just reacting—what's being built? What are we currently in today?

Katina Michael: We've always been, over the last two hundred years or so, since the Industrial Revolution, been stuck in profit maximisation and sales maximisation modes. Organisations generally have one of two goals: they either want to be a profit maximisation firm, or a sales maximisation firm. And if the goal is profit maximisation, it's about making money, it's about making your shareholders rich. Do I care about the externalities and the side-effects on the everyday consumer? No, I don't. Most people you ask who are building new engineering systems don't think about ethics. It's their job to build, to create, to push the boundaries, to build new applications that people will find a use value in, but these days we're not even concerned about the value of the product. What has Google done for example recently in launching their digital glass product? They've released Google glasses to 8,000 "explorers". "Go and explore. Tell us what you do with Google glass. Tell us your new applications. Thrill us with biometric recognition of your friends, and your address books. Augmediate your world so when you look through your digital glasses, only see the advertisements that you want to see on the billboards."

Jordan Brown: The bubble-

Katina Michael: Again, we're living in a screen bubble. We're protected by this forcefield of sorts, again, returning back to the self—it's about me, not you. And it's also finding comfort in the creation of inventions that lack positive utility for society.

Jordan Brown: Does this bubble also serve as the greatest surveillance grid ever constructed?

Katina Michael: Large service providers in the world today and we all know who they are, servicing so-called "free" applications, "free" email, "free" uploads of data, hold the key to unlocking who we are as a digital footprint, as a digital DNA. The more we give away freely to these free services, the more they will be used against us, to identify us, to categorise us, to segment us into a particular market type- the elderly, the more secure; the socialite; the worker bee, the teenager etc. However, we are going beyond these typical market segmentations that were created in the 1990s, for example, through the mobile revolution. It is about you—and not about the collective today. We've become so smart in our algorithms, in our neural network approaches, in our semantic analysis, in our sentiment analysis,—various types of approaches to analysing what data you publicly disclose voluntarily—that this data is then being used and repurposed to send you out more of the same.

Jordan Brown: Or even data that can be inferred or assumed in aggregate.

Katina Michael: So data that you provide through touch points on your mobile phone, on your laptop, on other devices, even your VoIP sessions through Skype—can be used to infer a great deal about you. And we no longer require with predictive analytics concrete historical evidence to place you in a situational awareness context. If you are at location A, for example, university, then I can infer that most likely you are a student or an academic. Taking this further, if you are in a particular location on a Sunday morning which is not a well-known location to be visiting on a Sunday morning or is a Church for arguments sake in the opposite spectrum, then certain assumptions are being made about you and about your likes and dislikes and about your character traits. No-one is immune.

Jordan Brown: So does society control technology or is it the other way around?

Katina Michael: Society creates new technologies. Initially…

Jordan Brown: Actually, wait. That's something I'd like to clarify. Do you think it (technological advancement) is driven by these people over here that are creating all of this stuff and everyone else in the bubble is reacting to that; or do you think that it's us as a society opting for further advancement, for this surveillance grid to perpetuate itself, by the choices that we're making in that space (the bubble)? Or do you think it's this force over here (the technology companies) acting on this force over here (the consumers) and it's all just playing out as a phenomenon? What do you think?

Katina Michael: Every member of society has a role to play in society. All of us are governed by our life-world—that which encompasses the motivations and drivers for how we go about living our life. For example, if I'm an engineer in an engineering community and work for one of the large ICT organisations, my life-world tells me and informs me to say "Create, design, build, collaborate, share knowledge. Strive for that next product innovation and incremental innovation which is better than the one before." I could be driven by ethical codes of conduct in engineering and design, but I may not be really interested about legal issues or how the media interprets this, or even how consumers might interpret the product that I prototype, patent and release to the world. If I'm a user, am I mindfully adopting new technologies? Or am I just doing this on autopilot because everybody else is doing it? And if I don't have this application then I become ostracised—my community refuse to contact me because they say I am making it difficult for them if I do not join Facebook.

Jordan Brown: The mobile phone is another example.

Katina Michael: Yes. So the mobile phone can become an inclusive device or an exclusive device whereby it may include you if you are a fellow mobile phone user or it may exclude you if you are not. I'm not the first person to have trialled Facebook and to have lost a whole bunch of university friends when I deactivated my account. Quite interestingly when I reactivated my account two years later, again, people I had lost contact with started to communicate with me again. So new technologies can be used to make the people part of a social network, or they can be used to exclude people by default if they don't wish to opt into such a new application. So this is quite normal in the new devices. You either upgrade and keep on with the Joneses and with everybody else upgrading and be part of the in-crowd, the clique; or be left behind and live off the grid. However, most people don't have time for those living off the grid. If you don't have an email address these days, you're probably a non-person. There are members of society that for example don't have a drivers licence and don't have email accounts, and these people are being left behind; they are fearful of the change that is occurring, and they are being somehow co-opted into having to change. Even if you're 65 and never driven before—get your licence, it will guarantee your passport, for example, if you want to go overseas. If you've never had an email address before and you don't see the use in having an email address, you might have to if you want to communicate about product updates with company X, otherise you fall off their radar. So we have a very hard time in dealing with exceptions. If I don't have an ID card, if I don't have an email address, if I don't have a mobile phone, if I don't have a Facebook account; then the question is do you really exist? Your normality is probably even questioned outright? How do you as a person living off the grid deal with this scenario? Is there a manipulation of sorts by service providers? Have business processes advanced so much that you've just got to get on board if you don't want to be left behind?

Jordan Brown: So it's the network effect you're describing?

Katina Michael: What we are doing is empowering various online applications such as social networking applications like Facebook. I read yesterday that 30 million dead people have accounts on Facebook, and however there are close to 1,000,000,000 people now on Facebook (and some people have multiple accounts) but what you have is this "get on board, get on the bandwagon", the domino effect, the network effect—"make sure you're there, otherwise you'll miss out." And the more people that go on board unquestioningly actually propagate this false conception of the screen culture.

Jordan Brown: So what does that look like for the coming next few years?

Katina Michael: About 20 years ago, I heard about something called the Follow-Me-Number that was published in an International Telecommunications Union (ITU) report. A lot of protocols at the time that used VoIP were being developed, were being discussed: how can we create a Follow-Me-Number for every single person on Earth, sort of like a universal lifetime identifier? So you don't have to worry about changing mobile phone numbers, or losing one as you're changing phones, losing a SIM card etc. You have one constant email address that follows you around etc. And to be honest this smacks of person-number systems that were introduced just post the World War II period for social security purposes and rationing, and for giving people money that required it for services like social welfare. A Follow-Me-Number just like a unique DNA is quite eerie when you think about it. Just like you have fingerprints that can't be changed, in the future your Follow-Me-Number will be unchangeable. We're currently finding it easy to change credit card numbers when someone has defrauded us, but what will happen when your person number or your Follow-Me-Number is defrauded? What better way to institute a universal lifetime identifier than a microchip implant worn on the body and/or embedded in the body. What is real creepy is that most people would be devastated at having to change their cell phone number today, so in a strange sort of way, this Follow-Me-Number is already here.

Jordan Brown: Okay, and so taking some examples from history, say the infamous rise of the fascist regime in Germany which was heavily dependent on profiling people, is there a risk then based on historical experiences, of recent memories and generations, coupled with the "I've got nothing to hide, I've got nothing to fear" sentiment? Is there a reasonable concern that the rise of this big data, this vast surveillance society, lends control to a small group of people which could potentially enable really intense abuses?

Katina Michael: Most people who talk about privacy protections and privacy principles have studied history very well. They understand the risk associated with amassing large stores of personal data: your date of birth, your name, syndromes that you may have, whether you're a life insurance member, whether you drive three cars, have five children, have had previous marriages—all of this data is highly personal. On its own, as individual pieces of information, they may not be telling anyone anything. Collectively however, particular patterns can be used to infer almost anything. If we believe that what happened during World War II is not possible in today's society then we have a narrow view of history. Anyone, at any place, under any particular government agency control may find themselves on the wrong side. Here is more of that asymmetry I talk about often. I may not fit the latest fashion of thought, how will I fair in that particular community or society at large?

Jordan Brown: In the future?

Katina Michael: It can happen any time. If there's data that is stored, depending on the particular regime at that particular time, anything is possible. Who would have thought what occurred during World War II would have happened? Even with the limited automated computing that was available at that time, which was pretty clunky based on punch cards? People's religious beliefs were used during that time to segregate them. The Nazis attempted to remove Jews from their homes to make them completely at their mercy, if not kill them in some aspects as we saw in the gas chambers of Auschwitz. And what we need to understand is while we don't have modern-day gas chambers that look like gas chambers, you can squeeze the life out of anyone by the knowledge that you have of their personal data.

Jordan Brown: Yes. And the way I see it is that this panopticon has been built—it's not some grand conspiracy, they're not all colluding with each other (to make this happen). It's just the temperament of technological advancement—like how you say, the engineers mindset is "we create". It's just happening. The panopticon has been built, and the people in the bubble that are affected by that are reacting. So what's next? Where is that going?

Katina Michael: In April, we saw the devastating bombs that were used to maim and kill three individuals during the Boston Marathon. This is a classic example of where surveillance technology absolutely failed. Initially the wrong two suspects were identified. And this went viral on social media. They didn't do it. It was asymmetric. The two individuals who were wrongly accused of having planted the bombs were defamed in effect. They were scared, in one college student's response, to actually leave their household. A few days later, the real suspects were identified but what happened was the asymmetry had already taken place. Two perpetrators were identified early on, it went viral, they were categorised, and it was as if probably to them their whole world was against them. So we could have found ourselves in the same situation—wearing the 'wrong' clothes, looking the 'wrong' way at that particular location.

What authorities have now begun to question is how much more surveillance they could have applied to find the perpetrators faster and bring them to justice. However, biometrics failed the authorities. Surveillance footage failed the authorities. Until one of the victims who was maimed said: "I saw the perpetrators look at me and this is what they look like." He was able to give adirect evidence account of what he saw. However, if we say to ourselves "we can get better at this, we can introduce new technologies, we can introduce better biometrics on mobile CCTV cameras, our smartphones can have particular sensors that can be recording", and if I'm a law enforcement officer, "what we want to do is proactively profile the community to identify potential terrorists in that community". Well, then, we are going the wrong way. These are anticipatory strategies. This is situational awareness and proactive profiling which means that you are going on potential inferred data, or big data analysis as it's called, without actually being able to verify that this person has any intent in their head to commit a crime. This is when it gets a little bit scary as a member of the community. If you find yourself in the 'wrong' place via smartphone and its GPS enabled chipset, you look 'wrong', so your behavioural or your physical attire depicts you as a potential terrorist when you're not.

Jordan Brown: Like the fact they had a bag...

Katina Michael: You have a bag, you have a hat, you are wearing a hood or concealing your eyes with glasses, etc. What we don't want is a society where there's a chilling effect and people actually don't want to go outside their homes. And I have received numerous emails over the last 10 years of individuals potentially who have been suffering from mental illness but are scared to exit their front door due to surveillance cameras. I'm not propagating that view of the world, people are feeling it. These people are actually feeling this pervasive computing and invasion of their privacy, especially when they live in an apartment that is under constant camera view from an adjoining building. What do you do then? Keep your blinds closed and live in the dark? So yes, most people have nothing to hide in society, but some people feel they have everything to protect.

Jordan Brown: So it's happening then? The effects—maybe in its infancy—but you can see the chilling effect you're talking about happening now?

Katina Michael: I think the chilling effect is happening now, and if we do say that some people who are mentally disturbed are disturbed by these additional use of technologies—and were not just talking CCTV cameras, but people feeling like they've been implanted with chips for example... then we have to take these concerns seriously, not because they are happening, but because people are feeling triggers towards paranoid capacities. So you could say that "We don't care about these people, they're the minority. Let them have their paranoid schizophrenic attacks and their mental illness and it is their choice if they feel they cannot step foot outside their house ... that we're not going to solve their problems, a psychiatrist will, and they will perhaps need more medication but in actual fact and effect, we are creating new technologies that in the near future may have most people concerned about who's watching, when. Can you imagine what that might feel like? You know, I know there are sensors that track me. I know I'm wearing a smart phone that I can get instant communications with. I feel stressed by going to work because I know my boss knows when I actually arrived at work, when I take my lunch hour, when I visit the bathroom—these ID cards can tell many organisations what is going on with employee clocking on and clocking off during particular sessions of time during the day. But what kind of society are we moving to when we need alarms, bells and whistles, for absolutely every action we take?

Jordan Brown: Is that a panopticon?

Katina Michael: To me personally, what's a panopticon? I think it's when I might arrive in an uber-veillance society where I can't even think because the thoughts in my head have actually been inferred by somebody else. So I feel like I'm enslaved, I'm trapped—not within a prison wall, for example but within myself. I can't have the freedom to be who I want to be, to act like the person I want to act like, and just being myself without feeling someone is scrutinising my every move. And that is a real issue.

Jordan Brown: Can you tell me a little more about technology addiction?

Katina Michael: Most people haven't realised how over-reliant they've become on their smartphone. It's not just a tool used these days for emergencies. I think many people falsely make themselves believe that they're low end users of mobiles, when in fact, they're glued to them. You know, they drop off their children at the daycare centre, and there's an inclination to pick up the phone and to check the messages that have arrived from 5 minutes ago. You arrive at work, "Oh I gotta check the messages as I'm walking down to my office." And the excuse often used is that "I'm trying to capture and become more productive during my work life, and so that I don't have a bank-up of messages when I get to my office or when I get home at night". But actually what people do is they go home and they filter through even more messages, and then they get up in the middle of the night because they hear the phone buzz, and they're within an arm length distance of the phone and they'll be awoken and pick up the handset, look at the phone, respond in the middle of the night and then go back to sleep. We are living a 24-hour cycle these days. The world has become an always on, always connected, online, global world. We haven't been able to distinguish the boundaries between our home life very well and our work life, despite this quantified self movement saying things are getting better. And the thing is that there is a natural force in writing an email- you send a message, you get one back. It becomes an endless trail.

Many employers know this and so they provide free smartphones and free laptops to their workforce, because they know they will get more productivity out of them during the working week—even the weekend. So we're expected in this world to somehow carry on with our everyday relationships, as well as be always connected. Somebody submitted to me a short article for *IEEE Technology and Society Magazine* and said it was a hopeless situation one Sunday morning when he walked into an Indian cafe and this husband and wife arrived shortly after to sit at a nearby table. Thehusband's phone rung, he answered it, he was on the phone and while the wife was looking at the husband on the phone, she took out her phone and started interacting with her messages because she felt ignored. Their food arrived, and they ate it while they were texting and talking. At the end, they finished their meals, the husband got off the phone, the wife put her phone back in the purse and they got up and left. What kind of interaction is occurring at that point? None. At least not with one another in the physical space. We are almost stuck in the online world and cannot distinguish between the off-line and online world, even when we are in the presence of other people.

Personally, I know at times when I've been working at home on my laptop, trying to finish off an editorial for a pressing deadline, my children will come up and chat to me and I respond to them, and they say "Mum, you're not listening mum—close the screen." And I say, "Five minutes, I'm almost done, just give me five minutes, I know you're really hungry." They come back in five minutes. "Mum, we want to eat". And I say to the kids, "I'm sorry I'll be with you in five minutes." In effect, something like 20 minutes has gone by. By that time the children are so hungry that they'll come over to the keyboard and start pressing the keys in order to make me make a mistake so that I am forced to get off the computer.

That reality for that split second that I'm engaged in an online activity, or even on a computer—it doesn't have to be online—is in a virtual space. For example, my head is in that editorial. My children are at my side, but I can't distinguish between the two. And the more imminent is that which I'm engaging with on the screen because it has my full attention than that which is in the physical space. That is a dilemma that is I think is pervading most homes these days, and somehow families are keeping up with taking the kids (if they have kids by the way), taking the kids to sports, to other afternoon activities, in amongst this jostling of time between the Internet, the smart phone, and a laptop computer.

Jordan Brown: So that's it—the disconnection from reality.

Katina Michael: That is it. There is a disconnect from the reality in front of you. So I can be looking at you in the eyes, but my mind is still engaged in that practice and if we keep propagating this to newer innovations that continue to draw us away from the physical (e.g. augmented reality, e.g. drone applications)…

Jordan Brown: Or even just exacerbating the bubble?

Katina Michael: Yes, we are actually exacerbating the bubble. So if we propagate this culture, this screen culture, this online time, this not recognising the imminent physical people that are around you, then we're just going to get more and more lost in an online space which is really somewhat unreal. What happens online sometimes with regards to relationships at least may not be as real as we think they are. Sending someone a virtual hug, for example, is not exactly like having a real hug, a real embrace. But we seem to be filling up our world with status reports, status updates, Twitter messages—I mean, I read yesterday that there is a tweet alert for Huggies nappies so you figure out when your baby has wet his/her. Do I need a tweet to understand when my child… Am I so disconnected from my child that I need to get a tweet on my handset? I think there are two reasons that this application has been introduced. The first is that parents are so engaged with online activity that they forget that the child has not been changed in several hours, and the child gets nappy rash. The second reason why this happens is because parents potentially and carers have lost touch with the physical, they don't want to touch the child or remove the nappy to see or to smell—so our senses are being dulled down and replaced by tweets. This is really linked to our ability to recollect and to prioritise what is important.

The next thing that's going to happen is that we'll have even more alarm clocks. Some of us already have an ergonomic alarm warning that reminds us we need to get up, stretch our legs and move away from our computers set between 7 o'clock in the morning and 7 o'clock at night. But there are still many who eat their breakfast, their lunch, their dinner, in front of the computer. Those with home offices can sometimes suffer greatly as there are limited disruptions from peers besides an email, or telephone call. So we're like stuck in the old mines. You know, you go down into the pit, it is dark when you wake up, you start typing, and you stay there all day. You want to go home and it isdark still. Really, have we got better living conditions than they did back in the times of the mines, those really dirty mines with bad working conditions? Of course we're allegedly better off these days because of the clear air we're breathing in our offices but we are still to some degree stuck in the mine mentality where you wake up, it's dark, in fact sometimes we don't even go to sleep! I shouldn't even be using that expression, "we wake up" because many professional workers are now always on call, always connected, always replying, always sending messages back and forth, so day and night is even difficult to delineate.

Jordan Brown: It's a strange thing being disconnected.

Katina Michael: Some people have admitted that they never disconnect and that they find long plane rides especially difficult.

Jordan Brown: So people are further removed from the actual happenings behind the interfaces (of screen culture)… so if most people just pick up a phone, make a call and there's no real knowledge of what is happening behind that screen to make it possible—the vast wired and wireless infrastructures, the programming of the phone, the interfaces—as screen culture perpetuates itself and as the user becomes more removed from those processes, is there a loss of understanding as to how those infrastructures work and the risks that presents?

Katina Michael: We're often told as consumers not worry about the black box. The black box is the inner workings of a particular network, of a particular application: how it works, how it's built from basic principles, etc. For example, today people have reusable software. They don't need to know how to program. They can get a few chunks of code from here and there. They can have some level of work experience, bring reusable software together to do things that seemingly work but with little knowledge of the inner workings of a single module. This is a way of building new systems. The technical things are not for everyone. Don't worry about what things are going on. Don't worry that a call can be carried from A to B and go through about 15 Internet hops between locations. Sure, data can be intercepted, but don't worry about that—who wants to read your email? So this knowledge and approach to simplicity and to creative design and critical making is this hacktivist kinds of hackathons where people come together, you have these crazy ideas, you trial them out, and you get an end-to-end process going. Every person is like an individual unit in that building of a new prototype or a new application—they don't need to know what the next unit is doing.

But there's a problem with that kind of approach in that you may know your own particular area very well, but not know what's occurring in the next phase of the development of that product. It's like asking a professor who has built a small component—a scientist who has built a small component of an implant for the heart for example—to describe their own component, and they can do that really well, but then ask them to describe a little bit further out, and they say "Sorry, that's not my area, I can't really tell you how that was built." This is a problem because we don't realise what goes into potential wireless interceptions, potential jurisdictional issues between data storage versus data ownership versus data sovereignty versus requiring compliance with particular laws. People almost don't care about these fundamentals these days. And if they do, they're tactfully placing these data storage centres in places that will provide them with the liberty of accessing that information.

So the companies have a major contribution here on ensuring that they set up systems and networks—not just for their own good, but for the good of the community at large. And not to save money in their pocket by storing documents and data offshore which contains personal information of citizens. But I know what's occurring, it has to do with money again, it's got to do with profit maximisation, and it also has to do with becoming elusive to jurisdictional issues and legislative issues. So if I want to continually evade this jurisdiction based on that particular act, what I do is just put my data centre in a place where there are no laws against how I store things and for how long and the physical lifetime etc.

And this opens up individuals to abuses. This is a big problem. Governments are continually cutting public-sector roles and we've seen this at the state level over the last 12 months especially with the shift to cloud computing practices.

Jordan Brown: And then the users are deferring that knowledge and responsibility to people that don't necessarily have their interests at heart?

Katina Michael: Service providers provide terms and agreements that they very well know that users will not read. We had Google, for example, going and squashing their X number of different applications into a single privacy policy statement. Now, that can work in two ways. The argument that Google provide is that of simplification when in fact the reality is that it will decreases Google's liability to particular attacks on particular individual's datasets. So, by removing liability as a service provider, the onus goes back on the user, and the users are not equipped to deal with any breaches in their privacy or security.

Jordan Brown: And may not even know that such breaches or liabilities exist?

Katina Michael: Yes, most users don't know for example that they've been hacked, they cannot distinguish between sites that are real and phishing sites, and are not educated, are not cyber aware even about virus protection because the hacking attempts and the breaches in security and interception are so advanced these days. Consider Raytheon's RIOT software http://www.youtube.com/watch?v=7mcVA_D3sAg that can check whether Joe Bloggs has logged in and checked into FourSquare and looked at whether they're at the gym at 6 o'clock every morning or are doing different things during the day. And that's when we get to individual targeting. And I can, if I am empowered by this knowledge of looking at your personal journey and tracking through the day, then I can have some influence over you, because I know about your movements. And it doesn't mean that you've done anything wrong.

Jordan Brown: So that leads into one of the big questions then: it's progress for whom? So as you've mentioned, if someone has power over you and that's playing itself out, say for example if someone runs the Raytheon software that can potentially be watching a lot of people, we have an 'us versus them' dichotomy. It is progress for whom then? Who benefits?

Katina Michael: I think about this in terms of the poverty cycle—the rich getting richer and the poor getting poorer. And the poor get poorer because they're stuck in that rut and they give birth to children and the children are brought up in the same environment. And unless something magical happens they will continue to be in that environment and stuck in that cycle, that endless cycle. The same thing happens when I'm stuck in an endless life of upgrades, whether that has to do with computing, whether that has to do with any particular application that I might buy into. And so those people who are building the applications who know the inner working of the applications they're building and understand how the infrastructure works are more empowered than those people that are allegedly adopting voluntarily the products that are being sold to them. A lot of my students, for example, and I found this out early on, would work a whole week to pay for their mobile phones in the early days when it was particularly expensive. I used to ask, "Why are you so tired coming into my lecture theatre?" And I'm thinking, "You mean, you've worked all week to pay off your mobile phone?" There was something wrong about that model. I told them I'd rather you have no laptop for your honours project, I'd rather you not buy into a mobile phone, and come to my class awake so you can learn something new or at least participate in the dialogue and provide an opinion for me with the rest of the class and share it. So, it's almost like we're stuck in a cycle that's not called the poverty cycle—I don't know what you want to call it, "the enslaved cycle of ICT"... I don't know what you want to call it, but I've got to have this and I've got to have that, and I've got to have the next thing, and if I don't become a member of Facebook and I don't start geotagging—I've missed out... the opportunity cost is too great. However, what are the trade-offs here? I'm more worried about status updates on Facebook than I am about living my life. And this is the problem. I forget about living. I'm just doing what is expected to be done. Replying to that email, replying to that Facebook wall post... where's the common sense thinking in all this gone?

Jordan Brown: And in the meantime, those small groups of companies are further closing in on their influence of the people who are taking on those technological advancements, those developments?

Katina Michael: Of course, service providers become more and more empowered with the more and more personal data they gather. What you don't want is churn. Churn is when an individual user goes from one application to another. You want your user to be presented with stickiness drivers—this is a technical term in customer relationship management, so that your user, your customer comes back to your portal, interacts some more, gives away more—this is the whole business model of customer relations management. Provide enough stickiness drivers, they come back and they provide more, they disclose more. And how can we capitalise on this social ensemble? On this information disclosure? What can we do? Let's analyse it then. Let's analyse what they're talking about. Let's analyse what they feel about Brand X or Y, and if they feel badly about Brand X, let's employ the right strategies to counter that feeling. So are we being manipulated? Of course we are. And by the very data we disclose. This is the problem—we don't realise we're at the beginning and end of that cycle. We provide the information, someone analyses it and it's fed back to us and we eat it. So we might not think we're being manipulated, but in actual fact the whole idea of customer relationship management is about this cycle. It's about a stickiness driver and preventing churn. How Facebook, for example, can have more users than G+, but these guys are not silly. I mean, at the highest level, organisations that have the largest market share, if brought together in kind of a sharing and merging relationship—imagine, for example, Facebook, Google and a number of other organisations like Twitter, decide to share their user data and they profile individuals. We have to realise that anything is possible and these tech giants will continue to push the envelope.

Jordan Brown: So what happens then when you've got these companies that are individual entities? What happens if someone, say the intelligence organisations, say the National Security Agency in the United States comes along, and collects all of those vast data stores from the one's you mentioned: Google, Facebook, Twitter and others?

Katina Michael: We are being sold that we have transparency, at least outwardly, of the number of requests that are made to for example Google, of a particular user's content or metadata. So for example, Google publishes quarterly the number of requests they get from law enforcement agencies. Google have also stated that if the law enforcement request comes in for a heinous crime, for example, a murder or a rape or what have you, that they will not tell the user that they are under surveillance or that their data has been provided or will be provided to the law enforcement authorities. This is quite different from say a secret intelligence organisation that may wish to investigate individuals. We don't have that transparency. We should have transparency. Why should these secret intelligence organisations be exempt from a warrant process? And this is where a number of Acts in different jurisdictions really don't hold up to the mark and what I'm afraid of in the next 10 years is that we dilute these warrant processes and have warrantless monitoring. Just like some malls have got particular equipment to track users and their customers through shopping malls; how long they've stood in front of a window and pondered about walking in and then made a transaction within a particular store. So although at the moment this data is being gathered anonymously in the shopping mall context, the question is what will happen when we start to dilute privacy principles, privacy Acts, and say well, if these private companies are surveilling others and we have Raytheon, for example, producing products that are able to track people behaviourally, using for example check-in points and check-out points, then why not just leave it open to anybody? And we can make data-driven innovations from this. We can provide shoppers with better quality experiences through shopping malls—there's always an excuse for why to dilute privacy. There's always an excuse to strengthen security.

Jordan Brown: Do you think that's really been exacerbated in the "post 9/11 world", where terrorism is a buzzword used to dilute those privacy principles, and to shift the balance of power further towards these secret intelligence organisations?

Katina Michael: I think greater visibility was always on the cards. Being able to access data without warrant processes despite these age old privacy legislation enactments and surveillance device and listening Acts, and whole a gambit of telecommunications data interception Acts and so forth—it was always on the cards. Things move faster and easier when there are no security roadblocks, when I access anything I want as I want it, wherever I am. And we've seen this starting to dilute slowly since the inception of geographic information systems, census data on CD, customised to your needs—you know, ring the Australian Bureau of Statistics and tell us what you require, provide for us, for example, an Australian Business Register and identify businesses at a collection district level. We've now got satellite imagery we can purchase as tiles. This was available 15 years ago when I was in industry. We could provide this and overlay and register our own images and our own photographs—a bit like Google Maps for private organisations. This has been an ongoing process. Let me create a Google Maps—what a great idea. I can map every administrative boundary in the world. I can map every street location. I can look at what we have topographically on our Earth. Isn't it a fantastic idea? But when I then go to the next stage and let me go into different cities and let's start photographing every cadastre plot, and let's go and do more. You know, if you've got a photo of somebody's home and you want to upload it, hey, upload it under Google Maps. Isn't it nice to have the visibility of visiting a place before you've gone there of what it looks like—I'll never get lost again. Navigation, fantastic for creative industries and new services, fantastic for open innovation, but then where are we going next? Let's use Google glasses and let's not just take a Street view, let's go into the house, let's go into the plot, let's record 24×7 and upload that up onto the Internet. What we're being asked to be is drones. We're "manned drones"- not "unmanned drones"—we're manned drones. And I I can tell you that in the near future, what we will have is people being paid to be drone-like recording devices, where they walk up and down malls, they walk up and down public streets capturing visual evidence of passers-by as they go about their private business. At what point are we going to say we should not be uploading this data to Google maps? We should not be videoing everything in sight, recording it, and uploading as if we own it or I own your image, or Google owns it because it's on Google maps? At what point do we say enough's enough and we stop even surveilling one another?

Jordan Brown: Makes me think of the case in Britain where people drove out Google Street View with pitchforks…

Katina Michael: Yes. They got Google out quick smart because they ambushed the vehicle and threatened to smash it! But what's worse is if we're going to be recording everything we see. Imagine, I'm recording you recording me right now. And that's okay, if we've got consent to interact with one another that way, but there is no way I'm going to get everyone's consent as I walk down the street. And some people may be having a bad day—you're entitled to have a bad day. If you don't take it out on anyone else for example, you may not be feeling well, you may be crying, you may be suffering, you may have had a relationship breakdown. Do you want that captured on video? That private moment as you're walking down the street—you've just been given the news that your child is about to die in a hospital. Or you've just been told by your husband, I'm sorry I don't love you anymore. Is that what we want to capture—all those bad moments? We've got to get serious and get real, because life is not hunky-dory. Life is not always smiling, and like we see on those Google Glass promotions—the airbrushed look, you know, at 6 o'clock in the morning, I know what my hair looks like. I know my kids are screaming for food. Do I want that publicised on television? No. Do I want that publicised on the Internet? Of course not. If I want to go and get my mail from my mailbox with my pyjamas on and my robe, I should be able to do that without feeling "Oh, should I be dressed like this? Should I brush my hair before I go outside?" You know, Sunday mornings for example. And what we're doing is we're about to say "Hey, that's okay, let's pervade everyone's life. Let's not care. Let's see where we're going to go with breaking down everyone's privacy. Let's not look back—look forward, advance." And this promise is a fake promise because of the other stuff that I mentioned a while ago happens. We have struggles, we have challenges, we have crises in our life. We don't want to be replaying those over and over again.

Jordan Brown: Can I ask again of the person that would say "I've got nothing to hide, nothing to fear"… How do you persuade them given that situation?

Katina Michael: I should just stay outside their home and start capturing their every move as they interact in their front lawn, their back lawn, anywhere I can see from the front of their yard, and then what I should do is get in my car, put a GPS device on theirs covertly and follow them down the street. And then I should get out at their workplace and say "Hi, it's me again. I'm wearing the camera. I'm recording you. Don't worry, I won't put it up on the Internet today." And then I should follow them home and then see how they feel the next day when I do the same thing. And the day after that, and the day after that... And I think they'll get really sick of me really quick.

Jordan Brown: So is it only then because a lot of those processes perhaps aren't so close to that person, say with pervasive CCTV doing just that? Is it because it's not part of that person's awareness potentially, that they may feel that "it's not a problem, it doesn't worry me"?

Katina Michael: Most people who go about their everyday life are oblivious to CCTV cameras—even mobile CCTV now on police cars. And what that's called is the novelty effect—it wears off. So if something is new, I look up and I think "Oh, it's new, it's invaded my space", just like when telegraphs were introduced and people saw terrestrial lines that carried voice calls: "Wow, what are these things?", you know? We see windmills today and we think: "Oh wow, a windmill", or we see other infrastructure and we think "Aren't those base stations at the top of the building looking ugly? Haven't they destroyed the landscape?" So we do notice these things initially, but we become oblivious to them over time. I don't notice base stations any more and I used to work very closely with where base stations were located.

Jordan Brown: For mobile phones?

Katina Michael: Yes, for mobile phones. I don't notice CCTV cameras as much as I used to, they've sort of become transparent to the industry design of most buildings. They now have an aesthetic quality about them. I notice that children notice them, because their world is new, everything is new to a child, as they go to a mall for the first time, they ask the questions: "What's that?" But the novelty effect wears off and adults and with that wearing off we become immune perhaps, and we forget to question what is going on. It is like being stuck in a fog, you cannot see all around you, and you hope for that car that you're tailing with the blinkers on, is headed the right way… otherwise it is the blind leading the blind…

Jordan Brown: So where are we headed then?

Katina Michael: So why are we headed on this trajectory? Where are we headed? Why is this happening? For a long time when I was studying ICT in my undergraduate years, I used to study tech-evangelists and this whole idea of technology evangelists was striking to me. Who are these guys with job descriptions called tech-evangelists? What was their role? And I remember being at a conference in Sydney of all places when I received a business card that said "I'm a tech evangelist" before the dot.com era. More recently, I looked at a job title from IBM that had the descriptor "chief storyteller." Oh yes, now "what do you do for a living Sir?" | "I am a storyteller." And that storyteller was similar to a tech-evangelist. They sat between applications development and solutions architecture. So I'm a storyteller. I tell you stories about how you can harness these products for your business.

Jordan Brown: So it's like a spin doctor? Marketing?

Katina Michael: Yes. So we've created organisational positions- if you want to talk about manipulation, a tech-evangelist is probably a great manipulator and wants you to buy a particular brand and wants you to think a particular way, possess a particular ideology but so does the storyteller. Stories and metaphor can evoke huge reactions in individuals. Now, the question is: who is proposing these new ways forward? Of course we can look at the patent database and claim that these individuals who have over a hundred patents each in these particular areas whether it's digital glass or any other innovations, smartphones, or wireless technologies- they're the ones driving innovation. But in actual fact when we start to theorise, and say who are those thought leaders? Who are those people in the think-tanks? Do they have diverse backgrounds? Are they representing me as I should be represented? And when you start to dig a little deeper, it's really a very small number of people that are driving these new innovations either by accident or by conscious decision making. For example, Facebook. You know, it was supposedly an accident, and it took off really well. I'm sure everything that has happened since the accident, since that coincidence, has not been an accident- it has been very deliberate in strategy. But I also believe that these very successful companies are co-opted by various government agencies to their own ends. Private business must always be within the grasp of government, otherwise the government does not have the ability to provide "security" to its citizens. And this is where the paradox is- for a government to claim that it has "national security" as a core interest, it must either have some control over private enterprise, or enforce "a watering down of company security profiles". There is a symbiosis between government and private enterprise for this very reason.

Jordan Brown: Is the idea of "storytellers" like a euphemism for advertising itself?

Katina Michael: Yes, and application developers, and business developers- they all develop 'things'. The question is whether we let ourselves believe what is being proposed by the futurists, or whether we say, "Hey, that sounds really dumb. I don't want to live in a society like that. I don't want my kids being raised in a society like that. I don't want to live forever"… or whatever is the latest high-tech fad.

Jordan Brown: And that's another one of those things too. Ray Kurzweil and other futurists like that such as Michio Kaku (and perhaps Kevin Warwick), see those points as downsides to the 'human condition' for want of a better phrase. Getting sick, feeling sad, having a finite life: "these are all things that are undesirable." Does that not in-and-of itself say how fundamentally disconnected those ideas are from reality? And also in conjunction with what we were talking about before about "The Bubble" and "screen culture", and having your brain in a space which isn't in the real world? Because to me, that says it all. If someone says, "I see being human, being alive, being a biological creature on a finite planet as undesirable." It encompasses all of it. It's basically saying we should be dead, we should be machines, not be human anymore, not live in reality anymore. And how is that going to happen? How's that going to work?

Katina Michael: There are lots of different arguments to that point of view, the point of view that says, "I don't want to die, I want to live forever, and I want to do away with my *sarx*—which is my body. I want to do away with my limbs because they have a physical lifetime. And I want to live forever, I want someone to flip a switch, make sure I'm always on, upload my mind to the Internet for example, and be free of physical spaces and dimensions." The question is how realistic is this when there are people dying of a lack of food every day. The reverse argument says that if we all were to have our minds uploaded onto the Internet for example, or a data storage device in some way, in some shape or form, then we would do away with hunger altogether. But to that I always say there are technical failures. There are smart grid failures, energy failures—what happens if accidentally your smart grid powers off? And who's going to be alive physically to turn off and on that switch? Don't tell me a fallible machine?! There's got to be someone always there, a human, using their mind, using their physical tactile fingers to actually do something to the physical, breathing, "storage network." But this also presupposes that we are not spiritual beings and that what makes us up is simply "biology" without "spirit". Yes, we might one day be able to tap into the mind, but there is something that makes us who we are, and that part cannot be replicated, no matter how hard we try!

Jordan Brown: And also that that possibility isn't available for everyone? It's only available to the few that can afford it. So the third world, for example—it's not for them.

Katina Michael: So are we creating an elitist society? Those who can afford actually, can adopt these new technologies, just like people who have invested in cryogenics and other means of potentially keeping themselves alive and leaving their estate to themselves in a legal sense. So, "I can keep being cloned and coming back to life, and I've got my estate and I live my life again another hundred years, and if we stretch it to 103—very good." But most people on Earth won't be able to afford these elite services, if they do come into existence in the future, as has been proposed by many futurists. And the question is- what kind of life would that be? I like my body. Although, I acknowledge that there are people who are entrapped within their body, e.g. disabled people. I can see *for* and *against* arguments for this kind of lifestyle. I can see how we could free people who are trapped within their wheelchairs, and even within their minds in some syndromes through the upgrading of their mind—if that is ever to become a practical capability.

But if we think of the here and now, and what people really need today, it's not more of that kind of thinking. We aren't machines. We're people. We've got blood rushing through our bodies. We've got veins, and we've got a heart that's beating and pumping blood. We've got a pulse rate. I can touch you by bringing out my hand and I can sense your touch, I can sense your feeling. Do I wish to augment my body? Hey, it's your body do with it what you want, but I should have the right to live how I wish as well. But if we don't look at what is occurring to us as individuals, we may slowly succumb to becoming technology without realising it. And it does start with basic principles. It does start with having my mobile phone within reach when I am asleep and the question is whether the phone is an extension of me or I'm an extension of the phone. The more machines that we build around us that are "always on" in this Internet of Things, the more I become subject to that machine, rather than the opposite. I am at the mercy of the machine. I am at the mercy of my own creation. Is that really a world we want our children to be raised in?

KEY TERMS AND DEFINITIONS

Bearables: Another term for implantables, for technology that is embedded beneath the skin.

Behavioral Tracking: Refers to a range of technologies and techniques used typically by online website publishers and advertisers but may also include smart phone usage patterns allowing service providers to increase the effectiveness of their campaigns. Users are very often oblivious to the goings on as no previous consent has been sought from individuals for the tracking to occur.

Bubble: A metaphor for pervasive consumerism. Consumers remain unaware not only of their high-tech usage patterns but also of the bigger picture issues affecting them with respect to technology adoption. They are stuck in a 'bubble' so to speak and that bubble can burst at any time.

Drones: Is a colloquiual term for unmanned aerial vehicles. Drones that carry ammunition are deployed predominantly for military and special operation applications.

ENIAC: Electronic Numerical Integrator And Computer was touted the first electronic general-purpose computer.

Implantables: Are microchips that can be injected into the body. Form factors vary but usually include tags or transponders.

Off the Grid: Being completed disconnected from any form of telecommunications, including land-line telephone, smartphone, email, and Internet more broadly.

Push Marketing: Is when a customised marketing alert comes to your smartphone based usually on your location. Usually these techniques offer purported discounts luring consumers to impulse buying. Traditional push marketing techniques include targeted mail order catalogues to your home, and email alerts based on data from online behavioural tracking.

Screen Culture: Is a culture which is dominated by screens of all types but particularly digital displays. It may include digital billboards, television, gaming consoles, digital cameras, computers, smartphones, wearables digital glasses, or anything else that introduces another layer between the naked eye and the natural world.

Smart Grid: Is a modernized electrical grid that uses information and communications technology to gather and act on information, such as information about the behaviors of suppliers and consumers. Smart grids are meant to improve the efficiency, reliability, economics, and sustainability of the production and distribution of electricity.

Smart Meter: Is usually an electrical meter that records consumption of electric energy in hourly intervals and communicates readings back to the utility base on a daily basis for monitoring and billing purposes.

Storytellers: A position title in some large technology companies. One step removed from a technology evangelist, the storyteller sits somewhere between the salesperson and the solutions architecht, attempting to convince the client of the benefits of a given solution to their business problems. Storytellers are technically astute and are strong advocates for their company's product/service lines.

Technology Addiction: The state of being enslaved to a habit or practice or to something that is psychologically or physically habit-forming, such as all things high-tech, to such an extent that its cessation causes severe trauma. Most people admit they cannot forgo the use of their mobile phone or iPad devices for very long.

Chapter 17
Uberveillance and Faith-Based Organizations:
A Renewed Moral Imperative

Marcus Wigan
Oxford Systematics, Australia & Edinburgh Napier University, UK

ABSTRACT

Uberveillance extends the responsibilities of faith-based organisations to the power imbalances now emerging. This is less a matter of governance and strategy, and more one of the core values of faith-based organisations. These might be regarded from an ethical or moral standpoint, but the approach taken is to focus on the constituencies of faith-based organisations and the imperatives that have been woven into their aims and values. The specific ways in which such disempowerments emerge and the functional importance of making organizational responses are considered. Acknowledgement is made of the Science and Society Council of the Churches of Scotland, who catalysed the expression and articulation of these issues.

INTRODUCTION

The perspectives of faith-based organizations encapsulate most of those in the broader community, but with a stronger representative role when power exertion by society on weaker members, not only of their own community, becomes evident. This is a historically important role, in which most share. The developments in large scale databases, government pressures for a single identity, and the merger of a wide variety of data and information holders create cumulative ethical and moral issues for the community at large, and it is arguable that the value systems used to resolve these are currently neither sufficiently diverse nor effective enough to moderate these cumulative effects in a humane and ethical manner.

The fundamental nature of uberveillance, surveillance and integration of multiple identities is to create power imbalances. The special feature

DOI: 10.4018/978-1-4666-4582-0.ch017

of the information and communications system's disintermediation of this process is removal of most of the mechanisms that could correct or undo the cumulative power shifts that information asymmetries establish.

Faith-based organisations are, in the main, religious bodies, but not all, as a range of humanist bodies also share many of the same characteristics. Faith-based organisations usually have a special mission to correct power imbalances that disadvantage the weaker and more vulnerable in the community, not only amongst their members, but in the community at large. While not widely known, the Roman Catholic Church states this explicitly (Pontifical Council for Justice and Peace, 2005) in the words of Pope John Paul II addressing the Bishops in Mexico in 1979:

This love of preference for the poor, and the decisions which it inspires in us, cannot but embrace the immense multitudes of the hungry, the needy, the homeless, those without health care and, above all, those without hope of a better future (p62).

This view, reiterated in several contexts by Pope Paul II[1], is typical of the formal underpinnings of the commitment of faith-based organisations to addressing asymmetries of power and possibilities for the weak. The new imbalances induced by the asymmetric application of state-based dataveillance and physical surveillance to accumulating vast data records on individuals are steadily becoming more comprehensive and widespread; they need to attract the attention and action of faith-based organisations, at the very least in the interests of their more vulnerable members, and more broadly by attesting to their political capacities in influencing these objectives for the broader community.

Such actions are not often visible: why? One reason could be the comparatively limited use of the Internet by religious individuals as compared to their secular counterparts (Armfield, Dixon, & Dougherty, 2006), although Armfield et al point to

the organizational structural power of the pastor in articulating, explicitly or implicitly, the desirability of certain uses. This Foucaultian perspective (O'Farrell, 2005) on a Christian sample community, where the pastor's powers derive from being expected to guide, feed and protect a spectrum of humanity unable to spiritually fend for themselves, resonates with the functional role of the Mullah in Islam. Thus the enunciator expresses much of the power of faith-based organizations – and may well not express the same range or emphasis of the overall organisation at the higher level exemplified by the text cited for the Catholic Church (op.cit).

While these mediations of power at the congregational level are understandable, articulation of the higher levels of the faith-based organisations is where the political influence materializes.

The scarcity of public commentary by the Churches and other faith-based organisations on the values implied by uberveillance was highlighted (Wigan, 2010a) at a panel run by the Church of Scotland Church and Society Council towards a policy for science and society for this church; the implied moral duty of the churches to take positions on the disadvantaged by these trends and express them was asserted to be a logical consequence.

Secular approaches also emphasize the imbalances, more from an equity of access standpoint (Celeste, DiMaggio, Schafer, & Hargittai, 2004), yet echo a similar stance as follows:

... the research we call for here is one front in what should be a larger effort to understand the causes and impacts of inequality in access to and use of information of many kinds. Information figures crucially in the generation of inequality in advanced industrial societies in myriad ways ...

The approaches so far recognised to be relevant by a wider professional society are still limited, and largely neglect the shared values and roles of faith-based and even secular civil liberties internet-sensitive organisations such as the Electronic

Frontiers Foundation (www.eff.org), Knowledge Ecology International (http://keionline.org/), Electronic Frontiers Australia (www.efa.org.au), and the Australian Privacy Foundation (www.privacy.org.au) to name but a few. The power imbalance issue has yet to be catalyzed as an issue of broad public interest, and does not form a significant part of the recent European Commission Framework 7 Project ETICA [2] on the Ethical Issues of Emerging ICT Applications

This chapter is framed to assist in the establishment of this necessary process of recognizing the proper role of faith-based organisations in what is clearly a secular political issue. Establishing issues is much harder than prosecuting them (Wigan, 1994) once a point has finally been broadly recognized as *being* an issue. The terms on which such engagement can occur are a critical component in this process.

WHO ARE THE VULNERABLE NOW?

Access by Internet users is not the area of greatest concern any more: it is now the power imbalances emerging between Internet users with access – and the organizations (such as Facebook) preying on them for personal and saleable information. Conventional wisdom might suggest that a problem for faith-based organizations would be in the area of the perceived digital divide, i.e. enabling their members to secure equitable access to these and other resources. This assumed distinction between less advantaged groups in society being disadvantaged by poor take up and access to the Internet was a live issue from as early as the late 1990s (DiMaggio & Hargittai, 2001; Mossberger, Tolbert, & Stansbury, 2003), and is still an active concern. However, the complementary issues of what faith-based and other organizations do about the negative results of access being secured and used has been buried below these same assumptions: that improving access is the goal and the power imbalance is to be addressed as a matter of

equity. Typical of these initiatives is the Seniors Kiosk program of the Australian Government, where central facilities for access to computers with broadband connection were set up in places where older people congregated and some volunteer support was likely to be available [3], and similar programs elsewhere.

There has been an emergence of massive information power in the hands of, first, governments exploiting anti-terror legislation to enable widespread and effectively unaccountable monitoring of the population, and secondly huge bodies of personal information made available by Facebook usage and other social networks. We argue that this emergence indicates that the moral and ethical issues of information technology are now as important both in the information itself as well as in the mechanisms of access and use of it.

This neglected area has a special resonance for faith-based organisations, with their commitment to the disadvantaged. Again, the obvious groups that are so disadvantaged are not necessarily those one might expect. The data shows that higher educational and higher income groups take disproportionate use of (and access to) the Internet, and so these groups, with their greater exposure, and now arguably the most vulnerable to the new omnivorous and never forgetting Panopticon that the social networks and active population monitoring have created.

The migration of intelligence and anti-terrorist thinking has now begun to infect civil law. This migration was encouraged by the paroxysm of civil liberty and accountability destruction of the Howard years in Australia, and more generally in the instant and widespread reaction to 9/11, which empowered politicians to enact such invidious Acts in several countries.

Examples include the comprehensive collection and matching of all vehicle movements (both via speed cameras and Automatic Number Plate Recognition systems deployed for the purpose), not limited to lists of persons of interest, but as comprehensive data capture "in case of need".

Consumer monitoring by camera and identifiable mobile phone integration and linking to loyalty cards and credit cards is now emerging, as the RFID system becomes prevalent in the supply chain.

The complementary aspects of vulnerability enhanced by social data system utilization is now magnified by the spread of location based services, and hybrids of social networks and location based services, such as FourSquare (Clarke & Wigan, 2011).

This development handles a remarkable degree of detailed individual surveillance on recording unmatched as yet even by CCTV, where records a few weeks old are often deleted as operational policy. It is closer to the expansion of DNA records in some countries, where the records are ever expanding, and where the law may not be updated to ban its use for insurance, employment or other purposes (or at least to be used in any detectable manner for these purposes).

Each of these areas are domains where those on the "positive" side of the digital divide are becoming the most vulnerable: these are exactly the groups least often considered as being in need of protection from asymmetric information power ... yet protection from access might be the issue that needs addressing. At least structures of monitoring and accountability need to be in place to maintain the dignity of individuals in the face of an increasingly over-weighted information power with the massive historical record that both the state and social networks are amassing.

These examples demonstrate that faith-based organisations, with their firm commitment to justice and equitable treatment of all persons, need to take up considered positions in the policy formation debates, and begin to engage in the policy process.

There are arguments that this role of faith-based organizations has been subverted, and indeed some of these organizations have become players in power plays for less high flown objectives:

Tocqueville's view of the civic role of religious associations as a vital element of "participative democracy" and brake on the centralizing tendency of democratic power has been turned upside down. Recent empirical studies converge in suggesting that religion has become an instrument for power-aggrandizement and collecting votes (Ungureanu, 2008: p406).

Urgureanu picks a middle course drawn from developments in Europe in particular, with the key point:

The advocacy of religious discourses can have, under certain conditions, an intrinsic (and not only functional) value for the construction of the democratic legitimacy. This can occur, for instance, by means of selective democratic interpretations taking place in the opinion-oriented public sphere; such interpretations can turn religious motives into secular democratic justifications through inclusive discursive practices (Ungureanu, 2008: p407).

This point is specifically and formally endorsed by a recent Catholic encyclical (Benedict XVI, 2005), which both endorses the autonomy of secular politics and also the public role of religion as an indirect contribution to democracy.

The moral responsibility of faith-based organisations to intervene in the debates on power asymmetries is clear: although the acceptance of this by the secular state may be less so.

THE ROLE OF INFORMATION ETHICS

The basis for negotiating positions on power imbalances and asymmetries, and guarding against the negative effects, often relies upon secular ethics as well as religious values. Professionals in most fields have developed formal statements of ethics, but these are usually framed in such a way that they protect the professional group rather than the community at large, as one would expect for groups formed to protect and propagate their own special interests and perspectives. The outcomes have in general been positive for the community none the less, but the assumption that such formal ethical statements are adequate requires a broader frame of reference for values and personal dignity, and these areas are the domain of both general politics and faith-based organisations. The secular and the religious share this space, and have done so for centuries with varying levels of success in the power and assertion of their positions.

The development of information ethics is reasonably recent, but focuses in the main on the duties and responsibilities of individuals in information professions, rather than on organizational values and auditable actions. The sustained weakness of the Australian Privacy Commissioner (now subsumed into an Information Commission, but now with an even more muffled voice) is the natural consequence of inadequate public space pressures to resource and give greater powers and sanctions to this important body. The political and ethical divide is all too apparent in this case.

WHERE TO TARGET EFFORTS TO ADDRESS THESE ISSUES?

Specific areas where power assertion imbalances are emerging are discussed elsewhere in a more developed paper (Wigan, 2010b), covering issues such as the protection of multiple identities and other areas where power aggregation is occurring but not necessarily obvious until examined. The latter issue is discussed in detail elsewhere (Wigan, 2010c), as are the less obvious aspects of location based services (Clarke & Wigan, 2011).

There are a range of measures that might be considered to head off the looming prospect of the power asymmetries now emerging from uberveillance programs becoming permanently embedded. However these are now becoming secondary to the considered and active participation of a wider range of civil society bodies in formulating the steps to be taken and agreeing on the principles to be applied. Because of the current emphasis on individual access, enforcement, and far from benign neglect of the secular governance aspects of the trends discussed, bodies such as the faith-based organisations in societies must articulate their concerns and demands in a clear and effective manner that chimes with their own values, but must now form a far more public set of voices. Improved governance of national security is an element that needs to be addressed (Wigan, 2012).

There are several specific and convergent themes that highlight the importance of the role of ethical bodies in society: a major role of the faith-based community of Churches (Wigan, 2010b).

- Lack of ownership of one's own identity.
- Pressures to remove the right to maintain context-dependent multiple identities.
- Subversion of the presumption of innocence to a presumption of possible harm.
- Growth of intelligence techniques enabling mass population surveillance: the case of Location Based Services.
- Progressive criminalization of intellectual property with State support for commercial interests.
- Asymmetries of information leading to a greater need for contestability.
- Progressive accretion of biometric data unmatched by appropriate legislation on their use.
- Exemptions from accountability for highly personal data aggregated by politicians (still specifically permitted under Australia law).

The progressive digitization of society and its implicit and explicit interconnections raise many unprecedented moral and power relationships that have yet to be fully addressed.

The trends in terms of privacy and representation (and misrepresentation) of identity are pressing individuals in different ways to modify their behavior. Omnipresent surveillance has reached remarkable levels in countries such as the UK, so that the distributed presence over time and activities of individuals is now progressively linkable. These issues are clearly both moral and ethical, and it is in the domain of the faith-based bodies to play a role in their mediation and debate.

Expanding on just one of these moral and ethical issues, the personal identity that we have is becoming legally a collection of digital tokens, which may be exchanged, matched, sold and collected in virtual space with cumulative value. This has two major effects:

1. These tokens, which are ourselves, are not owned by ourselves. In a very real sense we have lost ownership of our identity, which has become a commodity in which we have no property or moral rights. They can be – and are – traded in the marketplace, be lost, and we are left vulnerable with little redress. This loss of dignity is hardly the most alarming aspect: our property and legal rights are now at risk in exponentially rising ways as a result;
2. Ever increasing power asymmetries are created by the ever accreting function creep that is enabled by this ability to steadily draw in historical and other data. The lifetime student number, the lifetime health record, the national ID card are all aspects of these keys to our own identity ... which we do not own or control.

Google (+), Facebook and the Australian Government are all pressing to assert the right to only a single identity in frameworks with no means for the community to resist the latter, and even withdrawal from the former cannot correct the problem of extant linkable data. The cumulative effect is to deny the ability of individuals to forgiveness, learning and making a new start when mistakes are made. The law allows juveniles special privileges in this regard, but these are negated by social media, health information numbers and other initiatives emerging on both governmental and social media fronts.

The loss of history is incredibly important to individual development. Forgiveness and growth often depend entirely on the ability for errors, convictions, and negative events and treatments to be buried by time to enable new growth in a person. Without spent convictions rehabilitation is marginalized. The right to be forgiven and to start anew is a fundamental moral theme in personal growth and in most religious frameworks.

When unique identifiers are available, then history can become impossible to lose. and forgiveness and regrowth become impossible. To take an emotive but widely recognised example; even the most minor youthful sexual transgression has the lifetime potential now to destroy an adult life with no recourse. Harbingers of our emergent joint predicament are the pedophiles released after decades in prison.

The context of long ago mistakes is rarely kept with the types of records from which we are now at risk, and much is hearsay or opinion unverified. Yet we have no opportunity to see this, and correct it. Accuracy in events is not contextual perspective, and asymmetries in information access have always been sources of great power. These opportunities are growing, and raise great moral questions about ourselves, our society and the faith-based institutions that project moral values into these great administrative debates.

One form of self-protection is the concept of context-based identities and thus privacy and quarantining of these information sources and perhaps misadventures in cyberspace or real life. Context-based identities can also act as protection from identity theft, or misattribution of evil acts to one's records ... let alone the inappropriate use of access to collate damning information about people for blackmail or worse.

The bureaucratic trend to regard unique identity as the only identity, and use of more than one as a reason for suspicion amplifies these risks. The reality of terrorism and the expansion of intelligence-based approaches to anticipating bad events have further amplified the risks in these trends.

Intelligence is about anticipation; about using probabilities to prevent acts and events, and this approach has little or none of the civil law framework of proof and admissible evidence, let alone facing one's accuser. These are necessary conditions to prevent bad events occurring.

The ethics of such approaches are based on the principle of the least overall damage ... yet individuals are valued in faith-based organisations, especially the powerless who often have no other voice. Moral imperatives suggest that multiple identities, and contextual verification and limited linkage of "identities" should be a fundamental principle in a moral world. There is little sign of this.

The growth of intangible property as a major resource has led to huge pressures to exert control over intellectual property of many kinds. Exploring how censorship and control of information of all kinds is not only possible, but recently a major target of the virtual worlds is an important subject for future discussion.

Now is the time and place to address the near universal impact of intangible property (music, words, videos) and how criminalization of the most minor offences or perceived violations are now to be backed by the full power of the State in support of large scale commercial interests. If this is in any doubt check the concerned debates[4] on the largely secret provisions of the currently negotiated TransPacific Partnership Agreement (TPP), and the World Customs Association moves on Intellectual Property Rights.

However, as Lessig pointed out so vividly (Lessig 2007), instead of a burgeoning of creativity a dead hand is falling on this generation's creativity: yet this was the original aim of the handover of an access use and denial of use monopoly "copy right". Censorship and mass monitoring of individual access to intellectual property resources converge into censorship of various kinds. Is this not a moral and well as an economic issue? Censorship is one word, responsible limitation of access is a paternalistic equivalent, in many ways another, for the same effect on the Internet.

These are simply examples of unbalanced power, especially asymmetric information power, and the principle of contestability is badly needed to allow a more just equilibrium to emerge. Religions have always played a major role in expressing and unveiling asymmetric abuses of power; the information and virtual worlds now make this a newly central role for them.

CONCLUSION

The emergence of serious imbalances in information power needs to be discussed more widely by more bodies in civil society, if any effective moderation of the established trends is to be secured. As this is a moral issue for faith-based organisations, who have been remarkably silent in this area to date, their active engagement in these areas needs to be encouraged and facilitated, albeit that:

Democratic and religious discursive practices are prima facie at loggerhead. (Ungurenau, 2008, p405)

These essentially moral issues are being identified and presented by purely secular bodies such as the Australian Privacy Foundation, who share many of the governance concerns expressed here. It is past time that the faith-based bodies formulated clear and appropriate expressions of their concerns, and brought their high level lobbying and communication channels and capacity to play in what is so clearly within their moral and ethical domain.

REFERENCES

Armfield, G. G., Dixon, M. A., & Dougherty, D. S. (2006). Organizational power and religious individuals' media use. *Journal of Communication & Religion, 29*(2), 421–444.

Benedict, X. V. I. (2005). *Deus caritas est.* Retrieved 9-1, 2012, from www.scborromeo.org/docs/deus_caritas_est.pdf

Celeste, C., DiMaggio, P., Shafer, S., & Hargittai, E. (2004). Digital inequality: From unequal access to differentiated use. In K. Neckerman (Ed.), *Social inequality* (pp. 355–400). New York: Russell Sage.

Clarke, R. A., & Wigan, M. R. (2011). You are where you have been: The privacy Implications of location and tracking technologies. *The Journal of Location Based Services, 5*(3-4), 138. doi:10.1080/17489725.2011.637969

DiMaggio, P., & Hargittai, E. (2001). *From digital divide to digital inequality.* Woodrow Wilson School of Public and International Affairs.

Ethical Issues of Emerging ICT Applications. (n.d.). Retrieved from http://ethics.ccsr.cse.dmu.ac.uk/etica

Lessig, L. (2007). *Laws that choke creativity.* Retrieved from http://www.ted.com/talks/larry_lessig_says_the_law_is_strangling_creativity.html

Mossberger, K., Tolbert, C. J., & Stansbury, M. (2003). *Virtual inequality: Beyond the digital divide.* Washington, DC: Georgetown University Press.

O'Farrell, C. (2005). *Michel Foucault.* London: Sage.

Pontifical Council for Justice and Peace. (2005). *Compendium of the social doctrine of the church: To his holiness Pope John Paul II.* Rome: Libreria Editrice Vaticana.

Ungureanu, C. (2008). The contested relation between democracy and religion: Towards a dialogical perspective? *European Journal of Political Theroy, 7*(4), 405–429. doi:10.1177/1474885108094052

Wigan, M. R. (1994). Establishing issues in transport policy. In K. W. Ogden, E. W. Russell, & M. R. Wigan (Eds.), *Transport policies for the new millennium* (pp. 1–19). Melbourne, Australia: Monash University.

Wigan, M. R. (2010a). *Identity, contestability and ethics of unified virtualisation.* Paper presented at the Panel Presentation to the 'Moral Maze on Virtualisation and Society' Forum Meeting. Edinburgh, Scotland.

Wigan, M. R. (2010b). Identity, contestability and ethics of unified virtualisation of society. In K. Michael (Ed.), *IEEE international symposium on technology and society* (pp. 399–405). Wollongong, Australia: ISTAS. doi:10.1109/ISTAS.2010.5514616

Wigan, M. R. (2010c). Owning identity - One or many - Do we have a choice? *IEEE Technology and Society Magazine, 29*(2), 7. doi:10.1109/MTS.2010.937026

Wigan, M. R. (2012). Contestability, democracy, and trust in the anti-terror age. *IEEE Technology and Society Magazine, 31*(1), 26-32.

KEY TERMS AND DEFINITIONS

ANPR: Automatic number plate recognition is a mass surveillance method that uses optical character recognition on images to read vehicle registration plates. ANPR can be used to store the images captured by the cameras as well as the text from the license plate, with some configurable to store a photograph of the driver. Systems commonly use infrared lighting to allow the camera to take the picture at any time of the day.

Church: Is a Christian religious institution, place of worship, or group of worshipers.

Civil Liberty: Are civil rights and freedoms that provide an individual specific rights.

Contestability: A struggle for superiority or victory between rivals.

Dignity: Is a term used in moral, ethical, legal, and political discussions to signify that a being has an innate right to be valued and receive ethical treatment.

Ethics: Also known as moral philosophy that involves systematizing, defending and recommending concepts of right and wrong conduct.

Morality: A system of moral principles. Conformity, or degree of conformity, to conventional standards of moral conduct.

Vulnerable: Exposed to the possibility of being attacked or harmed, either physically or emotionally.

Wisdom: Disposition to perform the right action under given circumstances.

ENDNOTES

[1] John Paul II, Encyclical Letter *Sollicitudo Rei Socialis*, 42: *AAS* 80 (1988), 572-573; cf. John Paul II, Encyclical Letter *Evangelium Vitae*, 32: *AAS* 87 (1995), 436-437; John Paul II, Apostolic Letter *Tertio Millennio Adveniente*, 51: *AAS* 87 (1995), 36; John Paul II, Apostolic Letter *Novo Millennio Ineunte*, 49-50: *AAS* 93 (2001), 302-303

[2] http://ethics.ccsr.cse.dmu.ac.uk/etica

[3] www.bhlibrary.org.au/images/LinkClick.pdf accessed 15-4-2013

[4] Weatherall, K (2013) TPP-Australian Section by Section Analysis of the Enforcement Provisions (August 2013 Leaked Draft) Accessed 19 Nov 2013 at http://works.bepress.com/cgi/viewcontent.cgi?article=1032&context=kimweatherall"

Glossary

3G: Third generation

4G: Fourth generation

AAAA: American Association of Advertising Agencies

ACE: Adult Community Education

ACPO: Associate of Chief Police Officers

ACTA: Anti-Counterfeiting Trade Agreement

ACS: Australian Computer Society

ADA: Americans with Disabilities Act

A-LBS: Assisted Location Based Services

ALRC: Australian Law Reform Commission

ANAAA: Australian Association of National Advertisers

ANPR: Automatic Number Plate Recognition

APEC: Asia-Pacific Economic Cooperation

APF: Australian Privacy Foundation

AR: Augmented Reality

AFLF: Australian Flexible Learning Framework

AMVA: American Medical Veterinary Association

BCI: Brain Computer Interface

BEM: Black Ethnic Minority

BMI: Brain Machine Interfaces

BNCI: Brain/Neuronal Computer Interaction

BPR: Best Practice Recommendation

BRD: Bovine Respiratory Disease

BSAVA: British Small Animal Veterinary Association

BTB: Bovine Tuberculosis

BWV: Body Worn Video

CASPIAN: Consumers Against Supermarket Privacy Invasion and Numbering

CB Radio: Citizens' Band Radio

CCTV: Closed Circuit Television

CfP: Concern for Privacy

CIA: Central Intelligence Agency

CJPOA: Criminal Justice and Public Order Act

CODIS: Combined DNA Index System

CoE: Council of Europe

CPU: Central Processing Unit

DAA: Digital Advertising Alliance

DARPA: Defense Advanced Research Projects Agency

DBS: Deep Brain Stimulation

DER: Digital Education Revolution

DIY: Do-It-Yourselfer

DOD: Department of Defense

DOI: Diffusion of Innovation

DOJ: Department of Justice

DNA: Deoxyribonucleic Acid

DST: Digital Signature Transponder

EASA: European Advertising Standards Alliance

EC: European Commission

ECtHR: European Court of Human Rights

EEG: Electroencephalography

EFA: Electronic Frontiers Australia

EFF: Electronic Frontiers Foundation

EGE: European Group on Ethics

EHS: Extension of the Human Senses Group

EMV: Europay, MasterCard and Visa

ENIAC: Electronic Numerical Integrator and Computer

EPC: Electronic Product Code

EPIC: Electronic Privacy Information Center

418

ETICA: Ethical Issues of Emerging ICT Applications

EU: European Union

FDA: Food and Drug Administration

FIPS: Fair Information Practices

fMRI: Functional Magnetic Resonance Imaging

FPGA: Field-Programmable Gate Array

FSS: Forensic Science Service

FTC: Federal Trade Commission

GIS: Geographic Information Systems

GMO: Genetically Modified Organisms

GPS: Global Positioning Systems

GSM: Global System for Mobiles

H+: Humanity Plus

HDR: High Dynamic Range

HGP: Human Genome Project

IAO: Information Awareness Office

ICT: Information and Communication Technology

ID: Identification

IEEE: Institute of Electrical and Electronics Engineers

IoT: Internet of Things

ISO: International Standards Organization

ISP: Internet Service Provider

ISTAS: International Symposium on Technology and Society

IP: Internet Protocol

IPP: Information Privacy Principles

IPG: Implanted Pulse Generator

ITU: International Telecommunications Union

KEI: Knowledge Ecology International

KM: Knowledge Management

LAN: Local Area Network

LBS: Location-Based Services

LMS: Learning Management System

MBTI: Myers Briggs Type Indicator

MEA: Multielectrode Array

MEG: Magnetoencephalography

MND: Motor Neurone Disease

MLP: Multilayer Perceptron

MRI: Magnetic Resonance Imaging

MPL: Modular Prosthesis Limb

NAHMS: National Animal Health Monitoring System

NAIS: National Animal Identification System

NASA: National Aeronautic and Space Administration

NDNAD: National DNA Database

NGIS: Next Generation Identification System

NGO: Non-Government Organization

NIRT: Nanoscale Interdisciplinary Research Team

NMP: Neuromotor Prostheses

NORC: Naturally Occurring Retirement Communities

NSA: National Security Agency

NSF: National Science Foundation

NPP: National Privacy Principles

NVELS: National VET E-Learning Strategy

NZLC: New Zealand Law Commission

OECD: Organization for Economic Co-operation and Development

PACE: Police and Criminal Evidence Act

PD: Parkinson's Disease

PDA: Personal Digital Assistant

PerAda: Pervasive Adaptation

PI: Privacy International

PIPEDA: Personal Information Protection and Electronic Documents Act

PIN: Personal Identification Number

PMT: Protection Motivation Theory

POV: Point of View

QR Code: Quick Response Code

RBF: Radial Basis Function

ReNaChip: Rehabilitation Nano Chip

RFID: Radio-frequency identification

RNSA: Research Network for a Secure Australia

RSA: Ron Rivest, Adi Shamir and Leonard Adleman

RVTS: Remote Vocational Training Scheme

SAID: Social Auto-Immune Disease

SLP: Single Locus Probe

SOC: Scene of Crime

SSIT: Society on the Social Implications of Technology

STS: Science and Technology Studies
TIA: Total Information Awareness
TIPS: Terrorism Information and Prevention System
TOTeM: Tales of Things Electronic Memory
TPB: Theory of Planned Behavior
TRA: Theory of Reasoned Action
TTL: Transistor-to-Transistor Logic
USDA: United States Department of Agriculture

UbiComp: Ubiquitous Computing
US: United States
VET: Vocational Education & Training
VIP: Very Important Persons
VoIP: Voice over Internet Protocol
VLRC: Victorian Law Reform Commission
W3C: World Wide Web Consortium
WiFi: Wireless Fidelity
WearComp: Wearable Computing
WYSIATI: What You See Is All There Is

Compilation of References

(2009). Uberveillance. InButler, S. (Ed.), *Macquarie Dictionary* (5th ed.). Sydney, Australia: Sydney University.

(2011). Do-Not-Track Online Act of 2011. U.S. Senate., 913, 1.

Aarti, R. (2011). *Pros and cons of RFID technology.* Retrieved April 12, 2011, from http://www.buzzle.com/articles/pros-and-cons-of-rfid- technology.html

Abt, C. (1970). *Serious games.* New York, NY: Viking Press.

Ace Animals Inc. (n.d.). *Sprague Dawley.* Retrieved September 27, 2007 from http://aceanimals.com/SpragueDawley.htm

Ackerman, S. (2011). Senators ask spy chief: Are you tracking us through our iPhones? *Wired.* Retrieved on July 18, 2011 from http://www.wired.com/dangerroom/2011/07/senators-ask-spy-chief-are-you-tracking-us-through-our-iphones/

Act, P. 1988 (Cth) (Australia)

Action on Rights for Children. (2007). *How many innocent children are being added to the national DNA database?* Retrieved from http://www.archrights.org.uk/issues/dna/dnabrief.htm

Agar, N. (2007). Where to transhumanism? The literature reaches a critical mass. *The Hastings Center Report, 37*(3), 12–17. doi:10.1353/hcr.2007.0034 PMID:17649897

Albrecht, K. (2006). *Spychips: How major corporations and government plan to track your every purchase and watch your every move.* New York: Plume.

Ali, H. M., & Ahmad, N. H. (2006). Knowledge management in Malaysian banks: A new paradigm. *Journal of Knowledge Management Practice, 7*(3).

Allan, R. (2006). Wireless sensing spawns the connected world. *Electronic Design, 54*(7), 49–56.

Allen, A. L. (1988). *Uneasy access: Privacy for women in a free society.* Totowa, NJ: Rowman & Littlefield.

ALRC. (2008). *For your information: Australian privacy law and practice: Report no. 108.* Retrieved from www.alrc.gov.au/publications/report-108

Alsop R. (2008, November 2). Coddled kids hit corporate culture. *St. Petersburg Times*, p. F2.

Amato, C., & Amato, L. (2005). Enhancing student team effectiveness: Application of Myers-Briggs personality assessment in business courses. *Journal of Marketing Education, 4*(27), 41–51. doi:10.1177/0273475304273350

Amendment, P. (Private Sector) Act 2000 (Cth) (Australia)

American Association of Advertising Agencies. (2011). *Digital advertising alliance begins enforcing next phase of self-regulatory program for online behavioural advertising.* Retrieved from http://www.aaaa.org/news/press/Pages/052311_digital_next.aspx

American Friends of Tel Aviv University. (2010, June 29). Pacemaker for your brain: Brain-to-computer chip revolutionizes neurological therapy. *ScienceDaily.* Retrieved July 20, 2011, from http://www.sciencedaily.com/releases/2010/06/100628152645.htm

An act to add Section 22947.45 to the Business and Professions Code, relating to business. State of California Senate. 761. (2011)

Andrejevic, M., & Arnott, C. (2011). *Internet privacy research*. Retrieved from http://cccs.uq.edu.au/documents/privacy-report.pdf

Angeles, R. (2005). RFID technologies: Supply-chain applications and implementation issues. *Information Systems Management, 22,* 51–65. doi:10.1201/1078/44912.22.1.20051201/85739.7

Angwin, J. (2012, February 17). Google's iPhone tracking. *Wall Street Journal*. Retrieved from online.wsj.com/article_email/SB10001424052970 2048804045772253804565 99176- %0A%0AlMyQjAxMTAyMDEwNjExNDYyWj .html?mod=wsj_share_email#articleTabs%3Darticle %26project%3DSAFARITRACKINGCODE0212

Anon. (2012a). Australian flexible learning framework. *Australian Flexible Learning Framework (NVELS)*. Retrieved from http://www.flexiblelearning.net.au/

Anon. (2012b). Digital education revolution. *Australian Government Digital Education Revolution*. Retrieved from http://www.deewr.gov.au/Schooling/DigitalEducationRevolution/Pages/default.aspx

Applied Digital Solutions Inc. (n.d.). *Press release: Applied digital solutions' VeriChip corporation confirms that the attorney general of Mexico and some of his staff have been 'chipped'*. Retrieved October 24, 2007, from http://www.adsx.com/pressreleases/2004-07-14

Argyris, C. (1962). *Interpersonal competence and organizational effectiveness*. Homewood, IL: Irwin.

Armfield, G. G., Dixon, M. A., & Dougherty, D. S. (2006). Organizational power and religious individuals' media use. *Journal of Communication & Religion, 29*(2), 421–444.

Article 29 Working Party. (2011). Opinion 16/2011 on EASA/IAB Best Practice Recommendation on Online Behavioural Advertising Adopted on 8 December 2011, 02005/11/EN WP188

Ashton, K. (2009). That 'internet of things' thing. *RFID Journal, 22.*

Asif, Z., & Mandviwalla, M. (2005). Integrating the supply chain with RFID: A technical and business analysis. *Communications of the Association for Information Systems, 15*(1).

Attaran, M. (2006). The coming age of RFID revolution. *Journal of International Technology and Information Management, 15*(4), 77–87.

Atul, G. (2006). Critical care workforce: A policy perspective. *Critical Care Medicine, 34*(3), S7–S11. PMID:16477206

Australian Association of National Advertisers. (2011, March). *Australian best practice recommendation for online behavioural advertising*. Retrieved from http://www.easa-alliance.org/page.aspx/386

Australian Government. (2009). *First stage response to ALRC privacy report*. Retrieved from http://www.dpmc.gov.au/privacy/reforms.cfm

Australian Privacy Foundation. (2010). *Submission to senate committee on environment, communications and the arts' inquiry into the adequacy of protections for the privacy of Australians online*. Retrieved from http://www.privacy.org.au/Papers/Sen-OLP-100813.pdf

Australian Privacy Foundation. (2011a). *APF policy statement re- visual surveillance, including CCTV*. Retrieved May 9, 2011, from http://www.privacy.org.au/Papers/CCTV-1001.html

Australian Privacy Foundation. (2011b). *Nomination for the Australian privacy foundation 'big brother awards' 2011*. Retrieved May 9, 2011, from http://www.privacy.org.au/bba/2011/BBA2011_WORSTCORP_Biometrics_Published.rtf

Aziz, O., Lo, B., Yang, G., & Darzi, A. (2005). Pervasive healthcare: Clinical drive, technological innovations, and socio-economic benefits. In *Proceedings of the Perspective in Pervasive Computing*. Retrieved April 29, 2011, from http://ieeexplore.ieee.org/stamp/stamp.jsp?tp=&arnumber=5678079&isnumber=5678066

Bakir, V. (2010). *Sousveillance, media and strategic political communication*. London, UK: Continuum.

Baldwin, T. (2010, March 25). National healthcare will require RFID chips. *The New American*. Retrieved April 17, 2011, from http://beforeitsnews.com/story/28/042/National_Healthcare_Will_Require_National_RFID_Chips.html

Bales, R. (1950). *Interaction process analysis: A method for the study of small groups*. Reading, MA: Addison-Wesley.

Ball, D. J. et al. (1991). Evaluation of a microchip implant system used for animal identification in rats. *Laboratory Animal Science, 41*(2), 185–186. PMID:1658454

Bandura, A., & Menlove, F. (1968). Factors determining vicarious extinction of avoidance behavior through symbolic. *Journal of Personality and Social Psychology, 8*(2), 99–108. doi:10.1037/h0025260 PMID:5644484

Banfield: The Pet Hospital. (2005, November 3). *Banfield applauds legislation promoting open pet microchip technology where all scanners read all chips*. Author.

Barben, D., Fisher, E., Selin, C., & Guston, D. H. (2008). Anticipatory governance of nanotechnologies: Foresight, engagement and integration. In E. J. Hackett, O. Amsterdamska, M. Lynch, & J. Wajcman (Eds.), *The new handbook of science and technology studies*. Cambridge, MA: MIT Press.

Barclay, R. O., & Murray, P. C. (1997). *What is knowledge management*. Retrieved from www.media-access.com/whatis.html

Basden, A., & Burke, M. (2004). Towards a philosophical understanding of documentation: A Dooyeweerdian framework. *The Journal of Documentation, 60*(4), 352–370. doi:10.1108/00220410410548135

Basics, B. C. I. (2010). *Future BNCI (brain neural computer interface)*. Retrieved February 28, 2011, from http://future-bnci.org/index.php?option=com_content&view=article&id=58&Itemid=59

Bateman, C. (2004). Txt me: Supporting disengaged youth using mobile technologies, Australia. *Australian Flexible Learning Framework*. Retrieved from http://www.google.com/url?q=http://www.flexiblelearning.net.au/projects/media/txt_me_evaluation_report.pdf&sa=U&ei=3egTT_3OJezxmAX2m72BCg&ved=0CAYQFjAB&client=internal-uds-cse&usg=AFQjCNFcOSRnC92uRBw-bLBPqaub0A5ffYQ

Bauman, Z. (1992). *Intimations of postmodernity*. London: Routledge.

BBC. (1991). *Birmingham six freed after 16 years*. Retrieved from http://news.bbc.co.uk/onthisday/hi/dates/stories/march/14/newsid_2543000/2543613.stm

Beattie, K. (2009). S and Marper v UK: Privacy, DNA and crime prevention. *European Human Rights Law Review, 2*, 231.

Beauchamp, T., & Childress, J. (1979). *Principles of biomedical ethics*. New York, NY: Oxford University Press.

Beck, U. (1992). *Risk society: Towards a new modernity*. London: Sage.

Begley, S. (2004, July 9). Prosthetics operated by brain activity move closer to reality. *Wall Street Journal*. Retrieved July 22, 2011 from http://online.wsj.com/article/0,SB108931910809458926,00.html

Bell, G., & Gemmell, J. (2007). A digital life. *Scientific American*. Retrieved on July 18, 2011 from http://www.scientificamerican.com/article.cfm?id=a-digital-life

Benedict, X. V. I. (2005). *Deus caritas est*. Retrieved 9-1, 2012, from www.scborromeo.org/docs/deus_caritas_est.pdf

Benedikter, R., Giordano, & Fitzgerald. (2010). The future of the self-image of the human being in the age of transhumanism, neurotechnology and global transition. *Futures, 42*, 1102–1109. doi:10.1016/j.futures.2010.08.010

Bennett, C. J., & Raab, C. D. (2006). The governance of privacy: Policy instruments in global perspective (2nd ed.). Cambridge, MA: MIT Press.

Benn, S. I. (1971). Privacy, freedom, and respect for persons. In J. R. Pennock, & J. W. Chapman (Eds.), *NOMOS XIII: Privacy* (pp. 1–26). New York: Atherton Press.

Berger, C. R. (1975). Beyond initial interaction: Uncertainty, understanding and the development of interpersonal relationships. In H. Giles, & R. St Clair (Eds.), *Language and social psychology* (pp. 122–145). Oxford, UK: Blackwell.

Berger, T. et al. (2008). *Brain computer interfaces: An international assessment of research and development trends*. New York, NY: Springer. doi:10.1007/978-1-4020-8705-9

Berger, T. et al. (2011). A cortical neural prosthesis for restoring and enhancing memory. *Journal of Neural Engineering, 8*(11). PMID:21677369

Bickel, A. (2010, February 12). Ag department drops livestock tracing program. *Tribune Business News.*

Biran, S. et al. (2011). Development of carcinoma of the breast at the site of an implanted pacemaker in two patients. *Journal of Surgical Oncology, 11*(1), 7–11. doi:10.1002/jso.2930110103 PMID:219300

Blackall, L. (2007). *The future of learning in a networked world.* Retrieved from http://learningnetworkedworld.blogspot.com 1

Blackall, L. (2011). Leigh Blackall: What is a definition of networked learning? *Leigh Blackall.* Retrieved from http://leighblackall.blogspot.com/2011/06/what-is-definition-of-networked.html

Blanchard, K. T. et al. (1999). Transponder-induced sarcoma in the heterozygous p53+/- mouse. *Toxicologic Pathology, 27*(5), 519–527. doi:10.1177/019262339902700505 PMID:10528631

Bloomberg Business Week. (2006). The snooping goes beyond phone calls. *Bloomberg Business Week.* Retrieved on July 21, 2011 from http://www.businessweek.com/magazine/content/06_22/b3986068.htm

Bloustein, E. (1964). Privacy as an aspect of human dignity: An answer to Dean Prosser. *New York University Law Review, 39*, 962–1007.

Boran, M. (2011). *Thad Starner: Wearable computing for smarter living.* Retrieved on July 21, 2011 from http://newtechpost.com/2011/03/16/thad-starner-wearable-computing-for-smarter-living

Bostrom, N. (2004). Is transhumanism the world's most dangerous idea? *Betterhumans.* Retrieved from http://www.nickbostrom.com/papers/dangerous.html

Bostrom, N. (2012). *Transhumanist values.* Retrieved October 29, 2012 from http://www.nickbostrom.com/ethics/values.html

Bostrom, N. (2005). In defense of posthuman dignity. *Bioethics, 19*(3), 202–214. doi:10.1111/j.1467-8519.2005.00437.x PMID:16167401

Brady, B. (2008, January 13). Prisoners to be chipped like dogs. *The Independent.* Retrieved August 22, 2011 from http://www.independent.co.uk/news/uk/politics/prisoners-to-be-chipped-like-dogs-769977.html?pagewanted=all

Brand, K. G. et al. (1975). Etiological factors, stages, and the role of the foreign body in foreign body tumorigenesis: A review. *Cancer Research, 35*(2), 279–286. PMID:1089044

Brand, K. G. (1982). Cancer associated with asbestosis, schistosomiasis, foreign bodies or scars. In *Cancer—A comprehensive treatise: Etiology: Chemical and physical carcinogenesis* (pp. 671–676). New York: Plenum Press.

Branscomb, A. W. (1994). *Who owns information?* New York: Basic Books.

Brent, M. H., & Vitall, S. A. (2007). Knowledge sharing in large IT organisations: A case study. *VINE: The Journal of Information and Knowledge Management Systems, 37*(4), 421–439.

Brin, D. (1998). *The transparent society.* Boston: Perseus Books.

British Small Animal Veterinary Association. (2004). *Microchip report 2004.* Retrieved October 13, 2007 from http://www.bsava.com/VirtualContent/85185/Adverse_Reaction_Report_2004PDF.pdf

British Small Animal Veterinary Association. Microchip Report 2003. (2003). Retrieved October 13, 2007 from http://www.bsava.com/VirtualContent/85185/Adverse_Reaction_Report_2003PDF.pdf

Britt, P. (2007). New push to protect your identity. *Information Today, 24*(1), 1–46.

Brodkin, J. (2012a). Web privacy standards: Easy to break, hard to enforce. *ARS Technica.* Retrieved from http://arstechnica.com/tech-policy/news/2012/02/web-privacy-standards-easy-to-break-hard-to-enforce.ars

Brodkin, J. (2012b). Google privacy change taking effect today is illegal, EU officials say. *ARS Technica.* Retrieved from http://arstechnica.com/tech-policy/news/2012/03/google-privacy-change-taking-effect-today-is-illegal-eu-officials-sayers

Bronit, S., Harfield, C., & Michael, K. (2010). The social implications of covert policing. Wollongong, Australia: University of Wollongong Press - The Centre for Transnational Crime Prevention (CTCP) - Faculty of Law.

Brooking, A. (1996). *Intellectual capital.* London: International Thomson Business Press.

Brown, C., & Braun, K. (2008). Globalization, women's migration, and the long-term-care workforce. *The Gerontologist*, *48*(1), 16–24. doi:10.1093/geront/48.1.16 PMID:18381828

Brown, J. (2001). *Using surveys in language programs.* Cambridge, UK: Cambridge University.

Brown, J. S., & Duguid, P. (2000). *The social life of information.* Cambridge, MA: Harvard Business Press.

Burdon, M. (2011). Contextualising the tensions and weaknesses of information privacy and data breach notification laws. Santa Clara Computer and High-Technology Law Journal, 27(1), 63–129.

Burdon, M., & Telford, P. (2010). The conceptual basis of personal information in Australian privacy law. Murdoch Elaw Journal, 17(1), 1–27.

Bureau of Infrastructure, Transport and Regional Economics. (2001). *Australia.* Canberra, Australia: Australian Government Statistical Report.

Burgess, J. G., Warwick, K., Ruiz, V., Gasson, M. N., Aziz, T. Z., Brittain, J., & Stein, J. (2010). Identifying tremor-related characteristics of basal ganglia nuclei during movement in the Parkinsonian patient. *Parkinsonism & Related Disorders*, *16*(10), 671–675. doi:10.1016/j.parkreldis.2010.08.025 PMID:20884273

Burke, M. (2003). Philosophical and theoretical perspectives of organization structures as information processing systems. *The Journal of Documentation*, *59*(2), 131–142. doi:10.1108/00220410310463482

Burke, M. (2006). Achieving information fulfilment in the networked society: Part 1: Introducing new concepts. *New Library World*, *107*(9/10), 21–26. doi:10.1108/03074800610702624

Burke, M. (2007). Cultural issues, organizational hierarchy and information fulfilment: An exploration of relationships. *Library Review*, *56*(8), 236–245. doi:10.1108/00242530710818018

Burkhardt, J. (2008). The ethics of agri-food biotechnology: How can an agricultural technology be so important? In K. David, & P. B. Thompson (Eds.), *What can nanotechnology learn from biotechnology? Social and ethical lessons for nanoscience from the debate over agricultural biotechnology and GMOs.* Amsterdam, The Netherlands: Academic Press. doi:10.1016/B978-012373990-2.00003-0

Burton, C. (1999). The United Kingdom national DNA database. *Interpol.* Retrieved from http://www.interpol.int/Public/Forensic/dna/conference/DNADbBurton.ppt

Busch, L. (2008). Nanotechnologies, food, and agriculture: Next big thing or flash in the pan? *Agriculture and Human Values*, *25*(2), 215–218. doi:10.1007/s10460-008-9119-z

Busch, L. (2011). *Standards: Recipes for reality.* Cambridge, MA: MIT Press.

Busch, L., & Lloyd, J. R. (2008). What can nanotechnology learn from biotechnology? In K. David, & P. B. Thompson (Eds.), *What can nanotechnology learn from biotechnology? Social and ethical lessons for nanoscience from the debate over agricultural biotechnology and GMOs.* Amsterdam, The Netherlands: Academic Press. doi:10.1016/B978-012373990-2.00013-3

Business Wire. (2007, November 28). VeriChip corporation to unveil plans for self-contained implantable RFID glucose-sensing microchip at Grand Hyatt in New York on December 4. *Business Wire.* Retrieved August 19, 2011, from http://www.businesswire.com/news/home/20071128005305/en/VeriChip-Corporation-Unveil-Plans-Self-Contained-Implantable-RFID

Business Wire. (n.d.). *Vantaa Finland.* Retrieved from www.vtitechnologies

Butler, J. M. (2005). *Forensic DNA typing: Biology, technology, and genetic of STR markers.* Oxford, UK: Elsevier.

Callahan, D. (1999). The hastings center and the early years of bioethics. *Kennedy Institute of Ethics Journal*, *9*(1), 53–71. doi:10.1353/ken.1999.0001 PMID:11657314

Calvino, I. (1972). *Invisible cities*. Italy: Giulio Einaudi.

Campbell, Clark, Loy, Keenan, & Matthews, Winograd, & Zoloth. (2007). The bodily incorporation of mechanical devices: Ethical and religious issues (part 2). *Cambridge Quarterly of Healthcare Ethics*, *16*(3), 268–280. doi:10.1017/S0963180107070302 PMID:17695618

Canada, I. A. B. (2011) *Canada's advertising industry releases self-regulation framework for online behavioural advertising that ensures transparency, education, choice and accountability for consumers*. Retrieved from http://www.iabcanada.com/pr-news/oba-self-regulatory-framework

Capurro, R. (2010). *Ethical aspects of ICT implants in the human body*. Paper presented at the Meeting of the IEEE Symposium on Technology and Society (ISTAS10). New South Wales, Australia.

Carey, B. (2012, September 14). Brain implant improves thinking in monkeys, first such demonstration in primates. *The New York Times*, pp. A19.

Carlson, R. E., Silverman, S. R., & Mejia, Z. (2007). *Development of an implantable glucose sensor*. PositiveID Corp. Retrieved April 10, 2011 from http://www.positiveidcorp.com/files/Glucose-Sensor.pdf

Carmel, S., & Lowenstein, A. (2007). Addressing a nation's challenge: Graduate programs in gerontology in Israel. *Gerontology & Geriatrics Education*, *27*(3), 49–63. doi:10.1300/J021v27n03_04 PMID:17347110

Carpenter, S. R., Mooney, H. A., Agard, J., Capistrano, D., DeFries, R. S., & Díaz, S. et al. (2009). Science for managing ecosystem services: Beyond the millennium ecosystem assessment. *Proceedings of the National Academy of Sciences of the United States of America*, *106*(5), 1305–1312. doi:10.1073/pnas.0808772106 PMID:19179280

Cascino, G. D. et al. (2007). Breast cancer at site of implanted vagus nerve stimulator. *Neurology*, *68*(9), 703. doi:10.1212/01.wnl.0000256035.72892.a7 PMID:17325283

Celeste, C., DiMaggio, P., Shafer, S., & Hargittai, E. (2004). Digital inequality: From unequal access to differentiated use. In K. Neckerman (Ed.), *Social inequality* (pp. 355–400). New York: Russell Sage.

Charles River Laboratory. (2007). *Research models and services catalog 2007*. Author.

Charnick, J. (2012). *Upstart film collective*. Retrieved from http://www.upstartfilmcollective.com/portfolios/jcharnick/essays/rear-window.html

Chekhov, A. (1999). *Later short stories, 1888-1903*. New York: Modern Library.

Chen, C. et al. (1998). How can cooperation be fostered? The cultural effects of individualism-collectivism. *Academy of Management Review*, *23*(2), 285–304.

Cho, B., & Hooker, N. H. (2002). *A note on three qualities: Search, experience and credence attributes* (Working Paper: AEDE-WP-0027-02). Columbus, OH: The Ohio State University. Retrieved from http://aede.osu.edu/resources/docs/pdf/CC8A1957-93C9-42AF-AE91E1288DD1EEED.pdf

Choney, S. (2011). *NBC news*. Retrieved from http://www.nbcnews.com/technology/technolog/memory-card-mouth-saves-police-shooting-video-122903

Choo, C., et al. (2001). Environmental scanning as information seeking and organizational learning. *Information Research*, *7*(1).

Chorost, M. (2011). *World wide mind, the coming integration of humanity, machines and the internet*. New York, NY: Free Press.

Clarke, R. (2007). What 'überveillance' is, and what to do about it. In *Proceedings of 2nd RNSA Workshop on the Social Implications of National Security - From Dataveillance to Überveillance*. Retrieved from http://www.rogerclarke.com/DV/RNSA07.html

Clarke, R. (2012). The regulation of point-of-view surveillance in point of view technologies in law enforcement. In *Proceedings of the Sixth Workshop on the Social Implications of National Security*. Sydney, Australia: Wollongong University. Retrieved from http://www.rogerclarke.com/DV/PoVSR.html

Clarke, R. (1988). Information technology and dataveillance. *Communications of the ACM*, *31*(5), 498–512. doi:10.1145/42411.42413

Clarke, R. (1994). The digital persona and its application to data surveillance. *The Information Society, 10*(2), 77–92. doi:10.1080/01972243.1994.9960160

Clarke, R. (1999). *Introduction to dataveillance and information privacy, and definitions of terms*. Roger Clarke's Dataveillance and Information Privacy Pages.

Clarke, R. (2007). Appendix to what überveillance is and what to do about it: Surveillance vignettes. In K. Michael, & M. G. Michael (Eds.), *From dataveillance to überveillance (uberveillance) and the realpolitik of the transparent society*. Wollongong, Australia: University of Wollongong.

Clarke, R. (2010, Summer). What is überveillance? (And what should be done about it?). *IEEE Technology and Society Magazine*, 17–25. doi:10.1109/MTS.2010.937030

Clarke, R. A., & Wigan, M. R. (2011). You are where you have been: The privacy Implications of location and tracking technologies. *The Journal of Location Based Services, 5*(3-4), 138. doi:10.1080/17489725.2011.637969

Clymer, A. (2002, July 26). Worker corps to be formed to report odd activity. *New York Times*. Retrieved May 10, 2010 from http://www.nytimes.com/2002/07/26/us/traces-terror-security-liberty-worker-corps-be-formed-report-odd-activity.html

Coffey, B., Mintert, J., Fox, S., Schroeder, T., & Valentin, L. (2005). The economic impact of BSE: A research summary (Kansas State University Agricultural Experiment Station and Cooperative Extension Service). *Livestock Economics,* MF-2679.

Collins, D. (2002, May 10). Florida family takes computer chip trip: Tiny implants contain phone numbers, information on medications. *CBS News*. Retrieved April 11, 2011 from http://www.cbsnews.com/stories/2002/05/10/tech/main508641.shtml

Commercial Privacy Bill of Rights Act of 2011, U.S. Senate. 799, 1. (2011)

Computer Professionals for Social Responsibility. (n.d.). *Ten commandments of computer ethics*. Retrieved from http://cpsr.org/issues/ethics/cei/

Consumer Federation of America and Consumers' Union. (2008, April 11). *Comments to the FTC concerning the proposed online behavioural advertising self-regulatory principles*. Author.

Cox, A. (2005). What are communities of practice? A comparative review of four seminal works. *Journal of Information Science, 31*(6), 527–540. doi:10.1177/0165551505057016

Cranor, L., Leon, P., Ur, B., Balebako, R., Shay, R., & Wang, Y. (2011). Why Johnny can't opt out: A usability evaluation of tools to limit online behavioural advertising. *Carnegie Mellon University*. Retrieved from www.cylab.cmu.edu/research/techreports/2011/tr_cylab11017.html

Crittenden, P. (2008). *Changing orders: Scenes of clerical and academic life*. Sydney, Australia: Brandl & Schlesinger.

Davenport, T. H. (1993). *Process innovation: Reengineering work through information technology*. Cambridge, MA: Harvard Business School Press.

Davenport, T., & Prusak, L. (2000). *Working knowledge: How organisations manage what they know*. Boston: Harvard Business School Press.

David, K. (2008). Socio-technical analysis of those concerned with emerging technology, engagement and governance. In K. David, & P. B. Thompson (Eds.), *What can nanotechnology learn from biotechnology? Social and ethical lessons for nanoscience from the debate over agricultural biotechnology and GMOs*. Amsterdam, The Netherlands: Academic Press. doi:10.1016/B978-012373990-2.00001-7

Davidson, P. (2011). *Website*. Retrieved from www.vtitechnologies

Davies, S. (1992). *Big brother: Australia's growing web of surveillance*. Sydney: Simon & Schuster.

DeCew, J. W. (1997). *In pursuit of privacy: Law, ethics, and the rise of technology*. Ithaca, NY: Cornell University Press.

deCharms, R. C. (2008). Applications of real-time fMRI. *Nature Reviews. Neuroscience, 9*, 720–729. doi:10.1038/nrn2414 PMID:18714327

Deleuze, G., & Felix, G. (1987). *A thousand plateaus.* Minneapolis, MN: University of Minnesota Press.

Demographics of Aging. (2011). Retrieved April 4, 2011,from http://www.transgenerational.org/aging/demographics.htm#Characteristics

Denning, T., Matsuoka, Y., & Kohno, T. (2009). Neurosecurity: Security and privacy for neural devices. *Neurosurgical Focus, 27*(1), 1–4. doi:10.3171/2009.4.FOCUS0985 PMID:19569895

Dennis, K. (2008). Viewpoint: Keeping a close watch–the rise of self-surveillance and the threat of digital exposure. *The Sociological Review, 56*(3), 347–357. doi:10.1111/j.1467-954X.2008.00793.x

Devaraj, S., Easley, R., & Crant, J. (2008). How does personality matter? Relating the five-factor model to technology acceptance and use. *Information Systems Research, 19*(1), 93–105. doi:10.1287/isre.1070.0153

Digital Advertising Alliance. (2009). *Self-regulatory principles for online behavioural advertising.* Retrieved from http://www.iab.net/public_policy/behavioral-advertisingprinciples

Digital Advertising Alliance. (2011). *Self-regulatory principles for multi-site data.* Retrieved from http://www.aboutads.info/resource/download/Multi-Site-Data-Principles.pdf

Digital Advertising Alliance. (2012). *White House, DOC and FTC commend DAA's self-regulatory program to protect consumer online privacy: DAA announces plans to expand program consumer choice mechanisms.* Retrieved from http://www.aaf.org/default.asp?id=1322

DiMaggio, P., & Hargittai, E. (2001). *From digital divide to digital inequality.* Woodrow Wilson School of Public and International Affairs.

Dino, J. (2008). Extension of the human senses. *Ames Technology Capability and Facilities.* Retrieved July 22, 2011 from http://www.nasa.gov/centers/ames/research/technology-onepagers/human_senses.html

Do Not Track Me Online Act, U.S. House of Representatives. 654, 1. (2011)

Dobelle Institute. (2000, January 18). Artificial vision system for the blind announced by the Dobelle Institute. *ScienceDaily.* Retrieved July 20, 2011, from http://www.sciencedaily.com/releases/2000/01/000118065202.htm

Dogster's For the Love of Dogs Blog. (n.d.). *Comments posted in response to: VeriChip Corporation microchips linked to tumors, former presidential candidate Tommy Thompson linked to company.* Retrieved October 18 2007 from http://dogblog.dogster.com/2007/09/13/verichip-corporation-microchips-linked-to-tumors-former-presidential-candidate-tommy-thompson-linked-to-company/

Donnelly, S. B. (2006, May 17). A new tack for airport screening: Behave yourself. *Time.* Retrieved February 1, 2008 from http://www.time.com/time/nation/article/0,8599,1195330,00.html

Donoghue, J. (2006). A link between mind and machine that can turn thought into movement. *Nature, 442.*

Dostoevsky, F. (2000). *Notes from underground.* New York: W.W. Norton & Company.

Doyle, A., Lippert, R., & Lyon, D. (2011). *Eyes everywhere: The global growth of camera surveillance.* Retrieved from http://www.routledge.com/books/details/9780415668644/

Drivers Privacy Protection Act of 1994 18 USC § 2725

Drummond, K. (2009, June 14). Pentagon preps soldier telepathy push. *Wired.* Retrieved January 19, 2013 from http://www.wired.com/dangerroom/2009/05/pentagon-preps-soldier-telepathy-push/

Drummond, K. (2010, July 15). Human trials next for darpa's mind-controlled artificial arm. *Wired.* Retrieved July 20, 2011 from http://www.wired.com/dangerroom/2010/07/human-trials-ahead-for-darpas-mind-controlled-artificial-arm/

Drummond, K. (2012, September 27). DARPA's 'transient electronics' will disappear anywhere. *Forbes.* Retrieved November 19, 2012 from http://www.forbes.com/sites/katiedrummond/2012/09/27/darpa-transient-electronics/

Dubreuil, G. H., Bengtsson, G., Bourrelier, P. H., Foster, R., Gadbois, S., & Kelly, G. N. et al. (2002). A report of TRUSTNET on risk governance - Lessons learned. *Journal of Risk Research, 5*(1), 83–95. doi:10.1080/13669870110039916

Duhig, C. (2012, February 19). How companies learn your secrets. *New York Times*. Retrieved from http://www.nytimes.com/2012/02/19/magazine/shopping-habits.html?_r=1

Duhigg, C. (2012). *The power of habit—Why we do what we do in life and business*. New York: Random House.

Dutch Group Nijhof. (n.d.). *Wel of niet chippen?* Retrieved October 13 2007 at http://www.invisio.nl/antichip/

Dyer, J. H., & Nobeoka, K. (2000). Creating and managing a high-performance knowledge-sharing network: The Toyota case. *Strategic Management Journal, 21*, 345–367. doi:10.1002/(SICI)1097-0266(200003)21:3<345::AID-SMJ96>3.0.CO;2-N

EASA. (2011). *Best practice recommendation on online behavioural advertising*. Retrieved from http://www.easa-alliance.org/page.aspx/386

Elbakidze, L. (2007). Economic benefits of animal tracing in the cattle production sector. *Journal of Agricultural and Resource Economics, 32*(1), 169–180.

Elcock, L. E. et al. (2001). Tumors in long-term rat studies associated with microchip animal identification devices. *Experimental and Toxicologic Pathology, 52*, 483–491. doi:10.1016/S0940-2993(01)80002-6 PMID:11256750

Electronic Frontiers Australia. (2007). *Submission to the DHS access card consumer and privacy taskforce in response to discussion paper no. 3: Registration*. Retrieved November 2, 2010, from http://www.efa.org.au/Publish/efasubm-dhstf-regist-200704.html

Electronic Privacy Information Centre. (n.d.). *Online tracking and behavioural profiling*. Retrieved from http://epic.org/privacy/consumer/online_tracking_and_behavioral.html

Elenurm, T. (2007). *Entrepreneurial knowledge sharing about business opportunities in virtual networks*. Paper presented at the 8th European Conference on Knowledge Management. Barcelona, Spain.

Ellul, J. (1967). *The technological society* (J. Wilkinson, Trans.). New York: Vintage.

EPIC. (2012). *Google's circumvention of browser privacy settings*. Retrieved from http://epic.org/privacy/google/tracking/googles_circumvention_of_brows.html

Ericson, R. (2007). *Crime in an insecure world*. Cambridge, UK: Polity Press.

Escoffier, L. (2010). Radio frequency identification tags, memory spots, and the processing of personally identifiable information, and sensitive data: When there is no balance between right and wrong. In *Proceedings of 2010 Association for the Advancement of Artificial Intelligence Spring Symposium Series*. Retrieved April 28, 2011 from http://www.aaai.org/ocs/index.php/SSS/SSS10/paper/viewFile/1050/1484

Ethical Issues of Emerging ICT Applications. (n.d.). Retrieved from http://ethics.ccsr.cse.dmu.ac.uk/etica

Ethics Resource Center. (n.d.). Millennials, gen x and baby boomers: What do they think about ethics? In *2009 National Business Ethics Survey*. Retrieved from http://ethics.org/files/u5/Gen-Diff.pdf

European Union. (1995). *Directive 95/46/EC of the European parliament and of the council of 24 October 1995 on the protection of individuals with regards to the processing of personal data, and on the free movement of such data OJ L 281/31*. Author.

European Union. (2002). *Directive 2002/58/EC of the European parliament and of the council of 12 July 2002 concerning the processing of personal data and the protection of privacy in the electronic communications sector (directive on privacy and electronic communications) OJ L 201/37*. Author.

European Union. (2009). *Directive 2009/136/EC of the European parliament and of the council of 25 November 2009 amending directive 2002/22/EC on universal service and users' rights relating to electronic communications networks and services, directive 2002/58/EC concerning the processing of personal data and the protection of privacy in the electronic communications sector and regulation (EC) no 2006/2004 on cooperation between national authorities responsible for the enforcement of consumer protection laws. OJ L 337/11*. Author.

European Union. (2012). *Proposal for a commission regulation (EC) no 0011/2012 of 25 January 2012 on the protection of individuals with regard to the processing of personal data and on the free movement of such data (general data protection regulation) COM(2012)*. Author.

Evans, N., Forney, D., & Guido-DiBrito, F. (1998). *Student development in college: Theory, research, and practice.* San Francisco: Jossey-Bass.

Fauci, A. S. (2002). Bioterrorism: Defining a research agenda. *Food and Drug Law Journal, 57,* 413–421. PMID:12703508

Federal Trade Commission. (2009). *Self-regulatory principles for online behavioural advertising.* Retrieved from http://www.ftc.gov/os/2009/02/P085400behavad-report.pdf

Federal Trade Commission. (2010). *Protecting consumer privacy in an era of rapid change.* Retrieved from http://online.wsj.com/public/resources/documents/PrivacyReport_FINAL.pdf

Feinberg, J. (1980). The nature and value of rights. In *Rights, Justice, and the Bounds of Liberty: Essays in Social Philosophy* (pp. 143–158). Princeton, NJ: Princeton University Press.

Fereday, L. (1999). *Technology development: DNA from fingerprints.* Retrieved from http://www.ojp.usdoj.gov/nij/topics/forensics/events/dnamtgtrans6/trans-i.html

Feyerabend, P., & Hacking, I. (2010). *Against method.* London: Verso Books.

Fidler, D. P. (2003). Public health and national security in the global age: Infectious diseases, bioterrorism and realpolitik. *The George Washington International Law Review, 35,* 787–856.

Filipova-Neumann, L., & Welzel, P. (2010). Reducing asymmetric information in insurance markets: Cars with black boxes. *Telematics and Informatics, 27,* 394–403. doi:10.1016/j.tele.2010.03.003

Fletcher, G., Griffiths, M., & Kutar, M. (2011). *A day in the digital life: A preliminary sousveillance study.* Retrieved from ttp://ssrn.com/abstract=1923629

Ford, P., & Kuba, C. (2006). Stimulating debate: Ethics in multidisciplinary functional neurosurgery committee. *Journal of Medical Ethics, 32*(2), 106–109. doi:10.1136/jme.200X.013151 PMID:16446416

Foreign-Born Workers and Baby Boomers. (2010). *Baby-boomercaretaker.com.* Retrieved March 19, 2011, from http://www.babyboomercaretaker.com/baby-boomer/foriegn-born-workers-and-babyboomers.html

Forensic Science Service. (2009a). *Analytical solutions: DNA solutions.* Retrieved from http://www.forensic.gov.uk/html/services/analytical-solutions/dna/

Forensic Science Service. (2009b). *Sally Anne Bowman.* Retrieved from http://www.forensic.gov.uk/html/media/case-studies/f-47.html

Forrester, J. (1965). *Industrial dynamics.* Cambridge, MA: MIT Press.

Foster, K. R., & Jaeger, J. (2007). RFID inside. *IEEE Spectrum, 44,* 24–29. doi:10.1109/MSPEC.2007.323430

Foucault, M. (1975). *Surveiller et punir.* Paris, France: Gallimard.

Foucault, M. (2003). Governmentality. In P. Rabinow, & N. Rose (Eds.), *The essential Foucault* (pp. 229–245). New York: New Press.

Freckelton, I. (1989). DNA profiling: Forensic science under the microscope. In J. Vernon, & B. Selinger (Eds.), *DNA and criminal justice* (Vol. 2). Academic Press.

Frenzel, L. (2001). An evolving ITS paves the way for intelligent highways. *Electronic Design, 49*(1), 102.

Fried, C. (1984). Privacy: A moral analysis. In F. Schoeman (Ed.), *Philosophical dimensions of privacy* (pp. 203–222). Cambridge, UK: Cambridge University Press. doi:10.1017/CBO9780511625138.008

Fukuyama, F. (2004, September 1). Transhumanism. *Foreign Policy.*

Futures Company. (n.d.). *Millenials ahead.* Retrieved from http://www.lifebenefits.com/lb/pdfs/Millennials_Ahead_Report.pdf

Gallagher, S. (2012, February 23). White House announces new privacy 'bill of rights', do not track agreement. *ARS Technica.* Retrieved from http://arstechnica.com/tech-policy/news/2012/02/white-house-announces-new-privacy-bill-of-rightsdo-not-track-agreement.ars

Ganascia, J. (2010). The generalized sousveillance society. *Social Sciences Information. Information Sur les Sciences Sociales, 49,* 489–506. doi:10.1177/0539018410371027

García-Montes, J., Caballero-Muñoz, D., & Perez-Alvarez, M. (2006). Changes in the self resulting from the use of mobile phones. *Media Culture & Society, 28*(1), 67–82. doi:10.1177/0163443706059287

Gaskell, G. (2008). Lessons from the bio-decade: A social science perspective. In K. David, & P. B. Thompson (Eds.), *What can nanotechnology learn from biotechnology? Social and ethical lessons for nanoscience from the debate over agricultural biotechnology and GMOs.* Amsterdam, The Netherlands: Academic Press. doi:10.1016/B978-012373990-2.00012-1

Gasson, M. N. (2010). Human enhancement: Could you become infected with a computer virus? In *Proceedings of IEEE International Symposium on Technology and Society* (pp. 61-68). Wollongong, Australia: IEEE.

Gasson, M. N., Wang, S. Y., Aziz, T. Z., Stein, J. F., & Warwick, K. (2005). Towards a demand driven deep-brain stimulator for the treatment of movement disorders. In *Proceedings of 3rd IEE International Seminar on Medical Applications of Signal Processing* (pp.83-86). London, UK: IEE.

Gasson, M. (2008). ICT implants: The invasive future of identity? *Advances in Information and Communication Technology, 262*(2), 287–295.

Gasson, M. N., Hutt, B. D., Goodhew, I., Kyberd, P., & Warwick, K. (2005b). Invasive neural prosthesis for neural signal detection and nerve stimulation. *International Journal of Adaptive Control and Signal Processing, 19*(5), 365–375.

Gasson, M. N., Kosta, E., Royer, D., Meints, M., & Warwick, K. (2011). Normality mining: Privacy implications of behavioral profiles drawn from GPS enabled mobile phones. *IEEE Transactions on System, Man, Cybernetics. Part C, 41*(2), 251–261.

Gavison, R. (1984). Privacy and the limits of law. In F. Schoeman (Ed.), *Philosophical dimensions of privacy* (pp. 346–402). Cambridge, UK: Cambridge University Press. doi:10.1017/CBO9780511625138.017

Geary, J. (2003). *The body electric: An anatomy of the new bionic senses.* New Brunswick, NJ: Rutgers University Press.

Gellman, R. (1997). Does privacy law work? In P. Agre & M. Rotenberg (Eds.), Technology and privacy: The new landscape (pp. 193–218). Cambridge, MA: MIT Press.

Genewatch, U. K. (2009a). *A brief legal history of the NDNAD.* Retrieved from http://www.genewatch.org/sub-537968

Genewatch, U. K. (2009b). *Police and criminal evidence act (PACE) consultations.* Retrieved from http://www.genewatch.org/sub-551990

Genewatch, U. K. (2009c). *Whose DNA profiles are on the database?* Retrieved from http://www.genewatch.org/sub-539482

Geoghegan, J. (2009, October 12). Criticism for police over silence on DNA database. *Echo.* Retrieved from http://www.echo-news.co.uk/news/4673015.Criticism_for_police_over_silence_on_DNA_database/

Georgia General Assembly. (2010). *Senate bill 235.* Retrieved January 12, 2011, from http://www1.legis.ga.gov/legis/2009_10/versions/sb235_As_passed_Senate_5.htm

Geser, H. (2006). Is the cell phone undermining the social order? Understanding mobile technology from a sociological perspective. *Knowledge, Technology & Policy, 19*(1), 8–18. doi:10.1007/s12130-006-1010-x

Gilly, M., & Zeithaml, V. A. (1985). The elderly consumer and adoption of technologies. *The Journal of Consumer Research, 12*(3), 353–357. doi:10.1086/208521

Gilmer, D. F., & Aldwin, C. M. (2003). *Health, illness, and optimal ageing: Biological and psychosocial perspectives.* Thousand Oaks, CA: Sage Publications.

Glaberson, H. (2011, February 7). US chocolatier develops virtual factory world. *Confectionery News.* Retrieved from www.confectionerynews.com/content/view/print/356965

Goldsmith, A. (2005). Police reform and the problem of trust. *Theoretical Criminology, 9*(4), 443–470. doi:10.1177/1362480605057727

Goleman, D. (2006). *Social intelligence: A new science of human relationships.* New York: Bantam Dell.

Goleman, D., Boyatzis, R., & McKee, A. (2004). *Primal leadership: Learning to lead with emotional intelligence.* Boston: Harvard Business School Press.

Gollaher, D. L. (1998). Genetic discrimination: Who is really at risk? *Genetic Testing, 2*(1), 13. doi:10.1089/gte.1998.2.13 PMID:10464593

Good, N. S. (2008). *Designing for informed consent: A multi-domain, interdisciplinary analysis of the technological means to provide informed consent.* Retrieved April 5, 2011, from http://search.proquest.com/docview/3046 97082?accountid=7117

Google. (n.d.). *Privacy policy.* Retrieved from http://www.google.com/policies/privacy/

Goold, B., Loader, I., & Thumala, A. (2010). Consuming security? Tools for a sociology of security consumption. *Theoretical Criminology, 14*(1), 3–30. doi:10.1177/1362480609354533

Gorman, S. (2008, March 10). NSA's domestic spying grows as agency sweeps up data. *The Wall Street Journal.* Retrieved June 14, 2010 from http://homepage.mac.com/imfalse/chapel_annex/NSAs_Domestic_Spying_Grows_As_Agency_Sweeps_Up_Data_WSJ.pdf

Gourlay, S. (2001). Knowledge management and HRD. *Human Resource Development International, 4*(1), 27–46. doi:10.1080/13678860121778

Graafstra, A. (2007). Hands on. *IEEE Spectrum, 44*(3), 18–23. doi:10.1109/MSPEC.2007.323420

Graimann, B., Allison, B., & Pfurtscheller, G. (2010). Brain-computer interfaces: A gentle introduction. In B. Graimann, B.Z. Allison, & G. Pfurtscheller (Eds.), *Brain-computer interfaces: Revolutionizing human-computer interaction.* Berlin: Springer-Verlag. Retrieved April 29, 2011 from http://www.springerlink.com/content/um1k229348172m20/fulltext.pdf

Grange, J. M., & Yates, M. D. (1994). Zoonotic aspects of mycobacterium Bovis infection. *Veterinary Microbiology, 40*, 137–151. doi:10.1016/0378-1135(94)90052-3 PMID:8073621

Greely, H. (2004). Remarks at the Regan lecture. In *Prediction, litigation, privacy, and property: Some possible legal and social implications of advances in neuroscience.* Retrieved on July 24, 2011 from http://www.scu.edu/ethics/publications/submitted/greely/neuroscience_ethics_law.html

Greely, H. (2006). Neuroethics and ELSI: Similarities and differences. *Minn. J. LSCI. & Tech., 599.*

Greenemeier, L. (2008). Monkey think, robot do. *Scientific American.* Retrieved on July 21, 2011 at http://www.scientificamerican.com/article.cfm?id=monkey-think-robot-do

Greenemeier, L. (2011). Does the Murdoch hacking scandal signify the end of privacy? *Scientific American.* Retrieved on July 21, 2011 from http://www.scientificamerican.com/article.cfm?id=murdoch-phone-hack-scandal-privacy

Greenleaf, G. (2001). Tabula rasa: Ten reasons why Australian privacy law does not exist. The University of New South Wales Law Journal, 24(1), 262–269.

Greenleaf, G. (2007a). Australia's proposed ID card: Still quacking like a duck. *University of New South Wales Law Research Series.* Retrieved November 2, 2010, from http://papers.ssrn.com/sol3/papers.cfm?abstract_id=951358&download=yes

Greenleaf, G., Waters, N., & Bygrave, L. A. (2007). Implementing privacy principles: After 20 years its time to enforce the privacy act. *University of New South Wales Law Research Series.* Retrieved November 2, 2010, from http://papers.ssrn.com/sol3/papers.cfm?abstract_id=987763

Greenleaf, G. (2007b). Access all areas: Function creep guaranteed in Australia's ID card bill (no. 1). *Computer Law & Security Report, 23*(4), 332–341. doi:10.1016/j.clsr.2007.05.006

Griffin, D. (1997). Economic impact associated with respiratory disease in beef cattle. *The Veterinary Clinics of North America. Food Animal Practice, 13*, 367–377. PMID:9368983

Guizzo, E. (n.d.). Monkeys use brain interface to move and feel virtual objects. *IEEE Spectrum*. Retrieved from http://spectrum.ieee.org/automaton/robotics/medical-robots/monkeys-use-bidirectional-brain-machine-interface-to-feel-virtual-objects

Guston, D. H. (1999). Evaluating the first U.S. consensus conference: The impact of the citizens' panel on telecommunications and the future of democracy. *Science, Technology & Human Values, 24*(4), 451–482. doi:10.1177/016224399902400402

Guthrie Test (Heel Prick Test). (2009). *Discovery*. Retrieved from http://www.discoverychannel.co.uk/homeandhealth/article.jsp?section_id=7&theme_id=23&subtheme_id=80&article_id=81&site=uk

Hadfield, P., Lister, S., & Traynor, P. (2009). This town's a different town today: Policing and regulating the night-time economy. *Criminology & Criminal Justice, 9*(4), 465–485. doi:10.1177/1748895809343409 PMID:20401315

Haggerty, K. D., & Ericson, R. V. (2000). The surveillant assemblage. *The British Journal of Sociology, 51*(4), 605–622. doi:10.1080/00071310020015280 PMID:11140886

Haggerty, K., & Samatas, M. (2010). *Surveillance and democracy*. New York: Routledge-Cavandish.

Halfhill, T., Sundstrom, E., Lahner, J., Calderone, W., & Neilsen, T. (2005). Group personality composition and group effectiveness. *Small Group Research, 36*(1), 83–105. doi:10.1177/1046496404268538

Hall, D. (2010). Food with a visible face: Traceability and the public promotion of private governance in the Japanese food system. *Geoforum, 41*(5), 826–835. doi:10.1016/j.geoforum.2010.05.005

Halley, D. (2010). Computer chip implant to program brain activity, treat Parkinson's. *Singularity Hub*. Retrieved July 20, 2011, from http://singularityhub.com/2010/07/21/computer-chip-implant-to-program-brain-activity-treat-parkinsons/

Halliday, M. (2013). *Interviews with M.A.K. Halliday: Language turned back on himself*. London: J.R.Martin, Bloomsbury Academic.

Halperin, D., et al. (2008). *Pacemakers and implantable cardiac defibrillators: Software radio attacks and zero-power defenses*. Paper presented at the 2008 IEEE Symposium on Security and Privacy. Oakland, CA. http://doi.ieeecomputersociety.org/10.1109/SP.2008.31

Hameed, J., Harrison, I., Gasson, M. N., & Warwick, K. (2010). A novel human-machine interface using subdermal magnetic implants. In *Proceedings of IEEE International Conference on Cybernetic Intelligent Systems* (pp. 106-110). Reading, UK: IEEE.

Hancock, J. (2010, April 27). Install microchips in illegal immigrants, GOP candidate says. *Iowa Independent*. Retrieved April 20, 2011 from http://iowaindependent.com/32926/install-microchips-in-illegal-immigrants-gop-candidate-says

Hannah, W., & Thompson, P. B. (2008). Nanotechnology, risk and the environment: A review. *Journal of Environmental Monitoring, 10*, 291–300. doi:10.1039/b718127m PMID:18392270

Hansard. (1993). *Royal commission on criminal justice*. Retrieved from http://hansard.millbanksystems.com/commons/1993/jun/24/royal-commission-on-criminal-justice

Hansard. (2009). *DNA databases*. Retrieved from http://www.publications.parliament.uk/pa/cm200809/cmhansrd/cm091027/text/91027w0019.htm

Hansard. (2009). *Police: Databases*. Retrieved from http://www.publications.parliament.uk/pa/ld200809/ldhansrd/text/90505w0003.htm

Hargreaves, J. S. (2010). Will electronic personal health records benefit providers and patients in rural America?. *Telemedicine and e-Health, 16*(2), 167.

Härmä, A. (2009). Ambient human-to-human communication. In H. Nakashima, H. Aghajan, & J. Augusto (Eds.), *Handbook of ambient intelligence and smart environments* (pp. 795–823). New York: Springer.

Harper, S. (2006). *Ageing societies: Myths, challenges and opportunities*. London: Arnold Publishers.

Harrop, P., & Napier, E. (2006). *Food and livestock traceability*. London: IDTechEx.

Hart, C. (2007). Micro-chipping away at privacy: Privacy implications created by the new Queensland driver license proposal. Queensland University of Technology Law and Justice Journal, 7(2), 305-324.

Hayes, A. (2010). National snapshot: Current use of POV technologies in an Australian educational context. In *MobilizeThis Symposium*. Retrieved from http://www.alexanderhayes.com/curriculum-vitae/publications

Hayes, A. (2012). *Uberveillance triquetra*. Retrieved from http://archive.org/details/Uberveillancetriquetra

Hayes, A. (2013). Cyborg cops, googlers, and connectivism. *IEEE Technology and Society Magazine, 32*(1), 23–24. doi:10.1109/MTS.2013.2247731

Headlam, B. (2000, June 11). The mind that moves objects. *NY Times,* 63-64.

Healthline. (n.d.). *Privacy policy*. Retrieved from http://www.healthline.com/health/privacy-policy

Heidegger, M. (1977). *The question concerning technology, and other essays*. New York: Harper & Row.

Herbert, W., & Tuminaro, A. K. (2009). Symposium: The impact of emerging technologies in the workplace: Who's watching the man (who's watching me)? *Hofstra Labor & Employment Law Journal, 25*, 355–393.

Hill, T. J. (1984). Autonomy and benevolent lies. *The Journal of Value Inquiry, 18*, 251–297. doi:10.1007/BF00144766

Hochberg, L. R., Serruya, M. D., Friehs, G., Mukand, J. A., Saleh, M., & Caplan, A. et al. (2006). Neuronal ensemble control of prosthetic devices by a human with tetraplegia. *Nature, 442*, 164–171. doi:10.1038/nature04970 PMID:16838014

Hofstede, G., & Hofstede, G. J. (2005). *Cultures and organisations: Software of the mind*. London: McGraw Hill.

Holloman, K. R., & Ponder, D. E. (2007). Clubs, bars, and the driver's license scanning system. In K. R. Larsen, & Z. A. Voronovich (Eds.), *Privacy in a transparent world* (pp. 40–47). Boulder, CO: Ethica Publishing.

Holman, M. (2006). *The nanotech report* (4th ed.). Lux Research, Inc.

Home Office. (2004). *Coldcases to be cracked in DNA clampdown*. Retrieved from http://press.homeoffice.gov.uk/press-releases/'Coldcases'_To_Be_Cracked_In_Dna?version=1

Home Office. (2005). *DNA expansion programme 2000–2005: Reporting achievement*. Retrieved from http://police.homeoffice.gov.uk/publications/operational-policing/DNAExpansion.pdf

Homeland Security News Wire. (2012, November 19). *DARPA seeking surveillance technology to predict future behavior*. Retrieved November 19, 2012 from http://www.homelandsecuritynewswire.com/dr20121119-darpa-seeking-surveillance-technology-to-predict-future-behavior

Hoofnagle, C. J. (2006). Privacy self-regulation: A decade of disappointment. In J. K. Winn (Ed.), Consumer protection in the age of the 'information economy' (pp. 381–404). Burlington, UK: Ashgate.

Hooper, S. (2004). Brain chip offers hope for paralyzed. *CNN.com*. Retrieved July 20, 2011 from http://ed.cnn.com/2004/TECH/10/20/explorers.braingate/

Howarth, B., & Ledwidge, J. (2011). *A faster future - The future of broadband: What it means for business, society and you*. Sydney, Australia: Five Senses Education.

Howe, N., & Strauss, W. (2000). *Millennials rising*. New York: Vintage Books.

Hsu, J. (n.d.). Intel wants brain implants in its customers' heads by 2020. *POPSI*. Retrieved from http://www.popsci.com/technology/article/2009-11/intel-wants-brain-implants-consumers-heads-2020

Huff, L., Cooper, J., & Jones, W. (2002). The development and consequences of trust in student project groups. *Journal of Marketing Education, 24*(1), 24–34. doi:10.1177/0273475302241004

Human Genome Project. (2009). *Human genome project information: Ethical, legal and social issues*. Retrieved from http://www.ornl.gov/sci/techresources/Human_Genome/elsi/elsi.shtml

Humane Society. (2008). *Slaughterhouse investigation: Cruel and unhealthy practices*. Retrieved from http://www.youtube.com/watch?v=zhlhSQ5z4V4

Humanity +. (2012). *Humanity + mission*. Retrieved October, 29, 2012, from http://humanityplus.org/about/mission/

IAO (Information Awareness Office). (2003). *Report to congress regarding the terrorism information awareness program*. Retrieved June 17, 2010 from http://www.globalsecurity.org/security/library/report/2003/tia-di_report_20may2003.pdf

IEEE Code of Ethics. (n.d.). Retrieved from http://www.ieee.org/about/corporate/governance/p7-8.html

Inda, J. (2006). *Targeting immigrants: Government, technology and ethics*. Malden, MA: Blackwell Publishing. doi:10.1002/9780470776315

Inness, J. C. (1992). *Privacy, intimacy, and isolation*. New York, NY: Oxford University Press.

Intendix. (n.d.). Retrieved from http://www.intendix.com

International Wealth Solutions. (2008). *Ageing Demographics*. Retrieved March, 6, 2011, from http://www.iwslimited.com/ageingdemographics.html

IOL. (2007). *Big brother will be watching you fly*. Retrieved February 1, 2008 from www.iol.co.za/index.php?set_id=14&click_id=116&art_id=iol1171352398668C560

IPGP. (n.d.). *IP address lookup*. Retrieved from http://www.ipgp.net/

Iqbal, M. U., & Lim, S. (2007). Privacy implications of automated GPS tracking and profiling. In K. Michael, & M. G. Michael (Eds.), *From dataveillance to überveillance (uberveillance) and the realpolitik of the transparent society* (pp. 225–240). Wollongong, Australia: University of Wollongong.

Ireland, S. (1989). What authority should police have to detain suspects to take samples? In J. Vernon & B. Selinger (Eds.), *DNA and criminal justice*. Retrieved from http://www.aic.gov.au/media_library/publications/proceedings/02/ireland.pdf

Isomursu, M., Häikiö, J., Wallin, A., & Ailisto, H. (2008). Experiences from a touch-based interaction and digitally enhanced meal-delivery service for the elderly. *Advances in Human-Computer Interaction*. doi:10.1155/2008/931701

Jackson Laboratory. (n.d.). *The jax mice data sheet: Strain name CBA/J*. Retrieved October 10, 2007 from http://jaxmice.jax.org/strain/000656.html

Jain, K. K., Manjit, S. S., & Gurvinder, K. S. (2007). Knowledge sharing among academic staff: A case study of business schools in Klang Valley, Malaysia. *Journal of the Advancement of Science and Arts*, 2, 23–29.

Jansen, J. A. et al. (1999). Biological and migrational characteristics of transponders implanted into beagle dogs. *The Veterinary Record*, 145, 329–333. doi:10.1136/vr.145.12.329 PMID:10530881

Jarrett, K. (2006). DNA breakthrough. *National Black Police Association*. Retrieved from http://www.nbpa.co.uk/index.php?option=com_content&task=view&id=40&Itemid=58

Jenkins, H. W., Jr. (2010, August 14). Google and the search for the future. *The Wall Street Journal*. Retrieved August 20, 2010, from http://online.wsj.com/article/SB10001424052748704901104575423294099527212.html

Jennings, C. (2006). Battlefield between the ears. *Nature*, 443(326), 911. doi:10.1038/443911a

Jennings, T. et al. (1988). Angiosarcoma associated with foreign body material: A report of three cases. *Cancer*, 62(11), 2436–2444. doi:10.1002/1097-0142(19881201)62:11<2436::AID-CNCR2820621132>3.0.CO;2-J PMID:3052791

Jha, A. (2004, September 9). DNA fingerprinting no longer foolproof. *The Guardian*. Retrieved from http://www.guardian.co.uk/science/2004/sep/09/sciencenews.crime

Jobling, M. A., & Gill, P. (2004). Encoded evidence: DNA in forensic analysis. *Nature Reviews. Genetics*, 5(10), 745. doi:10.1038/nrg1455 PMID:15510165

Johnson, C. G. (2010, September 7). ACLU raises questions about microchip tracking of preschoolers. *California Watch*. Retrieved August 16, 2011, from http://californiawatch.org/dailyreport/aclu-raises-questions-about-microchip-tracking-preschoolers-4476

Johnson, K. (1996). Foreign-body tumorigenesis: Sarcomas induced in mice by subcutaneously implanted transponders. *Toxicologic Pathology*, 33(5), 619.

Johnston, C. (2012, January 26). *Google's new privacy policy could anger FTC*. Retrieved from http://arstechnica.com/gadgets/news/2012/01/pascals-wager-googles-new-privacy-policy-could-anger-ft.cars

Jung, C. (1923/1971). *Psychological types*. Princeton, NJ: Princeton University Press.

Junglas, I., Johnson, N., & Spitzmüller, C. (2008). Personality traits and privacy perceptions: An empirical study in the context of location-based services. *European Journal of Information Systems*, *17*(4), 387–402. doi:10.1057/ejis.2008.29

Kafka, F. (1999). *Letters to milena*. New York: Vintage Classics.

Kahneman, D. (2011). *Thinking, fast and slow*. London: Penguin.

Karinen, R., & Guston, D. H. (2010). Toward anticipatory governance: The experience with nanotechnology. *Sociology of the Sciences Yearbook*, *27*(4), 217–232.

Kasper, D. L. et al. (2004). *Soft tissue sarcomas. Harrison's principles of internal medicine* (16th ed.). New York: McGraw-Hill Professional.

Kay, R. (2007). Head mounted cameras for security operations: What the officer sees the jury sees. *Security Solutions*, *7*, 49–50.

Keel, S. B. et al. (2001). Orthopaedic implant-related sarcoma: A study of twelve cases. *Modern Pathology*, *14*(10), 969–977. doi:10.1038/modpathol.3880420 PMID:11598166

Keller, W. W., & Mitchell, G. R. (Eds.). (2006). *Hitting first: Preventive force in U.S. security strategy*. Pittsburgh, PA: University of Pittsburgh Press.

Kennedy, P., Andreasen, D., Ehirim, P., King, B., Kirby, T., & Mao, H. et al. (2004). Using human extra-cortical local field potentials to control a switch. *Journal of Neural Engineering*, *1*(2), 72–77. doi:10.1088/1741-2560/1/2/002 PMID:15876625

Kerr, I., & Mann, S. (2006). *Exploring equiveillance*. Retrieved from http://www.anonequity.org/weblog/archives/2006/01/exploring_equiv_1.php

Kim, S., & Lee, H. (2006, May/June). The impact of organizational context and information technology on employee knowledge-sharing capabilities. *Public Administration Review*, 370–385. doi:10.1111/j.1540-6210.2006.00595.x

Kleinman, D. L., Delborne, J. A., & Anderson, A. A. (2011). Engaging citizens: The high cost of citizen participation in high technology. *Public Understanding of Science (Bristol, England)*, *20*(2), 221–240. doi:10.1177/0963662509347137

Kleinman, D., Powell, M., Grice, J., Adrian, J., & Lobes, C. (2007). A toolkit for democratizing science and technology policy: The practical mechanics of organizing a consensus conference. *Bulletin of Science, Technology & Society*, *27*(2), 154–169. doi:10.1177/0270467606298331

Klimas v Comcast Cable Communications Inc. 2006 C.A.6 (Mich.)

Koblinsky, L., Liotti, T. F., & Oeser-Sweat, J. (Eds.). (2005). *DNA: Forensic and legal applications*. Hoboken, NJ: Wiley.

Kohlbrand, J., & Foster, J. (2000, August 13). Human ID implant to be unveiled soon--'Wearers' of digital angel®' monitored by GPS, internet. *World Net Daily*. Retrieved April 8, 2011 from http://www.wnd.com/news/article.asp?ARTICLE_ID=17601

Kop, R., & Hill, A. (2008). Connectivism: Learning theory of the future or vestige of the past? *International Review of Research in Open and Distance Learning*, *9*(3).

Kosta, E., & Bowman, D. M. (2011). Treating or tracking? Regulatory challenges of nano-enabled ICT implants. *Law & Policy*, *33*(2), 256–275. doi:10.1111/j.1467-9930.2010.00338.x

Kulthau, C. (1993). A principle of uncertainty for information seeking. *The Journal of Documentation*, *49*(4), 39–55.

Kurkovsky, S., Syta, E., & Casano, B. (2011). Continuous RFID-enabled authentication: Privacy implications. *IEEE Technology and Society Magazine*, *30*(3), 34–41. doi:10.1109/MTS.2011.942306

Kurzweil, R. (2005). *The singularity is near: When humans transcend biology*. New York: Penguin.

Kuzma, J., & VerHage, P. Woodrow Wilson International Center for Scholars, & The Pew Charitable Trusts. (2006). Nanotechnology in agriculture and food production: Anticipated applications. Washington, DC: Project on Emerging Nanotechnologies.

Landau, R., Werner, S., Auslander, G. K., Shoval, N., & Heinik, J. (2009). Attitudes of family and professional caregivers towards the use of GPS for tracking patients with dementia: An exploratory study. *British Journal of Social Work*, 39(4), 670–692. doi:10.1093/bjsw/bcp037

Lave, J., & Wenger, E. (1991). *Situated learning: Legitimate peripheral participation*. Cambridge, UK: Cambridge University Press. doi:10.1017/CBO9780511815355

Lawrence, S. (1999, February 24). What is institutional racism? *The Guardian*. Retrieved from http://www.guardian.co.uk/uk/1999/feb/24/lawrence.ukcrime7

Lawton, W. (2004). *Research group exploring limb loss hopes biohybrid will bridge gap between human and machine*. Retrieved from http://www.brown.edu/Administration/George_Street_Journal/vol29/29GSJ06e.html

Lazaros, E. J., & Ahmadi, R. (2008). Integration of supportive design features and technology. *Technology Teacher*, 67(7), 20–25.

Le Calvez, S., Perron-Lepage, M.-F., & Burnett, R. (2006). Subcutaneous microchip-associated tumours in B6C3F1 mice: A retrospective study to attempt to determine their histogenesis. *Experimental and Toxicologic Pathology*, 57, 255–265. doi:10.1016/j.etp.2005.10.007 PMID:16427258

Lee, M. (2011). Privacy commissioner pushes for new powers. *ZDNet*. Retrieved from http://www.zdnet.com.au/privacy-commissioner-pushes-for-powers-339319337.htm

Leinonen, T. (2010). *Digital learning tools: Methodological insights*. Aalto, Finland: Aalto University, School of Arts Design and Architecture.

Leman-Langlois, S. (2008). Privacy as currency: Crime, information and control in cyberspace. In S. Leman-Langlois (Ed.), *Technocrime: Technology, crime and social control* (pp. 112–138). Collumpton, UK: Willan Publishing.

Lemonick, M. (2003, October 21). Robo-monkey's reward. *Time*. Retrieved July 20, 2011 from http://www.time.com/time/magazine/article/0,9171,524491,00.html

Lessig, L. (2007). *Laws that choke creativity*. Retrieved from http://www.ted.com/talks/larry_lessig_says_the_law_is_strangling_creativity.html

Levi, P. (1989). *The drowned and the saved*. New York: Vintage.

Lewan, T. (2007, September 8). Chip implants linked to animal tumors. *Associated Press*.

Lindsay, D. (2005). An exploration of the conceptual basis of privacy and the implications for the future of Australian privacy law. *Melbourne University Law Review*, 29(1), 179–217.

Lippincott, J. K. (2010). Information commons: Meeting millennials' needs. *Journal of Library Administration*, 50(1), 27–37. doi:10.1080/01930820903422156

Lips, A., Miriam, B., Taylor, J. A., & Organ, J. (2009). Identity management, administrative sorting and citizenship in new modes of government. *Information Communication and Society*, 12(5), 715–734. doi:10.1080/13691180802549508

Lipschutz, R. D. (2008). Imperial warfare in the naked city—Sociality as critical infrastructure. *International Political Sociology*, 3(3), 204–218. doi:10.1111/j.1749-5687.2008.00045.x

Lipschutz, R. D. (2014). World war infinity: Total information awareness in the global garrison. In S. Hurt, & R. D. Lipschutz (Eds.), *The public-private hybridization of the 21st century state*. Academic Press.

Lipschutz, R. D., & Turcotte, H. (2005). Duct tape or plastic? The political economy of threats and the production of fear. In B. Hartmann, B. Subramaniam, & C. Zerner (Eds.), *Making threats—Biofears and environmental anxieties* (pp. 25–46). Lanham, MD: Rowman & Littlefield.

Locke, J. (2003). *Two treatises of government and a letter concerning toleration* (I. Shapiro, Ed.). New Haven, CT: Yale University Press.

Loconto, A., & Busch, L. (2010). Standards, techno-economic networks, and playing fields: Performing the global market economy. *Review of International Political Economy*, *17*(3), 507–536. doi:10.1080/09692290903319870

Loconto, A., Stone, J. V., & Busch, L. (2011). Standards, certifications, and accreditations. In G. Ritzer (Ed.), *Encyclopedia of globalization*. Hoboken, NJ: Wiley-Blackwell.

Look, G. (1998). *Auto ID makes tracks in livestock traceability*. ID Systems - The European source for auto ID. Retrieved 13 May 2013 from http://www.idsyseuro.com/cow0498.htm

Lovells, H. (n.d.). *EU modifies cookie rules*. Retrieved from http://ehoganlovells.com/rv/ff0001fbca7d-4cfee09193f6c241b40f4ea39f24/p=32

Luckin, R., Bligh, B., Manches, A., Ainsworth, S., Crook, C., & Noss, R. (2012). *Decoding learning: The proof, promise and potential of digital education*. London: NESTA.

Luoma, S. N. (2008). *Silver nanotechnologies and the environment*. Philadelphia, PA: The Pew Charitable Trusts.

Lupton, D. (1993). Risk as moral danger: The social and political functions of risk discourse in public health. *International Journal of Health Services*, *23*(3), 425–435. doi:10.2190/16AY-E2GC-DFLD-51X2 PMID:8375947

Lupton, D., & Petersen, A. (1996). *The new public health: Health and self in the age of risk*. London: Sage.

Lyell, M. (2010). *To fight crime and win* (pp. 29–37). Police Association News.

Lynch, M. et al. (2008). *Truth machine: The contentious history of DNA fingerprinting*. Chicago: Chicago University Press. doi:10.7208/chicago/9780226498089.001.0001

Lyon, D. (1994). *The electronic eye: The rise of surveillance society*. Minneapolis, MN: University of Minnesota Press.

Lyon, D. (2001). *Surveillance society*. Buckingham, UK: Open University Press.

Maguire, G. (1999). *Transforming humanity: Toward the implantable walkstation*. Retrieved July 20, 2011 from http://www.scribd.com/doc/27151750/1/Transforming-Humanity-Toward-the-Implantable-Walkstation

Maisel, W. H., & Kohno, T. (2010). Improving the security and privacy of implantable medical devices. *The New England Journal of Medicine*, *362*, 1164–1166. doi:10.1056/NEJMp1000745 PMID:20357279

Mann, S. (2002). Sousveillance, not just surveillance. *Metal and Flesh, 6*(1).

Mann, S. (2011). Learning by being: Thirty years of cyborg existemology. *Wearcam*. Retrieved from http://www.wearcam.org/existed/existed.ps

Manning, P. K. (2008). A view of surveillance. In S. Leman-Langlois (Ed.), *Technocrime: Technology, crime and social control* (pp. 209–242). Cullompton, UK: Willan Publishing.

Mann, S. (1997). Wearable computing: A first step toward personal imaging. *Computer, 30*(2). doi:10.1109/2.566147

Mann, S. (2001). Can humans being clerks make clerks be human? Exploring the fundamental difference between UbiComp and WearComp. *Oldenbourg Electronic Journals*, *43*(2), 97–106.

Mann, S. (2001). *Intelligent image processing*. New York: John Wiley-IEEE Press. doi:10.1002/0471221635

Mann, S., & Niedzviecki, H. (2001). *Cyborg: Digital destiny and human possibility in the age of the wearable computer*. Toronto, Canada: Doubleday.

Mann, S., Nolan, J., & Wellman, B. (2003). Sousveillance: Inventing and using wearable computing devices for data collection in surveillance environments. *Surveillance & Society*, *1*(3), 331–355.

Mann, T., & Blunden, A. (2012). Uberveillance. In *Australian law dictionary*. Oxford, UK: Oxford University Press.

Martin-Negrier, M.-L. et al. (1996). Primitive malignant fibrous histiocytoma of the neck with carotid occlusion and multiple cerebral ischemic lesions. *Stroke*, *27*, 536–537. doi:10.1161/01.STR.27.3.536 PMID:8610325

Martinovic, I., Davies, D., Frank, M., Perito, D., Ros, T. D., & Song, D. (n.d.). *On the feasibility of side-channel attacks with brain-computer interfaces conference: USENIX security '12*. Retrieved from https://www.usenix.org/conference/usenixsecurity12/feasibility-side-channel-attacks-brain-computer-interfaces

Masters, A., & Michael, K. (2005). Humancentric applications of RFID implants: The usability contexts of control, convenience and care. In *Proceedings of Mobile Commerce and Services*. IEEE. doi:10.1109/WMCS.2005.11

Masters, A., & Michael, K. (2007). Lend me your arms: The use and implications of humancentric RFID. *Electronic Commerce Research and Applications, 6*(1), 29–39. doi:10.1016/j.elerap.2006.04.008

Matthews, R. (2009). Beyond 'so what?' criminology: Rediscovering realism. *Theoretical Criminology, 13*(3), 341–362. doi:10.1177/1362480609336497

McCarthy, P. E. et al. (1996). Liposarcoma associated with glass foreign body in a dog. *Journal of the American Veterinary Medical Association, 209*, 612–614. PMID:8755980

McCartney, C. (2006). *Forensic identification and criminal justice: Forensic science justice and risk*. Cullompton: Willan Publishing.

McCartney, C. (2006). The DNA expansion programme and criminal investigation. *The British Journal of Criminology, 46*(2), 189. doi:10.1093/bjc/azi094

McCorkle, D., Reardon, J., Alexander, J., Kling, N., Harris, R., & Iyer, R. (1999). Undergraduate marketing students, group projects, and teamwork: The good, the bad, and the ugly? *Journal of Marketing Education, 21*(2), 106–117. doi:10.1177/0273475399212004

McCulloch, C. (2012). *About me - Baranduda blog*. Retrieved from http://coachcarole.wordpress.com/about/

McCullough, S. (2004). *Senior dogs for dummies*. New York: John Wiley & Sons.

McGee, E. (2008). Bioelectronics and implanted devices. In B. Gordijn & R. Chadwick (Eds.), Medical enhancement and posthumanity (pp. 207-224). Berlin: Springer Science + Business media B.V.

McGee, E. (2010). Toward regulating human enhancement technologies. *AJOB Neuroscience, 1*(2), 49–50. doi:10.1080/21507741003699405

McGee, E., & Maguire, G. (1999). Implantable brain chips? Time for debate. *The Hastings Center Report, 29*(1), 7–13. doi:10.2307/3528533 PMID:10052005

McGee, E., & Maguire, G. (2007). Becoming borg to become immortal: Regulating brain implant technologies. *Cambridge Quarterly of Healthcare Ethics, 16*(3), 291–302. doi:10.1017/S0963180107070326 PMID:17695620

McInerney, J. (1996). Old economics for new problems - Livestock disease: Presidential address. *Journal of Agricultural Economics, 47*(1-4), 295–314. doi:10.1111/j.1477-9552.1996.tb00695.x

McIntyre, L. (2007). *French bulldog is catalyst for investigation of microchip-cancer connection*. Retrieved September 8, 2007 from http://www.spychips.com/blog/2007/09/rfid_cancer_story_catalyst_lov.html

McNamee, M.J., & Edwards. (2006). Transhumanism, medical technology and slippery slopes. *Journal of Medical Ethics, 32*, 513–518. doi:10.1136/jme.2005.013789 PMID:16943331

Melson, K. E. (1990). Legal and ethical considerations. In L. T. Kirby (Ed.), *DNA fingerprinting: An introduction*. Oxford, UK: Oxford University Press.

Memoto. (n.d.). Retrieved from http://memoto.com

Michael, K., & Clarke, R. (2013). Location and tracking of mobile devices: Überveillance stalks the streets. *Computer Law and Security Review, 29*(2).

Michael, K., & McCathie, L. (2005). The pros and cons of RFID in supply chain management. In *Proceedings of International Conference on Mobile Business* (pp. 623-629). Los Alamitos, CA: IEEE Computer Society.

Michael, K., & Michael, M.G. (2004). The social, cultural, religious, and ethical implications of automatic identification. In *Proceedings of the Seventh International Conference in Electronic Commerce Research*, (pp. 433-450). IEEE.

Michael, K., & Michael, M. G. (2007). A note on uberveillance. In K. Michael & M. G. Michael (Eds.), *From dataveillance to überveillance (uberveillance) and the realpolitik of the transparent society*. Wollongong, Australia: University of Wollongong. Retrieved June 3, 2011 from http://works.bepress.com/mgmichael/24

Michael, K., & Michael, M.G. (2010, Summer). A note on uberveillance. *IEEE Science and Technology Magazine*, 9-16.

Michael, K., & Michael, M.G. (2011). The social and behavioral implications of location-based services. *Journal of Location-Based Services, 5*.

Michael, K., & Michael, M.G. (2007). Homo electricus and the continued speciation of humans. In M. Quigley (Ed.), *The encyclopedia of information ethics and security*. Hershey, PA: IGI Global.

Michael, K., & Michael, M.G. (2011). Implementing namebers using microchip implants: The black box beneath the skin. In *This pervasive day: The potential and perils of pervasive computing*. London: Imperial College Press.

Michael, K., McNamee, A., & Michael, M. G. (2006). The emerging ethics of humancentric GPS tracking and monitoring. In *Proceedings of IEEE International Conference on Mobile Business*. IEEE.

Michael, M.G. (2007). A note on uberveillance. In From Dataveillance to Uberveillance and the Realpolitik of the Transparent Society, pp. 9-26.

Michael, M. G. (1998). The number of the beast, 666 (revelation 13: 16-18), background, sources, and interpretation. New South Wales, Australia: Macquarie University.

Michael, M. G. (2010). Demystifying the number of the beast in the book of revelation: Examples of ancient cryptology and the interpretation of the 666 conundrum. In *Proceedings of IEEE International Symposium on Technology and Society*. IEEE.

Michaelis, R. C., Flanders, R. G., & Wulff, P. H. (2008). *A litigator's guide to DNA: From the laboratory to the courtroom*. Burlington, UK: Elsvier.

Michael, K., & Michael, M.G. (2006). A note on uberveillance. In K. Michael, & M. Michael (Eds.), *From dataveillance to uberveillance and the realpolitik of the transparent society* (pp. 9–25). Wollongong, Australia: University of Wollongong.

Michael, K., & Michael, M. G. (2007). *From dataveillance to überveillance and the realpolitik of the transparent society*. Wollongong, Australia: University of Wollongong.

Michael, K., & Michael, M. G. (2008). *Innovative automatic identification and location-based services: From bar codes to chip implants*. Hershey, PA: IGI Global.

Michael, K., & Michael, M. G. (2011a). Implementing namebers using microchip implants: The black box beneath the skin. In J. Pitt (Ed.), *This pervasive day: The potential and perils of pervasive computing* (pp. 1–51). London: Imperial College Press.

Michael, K., & Michael, M. G. (2011b). The social and behavioral implications of location-based services. *Journal of Location-Based Services, 5*(3-4), 121–137. doi:10.1080/17489725.2011.642820

Michael, K., & Michael, M. G. (2013). The future prospects of embedded microchips in humans as unique identifiers: The risks versus the rewards. *Media Culture & Society, 34*(3).

Michael, K., Roussos, G., Huang, G. Q., Gadh, R., Chattopadhyay, A., Prabhu, S., & Chu, P. (2010). Planetary-scale RFID services in an age of uberveillance. *Proceedings of the IEEE, 98*(9), 1663–1671. doi:10.1109/JPROC.2010.2050850

Michael, M. G. (1999). The genre of the apocalypse: What are they saying now? *Bulletin of Biblical Studies, 18*, 115–126.

Michael, M. G. (2000). For it is the number of a man. *Bulletin of Biblical Studies, 19*, 79–89.

Michael, M. G. (2002). *The canonical adventure of the apocalypse of John: An Eastern Orthodox perspective*. New South Wales, Australia: Australian Catholic University.

Michael, M. G., Fusco, S. J., & Michael, K. (2008). A research note on ethics in the emerging age of überveillance. *Computer Communications, 31*(6), 1192–1199. doi:10.1016/j.comcom.2008.01.023

Michael, M. G., & Michael, K. (2006). National security: The social implications of the politics of transparency. *Prometheus, 24*(4), 359–364. doi:10.1080/08109020601029912

Michael, M. G., & Michael, K. (2007). A note on überveillance. In *From dataveillance to uberveillance and the realpolitik of the transparent society*. Wollongong, Australia: Academic Press.

Michael, M. G., & Michael, K. (2009). Uberveillance: Microchipping people and the assault on privacy. *Quadrant*, *53*(3), 85–89.

Michael, M. G., & Michael, K. (2010a). Surveillance and uberveillance. *IEEE Technology and Society Magazine*, *29*(2).

Michael, M. G., & Michael, K. (2010b). Towards a state of uberveillance. *IEEE Technology and Society Magazine*, *29*(2), 9–16. doi:10.1109/MTS.2010.937024

Michael, M. G., & Michael, K. (2011). The fall-out from emerging technologies: On matters of surveillance, social networks and suicide. *IEEE Technology and Society Magazine*, *30*(3), 15–18. doi:10.1109/MTS.2011.942312

Michigan Department of Agriculture. (2011). Retrieved 10 February 2011 from http://www.michigan.gov/mda/0,1607,7-125--156829--,00.html

Milman, H. A. (1994). *Handbook of carcinogen testing* (2nd ed.). Park Ridge, NJ: Noyes Publications.

Ministry of Justice. (2009). *Population in custody*. Retrieved from http://www.justice.gov.uk/publications/populationincustody.htm

Mohammadian, M., & Jentzsch, R. (2008). Intelligent agent framework for secure patient-doctor profiling and profile matching. *International Journal of Healthcare Information Systems and Informatics*, *3*(3), 8–57. doi:10.4018/jhisi.2008070103

Moore, S. (2009). F.B.I. & states vastly expand DNA databases. *The New York Times*. Retrieved from http://www.nytimes.com/2009/04/19/us/19DNA.html?_r=1

Moore, A. D. (2010). *Privacy rights: Moral and legal foundations*. University Park, PA: Pennsylvania State University Press.

Morales, L. (2010, December 21). US internet users ready to limit online tracking for ads. *Gallup*. Retrieved from http://www.gallup.com/poll/145337/internet-users-ready-limit-onlinetracking-ads.aspx

Morandini, P. (2009). *Tell me who your friends are and I'll tell what gene you are*. Retrieved from http://www.siga.unina.it/SIGA2009/SIGA_2009/6_01.pdf

Morrison, B. A. (2003). Soft tissue sarcomas of the extremities.[Bayl Univ Med Cent]. *Proc*, *16*(3), 285–290. PMID:16278699

Mossberger, K., Tolbert, C. J., & Stansbury, M. (2003). *Virtual inequality: Beyond the digital divide*. Washington, DC: Georgetown University Press.

Mousavi, A., Sarhadi, M., Lenk, A., & Fawcett, S. (2002). Tracking and traceability in the meat processing industry: A solution. *British Food Journal*, *104*(1), 7–19. doi:10.1108/00070700210418703

Mulligan, C. M. (2008). Perfect enforcement of law: When to limit and when to use technology. *Richmond Journal of Law and Technology*, *14*(4), 1–49.

Murasugi, E. et al. (2003, September 13). Histological reactions to microchip implants in dogs. *The Veterinary Record*, 328. doi:10.1136/vr.153.11.328 PMID:14516115

MURI (Multi-Disciplinary University Research Initiative). (2008). *Synthetic telepathy*. Cognitive NeuroSystems Lab, University of California, Irvine. Retrieved February 28, 2011 from http://cnslab.ss.uci.edu/muri/research.html

Musil, S. (2012). Microsoft: Google bypassed ie privacy settings too. *CNET*. Retrieved from http://news.cnet.com/8301-1009_3-57381371-83/microsoft-google-bypassed-ie-privacy-settings-too/

Myers Briggs Research Foundation. (2010). *How frequent is my type?* Retrieved January 3, 2011, from http://www.myersbriggs.org/my-mbti-personality-type/my-mbti-results/how-frequent-is-my-type.asp

Myers, I., & McCaulley, M. (1985). *Manual: A guide to the development and use of the Myers-Briggs type indicator*. Palo Alto, CA: Consulting Psychologists Press.

Myers, K. K., & Sadaghiani, K. (2010). Millennials in the workplace: A communication perspective on millennials' organizational relationships and performance. *Journal of Business and Psychology*, *25*(2), 225–238. doi:10.1007/s10869-010-9172-7 PMID:20502509

Nagel, T. (1998). Concealment and exposure. *Philosophy & Public Affairs*, *27*(1), 3–30. doi:10.1111/j.1088-4963.1998.tb00057.x

NAHMS (National Animal Health Monitoring System). (2000). *Treatment of respiratory disease in U.S. feedlots.* Retrieved 13 February 2011 from http://www.aphis.usda. gov/animal_health/nahms/feedlot/downloads/feedlot99/ Feedlot99isTreatResp.pdf

Naone, E. (2009). RFID's security problem. *Technology Review, 112*(1), 72–74.

Napier, R., & Gershenfeld, M. (2004). *Groups: Theory and experience* (7th ed.). Boston: Houghton-Mifflin.

Narayanan, A. (2010). *Do not track explained - 33 bits of entropy.* Retrieved from http://33bits.org/2010/09/20/ do-not-track-explained/

National Nanotechnology Coordination Office (NNCO). (2010). *Nanotechnology-enabled sensing: Report of the national nanotechnology initiative workshop.* Arlington, VA: NNCO.

National Women's Health Resource Center. (n.d.). *Clinical trials overview.* Retrieved October 10 2007 from http:// findarticles.com/p/articles/mi_m0PWR/is_2005_August_9/ai_n17214944/p

New South Wales Law Reform Commission. (2009). *Invasion of privacy, 120.* Retrieved November 1, 2010, from http://www.lawlink.nsw.gov.au/lawlink/lrc/ll_lrc. nsf/pages/LRC_r120toc

New Zealand Law Commission. (2011). *Key recommendations.* Retrieved from http://www.lawcom.govt. nz/sites/default/files/publications/2011/08/key_recommendations_-_for_report_release.pdf

New Zealand Law Commission. (2011). *Review of privacy.* Retrieved from http://www.lawcom.govt.nz/project/ review-privacy?quicktabs_23=report

Nicolelis, M. (n.d.). *Laboratory of Miguel Nicolelis.* Retrieved, July 20, 2011 from http://www.nicolelislab.net/

Nietzsche, F. (1990). *The twilight of the idols and the anti-Christ* (R. J. Hollingdale, Trans.). London: Penguin Classics.

Nissenbaum, H. (2010). *Privacy in context: Technology, policy, and the integrity of social life.* Stanford, CA: Stanford Law Books.

NNI (National Nanotechnology Initiative). (2008). *Strategy for nanotechnology-related environmental, health, and safety research.* Subcommittee on Nanoscale Science, Engineering, and Technology Committee on Technology National Science and Technology Council.

Noack, T., & Kubicek, H. (2010). The introduction of online authentication as part of the new electronic national identity card in Germany. *Identity in the Information Society, 3*(1), 87–110. doi:10.1007/s12394-010-0051-1

Norton, B. G. (2005). *Sustainability: A philosophy of adaptive ecosystem management.* Chicago: University of Chicago Press. doi:10.7208/chicago/9780226595221.001.0001

Nov, O., & Chen, Y. (2008). Personality and technology acceptance: Personal innovativeness in IT, openness and resistance to change. In *Proceedings of the 41st Hawaii International Conference on System Sciences*, (pp. 433-450). IEEE.

Nuffield Council on Bioethics. (2009). *Forensic use of bioinformation: Ethical issues.* Retrieved from http:// www.nuffieldbioethics.org/bioinformation

O'Brien, M. (2010). *Consuming talk: Youth culture and the mobile phone.* (Unpublished doctoral dissertation). Department. of Sociology, National University of Ireland, Maynooth, Ireland. Retrieved March 2, 2011 from http:// eprints.nuim.ie/2253/1/MOB_PhD.pdf

O'Connor, D., & De Lint, W. (2009). Frontier government: The folding of the Canada-US border. *Studies in Social Justice, 3*(1), 39–66.

O'Neill, M. (2008, March 9). Hackers attack epileptic forum and make sufferers convulse. *Geeks are Sexy.* Retrieved April 27, 2011 from http://www.geeksaresexy. net/2008/03/29/hackers-attack-epileptic-forum-and-make-sufferers-convulse/

O'Farrell, C. (2005). *Michel Foucault.* London: Sage.

Office for National Statistics. (2007). *Mid-2006 UK, England and Wales, Scotland and Northern Ireland: 22/08/07.* Retrieved from http://www.statistics.gov.uk/ statbase/Product.asp?vlnk=15106

Office for National Statistics. (2009). *UK population grows to 61.4 million.* Retrieved from http://www.statistics.gov. uk/pdfdir/popnr0809.pdf

Office of the Information Commissioner. (n.d.). *IP addresses, Google analytics and the privacy principles.* Retrieved from http://www.oic.qld.gov.au/files/InformationSheets/Information%20Sheet%20-%20privacy,%20IP%20addresses%20and%20Google%20Analytics.pdf

Office of the Information Privacy Commissioner. (2006). *Explanatory statement: Approval of biometrics institute privacy code.* Retrieved June 3, 2010, from www.privacy.gov.au/materials/types/download/9197/6797

Office of the Privacy Commissioner of Canada. (2005). *Canadian federal privacy commissioner's findings in PIPEDA case summaries #2005-319.* Retrieved from http://www.priv.gc.ca/cf-dc/2005/319_20051103_e.cfm

Office of the Privacy Commissioner of Canada. (2009). *Privacy legislation in Canada.* Retrieved from http://www.priv.gc.ca/fs-fi/02_05_d_15_e.cfm

Office of the Privacy Commissioner of Canada. (2010a). *Report on the 2010 office of the privacy commissioner of Canada's consultations on online tracking, profiling and targeting, and cloud computing.* Retrieved from http://www.priv.gc.ca/resource/consultations/report_201105_e.pdf

Office of the Privacy Commissioner. (2010). *Submission to senate finance and public administration legislation committee inquiry into exposure drafts of Australian privacy amendment legislation.* Retrieved from http://www.privacy.gov.au/materials/types/download/9565/7125

Ohm, P. (2010). Broken promises of privacy: Responding to the surprising failure of anonymization. UCLA Law Review. University of California, Los Angeles. School of Law, 57(6), 1701–1778.

Ontario (Liquor Control Board) v. Keupfer. (1974). *Dominion Law Reports* (3d). 47, p. 326.

Orwell, G. (1983). 1984. New York: Plume.

Orwell, G. (1949). *Nineteen-eightyfour—A novel.* London: Secker & Warburg.

Osmond, C. (2010). Anti-social behavior and its surveillant inter-assemblage in New South Wales, Australia. *Surveillance & Society, 7*(3/4), 325–343.

Palmer, D., Warren, I., & Miller, P. (2010). ID scanners in the night-time economy. In K. Michael (Ed.), *Proceedings of the 2010 IEEE International Symposium on Technology and Society (ISTAS '10)* (pp. 234-241). Wollongong, Australia: University of Wollongong.

Palmer, T. E. et al. (1998). Fibrosarcomas associated with passive integrated transponder implants. *Toxicologic Pathology, 26,* 170.

Pal, S., & Alocilja, E. C. (2009). Electrically active polyaniline coated magnetic (EAPM) nanoparticle as novel transducer in biosensor for detection of bacillus anthracis spores in food samples. *Biosensors & Bioelectronics, 24*(5), 1437–1444. doi:10.1016/j.bios.2008.08.020 PMID:18823768

Parameswaran, M., & Whinston, A. B. (2007). Research issues in social computing. *Journal of the Association for Information Systems, 8,* 336–350.

Parent, W. A. (1983). Privacy, morality, and the law. *Philosophy & Public Affairs, 12*(4), 269–288.

Parliamentary Office of Science and Technology. (2006). *Postnote: The national DNA database.* Retrieved from http://www.parliament.uk/documents/upload/POSTpn258.pdf

Patel, P. (2009). The brain-machine interface, unplugged. *IEEE Spectrum.* Retrieved from http://spectrum.ieee.org/biomedical/devices/the-brainmachine-interface-unplugged

Patton, P. (1994). MetamorphoLogic: Bodies and powers in A Thousand Plateaus. *Journal of the British Society for Phenomenology, 25*(2), 157–169.

Patton, P. (1995). Caught-You used to watch television: Now it watches you. *Wired, 3*(1), 124–127.

Pearson, I. (2012). Future visions. *Future Human Evolution.* Retrieved July 23, 2011 from http://www.humansfuture.org/future_post_human_futures.php.htm

Perakslis, C., & Michael, K. (2012). Indian millennials: Are microchip implants a more secure technology for identification and access control? In *Proceedings of IEEE International Symposium on Technology and Society (ISTAS12).* IEEE.

Perakslis, C., & Wolk, R. (2006). Social acceptance of RFID as a biometric security method. *IEEE Symposium on Technology and Society Magazine, 25*(3), 34-42.

Perusco, L., & Michael, K. (2007). Control, trust, privacy, and security: Evaluating location-based services. *IEEE Technology and Society Magazine, 26*(1), 4–16. doi:10.1109/MTAS.2007.335564

Pettitt, R. E. (2001). Traceability in the food animal industry and supermarket chains. *Revue Scientifique et Technique-Office International Epizootics, 20*(2), 584–597. PMID:11548528

Pew Research Center for the People and the Press. (n.d.). *Millenials: A portrait of the generation next.* Retrieved from http://www.pewsocialtrends.org/files/2010/10/millennials-confident-connected-open-to-change.pdf

Phillips, N. (2010, October 5). Inside the cookie monster - Trading your online data for profits. *The Australian.* Retrieved from http://www.smh.com.au/technology/technology-news/inside-the-cookie-monster--trading-your-online-data-for-profits-20101004-164ee.html

Pinter, M. M., Alesch, F., Murg, M., Seiwald, M., Helscher, R. J., & Binder, H. (1999). Deep brain stimulation of the subthalamic nucleus for control of extrapyramidal features in advanced idiopathic Parkinson's disease. *Journal of Neural Transmission, 106,* 693–709. doi:10.1007/s007020050190 PMID:10907728

Platt, S. et al. (2006). Spinal cord injury resulting from incorrect microchip placement in a cat. *Journal of Feline Medicine and Surgery, 9*(2), 157–160. doi:10.1016/j.jfms.2006.07.002 PMID:16982206

Police Home Office. (2009). *Police and criminal evidence act 1984 (PACE) and accompanying codes of practice.* Retrieved from http://police.homeoffice.gov.uk/operational-policing/powers-pace-codes/pace-code-intro/

Pontifical Council for Justice and Peace. (2005). *Compendium of the social doctrine of the church: To his holiness Pope John Paul II.* Rome: Libreria Editrice Vaticana.

Popper, D. E. (2007). Traceability: Tracking and privacy in the food system. *Geographical Review, 97*(3), 365–388. doi:10.1111/j.1931-0846.2007.tb00511.x

Positive, I. D. (2009, September 8). *VeriChip corporation agrees to acquire steel vault corporation to form PositiveID corporation.* Retrieved April 15, 2011 from http://investors.positiveidcorp.com/releasedetail.cfm?ReleaseID=428367

Positive, I. D. (2010). *PositiveID corporation launches the wireless body at ID WORLD international congress in Milan, Italy, Nov. 17.* Retrieved February 28, 2010 from http://investors.positiveidcorp.com/releasedetail.cfm?ReleaseID=531278

Positive, I. D. (n.d.). *Glucose-sensing microchip.* Retrieved April 21, 2011 from http://www.positiveidcorp.com/glucose_sensing.html

Posts Tagged Transhumanism. (2012). Retrieved October 29, 2012, from http://discombobulaated.wordpress.com/tag/transhumanism/

Premier Inc. (2009, September 8). Awards dynamic computer corporation with a group purchasing agreement to provide RFID asset tracking and management solutions for its member network. *Science Letter.*

Priest, D., & Arkin, W. (2011). *Top secret America: The rise of the new American security state.* Boston: Little, Brown.

Project Lifesaver International. (2012). *Bringing loved ones home.* Retrieved October, 28, 2012, from http://www.projectlifesaver.org/

Prusak, L., & Cohen, D. (1997). Knowledge buyers, sellers and brokers: The political economy of knowledge. In *The economic impact of knowledge.* New York: Butterworth-Heinemann.

Pulse, A. Sydney. (2011). *VTI technologies: SCC1300 combined sensor improves positioning accuracy in map-matching applications.* Retrieved from www.vti-technologies

Queensland Legislative Assembly Law and Justice Safety Committee (QLALJSC). (2009). *Inquiry into alcohol-related violence, interim report, 72.* Retrieved June 3, 2010, from http://www.parliament.qld.gov.au/view/committees/documents/lcarc/reports/Report%2073.pdf

R. v. Webb. (1943). *Canadian Criminal Cases* (C.C.C), *80,* pp. 151.

Raab, C. (1999). From balancing to steering: New directions for data protection. In R. Grant & C. J. Bennett (Eds.), Visions of privacy: Policy choices for the digital age (pp. 68–96). Toronto, Canada: University of Toronto Press.

Rachels, J. (1975). Why privacy is important. *Philosophy & Public Affairs, 4*(4), 323–333.

Radio Frequency Identification (RFID) Systems. (2011). Retrieved April 12, 2011, from http://epic.org/privacy/rfid/

Rajasekaran, M. P., Radhakrishnan, S., & Subbaraj, P. (2009). Elderly patient monitoring system using a wireless sensor network. *Telemedicine and e-Health, 15*(1), 73-79.

Rao, G. N., & Edmondson, J. (1985). Tissue reaction to an implantable identification device in mice. *Toxicologic Pathology, 18*(3), 412–416. doi:10.1177/019262339001800308 PMID:2267501

Rasmussen, K. et al. (1985). Male breast cancer from pacemaker pocket. *Pacing and Clinical Electrophysiology, 8*(5), 761–763. doi:10.1111/j.1540-8159.1985.tb05891.x PMID:2414760

Raths, D. (2009). RFID is finding a home in the data center. *Computerworld, 43*(18), 21.

Rawal, A. (2009). RFID: The next generation auto ID technology. *Microwave Journal, 52*(3), 58–76.

Raz, J. (1988). *The morality of freedom*. Oxford, UK: Clarendon Press, Oxford University Press. doi:10.1093/0198248075.001.0001

Reggatieri, A., Gamberi, M., & Manzini, R. (2007). Traceability of food products: General framework and experimental evidence. *Journal of Food Engineering, 81*, 347–356. doi:10.1016/j.jfoodeng.2006.10.032

Reidenberg, J. (1999). The globalisation of privacy solutions: Movement towards obligatory standards for fair information practices. In R. Grant & C. J. Bennett (Eds.), Visions of privacy: Policy choices for the digital age (pp. 217–230). Toronto, Canada: University of Toronto Press.

Reidenberg, J. R. (1992). Privacy in the information economy: A fortress or frontier for individual rights? Federal Communications Law Journal, 44(2), 195–243.

Reiman, J. H. (1976). Privacy, intimacy, and personhood. *Philosophy & Public Affairs, 6*(1), 26–44.

Rheingold, H. (2011). Social media classroom. *Invitation to the Social Media Classroom and Collabatory*. Retrieved from http://socialmediaclassroom.com/

Roberts, A., & Taylor, N. (2005). Privacy and the DNA database. *European Human Rights Law Review, 4*, 373.

Rodgers, M. C. (2009). Diane Abbott MP and liberty hold DNA clinic in Hackney. *Liberty*. Retrieved from http://www.liberty-human-rights.org.uk/news-and-events/1-press-releases/2009/24-09-2009-diane-abbott-mp-and-liberty-hold-dna-clinic-in-hackney.shtml

Rodota, S., & Capurro, R. (2005). *Opinion n⁰20-16/03/02005: Ethical aspects of ICT implants in the human body*. Retrieved December 12, 2010, from http://ec.europa.eu/european_group_ethics/docs/avis20_en.pdf

Rogers, E. M. (1995). *Diffusion of innovations*. New York: The Free Press.

Rose, D. (2011). *Chip checkups and embedded beds: RFID and home monitoring*.

Rosenberg, I. B. (2007). Involuntary endogenous RFID compliance monitoring as a condition of federal supervised release - Chips ahoy. *Yale Journal of Law and Technology, 10*, 331–359.

Rosenberg, R. S. (2004). *The social impact of computers*. Los Angeles, CA: Elsevier.

Rothenberger-Janzen, K., Flueckiger, A., & Bigler, R. (1998). Carcinoma of the breast and pacemaker generators. *Pacing and Clinical Electrophysiology, 21*(4), 769–771. doi:10.1111/j.1540-8159.1998.tb00137.x PMID:9584311

Rotter, P., Daskala, B., & Compano, R. (2008). RFID implants: Opportunities and challenges for identifying people. *IEEE Technology and Society Magazine, 27*(2), 24–32. doi:10.1109/MTS.2008.924862

Rubel, A. (2007). Privacy and the USA patriot act: Rights, the value of rights, and autonomy. *Law and Philosophy, 26*(2), 119–159. doi:10.1007/s10982-005-5970-x

Rubel, A. (2011). The particularized judgment account of privacy. *Res Publica (Liverpool, England), 17*, 275–290. doi:10.1007/s11158-011-9160-4

Rubio, J. (2012, December 7). US begins testing brain implants with hopes of slowing Alzheimer's. *The Verge*. Retrieved from http://mobile.theverge.com/2012/12/7/3740988/us-deep-brain-stimulation-implant

Saenz, A. (2009). Braingate2: Your mind just went wireless. *Singularity Hub*. Retrieved on July 22, 2011 from http://singularityhub.com/2009/06/17/braingate2-your-mind-just-went-wireless/

Savage, C., & Shane, S. (2011, February 24). U.S. arrests Saudi student in bomb plot. *The New York Times*. Retrieved February 25, 2011 from http://www.nytimes.com/2011/02/25/us/25terror.html

Schneider, M. E. (2006). Technology helps prevent wandering, falls by elderly residents. *Internal Medicine News*, *39*(10), 37. doi:10.1016/S1097-8690(06)73524-7

Scholz, T. S. (2011). *Learning through digital media: Experiments on technology and pedagogy*. Institute for Distributed Creativity (IDC). Retrieved from http://www.learningthroughdigitalmedia.net

Schultze, U., & Leidner, D. (2002). Studying KM in IS research: Discourses and theoretical assumptions. *Management Information Systems Quarterly*, *26*(3), 213–242. doi:10.2307/4132331

Selamat, M. H., & Choudrie, J. (2007). Using meta-abilities and tacit knowledge for developing learning based systems: A case study approach. *The Learning Organization*, *14*(4), 321–344. doi:10.1108/09696470710749263

Semple, J. (1993). *Bentham's prison: A study of the panopticon penitentiary*. Oxford, UK: Clarendon Press. doi:10.1093/acprof:oso/9780198273875.001.0001

Senate Environment and Communications Reference Committee. (2011, April 7). *The adequacy of protections for the privacy of Australians online*. Retrieved from http://www.aph.gov.au/Parliamentary_Business/Committees/Senate_Committees?url=ec_ctte/online_privacy/report/index.htm

Senate Finance and Public Administration Legislation Committee. (2010). *Inquiry into exposure drafts of Australian privacy amendment*. Retrieved from http://www.privacy.gov.au/materials/types/submissions/view/7125

Senate Standing Committee on Finance and Public Administration. (2007). *Human services (enhanced service delivery) bill 2007*. Retrieved November 2, 2010, from http://www.aph.gov.au/senate/committee/fapa_ctte/completed_inquiries/2004-07/access_card/report/index.htm

Shadish, W., Cook, T., & Campbell, D. (2002). *Experimental and quasi-experimental designs for generalized causal inference*. Boston: Houghton Mifflin.

Shah, N. A. (2011). *RFID chip in humans*. Retrieved March 30, 2011, from http://www.buzzle.com/articles/rfid-chip-in-humans.html

Shakespeare. (n.d.). *Richard* (11 Act V Scene 5). Retrieved from www.opensourceshakespeare.org

Shay, L. A., Conti, G., Larkin, D., & Nelson, J. (2012). A framework for analysis of quotidian exposure in an instrumented world. In *Proceedings of IEEE Carnahan Conference on Security Technology*. IEEE.

Shoval, N., Auslander, G., Cohen-Shalom, K., Isaacson, M., Landau, R., & Heinik, J. (2010). What can we learn about the mobility of the elderly in the GPS era? *Journal of Transport Geography*, *18*(5), 603–612. doi:10.1016/j.jtrangeo.2010.03.012

Siemens, G. (2005). Connectivism: A learning theory for the digital age. *ITDL*. Retrieved from http://www.itdl.org/Journal/Jan_05/article01.htm

Silber, M. D., & Bhatt, A. (2007). *Radicalization in the West: The homegrown threat*. New York: New York City Police Department Intelligence Division. Retrieved April 29, 2011 from http://msnbcmedia.msn.com/i/msnbc/Sections/NEWS/PDFs/nypd_radicalization_report.pdf

Simonsen Laboratories. (n.d.). *Fischer 344*. Retrieved October 5 2007 from http://www.simlab.com/products/fischer344.html

Simpson, J. R. (2008). Functional MRI lie detection: Too good to be true? *The Journal of the American Academy of Psychiatry and the Law*, *36*(4), 491–498. PMID:19092066

Singer, E. (2010, July 29). Glucose monitors get under the skin: Implantable devices work in diabetic pigs for over a year – Human tests could be next. *Technology Review*. Retrieved April 14, 2011 from http://www.technology-review.com/biomedicine/25889/?a=f

Singer, N., & Duhigg, C. (2012, October 27). Tracking voters' clicks online to try to sway them. *The New York Times*. Retrieved November 19, 2012 from http://www.nytimes.com/2012/10/28/us/politics/tracking-clicks-online-to-try-to-sway-voters.html

Smithsonian. (2012, August 19). These people are turning themselves into cyborgs in their basement. *Smartnews*. Retrieved from http://blogs.smithsonianmag.com/smart-news/2012/08/these-people-are-turning-themselves-into-cyborgs-in-their-basement/

Smits, I., Dolan, C., Vorst, H., Wicherts, J., & Timmerman, M. (2011). Cohort differences in big five personality factors over a period of 25 years. *Journal of Personality and Social Psychology*, *100*(6), 1124–1138. doi:10.1037/a0022874 PMID:21534699

Snider, L. (2011). *Criminalizing the algorithm? Stock market crime in the 21st century.* Paper presented at the ANZ Conference of Criminology. Geelong, Australia.

Snider, L. (2001). Crimes against capital: Discovering theft of time social justice. *Law, Order, and Neoliberalism*, *28*(3), 105–120.

Snowder, G. D., Van Vleck, L. D., Cundiff, L. V., & Bennett, G. L. (2006). Bovine respiratory disease in feedlot cattle: Environmental, genetic, and economic factors. *Journal of Animal Science*, *84*, 1999–2008. doi:10.2527/jas.2006-046 PMID:16864858

Sobel, D. (1995). *Longitude*. London: Fourth Estate.

Solove, D. (2001). Privacy and power: Computer databases and metaphors for information privacy. Stanford Law Review, 53(6), 1393. doi:10.2307/1229546 doi:10.2307/1229546

Solzhenitsyn, A. (2003). *The Gulag archipelago*. London: The Harvill Press.

Song, S. (2002, Spring). An internet knowledge sharing system. *Journal of Computer Information Systems*, 25–30.

Spiekermann, S. (2009). RFID and privacy: What consumers really want and fear. *Personal and Ubiquitous Computing*, *13*(6), 423–434. doi:10.1007/s00779-008-0215-2

Spillman, B. C. (2004). Changes in elderly disability rates and the implications for health care utilization and cost. *The Milbank Quarterly*, *82*(1), 157–194. doi:10.1111/j.0887-378X.2004.00305.x PMID:15016247

Sponselee, A., Schouten, B., Bouwhuis, D., & Willems, C. (2008). Smart home technology for the elderly: Perceptions of multidisciplinary stakeholders. In M. Mühlhäuser, A. Ferscha, & E. Aitenbichler (Eds.), *Communications in computer and information science constructing ambient intelligence* (Vol. 11). New York: Springer. doi:10.1007/978-3-540-85379-4_37

Stanton, R. (2005). RFID: Ripe for informed debate. *Computer Fraud and Security*, *12*, 12-14. doi:10.1016/S1361- 3723(05)70285-2

Stecklow, S. (2006, June 21). Twisting trail: U.S. falls behind in tracking cattle to control disease, USDA plans voluntary system after the industry divides on making one mandatory, a mad cow's unknown origins. *Wall Street Journal*.

Steffen, T., Luechinger, R., Wildermuth, S., Kern, C., Fretz, C., & Lange, J. et al. (2010). Safety and reliability of radio frequency identification devices in magnetic resonance imaging and computed tomography. *Patient Safety in Surgery*, *4*(2). doi: doi:10.1186/1754-9493-4-2 PMID:20205829

Stephan, K. D., Michael, K., Michael, M. G., Jacob, L., & Anesta, E. (2012). Social implications of technology: Past, present, and future. *Proceedings of the IEEE*, *100*(13), 1752–1781. doi:10.1109/JPROC.2012.2189919

Storey, J., & Barnett, E. (2000). Knowledge management initiatives: Learning from failure. *Journal of Knowledge Management*, *4*, 145–156. doi:10.1108/13673270010372279

Stosny, S. (2008). Guilt vs. responsibility is powerlessness vs. power: Understanding emotional pollution and power. *Anger in the Age of Entitlement*. Retrieved from http://www.psychologytoday.com/blog/anger-in-the-age-entitlement/200805/guilt-vs-responsibility-is-powerlessness-vs-power

Strak, J. (2011). Walmart asks meat industry to break with tradition. *Meatingplace*. Retrieved 13 May 2013 from http://www.traincan.com/feb22a-2011.asp

Surveillance Devices Act (Victoria), 1999.

Swedberg, C. (2007). Alzheimer care center to carry out verichip pilot. *RFID Journal*. Retrieved November, 6, 2012 from http://www.rfidjournal.com/article/view/3340

Swire, P., & Litan, R. (1998). None of your business: World data flow, electronic commerce, and the European privacy directive. Washington, DC: Brookings Institution Press.

Tabak, F., & Smith, W. (2005). Privacy and electronic monitoring in the workplace: A model of managerial cognition and relational trust development. *Employee Responsibilities and Rights Journal, 17*(3), 173–189. doi:10.1007/s10672-005-6940-z

Taipale, K. A. (2003). Data mining and domestic security: Connecting the dots to make sense of data. *Columbia Science and Technology Law Review, 5*(2). Retrieved June 17, 2010 from http://ssrn.com/abstract=546782

Takahashi, D. (2008, August 8). Defcon: Excuse me while I turn off your pacemaker. *VentureBeat*. Retrieved April 28, 2011 from http://venturebeat.com/2008/08/08/defcon-excuse-me-while-I-turn-off-your-pacemaker/

Tan, R., McClure, T., Lin, C. K., Jea, D., Dabiri, F., & Massey, T. et al. (2009). Development of a fully implantable wireless pressure monitoring system. *Biomedical Microdevices, 11*(1), 259–264. doi:10.1007/s10544-008-9232-1 PMID:18836836

Taylor, J. A., Lips, M., & Organ, J. (2008). Identification practices in government: Citizen surveillance and the quest for public service improvement. *Identity and Information Society, 1*(1), 135–154. doi:10.1007/s12394-009-0007-5

Taylor, N. (2011). A conceptual legal framework for privacy, accountability and transparency in visual surveillance systems. *Surveillance & Society, 8*(4), 455–470.

Telemedicine and Advanced Technology Research Center. (2009). *Annual report*. Retrieved April 29, 2011 from http://www.tatrc.org/docs/TATRC_report_2009.pdf

The Video Privacy Protection Act of 1998, 18 USC § 2710.1994.

Thompson, M. (2008, September 18). The Army's totally serious mind control project. *Time*. Retrieved February 28, 2010 from http://www.time.com/time/nation/article/0,8599,1841108,00.html

Thompson, T., & Phillips, T. (2005). *Death in Stockwell: The unanswered questions*. Retrieved from http://www.guardian.co.uk/uk/2005/aug/14/july7.terrorism

Thompson, D., Muriel, P., Russell, D., Osborne, P., Bromley, A., & Rowland, M. et al. (2002). Economic costs of the foot and mouth disease outbreak in the United Kingdom. *Rev. Sci. Tech. Off. Int. Epiz., 21*(3), 675–687. PMID:12523706

Thompson, P. (2010). Agrifood nanotechnology: Is this anything? In U. Fiedeler, C. Coenen, S. R. Davies, & A. Ferrari (Eds.), *Understanding nanotechnology: Philosophy, policy and publics*. Amsterdam: IOS Press.

Thompson, P. B., & Hannah, W. (2008). Food and agricultural biotechnology: A summary and analysis of ethical concerns. *Advances in Biochemical Engineering/Biotechnology, 111*, 229–264. doi:10.1007/10_2008_100 PMID:18496654

Thompson, S., & Genosko, G. (2009). *Punched drunk: Alcohol, surveillance and the LCBO, 1927-1975*. Blackpoint, Canada: Fernwood Publishing.

Tieger, P., & Barron-Tieger, B. (1998). *The art of speed-reading people: Harness the power of personality type and create what you want in business and in life*. Boston: Little, Brown and Company.

Tillmann, T. et al. (1997). Subcutaneous soft tissue tumours at the site of implanted microchips in mice. *Experimental and Toxicologic Pathology, 49*, 197–200. doi:10.1016/S0940-2993(97)80007-3 PMID:9314053

Today Magazine, U. S. A. (2010, September 20). Are consumers being stalked by RFID tags? *USA Today Magazine*, p. 8.

Travis, A. (2009, August 8). Police told to ignore human rights ruling over DNA: Details of innocent people will continue to be held: Senior officers will not get new guidance for a year. *The Guardian*. Retrieved from http://www.theguardian.com/politics/2009/aug/07/dna-database-police-advice

Trevarthen, A. (2005). *The importance of utilising electronic identification for total farm management: A Case study of dairy farms on the south coast of NSW*. University of Wollongong Thesis Collections.

Trevarthen, A., & Michael, K. (2006). *Beyond mere compliance of RFID regulations by the farming community: A case study of the Cochrane dairy farm.* Paper presented at the The Sixth International Conference on Mobile Business. Toronto, Canada.

Trevarthen, A., & Michael, K. (2008). *The RFID-enabled dairy farm: Towards total farm management.* Paper presented at the 7th International Conference on Mobile Business. Barcelona, Spain.

Trevarthen, A. (2007). The national livestock identification system: The importance of traceability in e-business. *Journal of Theoretical and Applied Electronic Commerce Research, 2*(1), 49–62.

Trivedi, B. (2007). Toxic cocktail. *New Scientist Magazine, 2619,* 44–47. doi:10.1016/S0262-4079(07)62217-9

Turkle, S. (2009, February 2). Interview. *Frontline.* Retrieved April 28, 2011, from http://www.pbs.org/wgbh/pages/frontline/digitalnation/interviews/turkle.html

Turow, J. (2011). The daily you. New Haven, CT: Yale University Press.

UC Berkeley School of Information. (2009). *Info 290: Surveillance, sousveillance, coveillance, and dataveillance.* Retrieved from http://www.ischool.berkeley.edu/courses/i290-sscd

UC Irvine Today. (2008, August 13). *Scientists to study synthetic telepathy.* Retrieved February 28, 2011 from http://today.uci.edu/iframe.php?p=/news/release_detail_iframe.asp?key=1808

Ungureanu, C. (2008). The contested relation between democracy and religion: Towards a dialogical perspective? *European Journal of Political Theory, 7*(4), 405–429. doi:10.1177/1474885108094052

United States Congress. (2011). 111 P.L. 353: *The FDA Food Safety Modernization Act,* 1-4-2011 (124 Stat 3885)

University of Finland. (2011). Retrieved from www.tut.fi

University of Reading. (2010, May 26). Could humans be infected by computer viruses? *ScienceDaily.* Retrieved July 23, 2011, from http://www.sciencedaily.com/releases/2010/05/100526095830.htm

University of Southern California. (2011, June 17). Scientists turn memories off and on with flip of switch. *ScienceDaily.* Retrieved July 20, 2011, from http://www.sciencedaily.com/releases/2011/06/110617081543.htm

Van der Rijta, P. (2007). *Precious knowledge: Virtualness and the willingness to share knowledge in organisational teams.* University van Amsterdam.

Vascellari, M. et al. (2004). Liposarcoma at the site of an implanted microchip in a dog. *Veterinary Journal (London, England), 168,* 188–190. doi:10.1016/S1090-0233(03)00121-7 PMID:15301769

Vascellari, M., & Mutinelli, F. (2006). Fibrosarcoma with typical features of postinjection sarcoma at site of microchip implant in a dog: Histologic and immunohistochemical study. *Veterinary Pathology, 43,* 545–548. doi:10.1354/vp.43-4-545 PMID:16846997

Vastenburg, M., Visser, T., Vermaas, M., & Keyson, D. (2008). Designing acceptable assisted living services for elderly users. In *Proceedings of the European Conference on Ambient Intelligence* (pp. 1-12). Berlin: Springer-Vertag.

VeriChip Corp. (2006, March 1). *Press release: RFID gets under their skin.* Retrieved October 24, 2007 from http://www.verichipcorp.com/news/1141257941

Victorian Law Reform Commission (VLRC). (2010). *Surveillance in public places final report, 18.* Retrieved November 1, 2010, from http://www.lawreform.vic.gov.au/wps/wcm/connect/justlib/law+reform/home/completed+projects/surveillance+in+public+places/lawreform+-+surveillance+in+public+places+-+final+report

von Hirsch, A., & Simister, A. P. (Eds.). (2006). *Incivilities: Regulating offensive behavior.* Oxford, UK: Hart Publishing.

Wagner, D., & Berger, J. (1985). Do sociological theories grow? *American Journal of Sociology, 90*(4), 697–728. doi:10.1086/228142

Wall Street Journal. (2011, May 30). A short guide to cookies. *Wall Street Journal*. Retrieved from http://online.wsj.com/public/page/0_0_WP_3001.html?currentPlayingLocation=158&currentlyPlayingCollection=Tech&currentlyPlayingVideoId={92E525EB-9E4A-4399-817D-8C4E6EF68F93}

Wall, S. (2007). Perfectionism in moral and political philosophy. In E. Zalta (Ed.), *Stanford encyclopedia of philosophy*. Retrieved July 21, 2011 from http://plato.stanford.edu/entries/perfectionism-moral/

Walsh, S. P., White, K. M., Cox, S., & Young, R. McD. (2011). Keeping in constant touch: The predictors of young Australians' mobile phone involvement. *Computers in Human Behavior*, 27, 333–342. doi:10.1016/j.chb.2010.08.011

Warren, S., & Brandeis, L. (1890). The right to privacy. *Harvard Law Review*, 4(5). doi:10.2307/1321160

Warwick, K. (n.d.). *Te next step towards true cyborgs*. Retrieved from http://www.kevinwarwick.com/Cyborg2.htm

Warwick, K., & Gasson, M. N. (2006). A question of identity – Wiring in the human. In *Proceedings of IET Wireless Sensor Networks Conference* (pp. 4/1-4/6). London: IET.

Warwick, K. (2000). The chip and I. In *The political subject: Essays on the self from art, politics and science*. London, UK: Lawrence and Wishart.

Warwick, K. (2004). *I cyborg*. Chicago: University of Illinois Press.

Warwick, K., Gasson, M. N., Hutt, B., Goodhew, I., Kyberd, P., & Andrews, B. et al. (2003). The application of implant technology for cybernetic systems. *Archives of Neurology*, 60(10), 1369–1373. doi:10.1001/archneur.60.10.1369 PMID:14568806

Warwick, K., Gasson, M. N., Hutt, B., Goodhew, I., Kyberd, P., & Schulzrinne, H. et al. (2004). Thought communication and control: A first step using radiotelegraphy. *IEE Proceedings. Communications*, 151(3), 185–189. doi:10.1049/ip-com:20040409

Webster, S. C. (2011a, March 26). $1 billion FBI biometrics identification programme. *The Nation*. Retrieved April 29, 2011 from http://www.nation.com.pk/pakistan-news-newspaper-daily-english-online/International/26-Mar-2011/1-billion-FBI-biometrics-identification-programme

Webster, S. C. (2011b, March 24). FBI dedicates $1 billion to massive biometrics identification program. *The Raw Story*. Retrieved April 7, 2011 from http://www.rawstory.com/rs/2011/03/24/fbi-dedicates-1-billion-to-massive-biometrics-identification-program/

Weil, P. (2002). Towards a coherent policy of co–development. *International Migration (Geneva, Switzerland)*, 40(3), 41–56. doi:10.1111/1468-2435.00196

Weiss, J. H. (1982). *The making of technological man: The social origins of french engineering education*. Cambridge, MA: The MIT Press.

Westacott, E. (2011). Does surveillance make us morally better? *Philosophy Now*. Retrieved from http://www.philosophynow.org/issues/79/Does_Surveillance_Make_Us_Morally_Better

White House. (2012). *Consumer data privacy in a networked world: A framework for protecting privacy and promoting innovation in a networked world*. Retrieved from http://www.whitehouse.gov/sites/default/files/privacy-final.pdf

White, L. (2007). The chase to trace. *National Provisioner*, 1/2001.

Whitehouse, D. (1999). Looking through cats' eyes. *BBC Sci/Tech News*. Retrieved July 20, 2011 from http://news.bbc.co.uk/2/hi/science/nature/471786.stm

Wigan, M. R. (2010a). *Identity, contestability and ethics of unified virtualisation*. Paper presented at the Panel Presentation to the 'Moral Maze on Virtualisation and Society' Forum Meeting. Edinburgh, Scotland.

Wigan, M. R. (1994). Establishing issues in transport policy. In K. W. Ogden, E. W. Russell, & M. R. Wigan (Eds.), *Transport policies for the new millennium* (pp. 1–19). Melbourne, Australia: Monash University.

Wigan, M. R. (2010c). Owning identity - One or many - Do we have a choice? *IEEE Technology and Society Magazine, 29*(2), 7. doi:10.1109/MTS.2010.937026

Wigan, M. R. (2012). Governance and evidence based policy under a national security framework. *IEEE Technology and Society Magazine*, 1.

William, M. (2011, January 16). Thoreau's cellphone experiment. *The Chronicle of Higher Education*. Retrieved February 5, 2011 from http://chronicle.com/article/Thoreaus-Cellphone-Experiment/125962/

Williams, R., & Johnson, P. (2005). Inclusiveness, effectiveness and intrusiveness: Issues in the developing uses of DNA profiling in support of criminal investigations. *Medical Malpractice: U.S., & International Perspectives, 545.*

Winston, B. (1998). *Media and technology: A history from the telegraph to the internet*. New York: Routledge.

Wong, K. Y., & Aspinwall, E. (2005). An empirical study of the important factors for knowledge management adoption in the SME sector. *Journal of Knowledge Management, 9*(3), 64–82. doi:10.1108/13673270510602773

World Health Organization. (n.d.). *Wingspread statement on the precautionary principle*. Retrieved from www.who.int/ifcs/documents/forums/forum5/wingspread.doc

Wu, D., Warwick, K., Ma, Z., Gasson, M. N., Burgess, J. G., & Pan, S. et al. (2010). Prediction of Parkinson's disease tremor onset using radial basis function neural network based on particle swarm optimization. *International Journal of Neural Systems, 20*(2), 109–118. doi:10.1142/S0129065710002292 PMID:20411594

Yang, G.-Z. (2006). *Body sensor networks*. London: Springer Verlag. doi:10.1007/1-84628-484-8

Zedner, L. (2010). Security, the state, and the citizen: The changing architecture of crime control. *New Criminal Law Journal, 13*(2), 379–403.

Zhang, D., Carr, D. J., & Alocilja, E. C. (2009). Fluorescent bio-barcode DNA assay for the detection of salmonella enterica serovar enteritidis. *Biosensors & Bioelectronics, 24*(5), 1377–1381. doi:10.1016/j.bios.2008.07.081 PMID:18835708

Zheng, W. (2005). A conceptualisation of the relationships between organisational culture and knowledge management. *Journal of Information and Knowledge Management, 4*(2), 113–124. doi:10.1142/S0219649205001110

Zimmer, C. (2004, February 2). Mind over machine. *Popular Science*, 47-48.

Zmud, R. (1979). Individual differences and MIS success: A review of the empirical literature. *Management Science, 25*(1), 966–979. doi:10.1287/mnsc.25.10.966

About the Contributors

M. G. Michael, Ph.D. (ACU), M.A (Hons) (MacqUni), M.Theol (SydUni), B.Theol (SCD), B.A.(SydUni), DipProfCouns (AIPC), is an Honorary Associate Professor in the School of Information Systems and Technology at the University of Wollongong, NSW, Australia. Michael is a theologian and historian with cross-disciplinary qualifications in the humanities and who introduced the concept of überveillance into the privacy and bioethics literature. Michael brings with him a unique perspective to Information Technology. His formal studies include ancient history, theology, general philosophy, political sociology, ethics, linguistics, and government. He was previously the coordinator of Information and Communication Security Issues and since 2005 has guest-lectured and tutored in Location-Based Services, IT and Citizen Rights, Principles of eBusiness, and IT and Innovation. The focus of his current research extends to modern hermeneutics and the Apocalypse of John; the historical antecedents of modern cryptography; the auto-ID trajectory; data protection, privacy and ethics related issues; biometrics, RFID and chip implants; national security and government policy; dataveillance and überveillance; and more broadly the system dynamics between technology and society. Michael is a member of the American Academy of Religion (AAR). He has guest edited the December 2006 volume of *Prometheus*, several *IEEE Technology and Society Magazine* issues in 2010-11, an issue for *Information Technology Cases* (2011), and more recently, the *Journal of Location-Based Services*. He is also the proceedings editor of four national security workshops sponsored by the Australian Research Council's Research Network for a Secure Australia.

Katina Michael, Ph.D., M.TransCrimPrev, B.I.T, is associate professor in the School of Information Systems and Technology at the University of Wollongong. Katina presently holds a full-time appointment as the Associate Dean – International of the Faculty of Engineering and Information Sciences. She is the *IEEE Technology and Society Magazine* editor-in-chief and also serves on the editorial board of Elsevier's *Computers & Security* journal. Since 2008, she has been a board member of the Australian Privacy Foundation, and in 2011 elected Vice Chair. Michael researches on the socio-ethical implications of emerging technologies. She has also conducted research on the regulatory environment surrounding the tracking and monitoring of people using commercial Global Positioning Systems (GPS) applications in the area of dementia, mental illness, parolees, and minors for which she was awarded an Australian Research Council Discovery grant. Michael has written and edited six books, guest edited numerous special issue journals on themes related to Radio-Frequency Identification (RFID) tags, supply chain management, location-based services, innovation, surveillance/überveillance, and big data. She has published over 125 academic peer reviewed papers and was responsible for the creation of the human factors series of workshops hosted annually since 2006 on the "Social Implications of National Security," sponsored by the Research Network for a Secure Australia. Katina was the program chair for IEEE's International Symposium on Technology and Society (ISTAS) in 2010 at the University of Wollongong, and 2013 at the University of Toronto. She is a Senior Member of the IEEE. She is well known for her research into the social implications of microchipping people.

* * *

Katherine Albrecht, Ed.D., is an internationally recognized privacy expert with a Doctorate in Human Development and Psychology from the Harvard University Graduate School of Education, with emphasis in Consumer Education. Katherine also received a Masters in Education (Ed.M.) from Harvard, in Technology, Innovation, and Education. She also holds an undergraduate degree in Business Administration and International Marketing, graduating with magna cum laude honors. Katherine is the Director of CASPIAN Consumer Privacy, an 18,000-member grass-roots activist group she founded in 1999 to oppose retail surveillance and numbering. She is co-author of the award-winning book *Spychips: How Major Corporations and Government Plan to Track Your Every Purchase and Watch Your Every Move,* which was a top ten Amazon non-fiction bestseller in 2006. Among her many achievements, Katherine has authored privacy legislation and testified before the Federal Trade Commission, The Federal Reserve Bank, The European Commission, and state legislatures around the globe. The Governor of New Hampshire appointed her to a two-year position as the consumer privacy and policy expert on that state's RFID Study Commission. She has given more than 2,000 interviews, among them to *CNN*, *Good Morning America*, *NPR*, the *BBC*, *Fox News*, and the *Wall Street Journal*. She is a regular guest on *Coast to Coast AM* radio with George Noory, and was featured in the films, *America: Freedom to Fascism*, *Shadow Government*, and *Behold a Pale Horse*. Katherine is the talk show host of her own nationally syndicated radio program, has written numerous publications including *for Scientific American, Addison Wesley, Denver University Law Review*, and IEEE, and has worked as a producer and consultant on instructional videos and documentaries.

Randy (Randall) Basham, Ph.D., is coauthor of two texts, one on computing and data analysis, and another on published multi-national mixed methods survey research on micro-economic development and democracy. His published dissertation examines computing and data visualization in the evaluation of therapeutic multi-group comparative outcomes across units of analysis. He has authored or coauthored numerous peer reviewed journal articles, book chapters, conference proceedings, encyclopedia entries, and published book reviews. His original technology-focused contributions have been recognized nationally and internationally. Dr. Basham has received university, or professional school, awards for scholarship, research, teaching, community services, and also several external career recognitions. His specialty or substantive area(s) of study are concerned with emerging technologies and their application to human service delivery, or professional social service direct practice. He is Member of the Board of Directors for the Human Service Information Technology Association (HUSITA) international professional association. He is a Representative for the North American, Council on Social Work Education (CSWE), Commission on Accreditation (COA), as a Site Reviewer. Currently, he serves as the Director and Graduate Advisor, of one of the most highly ranked, and largest of graduate programs of professional Social Work education in the United States.

Hilda Bouri recently graduated from Michigan State University, USA.

Jordan Brown is a filmmaker, artist, and activist whose work focuses on the interface between the dominant culture of globalization and the impact on people and ecology. He is the creator and editor of thoughtmaybe.com, an independent online hub for streaming documentary films; and the radical writer's platform rwrite.org—both of which have a strong focus on challenging the foundations, trajectory, and consequences of technoculture in a society captivated by the spectacle.

Mark Burdon, Ph.D., is a Lecturer in the TC Beirne School of Law at the University of Queensland. Mark Burdon's primary research interests are privacy law and the regulation of information sharing technologies. Mark has researched on a diverse range of multi-disciplinary projects involving the reporting of data breaches, e-government information frameworks, consumer protection in e-commerce, and information protection standards for e-courts.

Maria Burke, Ph.D., is a Reader in Digital Media and Business at the University of Salford. She researches within the context of Information Systems. Her main area of research expertise concerns knowledge management and the application of new digital technology to economic, environmental and social systems. Maria leads at Salford a £1.4m Digital Economy Grant as a partner of the collaborative group known as TOTeM (Tales of Things and Electronic Memory). Her work within TOTeM explored the application of digital technology, in this case, Quick Read Codes (QR Codes) and RFID (Radio Frequency Identification) to tag items used in commercial concerns on the high street. She also undertakes research into leadership in Higher Education and has secured funding from the Leadership Foundation for Higher Education to explore the new concept of "Urban and Edgy" universities. Her research publications include books, journal papers, conference papers, and book chapters.

Lawrence Busch is a Distinguished Professor of Sociology and Director of the Center for the Study of Standards in Society at Michigan State University. He has been on the faculty at the Norwegian University of Science and Technology, Lancaster University (UK), and what is now the Institut de Recherche pour le Développement (IRD). He is (co)author or (co)editor of twelve books including *Plants, Power and Profit: Social, Economic, and Ethical Consequences of the New Biotechnologies* (Blackwell 1991), *Toward a New Political Economy of Agriculture* (Westview 1991), *From Columbus to Conagra: The Globalization of Agriculture* (University of Kansas Press 1994), *Making Nature, Shaping Culture: Plant Biodiversity in Global Context* (University of Nebraska Press 1995), *The Eclipse of Morality: Science, State, and Market* (Aldine DeGruyter 2000), *Agricultural Standards: The Shape Of The Global Food And Fiber System* (Springer 2006), *Universities in the Age of Corporate Science: The UC Berkeley–Novartis Controversy* (Temple University Press 2007), and *Standards: Recipes for Reality* (MIT Press 2011). He has also authored or coauthored more than 150 other publications. Dr. Busch's current interests include the use of standards in public and private policymaking, biotechnology and nanotechnology policy, agricultural science and technology policy, higher education in agriculture, and public participation in the policy process.

Daniel Buskirk, Ph.D., M.S., B.S., is an Associate Professor in the Department of Animal Science with Michigan State University, USA. Dr. Buskirk's program emphasizes the improvement of profitability and competitive advantages of stocker and cow/calf operations in Michigan. The focus of this extension and research program is to work on areas that advance a coordinated, sustainable beef production system in the Eastern Corn Belt.

Rafael Capurro was born 1945 in Montevideo (Uruguay). Dr.Phil. in Philosophy from Düsseldorf University (1978). Postdoctoral teaching qualification (Habilitation) in Practical Philosophy (Ethics) from Stuttgart University (1989). Professor (em.) of Information Science and Information Ethics at Hochschule der Medien Stuttgart and Lecturer in Ethics at the University of Stuttgart (1986-2009). Director of the International Center for Information Ethics (ICIE) (1999 to present). Editor-in-Chief of the *International Review of Information Ethics* (IRIE) (2004 to present). Distinguished Researcher in Information Ethics, School of Information Studies, University of Wisconsin-Milwaukee, USA. Distinguished Researcher at the African Centre for Information Ethics, Department of Information Science, University of Pretoria, South Africa. Former member of the European Group on Ethics in Science and New Technologies (EGE) to the European Commission (2000-2010).

Roger Clarke, Ph.D., is a consultant on strategic and policy implications of advanced information technologies, with particular reference to eBusiness, information infrastructure, and dataveillance and privacy. He performs this work through Xamax Consultancy Pty Ltd, based in Canberra. He is a Visiting Professor in Cyberspace Law and Policy at the University of N.S.W., and a Visiting Professor in the Research School of Computer Science at the Australian National University (ANU). He is a Fellow of the Australian Computer Society (ACS), and a Fellow of the international Association for Information Systems (AIS). He has Commerce (Information Systems) degrees from UNSW, and a doctorate from the ANU. In 2009, he was awarded only the second Australian Privacy Medal, after Justice Michael Kirby. He has spent many years on the Board of the Australian Privacy Foundation (APF), including as Chair 2006-13, and on the Advisory Board of Privacy International (PI). He has also served variously as a Director, Secretary and Chair of several companies, of Electronic Frontiers Australia (EFA), and of the Internet Society of Australia (ISOC-AU).

Brigette Garbin is a research assistant in the TC Beirne School of Law at the University of Queensland.

Mark Gasson, Ph.D., is a visiting research fellow at the School of Systems Engineering, University of Reading, UK. In 2010, he became the first human to be infected by a computer virus, using an RFID device implanted in his body. In 2009, he brought attention to the privacy issues of GPS-enabled smartphones by creating detailed profiles of users tracked across Europe. His interdisciplinary research interests focus on user-centric applications of emerging technologies and pushing the envelope of Human-Machine interaction. In 2005, he was awarded a Ph.D. for invasively interfacing the nervous system of a human to a computer system, and in 2010, he had the honor of being the General Chair for the IEEE International Symposium on Technology and Society (ISTAS 2010) in Australia. Dr. Gasson regularly engages with the international press and other media outlets, including as a guest expert for BBC News.

He is passionate about engaging the public in engineering science through a variety of activities and has delivered well over 100 invited public and academic lectures internationally including at TEDx. Dr. Gasson was awarded his first degree in Cybernetics and Control Engineering in 1998 from the Department of Cybernetics, University of Reading, and until 2010, was a senior research fellow at the School of Systems Engineering, University of Reading.

Stephen Gasteyer is an Assistant Professor of Sociology with Michigan State University, USA. Dr. Gasteyer's research focuses on the nexus between water, land, and community development. Specifically, his research currently addresses: 1) community capacity development and civic engagement through leadership training; 2) the political and social processes that enable or hinder community access to water and land resources, specifically (but not exclusively) in rural communities; 3) the class and race effects of access to basic services (water, sanitation, food, healthcare); 4) community capacity, community resilience and water systems management; 5) the impacts of greening in economically depressed small cities; 6) the community aspects of bioenergy development; 7) international social movements and community rights to basic services; and 8) facilitating cross-sectoral and interdisciplinary partnerships to address water and land resources management. Before coming to Michigan State University, Dr. Gasteyer was on faculty in the Department of Human and Community Development at the University of Illinois. Prior to that, he was Research and Policy Director at the Rural Community Assistance Partnership in Washington, DC, and a research consultant on issues of global water governance. Dr. Gasteyer was a Peace Corps Volunteer in Mali from 1987 through 1990, and worked with environmental non-governmental organizations from 1993 through 1998 in the Palestinian territories. He received a BA from Earlham College in 1987, and a Ph.D. in Sociology from Iowa State University in 2001.

Erica Giorda is a Ph.D. candidate in the F.E.A.S.T. concentration of the Department of Sociology with Michigan State University, USA. Her research interests gravitate around food, environment, fair trade and sustainability, and the way standards and certifications are affecting those areas. Erica is completing a MA in Interdisciplinary Studies at Wayne State University with a thesis on the Environmental Movement in Detroit, and holds a BA in Philosophy on Natural Language Analysis with the University of Turin.

Amal Graafstra is author of *RFID Toys* and is the owner of several technology and mobile communications companies and a double RFID implantee. Amal loves thinking up interesting ways to combine and apply various technologies in his daily life. Since learning about RFID technology used in cats and dogs for identification, Amal wanted to leverage that technology himself. Getting implants meant there was no need to carry an RFID access card around and he could implement his own RFID access control systems instead of buying expensive off-the-shelf products. Soon after getting his first implant and posting some pictures of the process for a few friends, word quickly spread over the Internet and soon he found himself talking to everyone from industry players to clergy to book publishers about RFID technology and its possibilities. Amal has been interviewed about his adventures in RFID by television, print, and online news media from around the world, including the *Discovery Channel's Daily Planet* program. He uses his RFID implants to log into his computer, access his front door and opening his car door. He is proprietor at dangerousthings.com.

Daniel Grooms is a Professor and Associate Chair for Food Animal Programs, in the College of Veterinary Medicine, Department of Large Animal Clinical Science Michigan State University, USA. Dr. Dan Grooms is a native of central Ohio where he grew up on a small commercial cow-calf operation. Dr. Grooms received his B.S. degree in Animal Science from Cornell University in 1985, his D.VM. degree from The Ohio State University in 1989 and was in private practice in central Ohio for 5 years. He returned to Ohio State and received his Ph.D. in Veterinary Preventive Medicine in 1996. He joined the faculty in the Department of Large Animal Clinical Sciences at Michigan State University in 1996 and was promoted to professor in 2011. He is a diplomat of the American College of Veterinary Micro-biologists. His extension and research activities have focused on the control and prevention of infectious disease in cattle, specifically BVDV, Johne's disease, bovine respiratory disease, and bovine tuberculosis. He has also been actively involved in beef quality assurance, bull breeding soundness exam programs, pre-harvest food safety, biosecurity, emergency preparedness, and farm security initiatives in the state.

Alexander Hayes is a Ph.D. candidate within the School of Information Systems and Technology at the University of Wollongong, NSW, Australia. His research is focused on the implications of emergent, location-aware body worn video technologies in an educational context, citing user case studies and an extensive series of interviews with lead innovators across this rich field of interdisciplinary endeavor. He is currently a Professional Associate at the INSPIRE Centre, University of Canberra, ACT Australia, and also Visiting Researcher, Aalto University, Helsinki, Finland. Alexander Hayes is also a Business Development Manager, with the Augmate Corporation, New York, USA, a next generation Internet software company, providing visual efficiency tools for businesses and consumers by merging the digital and physical worlds using eye-wear (AR Glasses) and mobile devices. With an extensive background in e-learning and professional development, Alexander project manages solutions for corporate, business and industry across Australia and New Zealand. He is a recognized expert on emerging technology and its impacts on society, regularly invited to present at strategic events.

William A. Herbert is a labor and employment attorney with a scholarly interest in the intersection of new technologies and labor law. His published scholarship includes articles on social media, human tracking technologies, e-mail and Internet use in the workplace, genetic testing, electronic workplace privacy protections in the European Union, public sector labor law and history, the application of the United States Constitution to public sector labor law issues, and statutory discrimination and retaliation protections. His most recent article, "Can't Escape from the Memory: Social Media and Public Sector Labor Law," was published by *Northern Kentucky Law Review*. Mr. Herbert is a former Co-Chair of the American Bar Association Labor and Employment Section's Technology in the Practice and Workplace Committee and a current Co-Chair of the New York State Bar Association Labor and Employment Section's Technology in the Workplace and Practice Committee. He graduated from the Benjamin N. Cardozo School of Law, Yeshiva University and the University at Buffalo.

Rebecca Hester is Assistant Professor of Medical Humanities, University of Texas Medical Branch, USA. A strong advocate of social justice, Dr. Hester's interdisciplinary scholarship draws from the social sciences and the humanities to focus on questions of culture in medicine, the political economy of the body, and minority health and health subjectivity. She completed her doctoral training in the Department of Politics at the University of California Santa Cruz with an emphasis in Latin American and Latino Studies. She received her Bachelor's degree in Spanish and Portuguese with an emphasis in Latin American Literature from the University of California Berkeley where she graduated with high honors. Dr. Hester also completed university course work in Mexico and France. She came to the University of Texas Medical Branch in 2010 after completing a post-doctoral fellowship at the University of Illinois Urbana-Champaign. Her dissertation, *Embodied Politics: Health Promotion in Indigenous Mexican Migrant Communities in California*, which received the 2010 best dissertation in Latino Studies from the Latin American Studies Association, illuminates the moral and symbolic systems that are at work in community health promotion programs for indigenous Mexican migrants in California. Her current research continues to explore cultural and social aspects of health and medicine through two projects. The first is a two-year ethnographic study of the ways that academic medicine teaches notions of culture and cultural diversity to medical students. Specifically, this NIH-supported study identifies components of medical training at the University of Texas Medical Branch, one of the most diverse public medical schools in the country, that include critical, self-reflective discussions of power, and ethical practice in medicine, especially as these are manifested in relation to racial and cultural difference. The second avenue of research focuses on the social, ethical, and political implications of microchips that have been embedded in humans for medical purposes. This line of research is specifically interested in the ways that microchips can be used for social control and tracking purposes, especially for minority populations.

Jann Karp, Ph.D., M.A., BSoc.Sc., is a lecturer in Criminology in the School of Behavioral, Cognitive, and Social Science at the University of New England, Armidale, NSW, Australia. Karp has lectured at a number of institutions including the University of Western Sydney, Charles Sturt University, and the University of Newcastle. Karp spent 23 years in the NSW Police Service, completing her service as a senior constable of police. Karp has published on a variety of themes, more recently completing her full volume titled: *Truckies: Life Behind the Wheel*. She is fascinated by how technologies, especially tracking technologies like Global Positioning Systems (GPS), are used to monitor both vehicles and drivers respectively.

Ronnie D. Lipschutz is Professor of Politics and Provost of College Eight at the University of California, Santa Cruz. Lipschutz received his Ph.D. in Energy and Resources from UC-Berkeley in 1987 and an SM in Physics from MIT in 1978. He has been a faculty member at UCSC since 1990. Lipschutz conducts research in and writes on a range of topics related to global political economy, including U.S. global economic and military policy and strategy, changing conceptions and practices of security, changing forms of war, global governance, global civil society and corporate social responsibility, environmental politics, energy and resources, sustainability and political economy and popular culture. His most recent books are *Political Economy, Capitalism, and Popular Culture* (Rowman & Littlefield, 2010) and *The Constitution of Imperium* (Paradigm, 2008) as well as a text co-authored with Mary Ann Tétreault, *Global Politics as if People Mattered* (Rowman and Littlefield, 2009, 2nd ed.) co-edited books (with K. Ravi Raman) *Corporate Social Responsibility: Comparative Critiques* (Palgrave Macmillan, 2010) and (with Gabriela Kütting) of *Global Environmental Governance—Power and Knowledge in a Local-Global World* (Routledge, 2009).

Monica List is a Ph.D. candidate in Philosophy with Michigan State University, USA. She earned a veterinary medicine degree from the National University of Costa Rica in 2002, and a MA degree in bioethics, also from the National University in 2011. Her research interests include environmental philosophy, sustainability, animal ethics, veterinary ethics, development ethics, and bioethics.

Steve Mann holds degrees from the Massachusetts Institute of Technology (Ph.D. in Media Arts and Sciences '97) and McMaster University, where he was also inducted into the McMaster University Alumni Hall of Fame, Alumni Gallery, 2004, in recognition of his career as an inventor and teacher. While at MIT, in then Director Nicholas Negroponte's words he "brought the seed" that founded the Wearable Computing group in the Media Lab. In 2004, he was named the recipient of the 2004 Leonardo Award for Excellence for his article "Existential Technology" published in *Leonardo* 36:1. Mann is a researcher and inventor best known for his work on computational photography, particularly wearable computing and high dynamic range imaging. Mann also works in the fields of computer-mediated reality. He is a strong advocate of privacy rights, for which he was an award recipient of the Chalmers Foundation in the fine arts. His work also extends to the area of sousveillance, a term he coined for "inverse surveillance." Mann is author of more than 200 publications, including a textbook on electric eyeglasses and a popular culture book on day-to-day cyborg living. See for instance, *Intelligent Image Processing* and *Cyborg: Digital Destiny and Human Possibility in the Age of the Wearable Computer*. Mann is presently a tenured professor at the Department of Electrical and Computer Engineering with cross-appointments to the Faculty of Arts and Sciences, and Faculty of Forestry, at the University of Toronto, and is a Professional Engineer licensed through Professional Engineers Ontario.

Ellen M. McGee, Ph.D., is presently an ethics consultant, offering ethics education to organizations and institutions, and lecturing widely. She also serves as a speaker for the New York State Council on the Humanities. She is a retired adjunct professor of philosophy at Long Island University, where she taught for over twenty-five years, having previously taught at Fordham University. She founded and directed the Long Island Center for Ethics, now *The Institute for Education for Social Justice*. She subsequently served as Associate for Bioethics at The Long Island Center for Ethics where she coordinated the Nassau-Suffolk Health Care Ethics Network. Dr. McGee received a Ph.D. and M.A. in Philosophy from Fordham University, New York, and a B.A. from Marymount College, Tarrytown, New York. She teaches medical ethics, computer ethics, social work ethics, and philosophy at the undergraduate and graduate levels, and researches, lectures, and publishes in the areas of enhancement technologies, particularly implantable brain chips, and end-of-life care, suicide intervention, human rights, and reproductive issues. Dr. McGee has been a member of both *The Hospice Project* and *The Hospice and Alzheimer Project at the Hastings Center*; she was a member of the Advisory Committee on *Nursing Homes: New York State Partnership to Improve End-Of-Life Care*, and has served on a *Hospice Ethics Committee* and both the IRB and Ethics Committees of area hospitals. She has appeared as an ethics consultant on the network news and radio and has developed, organized and presented many conferences including those on Enhancement Technologies, Good Dying, and Health Care for Diverse Communities.

Peter Miller, Ph.D., B.A. (Hons), is a Principal Research Fellow at the School of Psychology, Deakin University. He is also the Commissioning Editor of the journal *Addiction*. His research interests include: Alcohol-related violence in licensed venues; Alcohol/drug use in rural populations, and; the behavior of vested interests such as the global alcohol industry lobby. Peter has recently completed two of the largest studies ever conducted into licensed venues, comparing 6 Australian cities over 3 years and talking to more than 10,000 patrons. Peter and is an Executive board member of the *International Society of Addiction Journal Editors* and the *National Alliance for Action on Alcohol*.

Darren Palmer, Ph.D., M.A. Criminological Studies, B.A. Hons, is the past convener of the major sequence in Criminology in the B.A. at Deakin University. He has had many years teaching experience at La Trobe University (Law and Legal Studies) and Deakin University (Police Studies/Criminology). He has taught in a range of areas including criminal justice, criminal law, psychology and crime, policing, and criminology research methods. His current research interests include: policing and criminal justice histories, the professionalization of police practice, police education, police memorials, police pursuits, changing forms of policing and security in Australia and internationally and more. He presently has about $900,000 of research funding examining a variety of topics including: the evaluation of interventions addressing alcohol and violence in the nighttime economy; the use of "new" technology in the nighttime economy with an emphasis on ID scanners; Australian Sudanese immigrants and their perceptions and experience of criminal justice; and rural and regional justice. He has been a member of the Australian and New Zealand Society of Criminology since 1988. He is the editor of the *Socio-Legal Bulletin* and is a member of the editorial committee of the international journal *Crime Prevention and Community Safety* and *Police Practice and Research: An International Journal*.

Christine Perakslis holds a Doctor of Education in Educational Leadership (2009), as well as a Certificate of Advanced Graduate Studies in Educational Leadership (2008) and a Master of Science in Management (2005) with a focus on Organization Development. She is presently an Associate Professor in the MBA Program in the Alan Shawn Feinstein Graduate School at Johnson and Wales University, teaching such courses as: Operations Management, and Contemporary Issues and Strategies. She was previously employed as the Joint Chief Operating Officer of a privately held company in the hospitality industry. Perakslis' research focuses on group integration competencies, behavioral motivators, and the social implications of technology. Her involvement with industry has included consulting work in various industries, as well as work with the Executive Forum for Coleman Research Group, the Editorial Advisory Board for IGI Global, the IEEE Technical Program Committee for the International Symposium on Technology and Society, and the Program Committee for the RNSA National Security Workshop. She continues to operate in a consulting capacity in industry which serves to broaden her industry-relevant teaching.

Gary Retherford has fifteen years of management experience, followed by fifteen years of sales, account executive, and business development experience. He has held positions with Siemens Building Technologies, Sec-Tron, Inc., Tyco International, and Simplex Time Recorder. Retherford is presently the regional sales manager at International Systems of America. In 2008, Retherford established Six Sigma Security, Inc. and began a LinkedIn group under the same name bringing together a network of security professionals from around the world interested in utilizing the Six Sigma methodology for improving Security. In 2006, Retherford was responsible for the first human implantable microchips for access control in a private organization at Citywatcher.com in Cincinnati, Ohio. The project was profiled on CNBC's *Big Brother, Big Business*. Retherford graduated with a B.S. in Business from Indiana University, Kelley School of Business, with emphasis in Management and Finance.

Alan Rubel is an assistant professor in the School of Library and Information Studies and in the Program in Legal Studies at the University of Wisconsin-Madison (USA). He is also an affiliate of the University of Wisconsin Law School. Prior to joining the faculty at UW, he was a Greenwall Fellow in Bioethics and Health Policy and Johns Hopkins and Georgetown universities, and law clerk to Justice Ann Walsh Bradley of the Wisconsin Supreme Court. He works in the areas of information ethics and policy, moral and political philosophy, and privacy law.

Chris Speed is Chair of Design Informatics at the University of Edinburgh where his research focuses upon the Network Society, Digital Art and Technology, and The Internet of Things. Chris has sustained a critical enquiry into how network technology can engage with the fields of art, design, and social experience through a variety of international digital art exhibitions, funded research projects, books journals, and conferences. At present, Chris is working on funded projects that engage with the flow of food across cities, an Internet of cars, turning printers into clocks and a persistent argument that chickens are actually robots. Chris is a co-organizer and compere for the Edinburgh www.ThisHappened. org events and is co-editor of the journal *Ubiquity*. Chris was PI for the TOTeM project investigating social memory within the "Internet of Things" funded by the Digital Economy (£1.4m) and the related Research in the Wild grant: Internet of Second Hand Things; PI for the JISC funded iPhone app Walking Through Time that overlays contemporary Google maps with historical maps; PI for Community Web2.0: creative control through hacking, a £40K feasibility study that explores parallels between virtual society (Internet) and actual society (communities); Co-I to the Sixth Sense Transport RCUK funded Energy project (£900k) which explores the implications for the next generation of mobile computing for dynamic personalized travel planning. He is also PI for the Travel Behaviors network funded by the RCUK Energy theme (£140k) and Co-I to both the EPSRC Creating trust through digital traceability project (Hull) and Learning Energy Systems project (Edinburgh).

Kelly Staunton is a research assistant in the TC Beirne School of Law at the University of Queensland.

John V. Stone is a Manager of Research and Engagement, Global Innoversity at Michigan State University, USA. He is Co-Director and Senior Research Scientist at the Center for the Study of Standards in Society (CS3). He holds a Ph.D. in Applied Anthropology from the University of South Florida, with an emphasis on Social Impact Assessment (SIA). Dr. Stone has over 25 years experience in applied social research, authored more than 30 scientific publications and technical reports, and delivered more than 85 presentations to professional and scientific societies and government agencies. His current work addresses ethnographic approaches to risk perception mapping, public engagement, and standards in the SIA of emerging technologies. Within this context, he seeks to promote "participatory equity" among potentially affected populations and increase their social access to policy-making processes. Dr. Stone holds seats on the American National Standards Institute's "Nanotechnology Standards Panel" and "Committee on Standards Education and Training." He co-founded the Risk Assessment and Policy Association, is a Fellow of the Society for Applied Anthropology, a founding member of the Society for Nano and Emerging Technologies, and he holds occasional membership in the International Association for Public Participation and the International Association for Impact Assessment.

Paul B. Thompson is the W.K. Kellogg Chair in Agricultural Food and Community Ethics with Michigan State University, USA. Thompson is the author of 13 books and editions, such as *The Spirit of the Soil: Agriculture and Environmental Ethics*; *The Ethics of Aid and Trade; Food Biotechnology in Ethical Perspective*, and co-editor of *The Agrarian Roots of Pragmatism*. He has served on many national and international committees on agricultural biotechnology and contributed to the National Research Council report *The Environmental Effects of Transgenic Plants*. He is a Past President of the Agriculture, Food and Human Values Society and the Society for Philosophy and Technology, and is Secretary of the International Society for Environmental Ethics.

Serafin Vilaplana is the former Information Technology specialist at the Baja Beach Club, Barcelona, Spain. Serafin was responsible for implementing the RFID implant access control and epayment solution in May 2004 at the club under the guidance of manager Conrad Chase.

Ian Warren, PhD, M.A., LLB, is a Senior Lecturer in Criminology and member of the Centre for Citizenship and Globalisation at Deakin University, Geelong, Australia. His research examines the relationship between technology and surveillance in contemporary Australian and international crime control debates. He has written widely on socio-legal responses to various crime problems including illicit drug trafficking, the regulation of the private security industry, the use of surveillance technologies to promote order in the Australian nighttime economy, and the comparative dimensions of privacy regulation in Australia and Canada. His research also examines the limits of contemporary global justice measures to deal with comparative and transnational crime problems. His research on the relationship between surveillance and criminal justice questions the widespread acceptance of technology to prevent and control crime.

Kevin Warwick is Professor of Cybernetics at the University of Reading, England, where he carries out research in artificial intelligence, control, robotics and biomedical engineering. He is a Chartered Engineer (CEng.) and is a Fellow of The Institution of Engineering and Technology (FIET). He is the youngest person ever to become a Fellow of the City and Guilds of London Institute (FCGI). Kevin was born in Coventry, UK, and left school to join British Telecom, at the age of 16. At 22, he took his first degree at Aston University, followed by a PhD and a research post at Imperial College, London. He subsequently held positions at Oxford, Newcastle and Warwick universities before being offered the Chair at Reading, at the age of 33. He has been awarded higher doctorates (DScs) both by Imperial College and the Czech Academy of Sciences, Prague. He was presented with The Future of Health Technology Award from MIT (USA), was made an Honorary Member of the Academy of Sciences, St. Petersburg and received The Mountbatten Medal in 2008. In 2000 Kevin presented the Royal Institution Christmas Lectures, entitled "The Rise of The Robots." He has also been awarded Honorary (DSc) Degrees by the

Universities of Aston and Coventry. Kevin's research involves robotics and he was responsible (with Dr Jim Wyatt) for Cybot, a robot exported around the world as part of a magazine "Real Robots" – this resulted in royalties totalling over £1M for Reading University. Robots designed and constructed by Kevin's group (Dr Ian Kelly, Dr Ben Hutt) have been on permanent interactive display in the Science Museums in London, Birmingham and Linz. Kevin's recent research involves a collaborative project with the Oxford neurosurgeon, Prof. Tipu Aziz, using intelligent computer methods to predict the onset of Parkinsonian tremors such that they can be stopped by means of a deep brain implant. In 2007, this work was hailed in the *Mail* on Sunday as "the most significant recent advance in biomedical engineering."

Kyle Powys Whyte is an Assistant Professor of Philosophy at Michigan State University, USA. He is an enrolled member of the Citizen Potawatomi Nation in Shawnee, Oklahoma. Whyte writes on environmental justice, the philosophy of technology and American Indian philosophy. His most recent research addresses moral and political issues concerning climate change impacts on Indigenous peoples. His articles have appeared in journals such as *Climatic Change, Ecological Processes, Synthese, Human Ecology, Journal of Global Ethics, American Journal of Bioethics, Journal of Agricultural & Environmental Ethics, Philosophy & Technology, Ethics, Policy & Environment, Environmental Justice*, and *Continental Philosophy Review*. His research has been funded by the National Science Foundation, U.S. Fish and Wildlife Service, and Spencer Foundation. He is a member of the American Philosophical Association Committee on Public Philosophy, Michigan Environmental Justice Coalition, and volunteer for the annual Growing Our Food System conference in Lansing, Michigan.

Marcus Wigan, Ph.D., is Principal of Oxford Systematics, Professorial Fellow at the University of Melbourne, Professor of both Transport and of Information Systems at Napier University, Edinburgh, and Visiting Professor at Imperial College, London, and serves on the Ethics Task Force and the Economic Legal and Social Implications Committee of the Australian Computer Society, of which he is a Fellow. He has worked on the societal aspects of transport, surveillance and privacy both as an engineer and policy analyst and as an organizational psychologist. He has published for over 30 years on the interactions between intellectual property, identity and data integration in electronic road pricing and intelligent transport systems for both freight and passenger movements. Marcus Wigan has qualifications in a wide range of fields from physics to psychology, business to intellectual property law. He has served on Ministerial Advisory Councils, as expert advisor to Parliamentary Committees and chaired Standards Australia Committees. He has published, participated and researched in transport and communications issues for over forty years. A life member of Electronic Frontiers Australia and the Australian Privacy Foundation, he remains a long term advocate for civil liberties and equity with special reference to ICT, telecommunications, and the Internet. He was recently elected onto the board of ACCAN (Australian Communications Consumer Action Network).

Index